Biomarkers in Cancer Chemoprevention

International Agency For Research On Cancer

The International Agency for Research on Cancer (IARC) was established in 1965 by the World Health Assembly, as an independently financed organization within the framework of the World Health Organization. The headquarters of the Agency are in Lyon, France.

The Agency conducts a programme of research concentrating particularly on the epidemiology of cancer and the study of potential carcinogens in the human environment. Its field studies are supplemented by biological and chemical research carried out in the Agency's laboratories in Lyon and, through collaborative research agreements, in national research institutions in many countries. The Agency also conducts a programme for the education and training of personnel for cancer research.

The publications of the Agency contribute to the dissemination of authoritative information on different aspects of cancer research. A complete list is printed at the back of this book. Information about IARC publications, and how to order them, is also available via the Internet at: **http://www.iarc.fr/**

Cover illustrations:

Top left:	Mammography
Bottom left:	Confocal laser microscopy of an eosinophilic granulocyte from the lung
Top right:	Tubular adenoma of the colon (N.J. Carr)
Bottom right:	COX-2 expression in human hepatocellular carcinoma (S. Sakisaka)

Biomarkers in Cancer Chemoprevention

Edited by A.B. Miller, H. Bartsch, P. Boffetta, L. Dragsted and H. Vainio

IARC Scientific Publications No. 154

International Agency for Research on Cancer, Lyon,

2001

Published by the International Agency for Research on Cancer,
150 cours Albert Thomas, F 69372 Lyon cedex 08, France

Distributed by Oxford University Press, Walton Street, Oxford, UK OX2 6DP (Fax: +44 1865 267782) and in the USA by Oxford University Press, 2001 Evans Road, Carey, NC 27513, USA (Fax: +1 919 677 1303).
All IARC publications can also be ordered directly from IARC*Press*
(Fax: +33 4 72 73 83 02; E-mail: press@iarc.fr).

IARC Library Cataloguing in Publication Data

Biomarkers in Cancer Chemoprevention
editors, A.B. Miller... [et al.]
(IARC Scientific Publication; 154)
1. Biological Markers. 2. Neoplasms - prevention & control.
I. Miller, A.B. (Anthony B).
II. Title III. Series

ISBN 92 832 2154 0 (NLM Classification: W1)
ISSN 0300-5085

Achevé d'imprimer sur rotative par l'imprimerie Darantiere à Dijon-Quetigny en décembre 2000
Dépôt légal : décembre 2000 – N° d'impression : 20-1352

Contents

Foreword vii

List of participants viii

Workshop report 1

Surrogate end-point biomarkers in chemopreventive drug development
G.J. Kelloff, C.C. Sigman, E.T. Hawk, K.M. Johnson, J.A. Crowell and K.Z. Guyton . 13

Biomarkers and surrogacy: relevance to chemoprevention
T.W. Kensler, N.E. Davidson, J.D. Groopman and A. Munoz 27

Development of difluoromethylornithine and Bowman-Birk inhibitor as chemopreventive agents by assessment of relevant biomarker modulation: some lessions learned
F.L. Meyskens, Jr 49

Selection and validation of biomarkers for chemoprevention: the contribution of epidemiology
P. Vineis and F. Veglia 57

Skin cancer

Ultraviolet radiation-induced photoproducts in human skin DNA as biomarkers of damage and its repair
K. Hemminki, G. Xu and F. Le Curieux 69

Intermediate-effect biomarks in prevention of skin cancer
J.-F. Doré, R. Pedeux, M. Boniol. M.-C. Chignol and P. Autier 81

Genetically determined susceptibility markers in skin cancer and their application to chemoprevention
H. Hahn 93

Colorectal cancer

Biomarkers in colorectal cancer
R.W. Owen 101

Intermediate effect markers for colorectal cancer
J.A. Baron 113

Susceptibility markers in colorectal cancer
J. Burn, P.D. Chapman, D.T. Bishop, S. Smalley, I. Mickleburgh, S. West and J.C. Mathers 131

Breast cancer

Endogenous hormone metabolism as an exposure marker in breast cancer chemoprevention studies
R. Kaaks 149

Mammographic density as a marker of susceptibility to breast cancer: a hypothesis
N.F. Boyd, G.A. Lockwood, L.J. Martin, J.W. Byng, M.J. Yaffe and D.L. Tritchler 163

Intermediate histological effect markers for breast cancer
A.B. Miller and A.M. Borges 171

Inherited genetic susceptibility to breast cancer
J. Chang-Claude 177

Prostate cancer

High-grade prostatic intraepithelial neoplasia as an exposure biomarker for prostate cancer chemoprevention research
J.R. Marshall 191

Intermediate biomarkers for chemoprevention of prostate cancer
T.H. van der Kwast 199

The role of molecular genetics in chemoprevention studies of prostate cancer
R.K. Ross 207

Liver cancer

Exposure biomarkers in chemoprevention studies of liver cancer
C.P. Wild and P.C. Turner 215

Significance of hepatic preneoplasia for cancer chemoprevention
P. Bannasch, D. Nehrbass and A. Kopp-Schneider 223

Hepatocellular carcinoma: susceptibility markers
H.E. Blum 241

Lung and oral cavity cancer

Carcinogen biomarkers for lung or oral cancer chemoprevention trials
S.S. Hecht 245

Lung cancer: chemoprevention and intermediate effect markers
M.S. Tockman 257

Metabolic polymorphisms as susceptibility markers for lung and oral cavity cancer
U. Nair and H. Bartsch 271

Foreword

The evaluation of potential cancer chemopreventive agents requires large, long-term, randomized trials in humans. Biomarkers that would allow the identification of subjects exposed to carcinogenic agents or of individuals genetically susceptible to cancer would be of great value. Similarly, markers of intermediate events on the carcinogenic pathway to cancer would facilitate the identification of the stages of carcinogenesis at which such agents might exert their protective effect. This could allow selection of smaller numbers of high-risk subjects and a shorter time-scale for trials. However, to be useful for such purposes, a biomarker should be highly sensitive and specific and, if used as a marker of outcome, must closely relate to the relevant stage of the carcinogenic process that the chemopreventive agent is designed to inhibit.

This volume is the product of an attempt to assess our current knowledge and experience with the use of biomarkers in cancer chemoprevention, and is based on an international workshop held at the Deutsches Krebsforschungszentrum (DKFZ), Heidelberg, Germany during February 27–29, 2000. It comprises an expert report on the current state of the art on use of biomarkers in cancer chemoprevention, introductory sections covering general topics in cancer chemoprevention, and contributions directed to biomarkers of exposure, intermediate effect and susceptibility relating to six major cancer sites: skin, colorectum, breast, prostate, liver and upper aerodigestive tract.

Financial support was provided by the Deutsches Krebsforschungszentrum, the National Institute of Environmental Health Sciences, Research Triangle Park, United States, the Foundation for Promotion of Cancer Research, Tokyo, Japan, the Danish Institute of Food Toxicology, Copenhagen, Denmark, and the Deutsche Forschungsgemeinschaft, Bonn, Germany. All contributions are gratefully acknowledged. I also should like to thank the local organizers at the DKFZ, especially Dr Anthony B. Miller, Dr Helmut Bartsch and Dr Norbert Frank, for their efforts in ensuring the success of the workshop.

P. Kleihues
Director, IARC

Biomarkers in Cancer Chemoprevention

German Cancer Research Centre, Heidelberg, Germany

27-29 February 2000

List of participants

P. Bannasch
Division of Cell Pathology - C0100
Deutsches Krebsforschungszentrum
Im Neuenheimer Feld 280
D 69120 Heidelberg
Germany

J.A. Baron
7927 Rubin Building
DHMC
1 Medical Center Drive
Dartmouth Medical School
Lebanon, NH 03756
USA

H. Bartsch
Division of Toxicology and Cancer Risk
Factors - C0200
Deutsches Krebsforschungszentrum
Im Neuenheimer Feld 280
D 69120 Heidelberg
Germany

H.E. Blum
Klinikum der Albert-Ludwigs-Universität
Innere Medizin Abt. II
Hugstetter Strasse 55
79106 Freiburg
Germany

P. Boffetta
Unit of Environmental Cancer
Epidemiology
International Agency for Research on
Cancer
150 cours Albert Thomas
69372 Lyon Cedex 08
France

V.A. Bohr
Laboratory of Molecular Genetics
National Institute on Aging
NIH
5600 Nathan Shock Dr.
Baltimore, MD 21224
USA

N.F. Boyd
Division of Epidemiology and Statistics
Ontario Cancer Institute
610 University Avenue
Toronto, Ontario
M5G 2M7
Canada

J. Burn
Department of Human Genetics
University of Newcastle upon Tyne
19/20 Claremont Place
Newcastle upon Tyne NE2 4AA
United Kingdom

J. Chang-Claude
Division of Clinical Epidemiology -C0500
Deutsches Krebsforschungszentrum
Im Neuenheimer Feld 280
D 69120 Heidelberg
Germany

J. Cheney
Editor
International Agency for Research on
Cancer
150 cours Albert Thomas
69372 Lyon Cedex 08
France

J.-F. Doré
INSERM U453
Centre Léon Bérard
28 rue Laënnec
69373 Lyon Cedex 08
France

L.O. Dragsted
Danish Veterinary and Food
Administration
Institute of Food Safety and Toxicology
19 Morkhoj Bygade
DK 2860 Soborg
Denmark

H. Hahn
Institute of Pathology
GSF Research Center of Environment
and Health
Ingolstädter Landstr. 1
85758 Neuherberg
Germany

S.S. Hecht
Wallin Grant Professor of Cancer
Prevention
University of Minnesota Cancer Center
Box UMHC
420 Delaware St. SE
Minneapolis, MN 55455
USA

K. Hemminki
Centre for Nutrition and Toxicology
Karolinska Institutet
Novum
141 52 Huddinge
Sweden

R. Kaaks
Unit of Nutrition and Cancer
International Agency for Research on
Cancer
150 cours Albert Thomas
69372 Lyon Cedex 08
France

G.J. Kelloff
Chemoprevention Branch
Division of Cancer Prevention and
Control
National Institutes of Health
National Cancer Institute
Bethesda, MD 20892
USA

T.W. Kensler*
Department of Environmental Health
Sciences
Johns Hopkins School of Hygiene and
Public Health
615 North Wolfe Street
Baltimore, MD 21205
USA

*Unable to attend

J.R. Marshall
Professor of Public Health
Arizona Cancer Center College of Medicine
University of Arizona
Tucson, AZ 85724
USA

F.L. Meyskens Jr.
Clinical Cancer Center
Building 23 – RT 81
University of California-Irvine
Orange, CA 92868
USA

A.B. Miller
Division of Clinical Epidemiology-C0500
Deutsches Krebsforschungszentrum
Im Neuenheimer Feld 280
D 69120 Heidelberg
Germany

U.J. Nair
Division of Toxicology and Cancer Risk Factors – C0200
Deutsches Krebsforschungszentrum
Im Neuenheimer Feld 280
D 69120 Heidelberg
Germany

R.W. Owen
Division of Molecular Toxicology - C0300
Deutsches Krebsforschungszentrum
Im Neuenheimer Feld 280
D 69120 Heidelberg
Germany

R.K. Ross
Preventive Medicine
Catherine & Joseph Aresty Professor of Preventive Medicine and Urology
1441 Eastlake Avenue Rm 8302B
Los Angeles, CA 90089 9181
USA

M.S. Tockman
Molecular Screening Program
H. Lee Moffitt Cancer Center and Research Institute
12902 Magnolia Drive
Tampa, FL 33612
USA

H. Vainio
Unit of Chemoprevention
International Agency for Research on Cancer
150 cours Albert Thomas
69372 Lyon Cedex 08
France

T.H. van der Kwast
Department of Pathology
Josephine Nefkens Institute
PO Box 1738
3000 DR Rotterdam
The Netherlands

P. Vineis
University of Turin
Unit of Cancer Epidemiology
Dipartimento di Scienze Biomediche e Oncologia Umana
Via Santena 7
10126 Torino
Italy

C.P. Wild
Molecular Epidemiology Unit
School of Medicine
Algernon Firth Building
University of Leeds
Leeds LS2 9JT
United Kingdom

Observers

B. Bertram
Division of Toxicology and Cancer Risk Factors – C0200
Deutsches Krebsforschungszentrum
Im Neuenheimer Feld 280
D 69120 Heidelberg
Germany

Dr N. Becker
Division of Clinical Epidemiology - C0500
Deutsches Krebsforschungszentrum
Im Neuenheimer Feld 280
D 69120 Heidelberg
Germany

U. Bussas
Division of Clinical Epidemiology - C0500
Deutsches Krebsforschungszentrum
Im Neuenheimer Feld 280
D 69120 Heidelberg
Germany

H. Dally
Division of Toxicology and Cancer Risk Factors - C0200
Deutsches Krebsforschungszentrum
Im Neuenheimer Feld 280
D 69120 Heidelberg
Germany

N. Frank
Division of Toxicology and Cancer Risk Factors - C0200
Deutsches Krebsforschungszentrum
Im Neuenheimer Feld 280
D 69120 Heidelberg
Germany

C. Gerhäuser
Division of Toxicology and Cancer Risk Factors - C0200
Deutsches Krebsforschungszentrum
Im Neuenheimer Feld 280
D 69120 Heidelberg
Germany

R. Godschalk
Division of Toxicology andCancer Risk Factors - C0200
Deutsches Krebsforschungszentrum
Im Neuenheimer Feld 280
D 69120 Heidelberg
Germany

M. Hergenhahn
Division Toxicology and Cancer Risk Factors - C0200
Deutsches Krebsforschungszentrum
Im Neuenheimer Feld 280
D 69120 Heidelberg
Germany

J. Linseisen
Division of Clinical Epidemiology -C0500
Deutsches Krebsforschungszentrum
Im Neuenheimer Feld 280
D 69120 Heidelberg
Germany

C. Mayer
Division of Toxicology and Cancer Risk Factors - C0200
Deutsches Krebsforschungszentrum
Im Neuenheimer Feld 280
D 69120 Heidelberg
Germany

C. Martinsohn
Division of Clinical Epidemiology -C0500
Deutsches Krebsforschungszentrum
Im Neuenheimer Feld 280
D 69120 Heidelberg
Germany

J. Nair
Division of Toxicology and Cancer Risk Factors – C0200
Deutsches Krebsforschungszentrum
Im Neuenheimer Feld 280
D 69120 Heidelberg
Germany

G. Sabbioni
Walther-Straub-Institut für Pharmakologie und Toxikologie
Ludwig-Maximilians-Universität München
Nußbaumstr. 26
D 80336 München
Germany

B. Schlehofer
Environmental Epidemiology - C0600
Deutsches Krebsforschungszentrum
Im Neuenheimer Feld 280
D 69120 Heidelberg
Germany

P. Schmezer
Division Toxicology and Cancer Risk Factors - C0200
Deutsches Krebsforschungszentrum
Im Neuenheimer Feld 280
D 69120 Heidelberg
Germany

O. Schmitz
Division of Molecular Toxicology - C0300
Deutsches Krebsforschungszentrum
Im Neuenheimer Feld 280
D 69120 Heidelberg
Germany

K. Steindorf
Environmental Epidemiology - C0600
Deutsches Krebsforschungszentrum
Im Neuenheimer Feld 280
D 69120 Heidelberg
Germany

M. Wießler
Division of Molecular Toxicology – C0300
Deutsches Krebsforschungszentrum
Im Neuenheimer Feld 280
D 69120 Heidelberg
Germany

A. Nieters
Division of Clinical Epidemiology - C0500
Deutsches Krebsforschungszentrum
Im Neuenheimer Feld 280
D 69120 Heidelberg
Germany

H. Osswald
Division of Toxicology and Cancer Risk Factors - C0200
Deutsches Krebsforschungszentrum
Im Neuenheimer Feld 280
D 69120 Heidelberg
Germany

I. Persson
International Agency for Research on Cancer
150 cours Albert Thomas
69372 Lyon Cedex 08
France

O. Popanda
Division of Toxicology and Cancer Risk Factors - C0200
Deutsches Krebsforschungszentrum
Im Neuenheimer Feld 280
D 69120 Heidelberg
Germany

A. Risch
Division of Toxicology and Cancer Risk Factors - C0200
Deutsches Krebsforschungszentrum
Im Neuenheimer Feld 280
D 69120 Heidelberg
Germany

H. Wikman
Division of Toxicology and Cancer Risk Factors - C0200
Deutsches Krebsforschungszentrum
Im Neuenheimer Feld 280
D 69120 Heidelberg
Germany

Biomarkers in Cancer Chemoprevention
Miller, A.B., Bartsch, H., Boffetta, P., Dragsted, L. and Vainio, H. eds
IARC Scientific Publications No. 154
International Agency for Research on Cancer, Lyon, 2001

Biomarkers in Cancer Chemoprevention

Workshop report*

Introduction

Cancer chemoprevention is the inhibition or reversal of carcinogenesis (before invasion) by intervention with pharmacologically active agents. While minimizing exposure to carcinogens and changes in lifestyle will eventually reduce cancer incidence, chemoprevention offers an alternative approach with the potential for more immediate results, especially in subjects known to be at high risk of cancer.

Numerous oncogenes, tumour-suppressor genes and phenotypic preneoplastic cellular changes are being discovered, allowing improved definition of events in the process of carcinogenesis. Late events are critical in the actual occurrence of cancer, and interventions during these stages of carcinogenesis are thus theoretically very attractive. With better understanding of possible ways to perturb carcinogenesis, preventive intervention becomes increasingly practical for many cancer sites.

Objectives of the workshop

The objectives of the workshop on Biomarkers in Cancer Chemoprevention were:

(i) To summarize the current state of knowledge on biomarkers indicative of individual susceptibility or of carcinogenic exposure and on intermediate biomarkers predictive of invasive cancer occurrence that are relevant to studies on cancer prevention;

(ii) To prepare a consensus report on the role of biomarkers in studies of cancer prevention, covering:

- validation of already existing biomarkers for the purposes of cancer prevention,
- interpretation of the results of studies of cancer prevention using biomarkers,
- future activities for development of new biomarkers;

(iii) To improve the future use of biomarkers in development of new cancer chemopreventive agents.

The workshop built upon a previous IARC meeting on the application of biomarkers in cancer epidemiology (Toniolo *et al.*, 1997), but was more specifically oriented to the application of biomarkers to cancer chemoprevention.

Definition of biomarkers

The term 'biomarkers' refers to indicators of exposures and/or events in biological systems or samples. Biomarkers potentially relevant to cancer chemoprevention include exposure, intermediate-effect, drug effect, tumour and susceptibility markers. Of these, the workshop restricted its attention to exposure, intermediate-effect and susceptibility markers. Many of the characteristics and issues pertaining to each type of biomarker are different and the three types of biomarker are considered separately. However, there is no strict boundary between them, and several markers can be considered as belonging to two or even to all three categories (examples of this would be DNA damage and its repair and mutations in tumour-suppressor genes).

In chemoprevention, an exposure biomarker is any substance or structure that reflects endogenous or exogenous exposure to carcinogenic risk factors, which can be measured in the human body or its products and which may be

*This paper is the consensus report of the workshop *Use of Biomarkers in Chemoprevention of Cancer*, which was held at the Deutsches Krebsforschungszentrum, Heidelberg, Germany, in February 2000.

predictive of the incidence or outcome of disease. A biological marker of intermediate effect is an indicator of the development in an individual of a carcinogenic change short of invasive cancer. A biological marker of cancer susceptibility is an indicator of a heritable ability of an organism to respond to the challenge of carcinogenic agent(s) or event(s).

Biomarkers in chemoprevention

Much of the initial work on biomarkers has been performed in cellular or whole animal models, but there has been no systematic assessment of lessons learnt from such studies to guide the search for biomarkers that might be useful in humans.

Intermediate-effect biomarkers of potential value in chemoprevention are already available for most of the accessible human cancers (e.g., mouth, colon, lung, breast, prostate) but not for cancers at inaccessible sites such as the pancreas. There is need to capitalize on advances in molecular and cellular biology, imaging and microsurgery to reach the relevant organs and access cells and cellular products that may harbour biomarkers.

In practice, biomarkers already play an important role in the evaluation of chemopreventive agents in phase II trials. However, the intermediate-effect biomarkers used in phase II trials have not been validated in relation to subsequent cancer occurrence, and therefore are not being used in phase III trials.

Even large trials in humans are dependent on identifying individuals at high risk for the relevant disease. High-risk populations include those known to be heavily exposed to an important etiological agent (such as heavy smokers at high risk for tobacco-induced cancers); individuals with recognized genetic predisposition, from either high- or low-penetrance genes; and those with already detectable preneoplastic lesions. Biomarkers of both exposure and susceptibility are being developed that could reduce the need for very large samples of subjects, even in phase III chemoprevention trials.

Validity of biomarkers

Utilization of validated biomarkers would markedly facilitate the development and testing of potential chemopreventive agents. Validation of a biomarker involves the understanding of the molecular mechanisms related to chemopreventive action, external exposures, the outcome (cancer or preneoplastic alterations) and the biomarker itself. Several questions should be addressed in the validation of biomarkers:

1. Does the chemopreventive intervention affect the biomarker?

2. Is the biomarker consistently found in populations and/or the disease?

3. Is there molecular understanding as a basis for the use of the biomarker?

4. Is the biomarker associated with a chemopreventive agent or risk factors for cancer?

5. Is the biomarker associated with the outcome (cancer)?

6. Is the chemopreventive intervention effect on the main outcome (cancer) mediated by the biomarker?

7. Are the chemopreventive or risk factor effects mediated by the biomarker?

Many issues related to the validation of biomarkers were discussed in Toniolo *et al.* (1997) and the reader is referred to the relevant chapters (Schulte & Perera, 1997; Pearce & Boffetta, 1997; Vineis, 1997; White, 1997; Schatzkin *et al.*, 1997; Boone & Kelloff, 1997; McMichael & Hall, 1997).

The use of non-valid biomarkers in cancer chemoprevention can lead to invalid inferences and generalizations and ultimately to erroneous assessment and use of cancer-chemopreventive agents. Anything less than perfect validity of the biomarker will lead to imperfection in our assessment of chemoprevention. It would be helpful to specify appropriate targets for the levels of sensitivity, specificity, variability and reproducibility of biomarkers that may be applicable in chemoprevention research.

Exposure biomarkers

Exposure biomarkers may include endogenous or exogenous agents and their metabolites or adducts in tissues or body products, whether in physiological or pathological amounts. Structural changes in the cell or organism which reflect exposure are also included. Exposure is thus defined in a very broad sense and encompasses any influence that might predict the incidence or outcome of disease, including dietary factors, hormonal status, redox status, agent–gene interactions, and others. This is in line with, although still more narrow than, previous definitions, such as that of Armstrong *et al.* (1992) where exposure is referred to as 'any of a subject's attributes or any agent with which he or she may come in contact that may be relevant to his or her health'.

Whether a biomarker is considered an exposure marker may depend on its intended use. For example, HBV antibody seropositivity may be regarded as an exposure biomarker for viral exposure, or as an acquired predisposition marker used to select individuals who are at increased risk of hepatocellular cancer. Another example would be the presence of an HPV-related gene in a cervical biopsy, which may be regarded as an exposure biomarker, since it discriminates between HPV-exposed and non-exposed persons. It may, however, also be used as an intermediate-effect biomarker, since it is usually accompanied by a cellular effect (cervical dysplasia) along the pathway to cervical cancer.

Biomarkers of exposure are sometimes defined more narrowly as biomarkers of external carcinogen exposures. These have been subdivided into biomarkers of exposure, of target tissue dose, and of biologically effective dose. Although exposure biomarkers are meant here to encompass all of these, biomarkers which fall outside this definition are also embraced, for example, because they are related to protective exposures.

In chemoprevention, the most important exposure biomarkers concern exposures directly related to the intervention agent and to the target exposure to be modified by the intervention. It must be assumed that exposures to other risk factors or preventive factors are evenly balanced between the arms of the study, which should be the case provided randomization has been effective, but it may be considered prudent to monitor such equivalence by the use of exposure biomarkers.

The contribution of exposure biomarkers to chemopreventive interventions

Exposure biomarkers may contribute to intervention studies in several ways:

- In the selection of target populations: to identify subgroups of the population who have been exposed to an agent. Any exposure marker which could help in the selection of the relevant target population, making sure they have been exposed (e.g., patients excreting aflatoxin adducts in their urine, cotinine status, viral infection status, etc.), would qualify.

- In monitoring of the target population for confounders: to check that randomization has distributed evenly all important risk factors which are supposed not to change differentially in the different arms of the study during the intervention (e.g., use of nonsteroidal anti-inflammatory drugs (NSAIDs) in a colon cancer intervention trial, alcohol intake in a trial on lowering aflatoxin carcinogenicity in liver cancer, smoking status in a trial on intervention with antioxidants against lung cancer). Measurements may be made repeatedly during the intervention study.

- In the measurement of outcome:

(*a*) to measure quantitatively the efficacy of the intervention, determined as the effect on the targeted 'exposure' risk factor (if applicable), e.g. decreased adduct levels, decreased oxidative stress factors;

(*b*) to monitor chemopreventive agent compliance, uptake and/or distribution;

(*c*) to assess inter-individual and between-group differences in efficacy of the chemopreventive intervention.

- In the interpretation of results: e.g., to evaluate the influence of sub-group variables (as defined by biomarkers for other exposures) which may have influenced the outcome.

Etiological considerations and choice of additional exposure markers

Initially it is important to evaluate an exposure biomarker in relation to the following questions:

- Is the marker causally related to the disease or the intervention?
- What is the rationale for using the biomarker in question, more specifically:
  - Does the marker reflect the targeted 'exposure' (in a broad sense)? and
  - Is the exposure related to the natural history of the disease process?

For instance, if oxidative stress is targeted by intervention with an antioxidant, it is necessary to consider the evidence for the existence of such oxidative damage and for its causal relationship to the relevant disease. An exposure biomarker should in this case reflect the oxidative stress in the target organ, either as a direct measurement (e.g., in biopsies, if possible) or in a surrogate sample known to reflect the variation at the target. Other exposures influencing the targeted exposure should also be monitored, e.g., the target organ, plasma or serum status of dietary or endogenous antioxidants and biomarkers of more general reactive oxygen radical exposures of DNA, lipids or proteins.

Steps in the development of new exposure biomarkers

The development of a new exposure biomarker should take the following course:

1. Observations on risk and protective factors in human or experimental systems.
2. Mechanistic studies in experimental systems to establish the relationship of the biomarker to the causal chain of molecular and cellular events.
3. Development of experimental model systems such as transgenic mice.
4. Development of the biomarker methodology for exposure determination.
5. Testing of the exposure biomarker methodology in animal and human pilot studies.
6. Exploration of markers in large-scale intervention studies.

At any stage of this process, there can be feedback on etiology, prevention, diagnosis, prognosis, and thus an iterative return to step 1.

During steps 4 and 5, a range of characteristics have to be evaluated which constitute the best approximation of the validity of a biomarker. They may be expressed on a continuous scale. These should encompass the approaches to validation considered by Toniolo *et al.* (1997).

Examples of exposure biomarkers

Exposure biomarkers may be used for any of the following purposes:

1. Assessment of exposure to external carcinogens (e.g., body levels of external agent (carcinogen), DNA adducts, protein adducts)
2. Assessment of exposure to harmful endogenous agents (e.g., indices of oxidative damage to DNA or resulting from lipid peroxidation, hormonal levels and levels of hormone-regulated products)
3. As compliance markers, including markers for the intervention agent, for 'adherence' to the risk group, etc.
4. To assess endogenous or exogenous factors which are believed to interact with biological processes leading to cancer (e.g., preventive agents, markers of diet or lifestyle)

During the workshop, a number of examples of exposure biomarkers were discussed related to the six cancer sites considered.

Relevant exposure biomarkers for the skin should reflect exposure to sunlight and biomarkers of defence against radical-induced damage. Development of erythema on the skin is a good marker for excess sun exposure and is very consistent with the adducts formed in DNA. The number of naevi in less sun-exposed locations of the body is believed to reflect mainly childhood exposure to sunlight and if so, might be used as a childhood exposure marker.

For colorectal cancer, NAT2 fast acetylation, GSTM1 null and high CYP1A2 status are potential modifiers of exposure to heterocyclic aromatic amines from cooked meat.

For breast cancer, endogenous levels of hormones or hormone-regulating proteins seem to be relevant exposure markers. Further development of these biomarkers should have high priority. Exposure to certain carcinogens belonging to the polycyclic aromatic hydrocarbon group or to the heterocyclic aromatic amine group gives rise to breast cancer in rat models and might be investigated in humans by using available or developing exposure biomarkers.

For prostate cancer, exposure biomarkers for endogenous androgens, androgen-regulated proteins and antiandrogens are currently available or under development. Selenium speciation biomarkers and plasma tocopherols might also be important exposure biomarkers for studies on prostate cancer.

For liver cancer, good biomarkers for cumulative exposures to alcohol are not available at present, whereas markers of exposure to aflatoxin B_1 and hepatitis B and C viruses may be adequately assessed in blood samples. Exposure biomarkers for direct or indirect DNA damage (e.g., 8-hydroxydeoxyguanosine (8-OH-dG) and etheno adducts) resulting from radical-mediated damage including lipid peroxidation might also prove useful in intervention studies on liver cancer.

For aerodigestive tract cancers, several biomarkers for exposure to polycyclic aromatic hydrocarbons and tobacco-specific nitrosamines have been developed. A simple biomarker for tobacco exposure, plasma cotinine, has the advantage of not being influenced by chemopreventive treatments. The apparent contradiction between elevated plasma β-carotene level as a negative predictor of lung cancer risk in smokers and the direct risk increase caused by β-carotene supplements calls for the use of other biomarkers of fruit and vegetable intake. Markers for other carotenoids are already available. Exposure biomarkers for various groups of dietary polyphenols have also been developed and biomarkers for other potentially modulating factors like isothiocyanates, indoles or terpenoids are available or under development.

Further research needed

In the area of exposure biomarkers, there are several areas where more research is needed, either in biomarker validation or in the development of new biomarkers.

(a) Exposure biomarker validation

There is a need for better and more systematic validation before exposure biomarkers are applied, requiring more interlaboratory validation efforts. The development of more than one marker for each end-point, using different analytical techniques, can disclose some of the analytical problems, which are difficult to identify with common validation procedures. Some aspects of validation are difficult to assess in pilot human studies, for example interactions with host factors or with dietary habits, and the inclusion of validation programmes into large human intervention studies is therefore important. An important aspect of the biological validation of an exposure biomarker is evaluation of underlying hypotheses on the relationship between the exposure and the effect. An example is whether oxidative stress or any DNA adduct formation is truly detrimental in all cases; if such processes are also used in endogenous signalling pathways, for instance in the induction of defence and repair, measurement of such end-points may lead to misclassification. With increasing sensitivity of methods, dose ranges may be studied which are well below those used in animal studies of dose–response relationships for genotoxic compounds. High priority should be given to the conduct of short-term human intervention studies for this type of biological validation of the hypotheses often implicitly underlying the use of exposure biomarkers.

(b) Development of new exposure biomarkers

In order to identify dietary or endogenous components which might influence newly identified molecular targets for chemoprevention, there is a need to develop in vitro/in vivo screening assays (for example, for screening of COX-2 inhibitors, compounds possessing hormonal activity, etc.). Such screening methods could be developed for exposure factors influencing a range of known or anticipated steps in carcinogenesis. The use of sophisticated animal model systems such as transgenic mice or even higher eukaryotic systems should be encouraged.

High priority should also be given to the development of biomarkers of exposure to substances which help the body to defend itself. In particular, better markers of dietary exposure and their interaction with host factors are needed.

The development of better (validated) oxidative stress markers at the DNA level, encompassing both direct and indirect damage, should have high priority due to the implication of redox factors in carcinogenesis at several stages. Specifically, the evaluation of surrogate markers in blood and their correlation with target organ exposures should be investigated in animal models and in human pilot studies. This would cover part of the strong need for more robust methods for DNA adduct measurements in human samples. However, also for other types of adducts, including those with small or bulky alkylating agents, more robust methods are required which can be applied in clinical laboratories. In particular, the sensitivity of the available assays needs to be improved. Furthermore, there is a strong need for biomarker methodologies which could accurately assess exposures to factors that work through mechanisms other than DNA adduct formation. Methods for measuring potential protective dietary factors, oxidative stress factors, viral infections and hormonal levels, which may all be characterized to some extent as epigenetic factors in carcinogenesis, should be extended and refined. Another area of interest is protein or RNA modification, which could be more widely used for biomarkers.

Intermediate-effect biomarkers

An intermediate-effect biomarker is a detectable lesion or biological parameter with some of the histological or biological features of preneoplasia or neoplasia but without evidence of invasion, which either is on the direct pathway from the initiation of the neoplastic process to the occurrence of invasive cancer, has a high probability of resulting in the development of cancer, or is a detectable biochemical abnormality which is highly correlated with the presence of such a lesion. Thus intermediate-effect markers include (*a*) detectable precancerous changes in an organ (confirmed by histology), (*b*) an alteration of a gene that is considered to play a causative role, and (*c*) other indicators of carcinogenesis, such as the expression of a marker that represents the cause of a cancer (e.g., HPV DNA positivity). Causation is not a requirement for inclusion in this group, but the expectation is that a relevant biomarker can eventually be linked, through a biological mechanism, to the cancer.

A hierarchy of intermediate-effect biomarkers can be perceived. Those that are known to be on the causal pathway to cancer are at the top, and can be truly called intermediate-effect markers. Then there are those markers for which present knowledge indicates only a probability of association with cancer, but it is uncertain as to whether they are on the causal pathway; these can only be called intermediate markers. However, whether one is talking about 'tissue lesions' (e.g., liver foci, skin papillomas, dysplastic lesions) or molecular markers (e.g., alterations in *ras* or *p53* genes), it is essential to know the statistical relationship between the lesions and cancer incidence.

A subset of intermediate-effect markers, which can be modulated, have been called surrogate end-point biomarkers (Kelloff *et al.*, 2000). An expectation is that if reduction in, but not abolition of, surrogate end-point biomarkers is shown, it can be demonstrated that those that remain do not have markers of progression, i.e. the bad actors have been removed.

Intermediate-effect markers contribute to chemopreventive interventions:

- in the selection of target populations: identification of subgroups of the population in whom precancerous lesions are detectable (e.g., patients with adenomatous polyps)
- in the measurement of outcome: surveillance of the efficacy of the intervention (e.g., women who test HPV DNA-positive in whom the subsequent incidence of high-grade cervical lesions is changed)
- in the interpretation of results: assessment of inter-individual and inter-population differences in efficacy of chemopreventive agents
- in the design of relevant animal models: creating informative animal models at high risk for

development of precancerous lesions (e.g., transgenic systems).

- in the characterization of molecular pathways involved.

General considerations regarding use of intermediate-effect biomarkers in chemoprevention

It has not been convincingly shown that the use of chemopreventive agents in men and women with any type of preneoplastic lesion can substantially reduce the subsequent development of a truly invasive cancer. In general, not enough is known about the natural history of precancerous lesions to identify those that will progress to invasive cancer if allowed to do so. It would be valuable to establish biomarkers of potential for such progression that could be used in chemoprevention research. Conventional histopathology alone has not proved reliable for such purposes.

Examples of intermediate markers

Skin cancer *in situ* is generally considered irreversible, and qualifies as an intermediate-effect marker. Markers of sun damage to the epidermis (sunburn cells or *p53* mutations) or melanocytes are indicators of early effects which are reversible. Actinic keratoses are later events, but may spontaneously regress after cessation of exposure to ultraviolet radiation. Naevi are risk factors that may be precursors of melanoma. They are clonal. Large atypical naevi are likely to be precursors, but it has not yet been determined whether acquired naevi represent actual premalignant lesions.

Adenomas are the most commonly used intermediate-effect biomarker for colorectal cancer. In chemoprevention trials, adenomatous polyps are used as biomarkers of risk, but can also be treated as surrogate end-points. There is a need to consider if chemoprevention will remove those adenomas that are most likely to progress to cancer. Oxidative DNA base modifications may be usable as markers, but are really biomarkers of exposure, not of outcome. The value of aberrant crypt foci or crypt fission as risk markers or as surrogate end-points is uncertain, as is the role of *ras* mutations in stools and of mucosal proliferation.

The degree of mammary density as a proportion of breast, measured on mammograms, is associated with increased risk of breast cancer. Ductal carcinoma *in situ* (DCIS) may be a marker of risk but may not be a precursor in the classic sense. However, DCIS is being used as a marker of drug effect in phase IIa studies, when an agent is given for three weeks, with modulation of cellular progression as end-point.

The natural history of prostate intraepithelial neoplasia (PIN) is not clear. High-grade PIN is associated with the subsequent risk of prostate cancer, but it is difficult to be certain that invasive cancer was not present at the time of diagnosis of PIN, while some of the architectural variants of PIN may be confused with intraductal spread of a concomitant adenocarcinoma. Chemoprevention studies have been conducted by enrolling men before planned prostate surgery and assessing the effect of intervention on high-grade PIN in the six-week interval before surgery. People with PIN have also been enrolled in studies and followed for cancer.

For liver (hepatocellular) cancer, it is possible that tests for *p53* mutations may indicate long-term changes, but accumulation of the p53 protein is not an early event in hepatic carcinogenesis in areas with low exposure to aflatoxin. Preneoplastic phenotypically altered lesions (falling short of either benign or malignant neoplasia) are very small and cannot be detected by conventional biochemical approaches in tissue homogenates. However, their assessment may be useful in short-term chemoprevention trials in individuals scheduled for liver transplantation. Cell proliferation is not a good marker for the earliest emerging types of preneoplastic lesion. Glycogenotic-basophilic cell lineages are found in humans with hepatocellular cancer.

Atypical cells in sputum are only weakly predictive of lung cancer. Chemopreventive agents have been found to modulate (upregulate) nuclear retinoic acid receptor β (RARβ). Many of the non-calcified nodules identified by helical computerized tomographic (CT) scanning show atypical adenomatous hyperplasia. These may be related to bronchioalveolar carcinoma. Immunostaining of A2/B1 heterogeneous nuclear ribonucleoprotein (hnRNP) is not correlated with histological change. Metabolically active preneoplastic cells seem to most actively express this protein.

Other relevant considerations

- Prevention of invasion is the fundamental goal in cancer chemoprevention.
- There is an increasing role for computer-assisted imaging for morphometric analysis to assess intermediate biomarkers in chemoprevention research.
- The term intraepithelial neoplasia (IEN) is being used to describe lesions such as colorectal adenomas, breast ductal carcinoma *in situ*, prostatic intraepithelial neoplasia, cervical intraepithelial neoplasia and actinic keratoses. Lesions such as Barrett's oesophagus, bronchial dysplasia, bladder dysplasia and oral leukoplakia are not neoplastic and should not be described as IEN, but may qualify as intermediate-effect biomarkers. Early biological events that could be potential targets of chemoprevention include alterations in protein kinases, transcription factors, enzymes involved in carbohydrate and lipid metabolism, factors that control angiogenesis, and altered components of the immune system.
- Animal models, particularly those using transgenic mice, are useful in the development of intermediate-effect biomarkers, but caution is required in extrapolating the results to the human situation.
- It is necessary to determine how good DNA modifications or adducts are as biomarkers, and how best to assess oxidative stress: by DNA adducts, protein oxidation, or other methods. Methods for measuring oxidative DNA base alterations are probably not yet sensitive enough. The use of classical DNA repair assays such as unscheduled DNA synthesis is not sufficient as the repair system is very refined and there is a large degree of heterogeneity in DNA repair within the genome. Active genes are repaired much faster and with different enzymology than inactive genomic regions. Techniques are required to measure this kind of repair in populations. Methods to assess DNA repair in individual cells are now emerging, and should be applied in population studies.

Susceptibility markers in chemoprevention

A biological marker of cancer susceptibility is an indicator of a heritable ability of an organism to respond to the challenge of carcinogenic agent(s) or event(s). In particular with respect to cancer chemoprevention, susceptibility markers are also indicators of an ability to respond to the cancer-preventive action. The marker can indicate an increased susceptibility to chemoprevention as well as a resistance to it. Thus the response can take the form of an enhanced sensitivity or of an adverse effect.

The very limited spectrum of susceptibility biomarkers available in the past, such as the phenotypically obvious features sex and skin colour, is expanding dramatically with the advances in the field of genetics. The traditional distinction between high- and low-penetrance gene defects obscures a continuum of susceptibility at the biological level. At the operational level, however, it is possible to identify a small number of genes in which pathological mutations are sufficiently predictive of cancer risk to influence clinical management. Any allelic variant of the 'major' genes, alteration in other interactive genes or in environmentally sensitive polymorphisms which would not be sufficiently penetrant to determine clinical practice will be categorized as low-penetrance biomarkers. As knowledge expands and the capacity to test multiple genes simultaneously becomes commonplace, high-penetrance 'genotypes' comprising several low-penetrance genetic variations may be recognized.

Susceptibility markers require specific types of validation, for example phenotype–genotype correspondence (the lack of such correspondence for CYP2E1 casts doubt on the usefulness of this marker for chemopreventive trials) and expression in relevant tissues.

Validated biomarkers of genetic or acquired susceptibility can be conceptualized as 'effect modifiers' in epidemiological studies. From a biological perspective, effect modification conceptually answers the question as to why two similarly treated (exposed) individuals or groups of individuals respond differently.

Use of susceptibility biomarkers in chemoprevention trials and interventions

Markers of susceptibility can contribute at different levels of chemopreventive trials and interventions.

1. Selection of high-risk target populations

The strategy of chemoprevention may be targeted to high-risk individuals. Patients diagnosed with cancer are at increased risk of a second primary cancer and are highly motivated to take part in trials. Individuals in families with genetic syndromes have increased risk of developing certain cancers. For colorectal cancer, the two major syndromes are hereditary non-polyposis colorectal cancer (HNPCC) and familial adenomatous polyposis (FAP). HNPCC accounts for 5% and FAP for 1% of all persons with the disease. The lifetime risk of developing colorectal cancer is almost 100% for FAP and around 70% for HNPCC. Randomized chemoprevention trials (with NSAIDs and nutritional supplements) in FAP and HNPCC families are in progress.

Results of chemoprevention trials in genetically high-risk individuals may apply to the general population if the high-risk cohort represents common disease pathways and the relevant biological pathways through which the chemopreventive agent operates are the same in the susceptible and the non-susceptible. By the same token, if pathways differ, an agent ineffective in the genetically susceptible may, nonetheless, have efficacy in the general population.

This subject-selection strategy has advantages but also disadvantages.

Advantages

- Individuals at risk may be easily traced through clinical or genetic registers.
- Perceived benefits to the individual and the families should increase compliance.
- Expensive interventions such as colonoscopy are available as part of routine health care.
- High penetrance reduces the number of participants and duration of treatment needed to achieve statistical power.

Disadvantages

- The high-risk cohort may not represent common disease pathways.
- Large-scale genotyping may overload the capacity to provide adequate pre-test genetic counselling.
- Compliance may be reduced if there is a perception of threat to health insurance if an individual's genetic status is revealed.
- Fear of cancer may encourage drop-out or non-compliance before a randomized trial is complete, especially if the agent under consideration is readily available to the general population.

2. Target of chemoprevention

Susceptibility markers can be the target of chemopreventive interventions at the phenotypic level. This is the case when the chemopreventive agent is chiefly aimed to modify the expression (phenotype) of a susceptibility factor. Many phase I or phase II metabolic pathways are inducible by a number of potential chemopreventive agents such as components of fruits and vegetables. For example, cruciferous vegetables can be administered as inducers of CYP1A2 and other inducible enzymes.

3. Modification of effect of chemoprevention

It is plausible that genetic susceptibility factors may modify the effect of chemopreventive agents. Such effect modification would be responsible for inter-individual and inter-population differences in the efficacy of chemopreventive interventions. In humans, however, no results are currently available showing such effect modification. Nevertheless, evidence that some possible chemopreventive agents (e.g., carotenoids) operate through induction of metabolizing enzymes in experimental systems supports the notion of a role for metabolic polymorphisms in the modulation of the response to chemoprevention.

Animal models in the design of new susceptibility biomarkers

The recent development of mouse strains with overexpressed or inactivated cancer-related genes has provided researchers with new models for testing chemoprevention strategies that could counteract specific genetic susceptibilites to cancer. The multiple intestinal neoplasia (Min) mouse, which carries a fully penetrant dominant mutation of the *Apc* gene, was first reported in 1990 (Moser *et al.*, 1990). Mice that are heterozygous for the *Apc* mutation develop scores of grossly detectable adenomas throughout the small intestine, and less so in the colon. Studies with Min and Cox-2 knock-out mice have provided strong evidence that Cox-2 plays a major role in intestinal carcinogenesis, and that NSAIDs which target the Cox-2 protein have great potential as chemopreventive agents.

Mutation of the *p53* gene is the most commonly observed genetic lesion in human cancer; more than 50% of all human tumours examined have identifiable *p53* mutations or deletions. Donehower *et al.* (1992) reported that homozygous *p53*-knock-out mice were viable but highly susceptible to spontaneous tumorigenesis (particularly lymphomas) at an early age. Hursting *et al.* (1994) reported that calorie restriction significantly delayed the onset of spontaneous tumorigenesis in $p53^{+/-}$ mice. Heterozygous *p53*-knock-out mice may be analogous to humans susceptible to heritable forms of cancer due to decreased *p53* expression, such as individuals with Li–Fraumeni syndrome. These mice exhibit increased sensitivity to mutagenic carcinogens, and thus may be susceptible to low-dose chronic carcinogen regimens that more closely mimic human exposures. Thus these mice may have great potential for developing models for studying modulatable biomarkers relevant to human cancer chemoprevention.

Design issues when susceptibility markers are integrated into chemoprevention trials

1. Selection of study populations

Differences in the distribution of allelic variants of putative susceptibility genes across populations can be used to define the study population by ethnic or geographical origin, with the aim of selecting the study population with the highest allele frequency. When there is a strong age-dependence in the penetrance of the susceptibility marker, the age range of the study population can be chosen accordingly to target the chemoprevention to the subjects with highest risk.

2. Sample size and statistical analysis

Calculation of study sample size needs to take into account the gene penetrance and the prevalence of the susceptibility allele(s) when the susceptibility allele(s) is(are) used to identify the target study population. In the case of risk modification by other susceptibility allele(s) or risk factors, the prevalence of combined genetic and other risk factors needs to be considered. When a susceptibility allele is treated as a modifier of the effect of chemoprevention in the intervention groups, the prevalence of the susceptibility allele needs to be considered in addition to the risk of disease in the study population. In general, the sample size will be determined by the subgroup with the lowest expected proportion of subjects.

Susceptibility alleles considered to modify the disease risk of the study group or the effect of chemoprevention need to be accounted for in the analysis of the efficacy of the chemoprevention.

Ethical issues in relation to susceptibility markers

A large literature has developed dealing with the health implications of predictive testing for late-onset disease using high-penetrance biomarkers. Chemoprevention trials targeted at carriers of low-penetrance biomarkers raise new challenges. On the one hand, disclosure of genetic status in relation to metabolic polymorphisms may cause unnecessary anxiety and create difficulty for clinicians asked to explain the results. On the other hand, confidential genotyping to stratify populations before enrolment would involve non-disclosure of genetic information, a practice liable to generate objections from research ethical committees.

Recommendations

1. Studies should be conducted in homogeneous, well defined, high risk groups (i.e., subjects with dysplastic lesions).

2. Costly five-year studies are not the ideal start for programmes, especially if they involve heterogeneous populations (mixture of high- and low-risk responders, due to genetics, diet, etc.).

3. There is an urgent need to develop chemopreventive options for susceptible high-risk individuals.

4. Research is needed to validate intermediate markers as indicators of effect with a high probability that the marker is on the causal pathway to cancer.

5. Research should be encouraged into the cellular and molecular biology and pathogenesis of preneoplasia in order to identify and validate intermediate biomarkers relevant to chemoprevention.

6. Techniques, assays and scientific results need to be compared and validated among different laboratories to develop reliable methods.

7. Chemopreventive agents should be sought that will yield (*a*) clinical benefit from the arrest or reversal of surrogate lesions and (*b*) enhanced quality of life. Continued monitoring for adverse effects that might not be observed in short-term surrogate end-point studies is important.

8. There is a need to explore the ethical dimension of genetically targeted chemoprevention. In particular, there is a dilemma in relation to disclosure of genetic information about low-penetrance biomarkers. Should information about biomarkers of little or no relevance to the individual be disclosed and if not, what methods should be employed to protect confidentiality? If a biomarker or combination of biomarkers becomes predictive of a disease, how will the information be made available to the participants?

9. Reproducible high-risk genotypes comprising several lower-penetrance genetic variants should be identified so as to expand the potential to target high-risk individuals for chemoprevention trials.

10. Concentration on populations with identifiable genetic polymorphisms would help in assessing the public health impact of chemoprevention.

11. It is important to assess whether agents that are potentially beneficial in some individuals may be harmful in others, and whether chemoprevention can be restricted to only those who will benefit from the agent.

12. Studies should have large enough sample size to permit assessment of gene–environment and gene–gene interactions. Such large studies require adequate long-term funding. However, with appropriate combination of exposure and susceptibility markers, it may be possible to concentrate on subjects especially at risk and reduce the sample sizes needed.

13. Meta-analyses and pooled analyses should be used to combine small studies. However, caution should be applied in interpreting their results.

References

Armstrong, B.K., White, E. & Saracci, R. (1992) *Principles of Exposure Measurement in Epidemiology*, Oxford, Oxford University Press, p. 4

Boone, C.W. & Kelloff, G.J. (1997). Biomarker end-points in cancer chemoprevention trials. In: Toniolo, P., Boffetta, P., Shuker, D.E.G., Rothman, N., Hulka, B. & Pearce, N., eds, *Application of Biomarkers in Cancer Epidemiology* (IARC Scientific Publications No. 142), Lyon, IARC, pp. 273–280

Coggon, D. & Friesen, M.D. (1997). Markers of internal dose: chemical agents. In: Toniolo, P., Boffetta, P., Shuker, D.E.G., Rothman, N., Hulka, B. & Pearce, N., eds, *Application of Biomarkers in Cancer Epidemiology* (IARC Scientific Publications No. 142), Lyon, IARC, pp. 95–101

Donehower, L.A., Harvey, M., Slagle, B.L., McArthur, M.J., Montgomery, C.A., Jr, Butel, J.S. & Bradley, A. (1992) Mice deficient for p53 are developmentally normal but susceptible to spontaneous tumours. *Nature*, **356**, 215–221

Hursting, S. D., Perkins, S. N. & Phang, J.M. (1994) Calorie restriction delays spontaneous tumorigenesis in p53-knockout transgenic mice. *Proc. Natl Acad. Sci. USA*, **91**, 7036–7040

Kelloff, G.J., Sigman, C.C., Johnson, K.M., Boone, C.W., Greenwald, P., Crowell, J.A., Hawk, E.T. & Doody, L.A. (2000) Perspectives on surrogate end points in the development of drugs that reduce the risk of cancer. *Cancer Epidemiol. Biomarkers Prev.*, **9**, 127–137

McMichael, A.J. & Hall, A.J. (1997) The use of biological markers as predictive early-outcome measures in epidemiological research. In: Toniolo, P., Boffetta, P., Shuker, D.E.G., Rothman, N., Hulka, B. & Pearce, N., eds, *Application of Biomarkers in Cancer Epidemiology* (IARC Scientific Publications No. 142), Lyon, IARC, pp. 281–289

Moser, A.R., Pitot, H.C. & Dove, W.F. (1990) A dominant mutation that predisposes to multiple intestinal neoplasia in the mouse. *Science*, **247**, 322–324

Pearce, N. & Boffetta, P. (1997) General issues of study design and analysis in the use of biomarkers in cancer epidemiology. In: Toniolo, P., Boffetta, P., Shuker, D.E.G., Rothman, N., Hulka, B. & Pearce, N., eds, *Application of Biomarkers in Cancer Epidemiology* (IARC Scientific Publications No. 142), Lyon, IARC, pp. 47–57

Schatzkin, A., Freedman, L.S., Dorgan, J., McShane, L., Schiffman, M.H., & Dawsey, S.M. (1997). Using and interpreting surrogate end-points in cancer research. In: Toniolo, P., Boffetta, P., Shuker, D.E.G., Rothman, N., Hulka, B. & Pearce, N., eds, *Application of Biomarkers in Cancer Epidemiology* (IARC Scientific Publications No. 142), Lyon, IARC, pp. 265–271

Schulte, P.A. & Perera, F.P. (1997). Transitional studies. In: Toniolo, P., Boffetta, P., Shuker, D.E.G., Rothman, N., Hulka, B. & Pearce, N., eds, *Application of Biomarkers in Cancer Epidemiology* (IARC Scientific Publications No. 142), Lyon, IARC, pp. 19–29

Toniolo, P., Boffetta, P., Shuker, D.E.G., Rothman, N., Hulka, B. & Pearce, N., eds (1997) *Application of Biomarkers in Cancer Epidemiology* (IARC Scientific Publications No. 142), Lyon, IARC

Vineis, P. (1997) Sources of variation in biomarkers. In: Toniolo, P., Boffetta, P., Shuker, D.E.G., Rothman, N., Hulka, B. & Pearce, N., eds, *Application of Biomarkers in Cancer Epidemiology* (IARC Scientific Publications No. 142), Lyon, IARC, pp. 59–73

White, E. (1997) Effects of biomarker measurement error on epidemiological studies. In: Toniolo, P., Boffetta, P., Shuker, D.E.G., Rothman, N., Hulka, B. & Pearce, N., eds, *Application of Biomarkers in Cancer Epidemiology* (IARC Scientific Publications No. 142), Lyon, IARC, pp. 73–93

Biomarkers in Cancer Chemoprevention
Miller, A.B., Bartsch, H., Boffetta, P., Dragsted, L. and Vainio, H. eds
IARC Scientific Publications No. 154
International Agency for Research on Cancer, Lyon, 2001

Surrogate end-point biomarkers in chemopreventive drug development

G.J. Kelloff, C.C. Sigman, E.T. Hawk, K.M. Johnson, J.A. Crowell and K.Z. Guyton

Relevant and feasible surrogate end-points are needed for the evaluation of intervention strategies against cancer and other chronic, life-threatening diseases. Carcinogenesis can be viewed as a process of progressive disorganization. This process is characterized by the accumulation of genotypic lesions and corresponding tissue and cellular abnormalities, including loss of proliferation and apoptosis controls. Potential surrogate end-points for cancer incidence include both phenotypic and genotypic biomarkers of this progression. In the US National Cancer Institute chemoprevention programme, histological modulation of a precancer (intraepithelial neoplasia) has so far been the primary phenotypic surrogate end-point in chemoprevention trials. Additionally, high priority has been given to biomarkers measuring specific and general genotypic changes correlated with the carcinogenesis progression model for the targeted cancer (e.g., progressive genomic instability as measured by loss of heterozygosity or amplification at specific microsatellite loci). Other potential surrogate end-points include proliferation and differentiation indices, specific gene and general chromosome damage, cell growth regulatory molecules, and biochemical activities (e.g., enzyme inhibition). Serum biomarkers thought to be associated with cancer progression (e.g., prostate-specific antigen) are particularly appealing surrogate end-points because of accessibility. Potentially chemopreventive effects of the test agent may also be measured (e.g., tissue and serum estrogen levels in studies of steroid aromatase inhibitors). To establish chemopreventive efficacy, prevention of virtually all biomarker lesions, or of those lesions with particular propensity for progression, may be required. Ideally, the phenotype and genotype of any new or remaining precancers in the target tissue of chemopreventive agent-treated subjects would show less, and certainly no greater, potential for progression than those of placebo-treated subjects.

Introduction

Cancer chemoprevention can be defined as treatment of carcinogenesis — i.e., its prevention, inhibition or reversal (Hong & Sporn, 1997; Kelloff, 2000). In most epithelial tissues, accumulating mutations (i.e., genetic progression) and loss of cellular control functions are observed during the course of sequential histological changes that culminate in cancer. These changes are manifested as the transition from normal histology to early intraepithelial neoplasia, through increasingly severe intraepithelial neoplasia to superficial cancers and finally invasive disease. Although the carcinogenic process can be relatively aggressive (e.g., in the presence of a DNA-repair-deficient genotype or viral transformant such as human papillomavirus), these changes generally occur over a long time period (Table 1). Cancers generally develop over decades and intraepithelial neoplasia (e.g., prostatic intraepithelial neoplasia, colorectal adenomas) may also progress slowly.

The progressive nature of carcinogenesis underscores the advantage of chemoprevention — to intervene when the mutations are fewer, even before tissue-level phenotypic changes are evident. However, a major obstacle to chemopreventive drug development is the use of cancer incidence as the end-point for determining efficacy in clinical trials. Such studies entail huge sample sizes, lengthy follow-up periods and high cost (Hong & Sporn, 1997; Kelloff *et al.*, 1995, 2000; Kelloff, 2000). Typically, cancer incidence reduction trials have planned durations of 5–10 years with subject accrual in the tens of thousands. Surrogate end-

Table 1. Incidence and multi-year time course for progression of precancers in selected cancer targets

Target organ	Precancer (IEN)	Estimated incidence	Years for precancer formation	Years for progression from precancer to cancer	References
Prostate	PIN	40–50% of men aged 40–60 years	20	10 or more to latent cancer; 3–15 further years to cancer	Bostwick, 1992
Breast	DCIS	46 000 new cases in women in 2000	14–18 from atypical hyperplasia	6–10	Frykberg & Bland, 1993 Page *et al.*, 1985; Greenlee *et al.*, 2000
Colon	Adenoma	30–40% of the western population aged > 60 years	5–20	5–15	Bruzzi, 1995; Day & Morson, 1978; Zauber *et al.*, 1996
Bladder	Ta, T1, TIS	37 500 cases in USA for 1997	20	<5	Cotran *et al*, 1989; Scher *et al.*, 1997
Oesophagus	Barrett's metaplasia	0.4% of the western population	5–20	5–20 to severe dysplasia; 3–4 further years to cancer	Ovaska *et al*, 1989; Williamson *et al*, 1991; Miros *et al*, 1991; Cameron & Lomboy, 1992; Jankowski *et al*, 1993; Falk & Richter, 1996

DCIS, ductal carcinoma *in situ*; PIN, prostatic intraepithelial neoplasia; TIS, transitional-cell carcinoma *in situ*.

point biomarkers are an important aspect of the chemopreventive drug development process in that they provide a means for overcoming these obstacles (e.g., American Association for Cancer Research, 1999; Hong & Sporn, 1997; Kelloff *et al.*, 1995, 2000; Kelloff, 2000; Sporn & Suh, 2000). The use of phenotypic and genotypic biomarkers as surrogate end-points for cancer incidence would permit the evaluation of chemopreventive efficacy in most cancer targets in up to three years with no more than several hundred subjects. Use of surrogate end-points is possible only because of increasing knowledge of the genetic, histopathological and molecular basis of carcinogenesis. This expanding appreciation of the carcinogenic process will support the continuing efforts to identify, validate and apply biomarkers as surrogate end-points for cancer incidence.

Rationale for surrogate end-points of carcinogenesis: molecular progression models

Carcinogenesis is characterized by a progressive loss of proliferation and apoptosis controls and increasing disorganization, aneusomy and heterogeneity. The appearance of specific molecular and more general genotypic damage is associated with increasingly severe dysplastic phenotypes (Califano *et al.*, 1996, and other studies cited below). In many cases, critical early steps include inactivation of tumour-suppressor genes, such as those for adenomatous polyposis coli (*APC*) or breast cancer (*BRCA*) and activation of oncogenes such as *ras*. Carcinogenesis may follow multiple paths, and be multifocal; not all cancers in a given tissue nor all cells in a given cancer may ultimately contain the same lesions. Progression may also be influenced by factors specific to the host tissue's environment, such as the action of hormones produced in stroma around the developing epithelial tumour and changes in tissue and chromatin structure (Schipper *et al.*, 1996; Sporn, 1996; Bissell *et al.*, 1999; Sporn & Suh, 2000; Stein *et al.*, 2000). Genetic progression models have been established for many human cancers, including colon, brain, bladder, head and neck, non-small-cell lung cancer and cervical intraepithelial neoplasia (Fearon & Vogelstein, 1990; Sidransky & Messing, 1992; Sidransky *et al.*, 1992a,b; Simoneau & Jones, 1994; Kishimoto *et al.*, 1995; Rosin *et al.*, 1995; Thiberville *et al.*, 1995; Califano *et al.*, 1996; Mao *et al.*, 1996). These models indicate that the sequence of genetic damage leading to cancer can involve myriad combinations of targets in the array of pathways that govern proliferation and apoptosis. These genotypic lesions, and the corresponding tissue and cellular abnormalities, have high potential to serve as surrogate end-points when they are sufficiently stable to allow screening during carcinogenesis. Specific carcinogenesis-associated molecular lesions identified so far, while important, may not be the most informative among those that will be discovered as research continues. Most cancer is preceded by an abnormal histological precancer phenotype which integrates the progressive genetic and molecular changes. Thus, focusing on assessment of this abnormal phenotype and the accompanying genotypic changes within the target tissue appears at present to provide the best opportunity for validating surrogate end-points.

Phenotypic and genotypic surrogate end-points to establish chemopreventive efficacy

Intraepithelial neoplasia, the embodiment of the abnormal cancer phenotype, serves as a promising surrogate end-point for chemoprevention studies in epithelial tissues (Kelloff *et al.*, 1995, 2000; Kelloff, 2000). Although shorter than the period for developing cancer, the latency for progression of intraepithelial neoplasia can also be lengthy compared with the practical time frame for a chemopreventive intervention study. Importantly, the number of precancers may far exceed the number of cancers that subsequently develop in the target tissue. Additionally, behavioural (e.g., smoking history), environmental (e.g., hormonal status) and co-existing disease (e.g., immune system competence) factors may influence progression in individual subjects. Intraepithelial neoplastic lesions that will progress may also have particular characteristics predisposing them to develop into cancers. For example, the potential of colorectal adenomas to progress to cancer correlates with histological growth pattern, size and severity of dysplasia (Muto *et al.*, 1975; Hamilton, 1992, 1996).

For these reasons, histological determination of drug-induced prevention or regression of intraepithelial neoplasia alone may not be sufficient for assessing chemopreventive efficacy. The specific and general genotypic effects comprising the progression models for carcinogenesis, and the underlying molecular pathology of the lesions, should also be considered in the evaluation. A reduced incidence of new precancers in the target tissue in agent-treated subjects would ideally be accompanied by a genotype reflecting decreased, and certainly no greater, carcinogenic potential. In particular, when regression of existing precancers is incomplete, the remaining lesions in the agent-treated subjects should have genotypes with equivalent or lower propensity for progression than placebo control subjects.

Potential surrogate end-points at major cancer target organs

Cancers in at least 12 organ systems have been evaluated as targets for chemopreventive agents: prostate, breast, colon, lung, head and neck, bladder, oesophagus, cervix, skin (non-melanoma and melanoma), liver, ovary and multiple myeloma (Kelloff, 2000; Kelloff *et al.*, 2000). Many classes of agent, including retinoids, antioxidants, anti-inflammatories, antiestrogens and antiandrogens, have shown promising chemopreventive activity in one or more of these organ systems (Hong & Sporn, 1997; Kelloff, 2000; Sporn & Suh, 2000; Kelloff *et al.*, 2000); more than 40 candidate chemoprevention drugs are currently under clinical development in studies sponsored by the US National Cancer Institute chemoprevention program (Kelloff, 2000; Kelloff *et al.*, 2000). Among the cellular mechanisms of chemopreventive action of these drugs are inhibition of angiogenesis, mutagenesis, proliferation and apoptosis, as well as modulation of hormone activity. Often single agents exhibit multiple interrelated and/or independent mechanisms that may each contribute to the overall chemopreventive effect. For example, in addition to modulating estrogen receptor binding, antiestrogens can inhibit insulin-like growth factor-I (IGF-I), while cyclooxygenase (COX)-2 inhibitors can modulate the peroxisome proliferator-activated receptors and the pathways controlled by these nuclear receptors. The selection of appropriate biomarkers to monitor the efficacy of these agents should consider their purported mechanisms of action in the target organ of interest. For all the biomarkers, it is highly desirable to measure modulation quantitatively as the difference between the biomarker value at baseline and the end of treatment. The change in the surrogate end-point measures on chemopreventive treatment should also be compared with that seen in appropriate controls. Thus, biopsies or other tissue measurements at baseline are essential.

Table 2 provides a target-organ based listing of the types of biomarker currently being used to study chemopreventive efficacy in clinical trials sponsored by the US National Cancer Institute. Many of these biomarkers were previously or are currently being evaluated, fully characterized and validated in animal models (Boone *et al.*, 2000) as well as in archival human tissues (e.g., Bacus *et al.*, 1999; Sneige *et al.*, 1999). As mentioned above, intraepithelial neoplasia are tissue-level phenotypic biomarkers that, because they are on the causal pathway to and are direct precursors of cancer, are generally considered suitable for following carcinogenesis. Cellular biomarkers such as nuclear and nucleolar morphology, mitotic index and DNA ploidy are also being evaluated; they may be useful in characterizing the progression potential of intraepithelial neoplasia (Kelloff, 2000; Kelloff *et al.*, 2000). Other possibly useful genotypic biomarkers include loss of heterozygosity and gene amplification, either at specific gene loci (e.g., those for tumour-suppressors such as *p53* or tumour growth accelerators such as c-*erb*B2) or at panels of microsatellite loci where mutations indicate increasing genomic instability (Califano *et al.*, 1996). These biomarkers appear to be particularly applicable as surrogate end-points for head and neck cancer, and may also prove useful in other tissues where microsatellite instability is a predominant feature of carcinogenesis, as in hereditary non-polyposis colorectal cancer (HNPCC) (Marra & Boland, 1995; Lynch & Smyrk, 1996).

Both phenotypic and genotypic changes during carcinogenesis may also be manifested by molecular biomarkers (Kelloff *et al.*, 2000). For example, excess proliferation may be seen in increased levels of cellular antigens such as proliferating cell nuclear antigen (PCNA) or Ki-67/MIB-1 or overexpression of growth factors such as epidermal growth factor (EGF), transforming growth factor (TGF)-α and IGF-I; reduced propensity to undergo apoptosis may be detected by increased expression of *bcl*-2. Aberrant differentiation may result in changes in G-actin, cytokeratins and blood-group

Table 2. Potential surrogate end-point biomarkers for chemoprevention trials in breast, colon and prostate

Type of biomarker	Breast	Colon	Prostate
Histological	DCIS, LCIS, atypical hyperplasia, mammographic density, nuclear and nucleolar morphometry	Adenomatous polyps, aberrant crypts, microadenomas, nuclear and nucleolar morphometry	PIN, nuclear and nucleolar morphometry
Genotypic	Gene amplification (c-*erb*B-2)	LOH, gene amplification	Chromosomal loss or gain (8p, 9q, (16q), gene amplification (c-*erb*B-2)
Proliferation/ growth control	Ki-67, *bcl-2/bax*, p53, cyclin D1, TGF-β, EGFR, VEGF, IGF-1 expression, S-phase fraction, apoptotic index	PCNA, Ki-67, *bcl-2/bax* expression, S-phase fraction, BrdU uptake, apoptotic index	PCNA, Ki-67, p53, *bcl-2/bax*, pc-1, TGF-β,VEGF, IGF-1 expression, apoptotic index
Differentiation	Myoepithelial cell markers (S-100, keratin 17, vimentin), altered cytoplasmic glycoprotein expression, altered cell surface antigen expression	Altered blood group-related antigens, mucin core antigens (T, Tn, sialyl Tn antigens), apomucins (MUC 1,2,3 genes) cytokeratins, brush border membrane enzymes (sucrase, isomaltase)	Loss of high molecular weight cytokeratins (50–64 kDa), altered blood group related antigens, vimentin

BrdU, bromodeoxyuridine; DCIS, ductal carcinoma *in situ*; EGF, epidermal growth factor; EGFR, epidermal growth factor receptor; IGF, insulin-like growth factor; LCIS, lobular carcinoma *in situ*; LOH, loss of heterozygosity; PCNA, proliferating cell nuclear antigen; PIN, prostatic intraepithelial neoplasia; PSA, prostate-specific antigen; TGF, transforming growth factor; VEGF, vascular endothelial growth factor.

antigens. Other molecular biomarkers may reflect general changes in cell growth control. These include TGF-β, cyclins, p53 and other tumour suppressors, as well as mutations and overexpression of oncogenes associated with carcinogenesis such as *ras* and the transcription factors *myc, fos* and *jun*. Tissue- and drug-related biomarkers may also be useful. Examples of tissue-related biomarkers are the expression of estrogen receptors in breast and prostate-specific antigen (PSA) in prostate. Drug-related biomarkers associated with chemopreventive activity include inhibition of ornithine decarboxylase by 2-difluoromethylornithine and inhibition of prostaglandin biosynthesis by non-steroidal anti-inflammatory drugs (NSAIDs); while such biomarkers do not necessarily demonstrate a chemopreventive effect, they are useful in assessing whether a biologically active dose of the agent was present and in evaluating the chemopreventive mechanisms that are operating.

Cohorts for surrogate end-point chemoprevention studies

Another important challenge in chemoprevention research is the identification of appropriate cohorts for clinical trials. Patients at high risk for developing precancerous lesions and cancers often have the highest potential to benefit from chemopreventive interventions. In such cohorts where the time course of carcinogenesis is accelerated, shorter studies may be feasible. Additionally, these patients may afford the best opportunity to study intervention modalities, because the increased incidence and/or prevalence of disease permits the use of fewer subjects. Patients with previous cancers or precancers constitute one appropriate cohort for chemopreventive intervention, since they are at high risk for new primary cancers. For example, the lifetime risk for a second primary tumour of the aerodigestive tract following a squamous-cell cancer of the head or neck has been estimated at 20–40% (Benner *et al.*, 1992). Premalignant changes (surrogate end-points) can be followed at both phenotypic and genotypic levels in these subjects. Slaughter *et al.* (1953) coined the term "field cancerization" to describe the early evidence of carcinogenesis found in normal-appearing mucosa of patients with previous head and neck cancers. Many studies have confirmed this phenomenon (Hjermann *et al.*, 1981; Benner *et al.*, 1992; Hittelman *et al.*, 1996). In these studies, the degree of genetic change detected was correlated with histological progression of the lesion towards cancer. For example eight of 15 patients having high levels of genetic damage (3.5% or more of cells with three or more copies of chromosome 9) in premalignant lesions of the oral cavity subsequently developed aerodigestive tract cancer, compared with none among patients with lower levels. Similar results were found in relation to chromosome 9 in lung tissue from previous smokers (Hittelman *et al.*, 1996), chromosome 17 in breast tissue (Dhingra *et al.*, 1994) and chromosome 1 in cervical tissue from patients with various grades of cervical intraepithelial neoplasia (CIN) (Segers *et al.*, 1995). While none of these studies tracked the development of specific lesions into cancers, they all confirmed that carcinogenesis could be detected by genotypic changes in high-risk tissue.

The high incidence of new lesions in head and neck cancer patients suggests that a trial duration of up to three years would be appropriate for phase II and III studies using surrogate end-points; as little as three years may even be a feasible duration for detecting a reduction in the incidence of second primary cancers. Patients with superficial bladder cancer are also appropriate subjects for chemoprevention studies, because the recurrence rate is approximately 50% within 6–12 months (Soloway & Perito, 1992) and 60–75% within 2–5 years (Herr *et al.*, 1990; Harris & Neal, 1992). Similar high rates of recurrence or new lesions apply to colorectal adenomas (e.g., Winawer *et al.*, 1993). Studies in these settings would appear to be particularly promising for the validation of surrogate end-points, which may then be suitable for application in cohorts without previous precancers or cancers.

Germline mutations and other genetic and molecular evidence of susceptibility may also be used to define high-risk cohorts. For example, subjects with familial adenomatous polyposis (FAP), which is identified by loss of the *APC* tumour-suppressor gene, develop hundreds to thousands of colorectal adenomas (Burt, 1996). Fabian has described high-risk breast cancer subjects suitable for chemoprevention studies based on the presence of atypical hyperplasia, aneuploidy and overexpression of *p53* and EGF. These biomarkers could potentially serve as surrogate end-points for breast cancer prevention trials (Fabian *et al.*, 1996). Patients scheduled for surgical treatment of pre-cancer or early cancer also provide cohorts for obtaining early evidence of efficacy. Agents are administered to these patients during the weeks after diagnostic biopsy and before more definitive surgery, so that modulation of biomarkers in the precancer/cancerous, and, if possible, normal-appearing tissue in the target organ can be assessed. The National Cancer Institute is now using such protocols in phase I/early phase II studies of breast and prostate cancer prevention (Kelloff, 2000; Kelloff *et al.*, 2000).

Challenges in using surrogate end-points

Numerous issues must be addressed in both the preclinical and clinical phases of chemopreventive drug development efforts. For example, how long must treatment be continued (including whether treatment cessation results in recurrence of precancerous lesions)? Can chemoprevention be distinguished from regression of existing disease? Can lifestyle factors that may significantly influence trial outcomes (e.g., high-fat diets, total caloric consumption) be controlled? Are the results of trials in specific high-risk or undernourished populations applicable to other populations or the populace as a whole? Several additional philosophical and practical concerns specific to the application of surrogate end-points in the evaluation of chemopreventive efficacy must also be considered. Temple (1995, 1999) previously addressed many of these in the context of cardiovascular drug development. Particularly relevant to chemoprevention are issues of sampling, the clinical benefit of biomarker modulation, and whether adverse events prove limiting to long-term chemopreventive agent administration.

Sampling

A critical issue in the application and validation of surrogate end-points is the development of standardized, appropriate and quantitative techniques for sampling the target tissues. To date, the greatest progress has been made in tissues that can be directly observed: oral cavity, colon, larynx, bladder, oesophagus, cervix, bronchus, skin. In these tissues, the focal lesion can be identified and stained, and the area of cancerization can be defined and imaged (e.g., cervix). However, in more inaccessible tissues — prostate, ovary, breast, liver, pancreas—detection of the focal lesion is uncertain, and it is difficult to map and image the cancerization field. Advances in the basic sciences, particularly genomics and proteomics, and in biomedical technologies such as imaging, are providing tools for further growth in this area. New diagnostic methodologies such as gene-chip analyses, the confocal microscope, digital mammography, the LIFE scope for visualizing bronchial tissue and the magnifying endoscope for colorectal monitoring will enhance the possibilitites for monitoring of precancerous tissue. Such techniques can be used for the identification and evaluation of early molecular targets for intervention, as well as for quantitative assessment of cancer risks and tissue- and cell-based changes in these early stages of carcinogenesis. Brown and Botstein (1999) have reviewed the significant potential of functional genomics in biology — the utility ranges from identification of a mutant genotype by a single nucleotide polymorphism to sub-cellular localization of gene products to elucidation of gene expression patterns along signal transduction pathways. The sequencing and functional analysis efforts of the Cancer Genome Anatomy Project are a major contribution to this area. Gene-chip microarrays can be used to define and quantify contributors to risk once appropriate parameters for analyses have been defined. To this end, proven cases from archival specimens from properly designed tissue banks can be utilized to elucidate and validate relevant and useful end-points. The method for cluster analysis of genome-wide expression described by Eisen *et al.* (1998) could be applied to provide a generalized comparison of gene expression in baseline and post-treatment lesions, using known effective drugs and placebo.

Clinical benefit of surrogate end-point modulation

As described by Blue and Colburn (1996), surrogate end-points fall on a continuum from showing no particular clinical benefit but only correlation to the target disease end-point (e.g., drug effect markers), through demonstrating clinical benefit that is not a direct effect on the target disease (e.g., immunostimulation), to demonstrating clinical benefit directly related to the target disease (e.g., inhibiting colorectal adenomas). Initially, the criteria for selecting surrogate end-points support drugs with clinical benefit directly related to cancer incidence prevention. As understanding improves of the role of general genotypic and specific molecular changes in carcinogenesis, and with careful correlative studies, effects on surrogate end-points with antecedent impact on clinical outcome may also support chemopreventive drug efficacy.

There are several conditions in which treatment of precancerous lesions would appear to provide direct clinical benefit, irrespective of the potential for cancer prevention. These situations typically involve a change in standard of care based on regression or prevention of precancerous lesions that would engender reduced morbidity, enhanced quality of life, delayed surgery or reduced surveillance frequency. Subjects with genetic predisposition to cancer development (e.g., FAP) may achieve such benefits from chemopreventive interventions. FAP is characterized by germline mutations in the *APC* tumour-suppressor gene. Patients with FAP develop hundreds to thousands of colorectal adenomatous polyps beginning in their teen years, and in the absence of treatment will almost certainly develop colorectal cancer by the age of 50 years; they are also at risk for other lesions, particularly duodenal polyps and cancers, and desmoid tumours. Once adenomas begin to appear, these patients are monitored by periodic colonoscopy (at approximately six-month intervals), removal of existing polyps and cancer screening. When polyp burden becomes unmanageable, most patients have partial or total colectomies and undergo continued monitoring thereafter. Agents which prevent or slow the progression of the adenomas could benefit these patients by delaying or obviating the need for colectomy. A decrease in the frequency of surveillance colonoscopies and cancer screenings would also benefit patients with FAP, as it could those with sporadic colorectal adenomas. New adenomas occur within 1–3 years post-resection in approximately 30% of patients with sporadic colorectal adenomas or cancers (Hamilton, 1996). These patients routinely undergo colonoscopy with removal of new lesions at 1–5-year intervals. Preventive treatment could potentially increase the screening interval, thereby decreasing associated morbidity and lowering health care costs.

Other conditions in which organ removal or other major surgery with high morbidity is standard include Barrett's oesophagus and superficial bladder cancers. Barrett's oesophagus, a precursor of oesophageal cancer, is currently managed by endoscopy with biopsy of metaplastic and dysplastic lesions; severe dysplasia may mandate partial or total oesophagectomy (Roth *et al.*, 1997). Because of the high rate of their recurrence and potential for progression, treatment for superficial bladder cancers includes periodic surveillance (every three months) and removal of new lesions, and may include cystectomy (Linehan *et al.*, 1997). In both diseases, treatment has profound detrimental effects on quality of life. Both are examples of situations in which preventive agents could provide clinical benefit by reducing the frequency of surveillance and the need for surgery.

Quality of life

Chemopreventive drugs may ultimately be given to asymptomatic populations for years or decades. Therefore, minimal toxicity is essential. Determining standards in terms of allowable type and frequency of side-effects and impact on quality of life will be critical issues as chemopreventive drugs are introduced. It is possible that life-threatening toxicities compromising such long-term drug use would not be detected within the time-frame of surrogate end-point-based efficacy trials. In the meta-analysis of cholesterol-lowering interventions cited below, the investigators found that despite their cholesterol-lowering efficacy, fibrates such as gemfibrozil were associated with increases in non-coronary heart disease mortality by ~30% (p<0.01) and total mortality by ~17% (p<0.01) on long-term administration (Gould *et al.*, 1995). A different, but dramatic, example of unanticipated late toxicity is provided by the results of the Cardiac Arrhythmia Suppression Trial (Fleming & DeMets, 1996). This randomized, placebo-controlled trial of three type 1C antiarrhythmic drugs was designed to evaluate mortality reduction in patients experiencing ten ventricular premature beats per hour and few or no symptoms following a recent myocardial infarction. Entry into the trial required that the patients respond to antiarrhythmic therapy as measured by at least 70% reduction in ventricular premature beats as a surrogate for arrhythmia. This trial was stopped when it was found that drug treatment was associated with increased mortality or cardiac arrest despite lowering ventricular premature beats (Echt *et al.*, 1991; Cardiac Arrhythmia Suppression Trial II Investigators, 1992).

Cardiovascular disease prevention: precedent for application of surrogate end-point biomarkers in drug development

Cancer chemoprevention shares the interest and need for surrogate end-points in drug development with other chronic diseases of ageing and life-threatening diseases. To date, the best characterized surrogate end-points in drug development have been for AIDS (Mellors *et al.*, 1996; Saag *et al.*, 1996) and cardiovascular drugs (Fleming & DeMets, 1996). In particular, the use of blood lipid-lowering as a surrogate end-point for cardiovascular disease provides a model for and insight into the issues surrounding the use of surrogates for cancer incidence in chemoprevention studies. In terms of the long time required for disease development, the multiple paths by which the disease progresses and the chronic administration of preventive drugs, the course of cardiovascular disease closely parallels carcinogenesis. In the cardiovascular setting, a well established surrogate end-point is cholesterol level, which is a validated predictor of coronary heart disease (Expert Panel on Detection, Evaluation and Treatment of High Blood Cholesterol in Adults, 1993). Modulation of cholesterol levels has been used to gain marketing approval for 3-hydroxy-3-methylglutaryl coenzyme A (HMGCoA) reductase inhibitors such as lovastatin (Sahni *et al.*, 1991; Fail *et al.*, 1992), simvastatin, pravastatin (Crouse *et al.*, 1992; Pitt *et al.*, 1993) and gemfibrozil (Frick *et al.*, 1987). HMGCoA reductase catalyses a critical step in cholesterol biosynthesis, the formation of mevalonate. Gould *et al.* (1995) carried out a meta-analysis of 35 randomized clinical trials that essentially summarized the evidence supporting cholesterol-lowering as a surrogate end-point for coronary heart disease. This review of all primary or secondary intervention studies of >2 years' duration included single-drug studies such as the Helsinki Heart Study of gemfibrozil (Frick *et al.*, 1987), as well as dietary (Dayton *et al.*, 1968; Burr *et al.*, 1989), surgical (Buchwald *et al.*, 1990) and multifactorial interventions (Miettinen *et al.*, 1982; Wilhelmsen *et al.*, 1986). The results show that cholesterol-lowering is correlated with coronary heart disease, non-coronary heart disease and overall mortality. Specifically, it was found that for every 10% lowering of cholesterol, coronary heart disease mortality was reduced by 13% ($p < 0.002$) and total mortality by 10% ($p < 0.03$), while no effect was found on non-coronary heart disease mortality. A caveat applies here, as to all studies with biomarkers — the relationship between lower coronary heart disease and lower cholesterol is derived from the average of individual responses, and the same correlation is not seen in each individual. The presence of confounding factors (e.g., smoking history and diabetes mellitus) may influence the proportion of disease attributable to any specific parameter in a multifactorial disease process.

Chemoprevention of colorectal adenomas

The data supporting validation of cholesterol levels as a surrogate end-point for coronary heart disease include an association with disease risk, in addition to the ability to predict activity of a given drug against that disease (Kelloff *et al.*, 2000). Analogous data might be applied to support the validation of a surrogate for cancer incidence. For example, it is well established that the presence of colorectal adenomas increases colorectal cancer risk (Hamilton, 1992; Winawer *et al.*, 1993) and that adenoma number, size and severity of dysplasia are predictive factors for cancer incidence. It has been estimated that 2–5% of all colorectal adenomas progress to adenocarcinomas if not removed or treated, with increasing rates for large and severely dysplastic polyps (Day & Morson, 1978; Hamilton, 1992; Bruzzi *et al.*, 1995). Cancer risk is reduced by polyp removal, and a strong correlation exists between the relative prevalence of adenomas and cancers across populations (Winawar, 1993). More than 20 epidemiological and intervention studies have demonstrated that regular NSAID use is associated with reduced adenoma incidence and that this decrease is correlated with declines in both cancer incidence and mortality (Greenberg & Baron, 1996). These data support the validation of adenomas as a surrogate end-point for colon cancer incidence.

A recently conducted clinical trial sponsored by the US National Cancer Institute and G.D. Searle examined the effect of the COX-2 inhibitor celecoxib at two doses against colorectal polyps in subjects with FAP). Overexpression of prostaglandins and COX isoenzymes is observed in colorectal polyps and tumours from animals and humans with germline *APC* gene mutations. Early clinical evidence of polyp regression with the NSAID sulindac has been demonstrated in FAP patients. Additional support for the trial has come from preclinical efficacy studies with celecoxib, and substantial epidemiological evidence of a protective effect of NSAIDs against colorectal carcinogenesis. Preclinical and clinical studies demonstrating reduced gastro-intestinal toxicity of celecoxib compared with traditional NSAIDs support the use of a COX-2-specific inhibitor in a chemopreventive setting. In this randomized, double-blind, placebo-controlled study of 83 FAP patients, a six-month intervention with 800 mg celecoxib per day significantly reduced polyp number by 28%, with 53% of treated subjects showing a 25% or greater reduction. A blinded physicians' assessment indicated a qualitative improvement in the colon and rectum, and to a lesser extent in the duodenum, of treated subjects. This trial led to accelerated marketing approval of celecoxib by the US Food and Drug Administration, as an adjunct to standard care for the regression and reduction of adenomatous polyps in FAP subjects. Although it can be inferred from data supporting the correlation of polyp burden with colon cancer incidence, it remains to be demonstrated in a randomized, placebo-controlled clinical study that a reduction in cancer incidence will be engendered by a drug which prevents polyps. Nonetheless, this study was a landmark in chemoprevention research with surrogate end-points, demonstrating that polyp burden can serve as an appropriate end-point for quantitative and qualitative assessments of chemopreventive efficacy in FAP patients. Follow-up studies are planned to assess the relative effect of celecoxib on polyp regression and prevention, and to determine whether greater efficacy can be engendered by combination therapy of celecoxib with the antiproliferative agent 2-difluoromethylornithine.

Summary and perspectives on the use of surrogate end-points in gaining marketing approval for chemopreventive agents

The critical scientific aspects of developing surrogate end-points to characterize cancer chemopreventive efficacy should be applied to the design of clinical development strategies to gain marketing approval for chemopreventive drugs. The multi-path, multifocal and multi-year course of carcinogenesis suggests that, initially, the most successful strategies will use well defined precancers (intraepithelial neoplasia) as surrogate end-points for cancer incidence. Despite their close temporal and histological association with cancers, only a relatively small percentage of intraepithelial neoplastic lesions progress. Therefore, determination of chemopreventive efficacy will rely on assurance that the lesions most likely to progress are inhibited; the genotype of any post-treatment lesions should be indicative of an equivalent or lower progression potential than baseline lesions. The phenotypic changes seen in intraepithelial neoplasia during short-term studies are likely to be subtle, so that quantitative measurements such as computer-assisted image analysis are desirable. Similarly, the evaluation of genotypic changes requires sensitive, quantitative analysis of gene expression such as is afforded by the various DNA microarray techniques. Standardization to provide adequate sampling and handling of non-related biopsy effects (e.g., timing of breast cell proliferation assessment during the menstrual cycle) will be essential. The gold standard for validating surrogate end-points is correlation with cancer incidence reduction. However, the resources (e.g., time and number of subjects) required for this definitive validation are enormous. Continued discussion and research on alternative strategies among all interested parties are needed to ensure that surrogate end-point-based chemoprevention indications are feasible. Demonstration of the clinical benefit of prevention of intraepithelial neoplasia as described above for FAP, sporadic colorectal adenomas, superficial bladder cancers and Barrett's oesophagus is one possible strategy. A second approach would follow an accelerated pathway for gaining marketing approval, as defined in the United States Food & Drug Administration regulations based on strongly-supported surrogate end-points for disease incidence in the setting of life-threatening disease such as cancer.

References

American Association for Cancer Research (1999) Prevention of cancer in the next millenium: report of the Chemoprevention Working Group to the American Association for Cancer Research. *Cancer Res.*, **59**, 4743–4758

Bacus, J.W., Boone, C.W., Bacus, J.V., Follen, M., Kelloff, G.J., Kagan, V. & Lippman, S.M. (1999) Image morphometric nuclear grading of intraepithelial neoplastic lesions with applications to cancer chemoprevention trials. *Cancer Epidemiol. Biomarkers Prev.*, **8**, 1087–1094

Benner, S.E., Hong, W.K., Lippman, S.M., Lee, J.S. & Hittelman, W.M. (1992) Intermediate biomarkers in upper aerodigestive tract and lung chemoprevention trials. *J. Cell. Biochem.*, **16G**, 33–38

Bissell, M.J., Weaver, V.M., Lelievre, S.A., Wang, F., Petersen, O.W. & Schmeichel, K.L. (1999) Tissue structure, nuclear organization, and gene expression in normal and malignant breast. *Cancer Res.*, **59**, 1757s–1764s

Blue, J.W. & Colburn, W.A. (1996) Efficacy measures: surrogates or clinical outcomes? *J. Clin. Pharmacol.*, **36**, 767–770

Boone, C.W., Stoner, G.D., Bacus, J.V., Kagan, V., Morse, M.A., Kelloff, G.J. & Bacus, J.W. (2000) Quantitative grading of rat esophageal carcinogenesis using computer-assisted image tile analysis. *Cancer Epidemiol. Biomarkers Prev.*, **9**, 495–500

Bostwick, D.G. (1992) Prostatic intraepithelial neoplasia (PIN): current concepts. *J. Cell. Biochem.*, **16H**, 10–19

Brown, P.O. & Botstein, D. (1999) Exploring the new world of the genome with DNA microarrays. *Nature Gen.*, **21** (Suppl.), 33–37

Bruzzi, P., Bonelli, L., Costantini, M., Sciallero, S., Boni, L., Aste, H., Gatteschi, B., Naldoni, C., Bucchi, L., Casetti, T., Bertinelli, E., Lanzanova, G., Onofri, P., Parri, R., Rinaldi, P., Castiglione, G., Mantellini, P. & Giannini, A. (1995) A multicenter study of colorectal adenomas—rationale, objectives, methods and characteristics of the study cohort. *Tumori*, **81**, 157–163

Buchwald, H., Varco, R.L., Matts, J.P., Long, J.M., Fitch, L.L., Campbell, G.S., Pearce, M.B., Yellin, A.E., Edmiston, W.A., Smink, R.D., Jr, Sawin, H.S., Jr, Campos, C.T., Hansen, B.J., Tuna, N., Karnegis, J.N., Sanmarco, M.E., Amplatz, K., Castaneda-Zuniga, W.R., Hunter, D.W., Bissett, J.K., Weber, F.J., Stevenson, J.W., Leon, A.S., Chalmers, T.C. & the POSCH Group (1990) Report of the Program on the Surgical Control of Hyperlipidemias (POSCH): effect of partial ileal bypass surgery on mortality and morbidity from coronary heart disease in patients with hypercholesterolemia. *New Engl. J. Med.*, **323**, 946–955

Burr, M.L., Fehily, A.M., Gilbert, J.F., Rogers, S., Holliday, R.M., Sweetnam, P.M., Elwood, P.C. & Deadman, N.M. (1989) Effects of changes in fat, fish, and fibre intakes on death and myocardial reinfarction: diet and reinfarction trial (DART). *Lancet*, **2**, 757–761

Burt, R.W. (1996) Cohorts with familial disposition for colon cancers in chemoprevention trials. *J. Cell. Biochem. (Suppl.)*, **25**, 131–135

Califano, J., van der Riet, P., Westra, W., Nawroz, H., Clayman, G., Piantadosi, S., Corio, R., Lee, D., Greenberg, B., Koch, W. & Sidransky, D. (1996) Genetic progression model for head and neck cancer: implications for field cancerization. *Cancer Res.*, **56**, 2488–2492

Cameron, A.J. & Lomboy, C.T. (1992) Barrett's esophagus: age, prevalence, and extent of columnar epithelium. *Gastroenterology*, **103**, 1241–1245

Cardiac Arrhythmia Suppression Trial II Investigators (1992) Effect of the antiarrhythmic agent moricizine on survival after myocardial infarction. *New Engl. J. Med.*, **327**, 227–233

Cotran, R.X., Kumar, V. & Robbins, S.L. eds (1989) *Robbins Pathologic Basis of Disease,* 4th Edition, Philadelphia, W. B. Saunders, pp. 1090–1094

Crouse, J.R., Byington, R.P., Bond, M.G., Espeland, M.A., Sprinkle, J.W., McGovern, M. & Furberg, C.D. (1992) Pravastatin, lipids, and atherosclerosis in the carotid arteries: design features of a clinical trial with carotid atherosclerosis outcome. *Control. Clin. Trials*, **13**, 495–506

Day, D.W. & Morson, B.C. (1978) The adenoma-carcinoma sequence. In: Bennington, J.L., ed., *The Pathogenesis of Colorectal Cancer*, Philadelphia, W.B. Saunders, pp. 58–71

Dayton, S., Pearce, M.L., Goldman, H., Harnish, A., Plotkin, D., Shickman, M., Winfield, M., Zager, A. & Dixon, W.J. (1968) Controlled trial of a diet high in unsaturated fat for prevention of atherosclerotic complications. *Lancet*, **2**, 1060–1062

Dhingra, K., Sneige, N., Pandita, T.K., Johnston, D.A., Lee, J.S., Emani, K., Hortobagyi, G.N. & Hittelman, W.N. (1994) Quantitative analysis of chromosome *in situ* hybridization signal in paraffin-embedded tissue sections. *Cytometry*, **16**, 100–112

Echt, D.S., Liebson, P.R., Mitchell, L.B., Peters, R.W., Obias-Manno, D., Barker, A.H., Arensberg, D., Baker, A., Friedman, L., Greene, H.L., Huther, M.L., Richardson, D.W. & the CAST Investigators. (1991) Mortality and morbidity in patients receiving encainide, flecanide, or placebo. *New Engl. J. Med.*, **324**, 781–788

Eisen, M.B., Spellman, P.T., Brown, P.O. & Botstein, D. (1998) Cluster analysis and display of genome-wide expression patterns. *Proc. Natl Acad. Sci. USA.*, **95**, 14863–14868

Expert Panel on Detection, Evaluation, and Treatment of High Blood Cholesterol in Adults (1993) Summary of the Second Report of the National Cholesterol Education Program (NCEP) Expert Panel on Detection, Evaluation, and Treatment of High Blood Cholesterol in Adults (Adult Treatment Panel II). *J. Am. Med. Ass.*, **269**, 3015–3023

Fabian, C.J., Kamel, S., Zalles, C. & Kimler, B.F. (1996) Identification of a chemoprevention cohort from a population of women at high risk for breast cancer. *J. Cell. Biochem.*, Suppl., **25**, 112–122

Fail, P.S., Sahni, R.C., Maniet, A.R., Voci, G. & Banka, V.S. (1992) The long-term clinical efficacy of lovastatin therapy following successful coronary angioplasty. *Clin. Res.*, **40**, 400A

Falk, G.W. & Richter, J.E. (1996) Reflux disease and Barrett's esophagus. *Endoscopy*, **28**, 13–21

Fearon, E.R & Vogelstein, B. (1990) A genetic model for colorectal tumorigenesis. *Cell*, **61**, 759–767

Fleming, T.R. & DeMets, D.L. (1996) Surrogate end points in clinical trials: are we being misled? *Ann. Intern. Med.*, **125**, 605–613

Frick, M.H., Elo, O, Haapa, K., Heinonen, O.P., Heinsalmi, P., Helo, P., Huttunen, J.K., Kaitaniemi, P., Koskinen, P., Manninen, V., Mäenpää, H., Mälkönen, M., Mänttäri, M., Norola, S., Pasternack, A., Pikkarainen, J., Romo, M., Sjöblom, T. & Nikkilä, E.A. (1987) Helsinki Heart Study: primary prevention trial with gemfibrozil in middle-aged men with dyslipidemia. Safety of treatment, changes in risk factors, and incidence of coronary heart disease. *New Engl. J. Med.*, **317**, 1237–1245

Frykberg, E.R. & Bland, K.I. (1993) *In situ* breast carcinoma. *Adv. Surg.*, **26**, 29–72

Gould, A.L., Rossouw, J.E., Santanello, N.C., Heyse, J.F. & Furberg, C.D. (1995) Cholesterol reduction yields clinical benefit. A new look at old data. *Circulation*, **91**, 2274–2282

Greenberg, E.R. & Baron, J.A. (1996) Aspirin and other nonsteroid anti-inflammatory drugs as cancer-preventive agents. In: Stewart, B.W., McGregor, D. & Kleihues, P., eds *Principles of Chemoprevention* (IARC Scientific Publications No. 139), Lyon, IARC, pp. 91–98

Greenlee, R.T., Murray, T., Bolden, S. & Wingo, P.A. (2000) Cancer statistics, 2000. *CA Cancer J. Clin.*, **50**, 7–33

Hamilton, S.R. (1992) The adenoma-adenocarcinoma sequence in the large bowel: variations on a theme. *J. Cell. Biochem. (Suppl. G)*, **16**, 41–46

Hamilton, S.R. (1996) Pathology and biology of colorectal neoplasia. In: Young, G.P., Levin, B. & Rozen, P., eds, *Prevention and Early Detection of Colorectal Cancer. Principles and Practice*, London, W.B. Saunders, pp. 3–21

Harris, A.L. & Neal, D.E. (1992) Bladder cancer—field versus clonal origin. *New Engl. J. Med.*, **326**, 759–761

Herr, H.W., Jakse, G. & Sheinfeld, J. (1990) The T1 bladder tumor. *Semin. Urol.*, **8**, 254–261

Hittelman, W.N., Kim, H.J., Lee, J.S., Shin, D.M., Lippman, S.M., Kim, J., Ro, J.Y. & Hong, W.K. (1996) Detection of chromosome instability of tissue fields at risk: *in situ* hybridization. *J. Cell. Biochem. (Suppl.)*, **25**, 57–62

Hjermann, I., Velve Byre, K., Holme, I. & Leren, P. (1981) Effect of diet and smoking intervention on the incidence of coronary heart disease: report from the Oslo Study Group of a randomized trial of healthy men. *Lancet*, **2**, 1303–1310

Hong, W.K. & Sporn, M.B. (1997) Recent advances in chemoprevention of cancer. *Science*, **278**, 1073–1077

Jankowski, J., Hopwood, D., Pringle, R. & Wormsley, K.G. (1993) Increased expression of epidermal growth factor receptors in Barrett's esophagus associated with alkaline reflux: a putative model for carcinogenesis. *Am. J. Gastroenterol.*, **88**, 402–408

Kelloff, G.J. (2000) Perspectives on cancer chemoprevention research and drug development. *Adv. Cancer Res.*, **278**, 199–334

Kelloff, G.J., Johnson, J.R., Crowell, J.A., Boone, C.W., DeGeorge, J.J., Steele, V.E., Mehta, M.U., Temeck, J.W., Schmidt, W.J., Burke, G., Greenwald, P. & Temple, R.J. (1995) Approaches to the development and marketing approval of drugs that prevent cancer. *Cancer Epidemiol. Biomarkers Prev.*, **4**, 1–10

Kelloff, G.J., Sigman, C.C., Johnson, K.M., Boone, C.W., Greenwald, P., Crowell, J.A., Hawk, E.T. & Doody, L.A. (2000) Perspectives on surrogate endpoints in the development of drugs that reduce the risk of cancer. *Cancer Epidemiol. Biomarkers Prev.*, **9**, 127–134

Kishimoto, Y., Sugio, K., Hung, J.Y., Virmani, A.K., McIntire, D.D., Minna, J.D. & Gazdar, A.F. (1995) Allele-specific loss in chromosome 9p loci in preneoplastic lesions accompanying non-small-cell lung cancers. *J. Natl Cancer Inst.*, **87**, 1224–1229

Linehan, W.M., Cordon-Cardo, C. & Isaacs, W. (1997) Cancers of the genitourinary tract. In: DeVita, V.T., Jr, Hellman, S. & Rosenberg, S. A., eds, *Cancer: Principles and Practice of Oncology*, Fifth Edition Philadelphia, Lippincott-Raven, pp. 1253–1395

Lynch, H.T. & Smyrk, T. (1996) Hereditary nonpolyposis colorectal cancer (Lynch syndrome). *Cancer*, **76**, 1149–1167

Mao, L., Schoenberg, M.P., Scicchitano, M., Erozan, Y.S., Merlo, A., Schwab, D. & Sidransky, D. (1996) Molecular detection of primary bladder cancer by microsatellite analysis. *Science*, 271, 659–662

Marra, G. & Boland, C.R. (1995) Hereditary nonpolyposis colorectal cancer: the syndrome, the genes, and historical perspectives. *J. Natl Cancer Inst.*, **87**, 1114–1125

Mellors, J.W., Rinaldo, C.R., Jr, Gupta, P., White, R.M., Todd, J.A. & Kingsley, L.A. (1996) Prognosis in HIV-1 infection predicted by the quantity of virus in plasma. *Science*, **272**, 1167–1170

Miettinen, T.A., Huttunen, J.K., Naukkarinen, V., Strandberg, T., Mattila, S., Kumlin T. & Sarna, S. (1982) Multifactorial primary prevention of cardiovascular diseases in middle-aged men: risk factor changes, incidence, and mortality. *J. Am. Med. Ass.*, **255**, 2097–2102

Miros, M., Kerlin, P. & Walker, N. (1991) Only patients with dysplasia progress to adenocarcinoma in Barrett's oesophagus. *Gut*, **32**, 1441–1446

Muto, T., Bussey, H.J.R. & Morson, B.C. (1975) The evolution of cancer of the colon and rectum. *Cancer*, **36**, 2251–2270

Ovaska, J., Miettinen, M. & Kivilaakso, E. (1989) Adenocarcinoma arising in Barrett's esophagus. *Dig. Dis. Sci.*, **34**, 1336–1339

Page, D.L., Dupont, W.D., Rogers, L.W. & Rados, M.S. (1985) Atypical hyperplastic lesions of the female breast. A long-term follow-up study. *Cancer*, **55**, 2698–2708

Pitt, B., Ellis, S.G., Mancini, J., Rosman, H.S. & McGovern, M.E. for the PLAC 1 Investigators (1993) Design and recruitment in the United States of a multicenter quantitative angiographic trial of pravastatin to limit atherosclerosis in the coronary arteries (PLAC 1). *Am. J. Cardiol.*, **72**, 31–35

Rosin, M.P., Cairns, P., Epstein, J.I., Schoenberg, M.P. & Sidransky, D. (1995) Partial allelotype of carcinoma *in situ* of the human bladder. *Cancer Res.*, **55**, 5213–5216

Roth, J.A., Putnam, J.B., Jr, Rich, T.A. & Forastiere, A. (1997) Cancers of the gastrointestinal tract: cancer of the esophagus. In: DeVita, V.T., Jr, Hellman, S. & Rosenberg, S.A., eds, *Cancer: Principles and Practice of Oncology*, Fifth Edition, Philadelphia, Lippincott-Raven, pp. 970–1251

Saag, M.S., Holodniy, M., Kuritzkes, D.R., O'Brien, W.A., Coombs, R., Poscher, M.E., Jacobsen, D.M., Shaw, G.M., Richman, D.D. & Volberding, P.A. (1996) HIV viral load markers in clinical practice. *Nature Med.*, **2**, 625–629

Sahni, R., Maniet, A.R., Voci, G. & Banka, V.S. (1991) Prevention of restenosis by lovastatin after successful angioplasty. *Am. Heart J.*, **121**, 1600–1608

Scher, H.I., Shipley, W.U. & Herr, H.W. (1997) Cancer of the bladder. Cancers of the genitourinary tract. In: DeVita, V.T., Jr, Hellman, S. & Rosenberg, S. A., eds, *Cancer: Principles and Practice of Oncology*, Fifth edition, Philadelphia, Lippincott-Raven, pp. 1300–1322

Schipper, H., Turley, E.A. & Baum, M. (1996) A new biological framework for cancer research. *Lancet*, **348**, 1149–1151

Segers, P., Haesen, S., Castelain, P., Amy, J.-J., De Sutter, P., Van Dam, P. & Kirsch-Volders, M. (1995) Study of numerical aberrations of chromosome 1 by fluorescent *in situ* hybridization and DNA content by densitometric analysis on (pre)-malignant cervical lesions. *Histochem. J.*, **27**, 24–34

Sidransky, D. & Messing, E. (1992) Molecular genetics and biochemical mechanisms in bladder cancer. Oncogenes, tumor suppressor genes, and growth factors. *Urol. Clin. North Am.*, **19**, 629–639

Sidransky, D., Frost, P., Von Eschenbach, A., Oyasu, R., Preisinger, A.C. & Vogelstein, B. (1992a) Clonal origin bladder cancer. *New Engl. J. Med.*, **326**, 737–740

Sidransky, D., Mikkelsen, T., Schwechheimer, K., Rosenblum, M.L., Cavenee, W. & Vogelstein, B. (1992b) Clonal expansion of p53 mutant cells is associated with brain tumour progression. *Nature*, **355**, 846–847

Simoneau, A.R. & Jones, P.A. (1994) Bladder cancer: the molecular progression to invasive disease. *World J. Urol.*, **12**, 89–95

Slaughter, D.P., Southwick, H.W. & Smejkal, W. (1953) 'Field cancerization' in oral stratified squamous epithelium. Clinical implications of multicentric origin. *Cancer*, **6**, 963–968

Sneige, N., Lagios, M.D., Schwarting, R., Colburn, W., Atkinson, E., Weber, D., Sahin, A., Kemp, B., Hoque, A., Risin, S., Sabichi, A., Boone, C., Dhingra, K., Kelloff, G. & Lippman S. (1999) Interobserver reproducibility of the Lagios nuclear grading system for ductal carcinoma *in situ*. *Hum. Pathol.*, **30**, 257–262

Soloway, M.S. & Perito, P.E. (1992) Superficial bladder cancer: diagnosis, surveillance and treatment. *J. Cell. Biochem. (Suppl.)*, **16I**, 120–127

Sporn, M.B. (1996) The war on cancer. *Lancet*, **347**, 1377–1381

Sporn, M.B. & Suh, N. (2000) Chemoprevention of cancer. *Carcinogenesis*, **21**, 525–530

Stein, G.S., Montecino, M., van Wijnen, A.J., Stein, J.L. & Lian J.B. (2000) Nuclear structure–gene expression interrelationships: implications for aberrant gene expression in cancer. *Cancer Res.*, **60**, 2067–2076

Temple, R.J. (1995) A regulatory authority's opinion about surrogate endpoints. In: Nimmo, W.S. & Tucker, G.T., eds, *Clinical Measurement in Drug Evaluation*, New York, John Wiley, pp. 3–22

Temple, R. (1999) Are surrogate markers adequate to assess cardiovascular disease drugs? *J. Am. Med. Ass.*, **282**, 790–795

Thiberville, L., Payne, P., Vielkinds, J., LeRiche, J., Horsman, D., Nouvet, G., Palcic, B. & Lam, S. (1995) Evidence of cumulative gene losses with progression of premalignant epithelial lesions to carcinoma of the bronchus. *Cancer Res.*, **55**, 5133–5139

Wilhelmsen, L., Berglund, G., Elmfeldt, D., Tibblin, G., Wedel, H., Pennert, K., Vedin, A., Wilhelmsson, C. & Werkö, L. (1986) The multifactorial primary prevention trial in Göteborg, Sweden. *Eur. Heart J.*, **7**, 279–288

Williamson, W.A., Ellis, F.H., Jr, Gibb, S.P., Shahian, D.M., Aretz, H.T., Heatley, G.J. & Watkins, E., Jr (1991) Barrett's esophagus. Prevalence and incidence of adenocarcinoma. *Arch. Intern. Med.*, **151**, 2212–2216

Winawer, S.J., Zauber, A.G., O'Brien, M.J., Ho, M.N., Gottlieb, L., Sternberg, S.S., Waye, J.D., Bond, J., Schapiro, M., Stewart, E.T., Panish, J., Ackroyd, F., Kurtz, R.C. & Shike, M. (1993) Randomized comparison of surveillance intervals after colonoscopic removal of newly diagnosed adenomatous polyps. *New Engl. J. Med.*, **328**, 901–906

Zauber, A.G., Bond, J.H. & Winawer, S.J. (1996) Surveillance of patients with colorectal adenomas or cancer. In: Young, G.P., Rozen, P. & Levin, B., eds, *Prevention and Early Detection of Colorectal Cancer*, London, W.B. Saunders, pp. 195–216

Corresponding author:

G.J. Kelloff
National Cancer Institute,
Division of Cancer Prevention,
EPN 201,
MSC 7322, 9000 Rockville Pike,
Bethesda,
MD 20892-7322,
USA

Biomarkers in Cancer Chemoprevention
Miller, A.B., Bartsch, H., Boffetta, P., Dragsted, L. and Vainio, H. eds
IARC Scientific Publications No. 154
International Agency for Research on Cancer, Lyon, 2001

Biomarkers and surrogacy: relevance to chemoprevention

T.W. Kensler, N.E. Davidson, J.D. Groopman and A. Muñoz

Clinical cancer prevention trials that use disease as the end-point are of necessity large, lengthy and costly. While such trials will always remain the 'gold standard' for establishing efficacy, they are unwieldy and inefficient for the rapid translation of our accelerating understanding of the molecular basis of cancer into preventive strategies. The inclusion of biomarkers in the process of chemopreventive agent development is crucial for the advancement of the field. This overview highlights the types of approach that are being used in the development and application of biomarkers in chemoprevention studies. Biomarkers, which measure exposure, susceptibility or risk factors, can be used in selecting study cohorts, assessing participant compliance and/or determining agent efficacy. Key features of biomarkers include reliability, precision, accuracy and validity. Not all biomarkers are suitable for all purposes and are likely to be imperfect in any single setting. Judicious selection and matching of biomarkers with agents and study cohorts is required for their effective utilization. A critical but non-dichotomous element of risk biomarkers is their degree of surrogacy. A classification scheme is provided that relates the degree of surrogacy of risk biomarkers to their utility in preventive interventions.

Introduction

The past decade has witnessed the development of an impressive array of biomarkers reflecting specific exposures to environmental agents and/or predicting disease risk in individuals. A biomarker may be defined as a chemical (or infectious) agent in accessible body matrices, an *in vivo* response to an exposure or set of exposures, or a genotype or phenotype indicative of susceptibility to disease, all measurable in body fluids, cells or tissues. Biomarkers have the potential to make possible better assessment of ambient environmental exposures; better methods for risk estimation and classification of at-risk individuals, communities and populations; better definition of mechanisms of exposure–disease linkages and the underlying susceptibility factors; clearer definition of the interactions of multiple agents and exposures on disease outcomes; and, ultimately, better and faster methods for assessing the effect on disease outcomes of exposure remediation and preventive interventions (Hulka *et al.*, 1990; Schulte & Perera, 1993; Muñoz & Gange, 1998). Many of these goals directly affect the development and maturation of the discipline of cancer chemoprevention.

Clinical cancer prevention trials that use disease as the end-point are of necessity large, lengthy and costly. While such trials will always remain the 'gold standard' for establishing efficacy, they are unwieldy and inefficient for the rapid translation of our accelerating understanding of the molecular basis of cancer into preventive strategies. Thus, inclusion of biomarkers, despite some intrinsic limitations, in the process of chemopreventive agent development and application is of central importance for the advancement of the field. As discussed by Kelloff *et al.* (1996), the major structural triad that needs to be considered in unison for the development of chemopreventive agents is the 'ABC' of chemoprevention: agents, biomarkers, and cohorts. Biomarkers can be used in three distinct but complementary ways. First, biomarkers can be used in defining study populations by classifying individuals at risk among whom putative preventive interventions are to be evaluated. Second, biomarkers can be used to accelerate assessment of the efficacy of preventive interventions, both in terms of identification of active agents in humans and optimization of their use (e.g., dose and schedule). Third, biomarkers can be used to monitor compliance to the agent that forms the basis of the intervention.

The mere existence of a biomarker does not mean that it will be useful to the field. At present the possibilities for biased use of biomarkers probably outweigh prospects for their informed use. This concern arises simply because few of the biomarkers now being used in either preclinical or clinical settings have undergone anything approaching rigorous validation. Indeed, paradigms for the validation of biomarkers are still evolving (Freedman *et al.*, 1992; Schulte & Perera, 1993; Groopman & Kensler, 1999) and considerable effort will be required for the validation of current and future biomarkers of potential use in chemoprevention studies. This overview seeks to highlight the types of approach that are being used in the development and application of such biomarkers that, in turn, reflect different components of the multistage, multifactorial process of carcinogenesis. Of particular importance is the recognition of the concept that the utility of biomarkers in prevention studies is not dichotomous (i.e., good or bad), but rather continuous, with some markers more informative than others, depending upon how they are used. Figure 1 provides a conceptual basis for the application of biomarkers in preventive interventions.

Criteria for useful biomarkers

Not all biomarkers are suitable for all purposes. Some will be helpful in selecting study cohorts, others will find use in assessing participant compliance, and others can be applied to determining agent efficacy in prevention trials. There are a number of analytical and biological criteria that define the utility of any given biomarker for chemoprevention studies (Schatzkin *et al.*, 1990; Kelloff *et al.*, 2000).

The development of most biomarkers being considered for application to chemoprevention trials is driven by improvements in analytical methods. Our abilities to measure ever-smaller amounts of molecules in a complex biological milieu provide ever-greater insight into the key pathways of the carcinogenic process and the potential modulating effects of chemopreventive agents. These molecules can be environmental

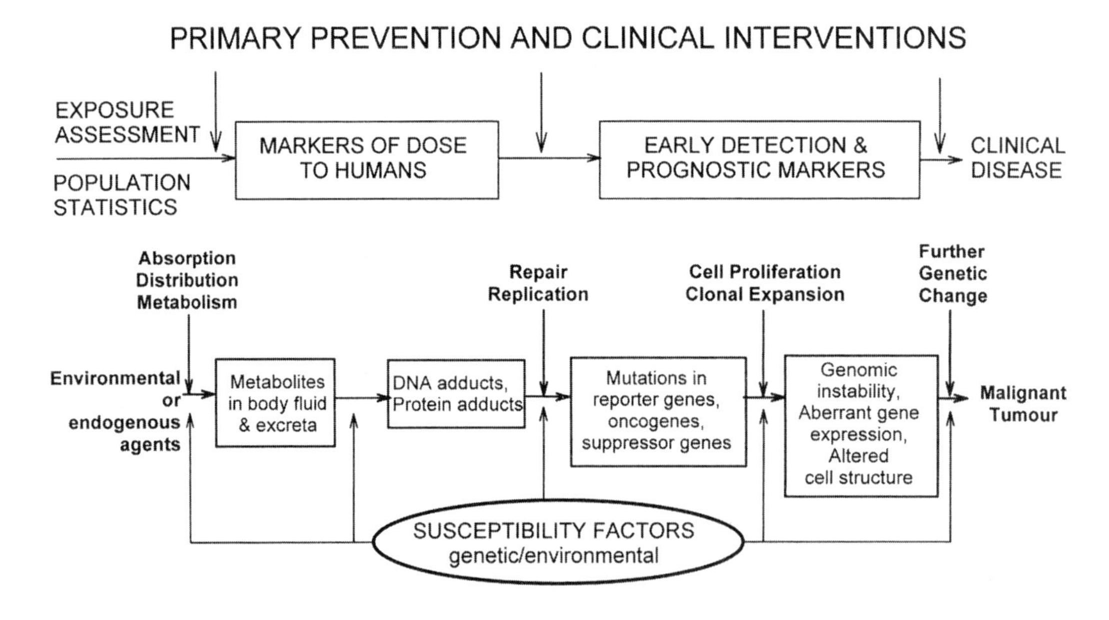

Figure 1. A conceptual basis for the application of biomarkers for use in prevention trials.

carcinogens, oligonucleotide sequences or expressed proteins. Simplicity and cost are important parameters. Complex laboratory-based assays of limited throughput may not be suitable for clinic- or field-based interventions where thousands of samples may be collected for analysis. Moreover, the biomarker needs to be obtained by non-invasive or minimally invasive procedures.

Regardless of intended application, there are fundamental analytical issues that need to be addressed for all biomarker assays. They include reliability, precision, accuracy and validity. Table 1 provides a description of each of these essential characteristics and indicates how they are measured.

Combination of the characteristics defined in Table 1 gives a full description of the properties of a given biomarker. A fully reliable ($S_w^2 = 0$), totally precise ($S^2 = 0$) and accurate ($m - \mu = 0$) biomarker is the desired goal, but this is extremely difficult to attain and almost never happens. It is also unusual to have a fully reliable and totally precise yet inaccurate ($m - \mu \neq 0$) biomarker, but this could be the case with a superb laboratory technique measuring the wrong analyte. From epidemiological and intervention points of view, the ideal biomarker will be fully reliable, reasonably precise ($S^2 > 0$ but not too large), and accurate. In practice, most population-based studies use biomarkers that are moderately reliable ($S^2 > 0$, but not large) and moderately precise ($S^2 > 0$ but not large), and accurate ($m - \mu = 0$).

Most reports document the reliability and precision of the biomarker measurements, but do not directly incorporate them into inferential summaries. Efforts should be made to integrate all aspects of the biomarker in inferences regarding their response to exposures and/or interventions. Another analytical issue that is important is tracking. In the context of a longitudinal study, if exposure is constant over time, the longitudinal measurements of a biomarker could be viewed as the x replicates in the experimental setting outlined in Table 1, so that ρ represents the tracking correlation. The degree of tracking of the biomarker will influence the needed frequency for repetitive sampling.

Of the three characteristics above, accuracy is the primordial one. In general, accuracy imparts validity to the biomarker, but inaccuracy does not preclude validity. As a matter of fact, in the context of chemoprevention trials, where the objective is to quantify the effect of an intervention on disease risk via modulation of one or more biomarkers, it is validity that is the most important feature of a biomarker. In the situation where the biomarker is inaccurate ($m - \mu \neq 0$, that is, biased), the hope is for the biases to operate in the same direction in both control and intervention groups, so that the validity of the study is preserved.

When the primary objective is to use a biomarker as the study end-point to monitor efficacy of the intervention, two biological features are also essential. The first feature is the degree of association between a risk biomarker and disease outcome — cancer. In general, the higher the association, the more useful the biomarker to chemoprevention trials. The second biological feature is that intermediate end-point biomarkers must be modulated by interventions in predictable, dose-dependent ways. These are necessary conditions for a biomarker to be a surrogate marker. Surrogacy is discussed in detail later in this chapter.

Biomarkers as measures of exposure: dose to humans

Humans are exposed to chemical, physical or biological carcinogens through contaminated air, water, soil, food or biological specimens (blood, semen, saliva). Thus, a person's exposure is the result of proximity to the agent superimposed upon many modifying factors. A biomarker of exposure may be the parent chemical itself, as exemplified by heavy metals. Frequently, however, it is a metabolic product of the agent formed in the body that serves as a marker of exposure and provides an internal dose measure. Carcinogen – DNA and carcinogen–protein adducts are also markers of exposure and are often referred to as measures of 'biologically effective' dose. Ideally, biomarkers of exposure should indicate the presence and magnitude of previous exposure to an environmental carcinogen. In the absence of a biomarker, assessment of exposure typically requires measurement of toxicant levels in the environment and characterization of the individual's presence in, and interaction with, that environment. The use of ambient measurements to determine exposure status of individuals is complicated because most etiological agents are not evenly distributed in the

Table 1. Essential characteristics for biomarkers

The experiment used to define measures of reliability, precision, accuracy and validity in this table consists of *x* replicate (i.e., same conditions) determinations on *n* homogeneous (i.e., same mean) individuals. The true common mean is denoted by μ; the observed overall mean (mean of individual means) by *m*; the within-individual error (average of the individual's variances of the sets of m replicates) by S_W^2; the error between individuals (variability of the individual's means around the overall mean) by S_B^2; the total error ($S_W^2 + S_B^2$) by S^2; the correlation between two measurements in an individual (S_B^2/S^2) by ρ; and the bias ($m-\mu$) by *b*. Measures with an * indicate those of the intervention group. The true difference in means between the intervention and control groups is denoted by Δ.

Concept	Description	Measures	Goals	
Reliability	Repeatability	S_W^2; ρ	$S_W^2 = 0$; $\rho = 1$	Fully reliable
Precision	Sharply measured	$S^2 = S_B^2 + S_W^2$	$S^2 = 0$	Totally precise
Accuracy	Measures 'true' level	$b = m - \mu$	$b = 0$	Unbiased
Validity	Measures 'true' change (or effect of intervention on outcome)	$b^* - b = (m^* - m) - \Delta$	$b^* - b = 0$ or $m^* - m = \Delta$	Internal validty

environment. Thus, the requirements for development of practical biomarkers of exposure must include an ability to integrate various routes and fluctuating exposures over time, to relate time of exposure to dose and to examine mechanisms in important biological targets (Muñoz, 1993). In turn, use of such accurate biomarkers of exposure will limit misclassification, which is often the major source of error in environmental epidemiology (Hulka, 1991). Proper identification of individuals at risk for exposure to cancer-causing agents offers strong potential for enriching the selection of study cohorts (and correspondingly reducing sample size requirements) for chemoprevention trials.

Urinary metabolites

In the early 1800s, Wöhler and his colleagues recognized that urine was a vehicle for the elimination of metabolites of xenobiotics, when they identified hippuric acid following dosing with benzoic acid (Young, 1977). Now, both oxidation and conjugation products of a multitude of drugs and environmental toxicants, including many carcinogens, have been identified and quantified in the urine of humans. Such measurements have become analytical staples for molecular epidemiologists seeking to identify causal linkages between carcinogen exposure and disease (Shuker *et al.*, 1993). Examples of such metabolites include mercapturic acids derived from glutathione conjugation of several carcinogens (De Rooij *et al.*, 1998) including aflatoxin (Wang *et al.*, 1999), benzene (Boogaard & van Sittert, 1995) and 1,3-butadiene (Hayes *et al.*, 2000); glucuronide and sulfate esters of heterocyclic amines such as 2-amino-1-methyl-6-phenylimidazo[4,5-b]pyridine (PhIP) (Lang *et al.*, 1999); 1-hydroxypyrene in workers exposed to polycyclic aromatic hydrocarbons (Bouchard & Viau, 1999); and oxidation products and glucuronides of 4-(methylnitrosamino)-1-(3-pyridyl)-1-butanone (NNK) in smokers (Hecht, 1997). These metabolites are obviously strong markers of exposure and in some instances have served as intermediate biomarkers in chemoprevention trials.

Hecht *et al.* (1995) have analysed the effects of consuming watercress, which is a rich source of phenethyl isothiocyanate (PEITC), on the metabolism of tobacco-specific carcinogens in smokers. They observed that watercress consumption increased urinary levels of two metabolites of NNK: 4-(methylnitrosamino)-1-(3-pyridyl)-1-butanol (NNAL) and its glucuronide, NNAL-Gluc. This increase was attributed to either inhibition of cytochrome P450 or induction of glucuronidation. Watercress consumption also affected nicotine metabolism in these individuals, with an elevation in levels of glucuronide of nicotine in urine samples collected during the intervention and levels returning to baseline after the watercress consumption period (Hecht *et al.*, 1999). Likewise, in a large, double-blinded, placebo-controlled trial of oltipraz conducted in the People's Republic of China, urinary markers were used to demonstrate pharmacodynamic action by the intervention agent. Measurements of phase I (aflatoxin M_1) and phase II (aflatoxin-mercapturic acid) metabolites were used to demonstrate that oltipraz inhibited oxidation and enhanced conjugation of aflatoxin relative to placebo (Wang *et al.*, 1999). Reductions in levels of circulating aflatoxin–protein adducts were also seen in participants receiving oltipraz (Kensler *et al.*, 1998).

DNA and protein adducts

While early-stage measurements provide unequivocal identification of chemical exposures, they do not provide evidence that toxicological damage has occurred. Measurements of carcinogen–DNA and carcinogen–protein adducts are of interest because they provide molecular, mechanism-based bridges between carcinogen exposures and disease end-points. These adducts are direct products of damage to critical macromolecular targets and reflect an integration of the toxicokinetic factors of absorption, distribution and metabolism. However, these toxicokinetic factors are not constant and can vary as a function of dose and duration of exposure. Replication of carcinogen-modified DNA is thought to result in the fixation of mutations that serve as initiating agents in transformation and, thus, formation of such adducts is assumed to be on the causal pathway. However, since the pattern and level of these adducts can also be profoundly influenced by repair processes of differing efficiency and fidelity, the usefulness of measurements of DNA adduct concentration to predict cancer incidence quantitatively remains unclear (Gaylor *et al.*, 1992). Indeed, several estimates of the overall contributions of carcinogen adducts to cancer risk, using animal models in which exposure can be carefully controlled, indicate that the attributable risk for these types of marker may be less than 10% (Travis *et al.*, 1996; Kensler *et al.*, 1997).

Given the complex interactive nature of the carcinogenic process, it is unreasonable to expect that a single, early marker can fully predict risk of cancer outcomes. Clearly, production of genetic damage by carcinogens is not a *sine qua non* for cancer. Nonetheless, like the metabolite markers for internal dose, the adduct biomarkers effectively delineate exposures and serve as modulatable end-points for judging the efficacy of certain classes of chemopreventive agents, notably those that protect cells by altering the metabolism and disposition of the reactive intermediates leading to DNA damage (Kensler, 1997). Indeed, modulation of carcinogen adduct levels has been used for a considerable period of time as a short-term end-point for the initial evaluation of chemopreventive agents *in vivo* (Kensler *et al.*, 1985). In general, the population-based predictive value is quite good. However, in some instances, especially when complex tumour induction regimens are used involving tumour initiators and promoters, correlations between adduct levels and ultimate tumour yields can be poor, or more perversely, negative (Hartig *et al.*, 2000). In the one case where levels of adducts were assessed for their predictive value of individual risk for developing cancer, no value was observed, despite the fact that a strong correlation was observed in the same experiment between adduct burden and treatment group risks for hepatocarcinogenesis (Kensler *et al.*, 1997). Once the individual animal was identified by treatment group assignment, the adduct biomarker provided no further information about cancer risk.

A variety of analytical methods are now available for detecting and quantifying covalent adducts formed between DNA and proteins and genotoxic chemicals. Methods for DNA adduct analysis include immunoassay, ^{32}P-postlabelling and physicochemical procedures based on such properties as fluorescence or involving mass spectrometry and electrochemical analysis. Protein adducts in haemoglobin and serum albumin can be analysed by physicochemical methods, principally gas chromatography/mass spectrometry, or by immunoassay. Collectively, methods are now available for the detection of DNA and/or protein adducts of many of the major classes of chemical carcinogens (Kaderlik *et al.*, 1992; Shuker & Farmer, 1992; Strickland *et al.*, 1993). These techniques have been used to measure composite and specific DNA adducts in cellular DNA isolated from peripheral lymphocytes, and from bladder and colonic tissues, as well as DNA adducts excreted in urine. Many of these methods are of sufficient sensitivity and specificity to detect ambient levels of exposure and are being applied to studies of tobacco use (polycyclic aromatic hydrocarbons, aromatic amines, tobacco-specific nitrosamines; dietary exposures (aflatoxins, *N*-nitrosamines, heterocyclic amines); medicinal exposures (cisplatin, alkylating agents, 8-methoxypsoralen, ultraviolet photoproducts); occupational exposures (aromatic amines, polycyclic aromatic hydrocarbons, 1,3-butadiene, oxides of styrene and ethylene, vinyl chloride); and oxidative damage (8-oxoguanine, thymine glycol, malondialdehyde) (Kensler *et al.*, 1996; Halliwell, 1998; Singer & Bartsch, 1999).

There are a few instances where adducts have been used as biomarkers in human intervention studies. Excretion of 8-oxodeoxyguanosine is associated with age, metabolic rate, caloric intake and antioxidant content of the diet (Fraga *et al.*, 1990; Simic, 1994). Simic & Bergtold (1991) investigated the effects of manipulation of the human diet on levels of urinary markers of DNA base damage, namely, thymidine glycol and 8-oxodeoxyguanosine. Excretion of biomarkers of oxidative DNA damage was suppressed when dietary composition was maintained but caloric intake was decreased. At isocaloric dietary intake, the level of damage depended upon diet composition. For diets containing carbohydrates, proteins and fats but lacking fruits and vegetables, the level of damage was higher than for diets including fruits and vegetables, which are rich in natural antioxidants. Feeding Brussels sprouts to healthy, nonsmoking volunteers also led to a small (28%) but statistically significant reduction in urinary excretion of 8-oxodeoxyguanosine (Verhagen *et al.*, 1995). Similarly, consumption of tomatoes was sufficient to alter levels of oxidative DNA base damage in white cells within 24 h (Rehman *et al.*, 1999). Clearly, levels of these biomarkers can be modulated in humans, making them attractive candidates for assessing the efficacy of antioxidant-based chemoprevention interventions. Dyke *et al.* (1994) have examined the effect of oral vitamin C supplementation on gastric mucosal DNA damage, as measured by ^{32}P-postlabelling in 43 patients. Gastric mucosal DNA damage was decreased in 28 of the patients after vitamin C supplementation ($p = 0.01$). Wallin *et al.* (1995) investigated the effect of phenobarbital (a phase I and II enzyme inducer) treatment of epileptic patients on levels of aromatic amine–haemoglobin adducts as a function of tobacco consumption. In comparison with patients receiving other anticonvulsants, a significant depression in adduct biomarker levels was observed with phenobarbital treatment.

Biomarkers as intermediate measures of disease: early detection and prognostic markers

The historical precursors to biomarkers in cancer research arose from the quest to discover cancer at an early and treatable stage. Numerous such 'tumour markers' are used currently to diagnose or confirm diagnosis for specific cancer types. Examples include carcinoembryonic antigen (CEA) for tumours at several sites, prostate-specific antigen (PSA) for carcinoma of the prostate, 5-hydroxy-indoleacetic acid in the urine for carcinoid tumours, α-fetoprotein for liver cancer and germ-cell tumours, and thyrocalcitonin for medullary carcinoma of the thyroid (Keefe & Meyskens, 2000). As reviewed by Schulte & Perera (1993), the history of tumour marker research, particularly in the areas of cancer cytology and cytogenetics, also provides examples of past attempts to validate markers and bring them into screening programmes. The use of Papanicolaou cytology as a marker of preclinical cervical cancer demonstrates how a good marker can lead to effective intervention, yet 27 years lapsed between the development and adoption of the Pap test (Greenwald *et al.*, 1990). In addition to the search for early detection markers, prognostic and predictive markers have been developed. These markers help guide decision-making about therapeutic options and opportunities.

Somatic mutations and genomic instability

Carcinogenesis is driven by an accumulation of genetic changes. These changes occur over time and lead to the evolution of extended clonal foci of neoplastic cells. A variety of detection methods have been developed to detect the presence of neoplastic cells in accessible samples of body fluids and tissues. Mutations are among the earliest lesions to occur following assault of the genome by endogenous or exogenous carcinogens.

Mutations can be detected in easily obtained cell types in reporter genes, whose modification is unrelated to the causal development of cancer, but which reflect exposures to carcinogens. The detection of mutations in the *HPRT* gene is currently the most extensively employed assay for detecting somatic mutations in human genes *in vivo*. *HPRT* mutations are examined in lymphocytes, and the standard assay involves T-lymphocyte cloning for phenotypic selection of 6-thioguanine-resistant mutant cells (Tates *et al.*, 1991). The location and type of mutations in a specific sequence of nucleotides defines the mutational spectrum and have been analysed in the *HPRT* locus of lymphocytes from humans exposed to a variety of genotoxins (Cole & Skopek, 1994). Another *in vivo* assay for the detection of somatic mutations is the glycophorin A assay. This assay is based on the autosomal glycophorin A locus that encodes the cell surface sialoglycoprotein expressed in the erythrocytic lineage and responsible for the M,N blood group (Grant & Bigbee, 1993). Most of the variants are derived from mutations in bone marrow stem cells and are, therefore, permanent, delineating lifetime exposures to mutagens and accumulated mutations. While rapid, facile and inexpensive, the assay is suitable for only one half of the human population, the M/N heterozygotes.

Significant attention has been focused in recent years on target genes for somatic mutations, oncogenes and tumour-suppressor genes. Such genes have been classified as gatekeeper and caretaker genes in terms of their control of net cellular proliferation or maintenance of genomic integrity, respectively (Kinzler & Vogelstein, 1997). The most prominent example of a gatekeeper is the *APC* gene in colorectal cancer. Alterations in *APC* lead to derangements of cellular proliferation pathways and mutation of *APC* is thought to be an early event in the process of colon carcinogenesis. Other gatekeeper genes frequently subject to mutation, such as K-*ras* and *p53*, appear to play important roles in later stages of carcinogenesis. The *p53* gene is well suited for mutational spectrum analyses for several reasons. First, it is commonly mutated in many human cancers. Second, the *p53* gene is small, permitting study of the entire coding region. Third, the point mutations that alter p53 function are distributed over a large region of the molecule, allowing extensive inferences of the mechanism of DNA damage involved (Hussain & Harris, 1998). While tumour-specific *p53* mutations have been identified in several human cancers, identification of individuals harbouring such mutations has been problematic. However, DNA can be isolated from the plasma (or serum) of patients with cancer and this plasma carries the same genetic mutations as DNA in the tumour (Nawroz *et al.*, 1996). Kirk *et al.* (2000) have analysed for a selective arginine-to-serine substitution in codon 249 of *p53* in DNA isolated from plasma. This codon is a hotspot for mutation in hepatocellular carcinoma occurring in populations that are exposed to aflatoxins and have a high prevalence of infection with hepatitis B virus (Hollstein, 1991). The 249-Ser mutation in *p53* was detected in DNA isolated from plasma by restriction endonuclease digestion of polymerase chain reaction products from exon 7 of the gene. Its presence is strongly associated with hepatocellular carcinoma in patients from The Gambia, a high-risk region, but not in patients with liver cancer from Europe. Such approaches allow earlier detection of liver cancer and provide possible intermediate end-points for assessing the impact of intervention programmes such as hepatitis B vaccination and chemoprevention. Mutations in gatekeeper genes can also be assessed in other settings such as *ras* gene mutations in stool (Sidransky *et al.*, 1992) and *p53* mutations in exfoliated bladder epithelial cells isolated from urine (Sidransky *et al.*, 1991). An exciting recent development is the measurement of mitochondrial DNA mutants in tumours and body fluids (Fliss *et al.*, 2000). By virtue of their clonal nature and high copy number, mitochondrial mutations may provide particularly sensitive markers for noninvasive detection of early neoplastic lesions.

It now appears that 4–10 events are necessary for the development of sporadic solid tumours (Fearon & Vogelstein, 1990); however, the normal baseline rate of mutation within a cell is insufficient to account for the required number of events (Loeb, 1998). Early inactivation of genes that maintain genomic stability (caretaker genes) could result in a mutator phenotype that would significantly destabilize the genome, increase the mutation rate and lead to tumour progression. Such genomic instability, reflecting the propensity and susceptibility of the genome to acquire multiple alterations, is believed to be the driving force behind multistage carcinogenesis. Genomic instability is manifest in several forms: aneuploidy, microsatellite instability and intrachromosomal instability. Probably the best characterized form of instability is microsatellite instability. It involves the insertion or deletion of one or two base pairs in simple repeat sequences (Perucho, 1996) and can result from inherited or somatic defects in DNA mismatch repair genes (e.g., *hMSH2* and *hMLH1*). Diagnostic assays have been developed for microsatellite instability in body fluids. Squamous cell carcinoma of the aerodigestive tract and bladder cancer can be detected through microsatellite analysis of saliva and urine, respectively (Mao *et al.*, 1996; Steiner *et al.*, 1997; Spafford *et al.*, 1998). Renal cancers can also be detected by molecular urinalysis (Eisenberger *et al.*, 1999), while early-stage lung cancer has been detected in tumour DNA isolated through bronchoalveolar lavage (Ahrendt *et al.*, 1999; Field *et al.*, 1999) and sputum (Mao *et al.*, 1994). Moreover, collateral microsatellite analysis of serum samples in some of these studies reveals evidence of circulating tumour DNA and may portend poorer prognosis. Nipple aspirates provide avenues for cytological and molecular analyses in breast cancer prevention trials (Fabian *et al.*, 1997). FISH assays are also available to diagnose and monitor the treatment of field cancerization, i.e., of diffuse genomic instability, even before the onset of intraepithelial neoplasia in patients with proven high risk of cancer (e.g., previous surgery for head and neck cancer). Resection, therapy and/or prophylaxis may all be appropriate in individuals manifesting markers of genomic instability.

Aberrant gene expression

Altered patterns of DNA methylation are among the earliest molecular changes to occur in the evolution of neoplastic cells. In particular, aberrant methylation of CpG dinucleotides that are clustered in the 5′ flanking and first exonic regions of many genes (CpG islands) appear to occur very early in tumour progression for several tumour types and could alter chromatin structure and/or play a role in the loss of tumour-suppressor or differentiation gene functions. Indeed, aberrant hypermethylation of CpG islands has been implicated in the transcriptional inactivation of many genes including those for Rb, p15, p16, p73, VHL, E-cadherin, TIMP3, glutathione *S*-transferase (GST) Pi, MLH1, BRCA1, estrogen receptor α, progesterone receptor, retinoic acid receptor β and androgen receptor. A critical finding is that aberrant promoter methylation is seldom seen in normal tissues except for imprinted genes and genes on the inactive X chromosome. In addition, hypermethylation changes are fairly constant among tumours and occur within the same regions, that is the promoter region of the target genes. Also, these changes can be assessed in a relatively stable molecule, DNA. PCR-based strategies to assess DNA promoter hypermethylation now exist, providing a sensitive method for detection using minute amounts of biological samples. All these features make detection of promoter region hypermethylation an attractive marker for detection of tumour cells (Laird & Jaenisch, 1996).

The potential clinical utility of this approach has been demonstrated by several pilot studies. For example, in non-small-cell lung cancer, hypermethylation of *p16* was detectable in bronchoalveolar lavage samples from patients with lung cancer whose tumours also had methylation of *p16*, but not from those collected from patients whose tumours did not show this change (Ahrendt *et al.*, 1999). Further, evidence of *p16* methylation has been detected in sputum from patients with lung cancer or those at high risk for lung cancer development (Belinsky *et al.*, 1998). A similar analysis of *p16* hypermethylation in patients with hepatocellular carcinoma showed *p16* methylation in 16/22 liver cancers and similar changes were detected in the plasma or serum of 13 of the 16 cases (Wong *et al.*, 1999). These studies have now been extended to include panels of methylated genes. A prototype analysis examined gene methylation patterns in normal lung, lung cancer and serum from non-small-cell lung cancer patients at the time of surgery. Overall, 15 of 22 tumours showed methylation of one or more of four genes in the tumour and the same alteration was found in the serum of 11 of these 15 patients (Esteller *et al.*, 1999). Similar approaches have been successfully applied to the study of patients with head and neck cancer.

Recent technological advances potentially provide powerful tools for direct analysis of the expression of multiple genes simultaneously in normal and abnormal tissues. These include open systems such as SAGE and closed systems such as microarrays and oligonucleotide chips. This field is currently in its infancy and initial studies are focused upon molecular classification of established tumours as a proof of principle. For example, a preliminary application of a microarray strategy to molecular classification of leukaemias has been described (Golub *et al.*, 1999). A long-range goal of this type of approach could be to predict clinical outcome in both the treatment and prevention settings. In the shorter term, microarray analyses can provide mechanistic readouts on the pharmacodynamic actions of chemopreventive agents.

Intraepithelial neoplasia

Intraepithelial neoplasia (IEN) is a precancerous lesion directly on the causal pathway to cancer and has traditionally been detected by histopathological methods. Two basic processes underlie the onset and development of IEN (Boone *et al.*, 1997). The first is genomic instability, the second is development within an epithelium having genomic instability of multicentric neoplastic lesions that independently progress through each of the following processes at an accelerating rate: clonal evolution, hyperproliferation, production of genomic structural variants, and apoptosis. IEN is the most common intermediate end-point currently applied in chemoprevention trials. It is used both in selection of study cohorts and as an end-point to assess efficacy. For the latter purpose, regression and prevention of recurrence of IEN are assessed. A wide range of IEN have been used for cohort selection for chemoprevention trials, including ductal carcinoma *in situ* (DCIS) and lobular carcinoma *in situ* (LCIS) for breast cancer; prostatic intraepithelial neoplasia (PIN); cervical intraepithelial neoplasia (CIN); adenomatous polyps for colon cancer; and dysplastic lesions of the stomach. Detailed discussions of the use of IEN in defining study cohorts and as outcome measures can be found in other chapters in this volume and in several reviews (Boone *et al.*, 1997; Keefe & Meyskens, 2000). Because IEN are at-risk foci of neoplastic cells, they are fertile regions for applications of molecular laboratory medicine that allow measurement of genomic instability and altered gene expression. Highly quantitative methods for assessing altered cell morphometry have also been developed. Quantitative computer-assisted image analysis systems can be used to measure features of nuclear morphometry such as increased size, altered shape, pleomorphism, altered chromatin texture, DNA ploidy and proliferative index, and are beginning to be applied to measure potential effects of antiproliferative agents in chemoprevention trials (Bacus *et al.*, 1999).

Biomarkers as measures of susceptibility

Epidemiological and human genetic studies have identified different types of at-risk individuals within populations (Harris, 1989). These interindividual differences in susceptibility to carcinogenesis or other diseases may be either acquired or inherited. Some individuals have heavy exposure to environmental carcinogens, while others are carriers of cancer-predisposing germline mutations in genes that, because of high penetrance, confer a very high risk for development of cancer (Dove, 2000). There is also another group of predisposing polymorphic, low-penetrance genes that more modestly increase the risk for cancer in individuals exposed to carcinogens. These genes can be involved in carcinogen metabolism, DNA repair, intracellular signalling (receptors) and immunosurveillance. The proportion of cancers attributable to such genetic traits may be high, because the frequency of these risk-modifying alleles is high in the overall population. In addition, there may be strong interactions between low-penetrance genes, that in the aggregate confer considerable risk for an individual to develop cancer (Hussain & Harris, 1998).

Low-penetrance susceptibility genes

Many enzymes are involved in the oxidative metabolism and conjugation of carcinogens in humans. Some of the genes that control expression of these enzymes are polymorphic and are not expressed in significant percentages of a population. Even for genes that are monomorphic, there can be huge variations in levels of expression and subsequent enzymatic activity. Thus, both intrinsic (e.g., genetic) and extraconstitutional factors (e.g., diet, hormonal status, occupation) can strongly influence the expression of xenobiotic (and endobiotic) metabolizing enzymes. The molecular basis for the genetic factors leading to variations in activity includes: nucleotide variations in the coding region of the gene (altered substrate binding or turnover rates); deletions in the coding region (inactive enzyme); polymorphisms in the regulatory regions of genes (altered basal or inducible expression); variations in polyadenylation signals (altered transcript half-life and enzyme levels); and gene amplification (increased enzyme levels) (Bartsch *et al.*, 2000). The difficult task lies in identifying individuals who harbour altered capacities for carcinogen metabolism evoked by these mechanisms. Function-based assays such as phenotyping by metabolite analyses of endogenous or exogenous substrates can be informative, but often are analytically laborious. Moreover, metabolic phenotyping is easily affected by confounders such as food or drug intake before testing, which do not affect genotyping assays (Barrett *et al.*, 1997). High-throughput gene analysis by DNA microarray techniques offers prospects for rapid identification of new mutations, while polymerase chain reaction (PCR) and restriction fragment length polymorphism (RFLP) methods provide easy approaches for characterization of known polymorphisms. However, in these instances, the analytical ease of measurement often outstrips the ability to appreciate the functional significance of these gene variants in humans. Measurements in the absence of understanding of the contributions of specific genetic variations to susceptibility modification are not likely to lead to improvements in the design, conduct and interpretation of chemopreventive interventions.

Genetic polymorphism is probably the single most important determinant of enzyme multiplicity in man and considerable inter-individual variation in drug oxidation and conjugation has been long recognized. Polymorphism refers to a monogenic variation that occurs with at least two phenotypes with sufficient frequency (>1%) to cause population differences. Polymorphisms in many, but not all, phase I (cytochrome P450 (CYP)) and phase II (conjugating) enzymes have been described. Variations in some *CYP* genes have been associated with increased risk for cancers of the lung, oesophagus, and head and neck in smokers (Bartsch *et al.*, 2000; Nair & Bartsch, this volume). Polymorphisms in other *CYP* genes elevate risk for breast cancer, presumably through effects on estrogen metabolism (Feigelson *et al.*, 1998). Polymorphisms in phase II enzymes can also influence cancer risk. In some instances, these enzymes contribute to the metabolic activation of procarcinogens, while in others, their role is in detoxication of reactive intermediates. As examples of this latter case, risk for smoking-related cancers can increase in individuals deficient in GSTM1 (Houlston, 1999; Bartsch *et al.*, 2000). In a more complicated scenario, polymorphisms in *N*-acetyltransferase appear to do both. The rapid acetylator phenotype for acetylation of aromatic and heterocyclic amine carcinogens is associated with increased risk for colon cancer and the slow-metabolizer phenotype with increased risk for urinary bladder cancer (Lang, 1997). Thus, the use of this biomarker as a predictor of individual risk will be dependent upon the context for its use. Moreover, gene–gene interactions between polymorphic phase I and II genes have been observed (Bartsch *et al.*, 2000). Fittingly, the manifestation of the contributions of these susceptibility genes is driven by levels of exposure to carcinogenic substances (Hietanen *et al.*, 1997). Thus, gene-$(n)_x$–gene–environment interactions are the true mediators of risk modification. Metabolic susceptibility genes in the absence of exposure are of little consequence. Development of study cohorts for chemoprevention in this context requires monitoring of biomarkers for both susceptibility and exposure.

DNA repair capacity represents another important susceptibility factor. Patients with the rare, cancer-prone inherited disorder xeroderma pigmentosum experience a greater than 1000-fold excess frequency of sunlight-related skin cancers (Kraemer & Slor, 1985). Laboratory studies indicate that cells from such patients are defective in repairing DNA damage induced by ultraviolet radiation and chemical carcinogens. Although less pronounced, variations in DNA repair capacity have been observed in the general population and may be important susceptibility determinants (Grossman & Wei, 1994). Several assays, notably host-cell reactivation for measuring cellular DNA repair capacity and an *in vitro* mutagen sensitivity assay, have been developed for application in population-based studies. Correlations between these assays have also been established (Wei *et al.*, 1996). Case–control studies indicate that diminished DNA repair capacity is a risk factor for upper aerodigestive-tract cancers, including lung cancer (Spitz *et al.*, 1996) and basal cell carcinoma of the skin (Grossman, 1997).

In recent years, several genes involved in the repair of mispaired nucleotides (mismatch repair) have been characterized (Bronner *et al.*, 1994). Mutations in these genes are particularly linked to an elevated risk of colon cancer. The mutations occur as heterozygotes and tumours are induced as a result of the loss of wild-type allele. It has been estimated that such mutations are carried by 1 in 200 people, thus constituting one of the most prevalent human disorder mutations (Barrett *et al.*, 1997). While mutations in DNA repair genes can result in loss of DNA repair protein, DNA polymorphisms may alter the structure of the DNA repair enzyme and modulate catalytic activity and efficiency. A recent study evaluated the effects of polymorphisms in the repair enzyme *XRCC1* (X-ray repair cross-complementing 1) in two populations by measuring levels of aflatoxin B_1–DNA adducts in placenta of Taiwanese maternity subjects and somatic glycophorin A variants in erythrocytes from smokers and nonsmokers. In both groups, a Arg399Gln amino acid change appeared to alter the phenotype of the protein, resulting in lowered DNA repair capacity (Lunn *et al.*, 1999).

High-penetrance susceptibility factors

Penetrance is 100% when every individual who carries the mutated gene develops the disease. In the most pronounced cases of familial cancer such as retinoblastoma, affected individuals transmit cancer predisposition to approximately 50% of their offspring (Dove, 2000). In these situations, one mutant allele at a single locus is sufficient for predisposition of individuals for cancer. The set of fully penetrant, dominantly transmitted familial cancers is expanding rapidly. In many cases such familial syndromes yield neoplasms of distinct histological origin and reflect loss of function in the mutated gene.

DNA repair genes provide several examples of loss-of-function familial cancer syndromes. Ataxia telangiectasia, Bloom's syndrome, xeroderma pigmentosum and Fanconi's anaemia lead to dramatically increased risk for lymphoma, solid tumours, skin cancer and acute myelogenous leukaemia, respectively. The aforementioned mismatch repair defects contribute to colorectal, endometrial and gastric carcinoma in patients with hereditary nonpolyposis colorectal cancer (HNPCC). Only a small proportion of colon tumours can be ascribed to members of high-risk families. However, the same gene in which germline mutations are found in high-risk families is often found to be mutated somatically in sporadic tumours at that site. For example, the adenomatous polyposis coli gene, *APC*, is mutated in the germline of familial adenomatous polyposis (FAP), or somatically in HNPCC, and in sporadic colon cancer. Mutations in other tumour-suppressor genes can be observed in multiple types of human tumours and are linked to familial syndromes (e.g., *p53* in Li–Fraumeni syndrome; *BRCA1* and *BRCA2* in familial breast and ovarian cancer).

Some of these genetic syndromes may represent suitable cohorts for inclusion in chemoprevention trials, although for the rarer forms, the need to accrue sufficient numbers of participants into trials may limit this approach. Although FAP comprises only 1% of colon cancer patients, there are some notable examples of chemoprevention trials in this cohort. FAP patients tend to develop thousands of adenomatous polyps, which are evenly distributed throughout the colon and rectum by the third decade of life. In the absence of surgical treatment, affected individuals are at high risk for development of colon cancer by the age of 40 years (Erbe, 1976). Numerous trials have been conducted in patients with polyps to prevent polyp recurrence using pharmacological (sulindac, celecoxib, aspirin, difluoromethylornithine, butyrate) as well as nutritional (fibre, calcium) approaches. These interventions have targeted the proliferative cancer phenotype of the polyps, rather than the underlying predisposing genetic defects. Use of a pharmacological agent to replace the function of a lost or mutated allele is beyond the bounds of current chemoprophylaxis, but within the promise of molecular medicine.

Interactions of susceptibility factors with interventions

Presence or expression of some of the low-penetrance susceptibility genes not only affects risk of carcinogenesis following exposure to genotoxins, but also can modify the potential efficacy of chemopreventive interventions. Many chemopreventive agents undergo metabolism, that may either activate or inactivate the agent. An interesting example of a gene–intervention interaction comes from the major chemopreventive components of cruciferous vegetables, isothiocyanates. These are potent anticarcinogens that act, in part, to induce levels of expression of conjugating enzymes, thereby detoxifying electrophilic forms of carcinogens. Many isothiocyanates are conjugated by GSTs, which facilitate their accumulation in cells (Zhang & Talalay, 1998), but which may also impede the manifestation of their pharmacodynamic actions as enzyme inducers. A case–control study by Lin *et al.* (1998) indicated that intake of broccoli (a cruciferous vegetable) was positively associated with reduced risk for colorectal adenomas. However, when stratified by *GSTM1* genotype, a significant protective effect of broccoli was observed only in subjects with the *GSTM1*-null genotype. In a similar vein, analysis of the interaction between dietary isothiocyanates (measured in urine) and genetic polymorphisms for GSTs in a prospective study conducted in Shanghai indicated that isothiocyanates seemed to decrease lung cancer risk, particularly among persons genetically deficient in the GST isoforms GSTM1 and GSTT1 that may inactivate these chemopreventive compounds (London *et al.*, 2000). We have observed in the chemoprevention trial of another inducer of carcinogen detoxifying enzymes (oltipraz) that the pharmacodynamic action was greatest in individuals who were *GSTM1*-null. In this instance, GSTM1, which is poorly induced by oltipraz, is thought to be the primary constitutive catalyst for the conjugation of aflatoxin with glutathione and likely masks induction of other isoforms of GST. However, in *GSTM1*-null individuals receiving oltipraz, the apparent induction of GSTA1 was unmasked and led to increased excretion of aflatoxin-mercapturic acid (Wang *et al.*, 1999). Clearly, the influence of pharmacogenetics on the actions of chemopreventive agents needs full consideration, as in other aspects of pharmacology such as chemotherapy (Balis & Adamson, 1999).

Degree of surrogacy: a classification paradigm for risk biomarkers

The first step towards the identification of a good early detection/prognostic or 'risk' biomarker is to document the prognostic value of the biomarker (B) on occurrence or incidence of disease (D). Hereafter, we denote this relationship by $B \rightarrow D$. The data to document this relationship often come from observational (cohort) studies which describe that natural history of the disease in humans (e.g., cholesterol for cardiovascular disease (Dawber, 1980); human immunodeficiency virus (HIV) load for AIDS (Mellors *et al.*, 1997); persistence of human papillomavirus (HPV) infection for cervical cancer (Ahdieh *et al.*, 2000)) and/or from disease models in animals. The relationship $B \rightarrow D$ is a prerequisite for a biomarker to have utility in the evaluation of efficacy of a chemoprevention agent (A). Since the primary objective in short-term trials is to use modulation of biomarkers as a measure of efficacy, it is also assumed that evidence is available documenting that the agent modifies the biomarker (i.e., $A \rightarrow B$). Such demonstrations often come from animal models.

Under the premises that $B \rightarrow D$ and $A \rightarrow B$, if the agent were not to have an effect on disease ($A \nrightarrow D$), this will indicate that the biomarker is useless as an evaluator of the lack of effect of the agent on disease. Furthermore, an effect on this type of biomarker can actually be misleading, as it may suggest a spurious efficacy of the agent. Therefore, a third criterion is that the agent must have an effect on disease occurrence (i.e., $A \rightarrow D$). While this criterion is often assumed, establishing this relationship through experimental or clinical studies is in fact very difficult.

Under the abovementioned criteria, a key characteristic of a biomarker is the determination of how much of the effect that an intervention has on disease is captured by the modification of the biomarker during the intervention. This capturing of information is the essence of surrogacy. In the best case, modulation of the marker by the intervention fully captures its effect on disease outcome (full surrogacy). In other words, once the modified (by the intervention) level of the biomarker is determined, no additional information on the intervention is needed to determine the risk of disease. An important feature of these criteria is to safeguard against using biomarkers that are modified by interventions but have no predictive value for effects on disease onset.

In the 1980s, Prentice (1989) suggested criteria to characterize this full surrogacy, such that, given (or, in statistical terms, conditional on) the biomarker level, there is no residual association between the agent and the disease ($A \not\to D|B$). Unfortunately, there are not many biomarkers that fulfil these stringent criteria. In part, this is because often there are other pathways by which an intervention affects disease that lie beyond the effect on a biomarker (Schatzkin *et al.*, 1990). Many potentially useful biomarkers do not lie directly and exclusively on the causal pathway(s) to disease. A less stringent classification providing a flexible scheme is obtained by quantifying the degree of surrogacy of a biomarker on a continuous rather than dichotomous scale. In terms of simple statistical models, the key comparison is to estimate the relative predictive value of the agent on disease when the biomarker is included as another predictor. Specifically, if the model to relate agent to disease is:

$$\text{Disease} = \alpha + \beta \bullet \text{Agent}$$

where β quantifies the change in disease due to changes in the agent (i.e., β measures $A \to D$), the degree of surrogacy can be determined by comparing β to β^* in the model

$$\text{Disease} = \alpha^* + \beta^* \bullet \text{Agent} + \gamma^* \bullet \text{Biomarker}.$$

Here, the coefficient β^* quantifies the effect of the agent after controlling for the level of the biomarker that, in itself, is modified by the agent. Freedman *et al.* (1992) proposed the use of $(\beta-\beta^*)/\beta$ as a measure of the proportion of the effect of an agent on disease explained by the biomarker. In the more stringent case of the biomarker fully capturing the effect of the agent on disease, one would expect the proportion explained to be equal to 1 (i.e., $\beta^*= 0$). The proportion explained is a direct measure of the degree of surrogacy. The further the proportion explained is from zero, the stronger the degree of surrogacy. As was indicated above, it is very rare to have the proportion explained equal to 1. In turn, the investigator could determine whether a 95% confidence interval for $(\beta-\beta^*)/\beta$ contains one. If so, this result would indicate an ideal degree of surrogacy. Unfortunately, the standard errors for the ratio $(\beta-\beta^*)/\beta$ are typically very large and the inclusion of one by the confidence interval is more likely to be due to the lack of precision in this estimate. The compensatory safeguard is to increase the sample size of the study to a level that defeats the advantages of biomarker studies. Another drawback of estimating the proportion explained is that it is not restricted to be always between zero and one. Biomarker modulation could be bidirectional. Furthermore, the possible residual effects of the agent on disease (after controlling for the attained biomarker levels) may vary by biomarker levels (i.e., interactions between biomarker and agent).

Alternative measures of the degree of surrogacy include the ratio of the measures of the effect of the agent on disease and the effect of the agent on the biomarker $(A \to D)/(A \to B)$ (Buyse & Molenberghs, 1998). A more epidemiologically based measure would be to require that the change in B due to A is of a magnitude that will correspond to a change in incidence of disease with a strength of a relative incidence below a prespecified level (e.g., 0.80). In other words, if B(A) is the level of the biomarker under the effect of the agent and B(not A) is the level of the biomarker for the control group, one would require that the protective effect of A on D be of a magnitude such that D[B(A)]/D[B(not A)] is less than 0.8. In this case, the reduction of the biomarker by the agent will translate in a reduction of more than 20% of disease occurrence. The more extreme this threshold, the more room there is for an outcome in which, even if not all the change induced by the agent on B translates into change on D, it is likely that the modulation of the biomarker does capture a beneficial effect of the agent.

In the context of HIV epidemiology, the level of HIV RNA in copies/ml provides an example of a biomarker with such a strong prognostic value on AIDS that an intervention modifying HIV viral level was correctly predicted to have an impact on AIDS. Cohort studies documented the strong association between HIV RNA and AIDS (i.e., $B \to D$), so that the relative hazards for AIDS were 1, 2.4, 4.3,

7.5, and 12.8 for HIV RNA <500, 500–3000, 3000–10 000, 10 000–30 000 and >30 000 copies/ml, respectively (Mellors *et al.*, 1997). In parallel, clinical trials showed that the use of a protease inhibitor-containing combination therapy dramatically reduced the levels of HIV RNA to undetectable levels (A→ B) in a large proportion (~2/3) of individuals (Hammer *et al.*, 1997). The expectation that modulation of HIV RNA was a good surrogate for the effect of protease inhibitor containing combination therapy against AIDS has been realized, and, indeed, therapies have been approved and recommended using HIV RNA as the end-point in clinical trials. After the introduction of these therapies in HIV-infected individuals, cohort studies have shown their effectiveness at the individual level (Philips *et al.*, 1999) and at the population level (Detels *et al.*, 1998; Muñoz *et al.*, 2000).

Classification paradigm for risk biomarkers

The preceding section has described methods to quantify the degree of surrogacy under the assumptions that B→D, A→B and A→D. These three conditions have been referred to as marginal and at-group-level relationships in the statistical and epidemiological literature, respectively. While these three marginal (at-group-level) relationships are necessary, their sufficiency for a biomarker to be a proper evaluator of chemopreventive strategies heavily depends on the degree of surrogacy.

To provide a classification paradigm of risk biomarkers, it is useful to quantify the conditional (at-individual-level) relationships. Specifically, measures should be provided of the conditional relationship of A to D given B ((A→D)|B) and of the conditional relationship of B to D given A ((B→D)|A). In other words, after knowing the attained value of the biomarker, to what extent does one need to also know the intervention assignment to appropriately describe those who developed disease?; and conversely, after knowing whether individuals were treated or not, does one need to also know the biomarker level to characterize disease incidence? Table 2 provides a classification of risk biomarkers according to the existence of these conditional relationships. The Type I biomarkers are those that have prognostic information for disease, that are modulated by the agent and for which the effect of the agent on disease is present at all levels of the biomarker, but, conditional on the agent, the biomarker levels do not predict disease. This type of biomarker, although useful for group comparisons and thus of some utility for evaluation of chemoprevention trials, does not provide information at the individual level about risk modification by the biomarker. The Type II biomarkers are those for which both conditional relationships are present. Namely, the agent modifies disease at all levels of the biomarker and the biomarker predicts disease among those receiving the agent as well as those not receiving the agent. This is likely to be the case for the major-

Table 2. Classification of biomarkers with at-group relationships according to the at-individual relationships

Type	Conditional/individual (A→D)|B	(B→D)|A	β*	γ*	Utility
I	Yes	No	≈β ≠0	≈0	Appropriate for group, but not for individual comparisons
II	Yes	Yes	≠0	≠0	Moderately useful for group and individual comparisons
III	No	Yes	≈0	≠0	Full surrogacy; ideal situation

ity of the biomarkers fulfilling the three necessary marginal (at-group-level) relationships. Biomarkers in this class have predictive value for at-group and at-individual levels and interventions influence disease both through the biomarker and through other means. The Type III biomarker corresponds to full surrogacy, whereby conditional on the biomarker, the agent has no residual effect on disease. In this case, the conditional relationship of the biomarker with disease, given the agent, equates to the marginal prognostic values on disease.

Biomarkers of the Type III or II categories will almost certainly derive from late events in the progression models of human carcinogenesis. High-penetrance genetic susceptibility syndromes and some forms of IEN are good candidates. (The example given above regarding HIV viral load and AIDS is certainly applicable here as well.) However, these markers by no means define the full extent of the population that will actually develop cancer. Additional biomarkers will need to be identified, developed and validated to capture the residual, seemingly low-risk individuals who still develop cancer in the absence of chemoprevention. Low-penetrance susceptibility genes, biomarkers of dose of environmental agents to humans, and some of the newer markers for genomic instability and altered gene expression are potential candidates in this setting. However, many of these biomarkers are likely to have characteristics of Type I, rendering their utility imperfect.

The limitations of Type I biomarkers are briefly highlighted by a study of the value of aflatoxin–albumin adducts for predicting the chemopreventive efficacy of oltipraz against hepatocellular carcinoma in an animal model. Studies in animals and humans have established serum aflatoxin–albumin adducts as biomarkers of exposure to aflatoxin B_1, a food-borne hepatocarcinogen (Wild *et al.*, 1990). To assess the utility of measurements of aflatoxin–albumin adducts in assessing the efficacy of oltipraz for prevention of hepatocellular carcinoma, 82 male F344 rats were dosed with 20 μg aflatoxin B_1 daily for five weeks after randomization into groups given no intervention or intervention (500 ppm oltipraz, during weeks –1 to 5 relative to aflatoxin B_1) (Kensler *et al.*, 1997). In this context, A is oltipraz, B is aflatoxin–albumin adducts and D is hepatocellular carcinoma. Serial blood samples were collected from each animal at weekly intervals throughout the period of aflatoxin B_1 exposure and were assayed for levels of aflatoxin–albumin by radioimmune assay. As shown in Figure 2 (panel a), the area under the curve (AUC) values for overall burden of aflatoxin–albumin adducts decreased by 39% in the oltipraz intervention group compared with no intervention (i.e., A→B). Similarly, total incidence of liver cancer dropped from 83% to 48% ($p < 0.01$) in these groups (i.e., A→D) (panel b). Overall, as shown in Figure 2 (panel c), a significant association ($p = 0.01$) was seen between biomarker AUC and risk of hepatocellular carcinoma (i.e., B→D). However, as shown in Figure 2 (panel d), when the predictive value of aflatoxin–albumin adducts was assessed within treatment groups, there was no association ($p = 0.56$) between AUC and risk of hepatocellular carcinoma (i.e., (B↛D)|A but the association of A to D remained in categories of the biomarker level (A→D)|B). In this case, once the intervention assignment was known, knowledge of the modulated biomarker level provided no further significant information regarding the likelihood of developing cancer for each individual animal. Thus, aflatoxin–albumin adducts can be useful in identifying potential study populations and for monitoring population-based changes induced by interventions, such as in chemoprevention trials, but have, in oltipraz-treated populations, very limited utility in identifying individuals destined to develop hepatocellular carcinoma. Figure 3 graphically depicts the general loci of biomarker types according to the at-individual relationships starting from the at-group relationships and highlights the positioning of the aflatoxin–albumin adduct biomarker as a Type I biomarker.

Conclusions

We have outlined and discussed the properties of biomarkers and challenges faced for their use in evaluating the putative effects of chemopreventive agents on disease and health improvement. These challenges require carefully conducted studies on animals under controlled conditions and studies in humans in which comprehensive data on agents, biomarkers and disease are collected. On top of these challenges, there is the almost universal situation where the determinants of a disease, and therefore, the means for preventing it, are

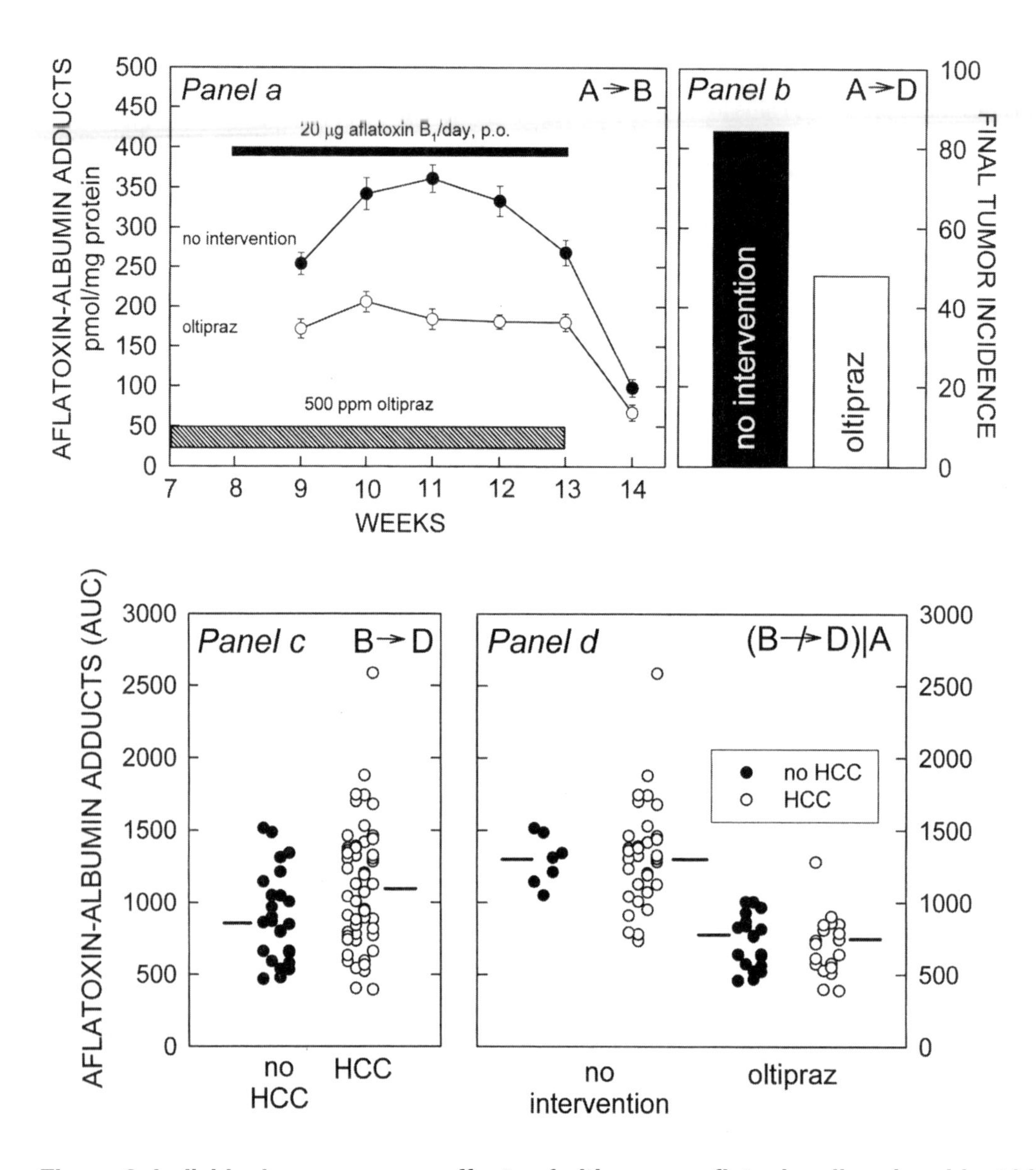

Figure 2. Individual versus group effects of oltipraz on aflatoxin–albumin adduct biomarkers and risk of hepatocellular carcinoma (HCC) in rats.

Panel a, mean serum levels of serial aflatoxin–albumin adducts in rats receiving no intervention or 500 ppm dietary oltipraz. The solid black bar indicates the period of aflatoxin exposure, whereas the striped bar displays the period of oltipraz administration. *Panel b*, effect of oltipraz intervention on incidence of hepatocellular carcinoma (HCC). *Panel c*, univariate association of biomarker burden (AUC: area under curve) with HCC. Biomarker burden was significantly lower in animals that did not develop HCC ($p < 0.01$). Bars, the median of the respective distributions. *Panel d*, bivariate association of AUC with HCC and intervention group.

Adapted from Kensler *et al.*, 1997 with permission.

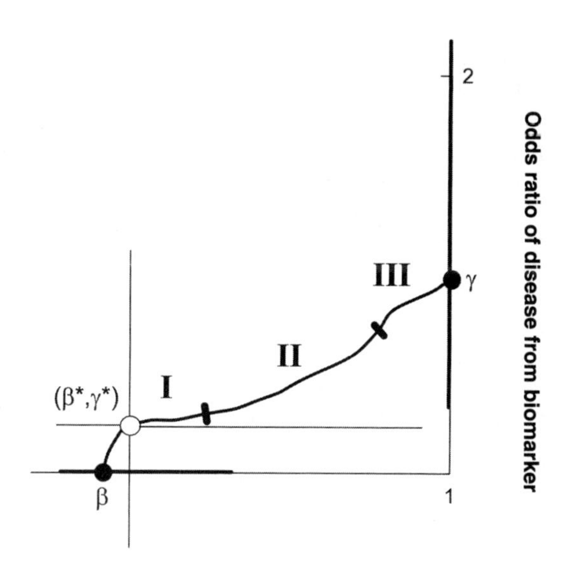

Figure 3. Theoretical distribution of biomarker types for a single agent–disease interaction.

The biomarker occupies one point on this curve. The abscissa corresponds to the value of β* and the ordinate to the value of γ*. The intercept on the abscissa corresponds to the value β capturing the at-group relationship of A→D. Similarly, the intercept of the ordinate corresponds to the value γ capturing the at-group relationship of B→D. β and γ can be thought of as measures of the protection conveyed by the (A)gent unadjusted by the (B)iomarker and of the risk of (D)isease predicted by B unadjusted by A, respectively. β* and γ* are the corresponding measures following adjustment for B and A, respectively. Values for β and γ and for β*, γ* are derived from the data-set depicted in Figure 2. Lines indicate 95% confidence intervals. From the perspective of statistical significance, for Type I biomarkers the values of β* are significant while those of γ* are not. Conversely, for Type III biomarkers, the values of γ* are significant while those of β* are not. The middle group, where both β* and γ* may not be significant, reflects Type II biomarkers.

multifactorial. It is unrealistic to expect that an agent affecting only a specific biomarker will have a major impact in terms of disease prevention. Agents, singly or in combination, that affect multiple components in the process of carcinogenesis are better poised to be effective. Failures of chemopreventive agents and apparent discrepancies between clinical trials and observational studies have more often been due to a lack of a multifactorial approach than to limitations of specific biomarkers or lack of randomization in observational studies. Judicious selection and matching of agents, biomarkers and study cohorts is required. Use of multiple biomarkers that incrementally and collectively enhance surrogacy will be important. In this way, individual biomarkers do not have to be perfect to be useful in advancing the development and evaluation of chemopreventive strategies.

References

Ahdieh, L., Muñoz, A., Vlahov, D., Trimble, C.L., Timpson, L.A. & Shah, K. (2000) Cervical neoplasia and repeated positivity of HPV infection in HIV seropositive and HIV seronegative women. *Am. J. Epidemiol.*, **151**, 1148–1157

Ahrendt, S.A., Chow, J.T., Xu, L.H., Yang, S.C., Eisenberger, C.F., Esteller, M., Herman, J.G., Wu, L., Decker, P.A., Jen, J. & Sidransky, D. (1999) Molecular detection of tumor cells in bronchoalveolar lavage fluid from patients with early stage lung cancer. *J. Natl Cancer Inst.*, **91**, 332–339

Bacus, J.W., Boone, C.W., Bacus, J.V., Follen, M., Kelloff, G.J., Kagan, V. & Lippman, S.M. (1999) Image morphometric nuclear grading of intraepithelial neoplastic lesions with applications to cancer chemoprevention trials. *Cancer Epidemiol. Biomarkers Prev.*, **8**, 1087–1094

Balis, F.M. & Adamson, P.C. (1999) Application of pharmacogenetics to optimization of mercaptopurine dosing. *J. Natl Cancer Inst.*, **91**, 1983–1984

Barrett, J.C., Vainio, H., Peakall, D. & Goldstein, B.D. (1997) 12th Meeting of the scientific group on methodologies for the safety evaluation of chemicals: susceptibility to environmental hazards. *Environ. Health Perspect.*, **105** (Suppl. 4), 699–737

Bartsch, H., Nair, U., Risch, A, Rojas, M., Wikman, H. & Alexandrov, K. (2000) Genetic polymorphism of CYP genes, alone or in combination, as a risk modifier of tobacco-related cancers. *Cancer Epidemiol. Biomarkers Prev.*, **9**, 3–28

Belinsky, S.A., Nikula, K.J., Palmisano, W.A., Michels, R., Saccomanno, G., Gabrielson, E., Baylin, S.B. & Herman, J.G. (1998) Aberrant methylation of p16(INK4a) is an early event in lung cancer and a potential biomarker for early diagnosis. *Proc. Natl Acad. Sci. USA*, **95**, 11891–11896

Boogaard, P.J. & van Sittert, N.J. (1995) Biological monitoring of exposure to benzene: a comparison between S-phenylmercapturic acid, trans,trans-muconic acid, and phenol. *Occup. Environ. Med.*, **52**, 611–620

Boone, C.W., Bacus, J.W., Bacus, J.V., Steele, V.E. & Kelloff, G.J. (1997) Properties of intraepithelial neoplasia relevant to the development of cancer chemopreventive agents. *J Cell Biochem.*, **28S/29S**, 1–20

Bouchard, M. & Viau, C. (1999) Urinary 1-hydroxypyrene as a biomarker of exposure to polycyclic aromatic hydrocarbons: biological monitoring strategies and methodology for determining biological exposure indices for various work environments. *Biomarkers*, **4**, 159–187

Bronner, C.E., Baker, S.M., Morrison, P.T., Warren, G., Smith, L.G., Lescoe, M.K., Kane, M., Earabino, C., Lipford, J., Lindblom, A., Tannergård, P., Bollag, R.J., Godwin, A.R., Ward, D.C., Nordenskjøld, M., Fishel, R., Kolodner, R. & Liskay, R.M. (1994) Mutation in the DNA mismatch repair gene homologue hMLH1 is associated with hereditary non-polyposis colon cancer. *Nature*, **368**, 258–261

Buyse, M & Molenberghs, G. (1998) Criteria for the validation of surrogate end points in randomized experiments. *Biometrics*, **54**, 1014–1029

Cole, J. & Skopek, R.T. (1994) Somatic mutation frequency, mutation rates, and mutational spectra in the human population. *Mutat. Res.*, **304**, 33–105

Dawber, T.R. (1980) *The Framingham Study: The Epidemiology of Atherosclerotic Disease*, Cambridge, Harvard University Press

DeRooij, B.M., Commandeur, J.N.M. & Vermeulen, N.P.E. (1998) Mercapturic acids as biomarkers of exposure to electrophilic chemicals: applications to environmental and industrial chemicals. *Biomarkers*, **3**, 239–303

Detels, R., Muñoz, A., McFarland, G., Kingsley, L.A., Margolick, J.B., Giorgi, J., Schrager, L.K. & Phair, J.P. (1998) Effectiveness of potent antiretroviral therapy on time to AIDS and death in men with known HIV infection. *J. Am. Med. Assoc.*, **280**, 1497–1503

Dove, W.F. (2000) Genes and cancer: risk determinants and agents of change. In: Abeloff, M.A., Armitage, J.O., Lichter, A.S. & Niederhuber, J.E., eds, *Clinical Oncology*, Second Edition, Churchill Livingstone, New York, pp. 54–76

Dyke, G.W., Craven, J.L., Hall, R. & Garner, R.C. (1994) Effect of vitamin C supplementation on gastric mucosal DNA damage. *Carcinogenesis*, **15**, 291–295

Eisenberger, C.F., Schoenberg, M., Enger, C., Hortopan, S., Shah, S., Chow, N.H., Marshall, F.F. & Sidransky, D. (1999) Diagnosis of renal cancer by molecular urinalysis. *J. Natl Cancer Inst.*, **91**, 2028–2032

Erbe, R.W. (1976) Inherited gastrointestinal-polyposis syndromes. *New Engl. J. Med.*, **294**, 1101–1104

Esteller, M., Sanchez-Cespedes, M., Rosell, R., Sidransky, D., Baylin, S.B. & Herman, J.G. (1999) Detection of aberrant promoter hypermethylation of tumor suppressor genes in serum DNA from non-small cell lung cancer patients. *Cancer Res.*, **59**, 67–70

Fabian, C.J., Zalles, C., Kamel, S., Zeiger, S., Simon, C. & Kimler, B.F. (1997) Breast cytology and biomarkers obtained by random fine needle aspiration: use in risk assessment and early chemoprevention trials. *J Cell. Biochem. Suppls.*, **28/29**, 101–110

Feigelson, H.S., Ross, R.K., Yu, M.C., Coetzee, G.A., Reichardt, J.K. & Henderson, B.D. (1998) Sex steroid hormones and genetic susceptibility to breast and prostate cancer. *Drug Metab. Rev.*, **30**, 421–34

Fearon, E.R. & Vogelstein, B. (1990) A genetic model for colorectal tumorigenesis. *Cell*, **61**, 759–767

Field, J.K., Liloglou, T., Xinarianos, G., Prime, W., Fielding, P., Walshaw, M.J. & Turnbull, L. (1999) Genetic alterations in bronchial lavage as a potential marker for individuals with a high risk of developing lung cancer. *Cancer Res.*, **59**, 2690–2695

Fliss, M.S., Usadel, H., Caballero, O.L., Wu, L., Buta, M.R., Eleff, S.M., Jen, J. & Sidransky, D. (2000) Facile detection of mitochondrial DNA mutations in tumors and bodily fluids. *Science*, **287**, 2017–2019

Fraga, C.G., Shigenaga, M.K., Park, J.-W, Degan, P. & Ames, B.N. (1990) Oxidative damage to DNA during aging: 8-hydroxy-2′-deoxyguanosine in rat organ DNA and urine. *Proc. Natl Acad. Sci. USA*, **87**, 4533–4537

Freedman, L.S., Graubard, B.I. & Schatzkin, A. (1992) Statistical validation of intermediate end points for chronic diseases. *Stat. Med.*, **11**, 167–178

Gaylor, D.W., Kadlubar, F.F. & Beland, F.A. (1992) Application of biomarkers to risk assessment. *Environ. Health Perspect.*, **98**, 139–141

Golub, T.R., Slonim, D.K., Tamayo, P., Huard, C., Gaasenbeek, M., Mesirov, J.P., Coller, H., Loh, M.L., Downing, J.R., Caligiuri, M.A., Bloomfield, C.D. & Lander, E.S. (1999) Molecular classification of cancer: class discovery and class prediction by gene expression monitoring. *Science*, **286**, 531–537

Grant, S.G. & Bigbee, W.L. (1993) *In vivo* mutation and segregation at the human glycophorin A (GPA) locus: phenotypic variation encompassing both gene-speficic and chromosomal mechanisms. *Mutat. Res.*, **288**, 163–172

Greenwald, P., Cullen, J.W. & Weed, D. (1990) Introduction: cancer prevention and control. *Sem. Oncol.*, **17**, 383–390

Groopman, J.D. & Kensler, T.W. (1999) The light at the end of the tunnel for chemical-specific biomarkers: daylight or headlight: *Carcinogenesis*, **20**, 1–11

Grossman, L. & Wei, Q. (1994) DNA repair capacity (DRC) as a biomarker of human variational responses to the environment. In: Vos, J.-M. H., ed., *DNA Repair Mechanisms: Impact on Human Diseases and Cancer*, Georgetown, TX, R.G. Landes, pp. 329–347

Grossman, L. (1997) Epidemiology of ultraviolet-DNA repair capacity and human cancer. *Environ. Health Perspect.*, **105** (Suppl. 4), 927–930

Halliwell, B. (1998) Can oxidative DNA damage be used as a biomarker of cancer risk in humans? Problems, resolutions and preliminary results from nutritional supplementation studies. *Free Radic. Res.*, **29**, 469–486

Hammer, S.M., Squires, K.E., Hughes, M.D., Grimes, J.M., Demeter, L.M., Currier, J.S., Eron, J.J., Jr, Feinberg, J.E., Balfour, H.H., Jr, Deyton, L.R., Chodakewitz, J.A. & Fischl, M.A. (1997) A controlled trial of two nucleoside analogues plus indinavir in persons with human immunodeficiency virus infection and CD4 cell counts of 200 per cubic millimeter or less. AIDS Clinical Trials Group 320 Study Team. *New Engl. J. Med.*, **337**, 725–733

Harris, C.C. (1989) Interindividual variation among humans in carcinogen metabolism, DNA adduct formation and DNA repair. *Carcinogenesis*, **10**, 1563–1566

Hartig, U., Loveland, P., Spitsbergen, H. & Bailey, G.S. (2000) Dose- and time-dependent promotion, suppression or blocking of 7,12-dimethylbenz[a]pyrene induced multi-target organ tumorigenesis by dietary indole-3-carbinole. *Proc. Am. Assoc. Cancer Res.*, **41**, 661

Hayes, R.B., Zhang, L., Yin, S., Swenberg, J.A., Xi, L., Wiencke, J., Bechtold, W.E., Yao, M., Rothman, N., Haas, R., O'Neil, J.P., Zhang, D., Wiemels, J., Dosemeci, M, Li, G & Smith, M.T. (2000) Genotoxic markers among butadiene polymer workers in China. *Carcinogenesis*, **21**, 55–62

Hartig, U., Loveland, P., Spitsbergen, J. & Bailey, G.S. (2000) Dose- and time-dependent promotion, suppression or blocking of 7,12-dimethylbenz[a]anthracene induced multi-target organ tumorigenesis by dietary indole-3-carbinol. *Proc. Am. Assoc. Cancer Res.*, **41**, 661

Hecht, S.S. (1997) Approaches to chemoprevention of lung cancer based on carcinogens in tobacco smoke. *Environ. Health Perspect.*, **105** (Suppl. 4), 955–963

Hecht, S.S., Chung, F-L., Richie, J.P., Jr, Akerkar, S.A., Borukhova, A., Skowronski, L. & Carmella, S.G. (1995) Effects of watercress consumption on metabolism of a tobacco-specific lung carcinogen in smokers. *Cancer Epidemiol. Biomarkers Prev.*, **4**, 877–884

Hecht, S.S., Carmella, S.G. & Murphy, S.E. (1999) Effects of watercress consumption on urinary metabolites of nicotine in smokers. *Cancer Epidemiol. Biomarkers Prev.*, **8**, 907–913

Hietanen, E., Husgafvel-Pursiainen, K. & Vainio, H. (1997) Interaction between dose and susceptibility to environmental cancer: a short review. *Environ. Health Perspect.*, **105** (Suppl. 4), 749–754

Hollstein, M., Sidransky, D., Vogelstein, B. & Harris, C.C. (1991) *p53* mutations in human cancers. *Science*, **253**, 1156–1160

Houlston, R.S. (1999) Glutathione S-transferase M1 status and lung cancer risk: A meta-analysis. *Cancer Epidemiol. Biomarkers Prev.*, **8**, 675–682

Hulka, B.S., Wilcosky, T.C. & Griffith, J.D. (1990) *Biological Markers in Epidemiology*, New York: Oxford University Press

Hulka, B.S. (1991) Epidemiological studies using biological markers: issues for epidemiologists. *Cancer Epidemiol. Biomarkers Prev.*, **1**, 13–19

Hussain, S.P. & Harris, C.C. (1998) Molecular epidemiology of human cancer: contribution of mutation spectra studies of tumor supressor genes. *Cancer Res.*, **58**, 4023–4037

Kaderlik, R.K., Lin, D.X., Lang, N.P. & Kadlubar, F.F. (1992) Advantages and limitations of laboratory methods for measurement of carcinogen-DNA adducts for epidemiological studies. *Toxicol. Lett.*, **64–65**, 469–475

Keefe, K.A. & Meyskens, Jr, F.L. (2000) Principles of prevention, diagnosis, and therapy. In: Abeloff, M.D., Armitage, J.O., Lichter, A.S. & Niederhuber, J.E., eds, *Clinical Oncology*, Second Edition, New York, Churchill Livingstone, pp. 318–365

Kelloff, G.J., Boone, C.W., Crowell, J.A., Nayfield, S.G., Hawk, E., Malone, W.F., Steele, V.E., Lubet, R.A. & Sigman, C.C. (1996) Risk biomarkers and current strategies for cancer chemoprevention. *J. Cell. Biochem.*, **25S**, 1–14

Kelloff, G.J., Sigman, C.C., Johnson, K.M., Boone, C.W., Greenwald, P., Crowell, J.A., Hawk, E.T. & Doody, L.A. (2000) Perspectives on surrogate end points in the development of drugs that reduce the risk of cancer. *Cancer Epidemiol. Biomarkers Prev.*, **9**, 127–137

Kensler, T.W. (1997) Chemoprevention by inducers of carcinogen detoxication enzymes. *Environ. Health Perspect.*, **105** (Suppl. 4), 965–970

Kensler, T.W., Egner, P.A., Trush, M.A., Budeing, E. & Groopman, J.D. (1985) Modification of aflatoxin B_1 binding to DNA in vivo in rats fed phenolic antioxidants, ethoxyquin and a dithiolthione. *Carcinogenesis*, **6**, 759–763

Kensler, T.W., Groopman, J.D. & Wogan, G.N. (1996) Use of carcinogen-DNA and biomarkers for cohort selection and as modifiable end points in chemoprevential trials. In: Stewart, B.W., McGregor, D. & Kleihues, P., eds, *Principles of Chemoprevention* (IARC Scientific Publications No. 139), Lyon, IARC, pp. 237–248

Kensler, T.W., Gange, S.J., Egner, P.A., Dolan, P.M., Muñoz, A., Groopman, J.D., Rogers, A.E. & Roebuck, B.D. (1997) Predictive value of molecular dosimetry: individual versus group effects of oltipraz on aflatoxin-albumin adducts and risk of liver cancer. *Cancer Epidemiol. Biomarkers Prev.*, **6**, 603–610

Kensler, T.W., He, X., Otieno, M., Egner, P.A., Jacobson, L.P., Chen, B., Wang, J.-S., Zhu, Y.-R., Zhang, B.-C., Wang, J.-B., Wu, Y., Zhang, Q.-N., Qian, G.-S., Kuang, S.-Y., Fang, S., Li, Y.-F., Yu, L.-Y., Prochaska, H.J., Davidson, N.E., Gordon, G.B., Gorman, M.B., Zarba, A., Enger, C., Muñoz, A., Helzlsouer, K.J. & Groopman, J.D. (1998) Oltipraz chemoprevention trial in Qidong, People's Republic of China: modulation of serum aflatoxin albumin adduct biomarkers. *Cancer Epidemiol. Biomarkers Prev.*, **7**, 127–134

Kinzler, K.W. & Vogelstein, B. (1997) Gatekeepers and caretakers. *Nature*, **386**, 761–763

Kirk, G.D., Camus-Randon, A.-M., Mendy, M., Goedert, J.J., Merle, P., Trepo, C., Brechot, C., Hainaut, P. & Montesano, R. (2000) Ser-249 mutations in plasma DNA of patients with hepatocellular carcinoma from The Gambia. *J. Natl Cancer Inst.*, **92**, 148–153

Kraemer, K.H. & Slor, H. (1985) Xeroderma pigmentosum. *Clin. Derm.*, **3**, 33–69

Laird, P.W. & Jaenisch, R. (1996) The role of DNA methylation in cancer genetic and epigenetics. *Ann. Rev. Genet.*, **30**, 441–464

Lang, N.P. (1997) Acetylation as an indicator of risk. *Environ. Health Perspect.*, **105** (Suppl. 4), 763–766

Lang, N.P., Nowell, S., Malfatti, M.A., Kulp, K.S., Knize, M.G., Davis, C., Massengill, J., Williams, S., MacLeod, S., Dingley, K.H., Felton, J.S. & Turteltaub, K.W. (1999) *In vivo* human metabolism of [2-^{14}C]2-amino-1-methyl-6-phenylimidazo[4,5-b]pyridine (PhIP). *Cancer Lett.*, **143**, 135–138

Lin, H.J., Probst-Hensch, N.M., Louie, A.D., Kau, K.H., Witte, J.S., Ingles, S.A., Frankl, H.D., Lee, E.R. & Haile, R.W. (1998) Glutathione transferase null genotype, broccoli, and lower prevalence of colorectal adenomas. *Cancer Epidemiol. Biomarkers Prev.*, **7**, 647–652

Loeb, L.A. (1998) Cancer cells exhibit a mutator phenotype. *Adv. Cancer Res.*, 72, 25–56

London, S.J., Yuan, J.-M., Chung, F.-L., Gao, Y.-T., Coetzee, G.A., Ross, R.K. & Yu, M.C. (2000) Diet-gene interaction in relation to lung cancer: prospective study of isothiocyanates and glutathione S-transferase M1 and T1 genetic polymorphisms among middle-aged in Shanghai, China. *Proc. Am. Assoc. Cancer Res.*, **41**, 552

Lunn, R.M., Langlois, R.G., Hsieh, L.L, Thompson, C.L. & Bell, D.A. (1999) XRCC1 polymorphisms: effects on aflatoxin B_1-DNA adducts and glycophorin a variant frequency. *Cancer Res.*, **59**, 2557–2561

Mao, L., Lee, D.J., Tockman, M.S., Erozan, Y.S., Askin, F. & Sidransky, D. (1994) Microsatellite alterations as clonal markers for the detection of human cancer. *Proc. Natl Acad. Sci. USA*, **91**, 9871–9875

Mao, L., Schoenberg, M.P., Scicchitano, M., Erozan, Y.S., Merlo, A., Schwab, D. & Sidransky, D. (1996) Molecular detection of primary bladder cancer by microsatellite analysis. *Science*, **271**, 659–662

Mellors, J.W., Muñoz, A., Giorgi, J.V., Margolick, J.B., Tassoni, C.J., Gupta, P., Kingsley, L.A., Todd, J.A., Saah, A.J., Detels, R., Phair, J.P., & Rinaldo, C.R. Jr (1997) Plasma viral load and CD4+ lymphocytes as prognostic markers of HIV-1 infection. *Ann. Intern. Med.*, **126**, 946–954

Muñoz, A. (1993) Design and analysis of studies of the health effects of ozone. *Environ. Health Perspect.*, **101** (Suppl. 4), 231–235

Muñoz, A. & Gange, S.J. (1998) Methodological issues for biomarkers and intermediate outcomes in cohort studies. *Epidemiol. Rev.*, **20**, 29–42

Muñoz, A., Gange, S.J. & Jacobson, L.J. (2000) Distinguishing efficacy, individual effectiveness and population effectiveness of therapies. *AIDS*, **14**, 754–756

Nawroz, H., Koch, W., Anker, P., Stroun, M. & Sidransky, D. (1996) Microsatellite alterations in serum DNA of head and neck cancer patients. *Nature Med.*, **2**, 1035–1037

Perucho, M. (1996) Cancer of the microsatellite mutator phenotype. *J. Biol. Chem.*, **377**, 675–684

Philips, A.N., Grabar, S., Tassie, J.M., Costagliola, D., Lundgren, J.D. & Egger, M. (1999) Use of observational databases to evaluate the effectiveness of antiretroviral therapy for HIV infection: comparison of cohort studies with randomized trials. *AIDS*, **13**, 2075–2082

Prentice, R.L. (1989) Surrogate end points in clinical trials: definitions and operational criteria. *Stat. Med.*, **8**, 431–440

Rehman, A., Bourne, L.C., Halliwell, B. & Rice-Evans, C.A. (1999) Tomato consumption modulates oxidative DNA damage in humans. *Biochem. Biophys. Res. Comm.*, **262**, 828–831

Schatzkin, A., Freedman, L.S., Schiffman, M.H. & Dawsey, S.M. (1990) Validation of intermediate end points in cancer research. *J. Natl Cancer Inst.*, **82**, 1746–1752

Schulte, P.A. & Perera, F.P. (1993) *Molecular Epidemiology: Principles and Practices*, San Diego, Academic Press

Shuker, D.E.G. & Farmer P.B. (1992) Relevance of urinary DNA adducts as markers of carcinogen exposure. *Chem. Res. Toxicol.*, **5**, 450–460

Shuker, D.E.G.., Prevost, V., Friesen, M.D., Lin, D., Ohshima, H. & Bartsch, H. (1993) Urinary markers for measuring exposure to endogenous and exogenous alkylating agents and precursors. *Environ. Health Perspect.*, **99**, 33–37

Sidransky, D., Von Eschenbach, A., Tsai, Y.C., Jones, P., Summerhayes, I., Marshall, F., Paul, M., Green, P., Hamilton, S.R., Frost, P. & Vogelstein, B. (1991) Identification of p53 gene mutations in bladder cancer and urine samples. *Science*, **252**, 706–709

Sidransky, D., Tokino, T., Hamilton, S.R., Kinzler, K.W., Levin, B., Frost, P. & Vogelstein, B. (1992) Identification of *ras* oncogene mutations in the stool of patients with curable colorectal tumors. *Science*, **256**, 102–105

Simic, M.G. (1994) DNA markers of oxidative processes *in vivo*: relevance to carcinogenesis and anticarcinogenesis. *Cancer Res.*, **54**, 1918s–1923s

Simic, M.G. & Bergtold, D.S. (1991) Dietary modulation of DNA damage in humans. *Mutat. Res.*, **250**, 17–24

Singer, B. & Bartsch, H., eds (1999) *Exocyclic DNA Adducts in Mutagenesis and Carcinogenesis* (IARC Scientific Publications No. 139), Lyon, IARC

Spafford, M.F., Reed, A.L., Xu, L., Koch, W., Westra, W., Califano, J.A. & Sidransky, D. (1998) Detection of head and neck squamous cell carcinoma in saliva using microsatellite analysis. *Proc. Am. Assoc. Cancer Res.*, **39**, 100

Spitz, M.R., Wu, X., Jiang, H. & Hsu, T.C. (1996) Mutagen sensitivity as a marker of cancer susceptibility. *J. Cell. Biochem.*, **25S**, 80–84

Steiner, G., Schoenberg, M.P., Linn, J.F., Mao, L. & Sidransky, D. (1997) Detection of bladder cancer recurrence by microsatellite analysis of urine. *Naure Med.*, **3**, 621–624

Strickland, P.T., Routledge, M.N. & Dipple, A. (1993) Methodologies for measuring carcinogen adducts in humans. *Cancer Epidemiol. Biomarkers Prev.*, **2**, 607–619

Tates, A.D., van Dam, F.J., van Mossel, H., Schoemaker, H., Thijssen, J.C.P., Woldring, V.M., Zwinderman, A.H. & Natarajan, A-T. (1991) Use of the clonal assay for the measurement of frequencies of HPRT mutants in T-lymphocytes from five control populations. *Mutat. Res.*, **253**, 199–213

Corresponding author:

T.W. Kensler
Department of Environmental Health Sciences,
Johns Hopkins School of Hygiene and Public Health,
Baltimore,
MD 21205,
USA

Travis, C.C., Zeng, C. & Nicholas, J. (1996) Biological model of ED01 hepatocarcinogenesis. *Toxicol. Appl. Pharmacol.*, **140**, 19–29

Verhagen, H., Poulsen, H.E., Loft, S., van Poppel, G., Willems, M.I. & van Bladeren, P.J. (1995) Reduction of oxidative DNA-damage in humans by brussels sprouts. *Carcinogenesis*, **16**, 969–970

Wallin, H., Skipper, P.L., Tannenbaum, S.R., Jensen, J.P.A., Rylander, L. & Olsen, J.H. (1995) Altered aromatic amine metabolism in epileptic patients treated with phenobarbital. *Cancer. Epidemiol. Biomarkers Prev.*, **4**, 771–773

Wang, J.-S., Shen, X., He, X., Zhu, Y.-R., Zhang, B.-C., Wang, J.-B., Qian, G.-S., Kuang, S.-Y., Zarba, A., Egner, P.A., Jacobson, L.P., Muñoz, A., Helzlsouer, K.J., Groopman, J.D. & Kensler, T.W. (1999) Protective alterations in phase 1 and 2 metabolism of aflatoxin B_1 by oltipraz in residents of Qidong, People's Republic of China. *J. Natl Cancer Inst.*, **91**, 347–354

Wei, Q., Spitz, M.R., Gu, J., Cheng, L., Xu, X., Strom, S.S., Kripke, M.L. & Hsu, T.C. (1996) DNA repair capacity correlates with mutagen sensitivity on lymphoblastoid cell lines. *Cancer Epidemiol. Biomarkers Prev.*, **5**, 199–204

Wild, C.R. Jiang, Y.Z., Allen, S.J., Jansen, L.A., Hall, A.J. & Montesano, R. (1990) Aflatoxin-albumin adducts in human sera from different regions of the world. *Carcinogenesis*, **11**, 2271–2274

Wong, I.H.N., Lo, Y.M.D., Zhang, J., Liew, C.-T., Ng, M.H.L., Wong, N., Lai, P.B.S., Lau, W.Y., Hjelm, N.M. & Johnson, P.J. (1999) Detection of aberrant *p16* methylation in the plasma and serum of liver cancer patients. *Cancer Res.*, **59**, 71–73

Young, L. (1977) The metabolism of foreign compounds – history and development. In: Parke, D.V. & Smith, R.L., eds, *Drug Metabolism: from Microbe to Man*, London, Taylor & Francis, pp. 1–11

Zhang, Y. & Talalay, P. (1998) Mechanism of differential potencies of isothiocyanates as inducers of anticarcinogenic phase 2 enzymes. *Cancer Res.*, **58**, 4632–4639

Biomarkers in Cancer Chemoprevention
Miller, A.B., Bartsch, H., Boffetta, P., Dragsted, L. and Vainio, H. eds
IARC Scientific Publications No. 154
International Agency for Research on Cancer, Lyon, 2001

Development of difluoromethylornithine and Bowman–Birk inhibitor as chemopreventive agents by assessment of relevant biomarker modulation: some lessons learned

F.L. Meyskens, Jr

A major goal in the development of chemopreventive agents has been to develop markers that reflect the underlying process of carcinogenesis and which are modulatable by the agent under study. An important application of such markers will be to select cohorts that are at elevated risk for cancer development, which should allow use of smaller sample sizes in definitive phase III trials as well as shorter duration (and lower cost), without loss of statistical power. Susceptibility and surrogate end-point biomarkers are particularly important in this respect. Intermediate markers are probably best assessed in terms of proportionate rather than relative risk.

The systematic development of difluoromethylornithine for use in chemoprevention against human cancer has involved pilot, phase IIa and IIb trials using participants with prior colonic polyps as the study group. A unique feature of the phase IIa study was the use of a dose de-escalation design which allowed selection of the lowest effective non-toxic dose of difluoromethylornithine. The phase IIb trial now in progress is using a combination of sulindac with difluoromethylornithine; the rationale for selection of markers for this study and for a randomized phase III registration trial is discussed. We also review the findings in phase I and IIa trials of Bowman–Birk inhibitor concentrate, in which patients with measurable oral leukoplakia are the study group.

Surrogate end-point biomarkers and chemoprevention: some conceptual thoughts and applied observations

There are several discrete types of measurement that should be considered separately during the design of chemoprevention trials: susceptibility (predisposition/hereditable), exposure, intermediate marker (non-causal and causal), drug-modulatable event (related to a carcinogenesis process or not) and tumour marker (Meyskens, 1992a,b). There has been a tendency among investigators to call different types of 'marker' by the same name. The development of a relative risk profile should be the first step, rather than the last (as currently tends to be done). This strategy has been adopted only in the National Surgical Adjuvant Breast and Bowel Project breast cancer prevention trial of tamoxifen, in which an increased relative risk for breast cancer was required for study entry (Fisher *et al.*, 1998). Two different major levels of risk should be assessed: genetic and epigenetic. Genetic risk should be considered in terms of both defined molecular abnormalities (e.g., tumour-suppressor genes, oncogenes, microsatellite stability, DNA repair, metabolic polymorphisms) and familial risk by genealogical analysis only. Epigenetic assessment can be broad, but should include consideration of at least the major known risk factors in general and those specific for the organ site being studied.

The development of validated surrogate end-point biological markers is a difficult task and to date no marker for carcinogenesis equivalent to cholesterol for atherogenesis-related disease has been validated. In assessment of the value of an intermediate marker for use in chemoprevention, its predictive value as an estimator of cancer risk needs to be stated, at least qualitatively. Most investigators accept a histologically defined end-point as a surrogate end-point biomarker with a high risk and modulation of such an end-point in a favourable manner as being indicative of chemopreventive activity. Evidence of alteration in the natural history of a preneoplastic lesion such as intraepithelial neoplasia, an adenoma or metaplasia is probably *de facto* sufficient to call an agent efficacious. It is much more difficult to relate pre-histological markers of cancer to risk, and even more so, those markers that may be associated or correlated with the true marker of risk, and not directly on the causal pathway.

To date, the concept of risk has been designated as relative risk. Since the maximum risk is 100%, use of the concept of proportionate risk, that is the proportion of total risk explained, could be more useful in the development of intermediate markers. An example of the principles underlying the concept of a hierarchy of markers is presented in Figure 1. In the first example, the germline absence or mutation of a tumour-suppressor gene leads to the inevitable or substantial likelihood of cancer development. Presence of the appropriate marker (M_1) would predict with high frequency the development of cancer and therefore a high proportionate risk (i.e., close to 1.0) could be assigned. Examples of this situation would include hereditary retinoblastoma and Li–Fraumeni syndrome. In such a case, an appropriate marker may be highly predictive of the development of cancer and its modulation will be predictive of a beneficial result. At the other end of the time-line of the process of carcinogenesis (M_5), a histological preneoplastic lesion also has a high chance of malignant conversion and therefore a patient with the biomarker has a high proportionate risk (i.e., close to one). There clearly is a spectrum of lesions;

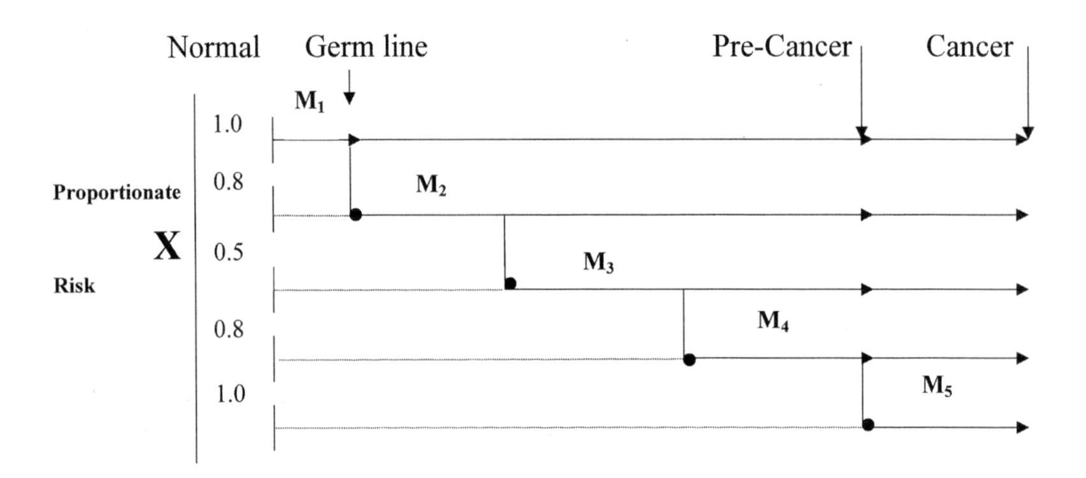

Figure 1. Development of a proportionate risk hierarchy of markers

The maximum likelihood of developing a cancer is 100% (1.0). Therefore, the risk assigned to any marker should be a proportion of this risk rather than portrayed as a relative risk. If the predictive properties of all markers were known, then choosing a few would likely result in a high prediction rate. Time to event would influence this selection process as well. A scoring system for markers related to risk of development of cancer based on this principle would be:

For x=0 to 1; $y = x_1M_1 + x_2M_2 + x_3M_3 + x_4M_4 + x_5M_{5=}$ 1; $M_{1,2,3,4,5}$= degrees of certainty for example, oral leukoplakia has a relatively low conversion rate, while erythroplakia with dysplasia has a high rate of malignant transformation.

However, in most cases the extent to which an intermediate marker predicts the development of cancer is not as great (e.g., M_2, M_3). Assessing the proportionate risk associated with a marker that falls somewhere between a germline mutation and a histological change has been considerably more difficult. At some point late in the carcinogenesis cascade, the value of a marker may again rise as the state of histological preneoplasia is approached; an example might be optically-measured nuclear morphometry (e.g., M_4).

Stringent criteria have previously been proposed to validate an intermediate marker as a surrogate end-point biomarker. These correspond to several steps of an algorithm:

- The identified marker must represent a step on the causal pathway to carcinogenesis
- The marker must be modulated by a chemoprevention agent
- Modulation of the marker must correlate with reduction of cancer incidence

Fulfilling these criteria requires a lengthy and expensive process that in the worst case might be marker- and/or chemoprevention agent-specific. In the rational development of surrogate end-point biomarkers, it might be more productive to link biomarkers to predisposition for risk rather than to attempt to assign a relative risk based on biological plausibility. An attempt to link these various features is presented in Table 1, in which the class of risk, relative risk and attributable risk in a population are assessed. Quantification (or even intelligent qualification) of these factors may allow identification of cohorts for chemoprevention trials that provide individuals that are at relatively high relative risk, but nevertheless, fairly representative of the more general population. For example, individuals with a strong family history for a cancer in which there are organ-specific polymorphisms predictive of risk might constitute a particularly useful cohort to study. Hopefully, identification of such cohorts will lead to more efficient trials that are of shorter duration and smaller size and hence less costly and of greater feasibility.

Development of difluoromethylornithine as a chemoprevention agent

Experimental data

Difluoromethylornithine (DFMO) is an irreversible enzyme-activated inhibitor of ornithine decarboxylase (Meyskens & Gerner, 1999). This enzyme catalyses the first step in the synthesis of polyamines in eukaryotes, the decarboxylation of

Table 1. Identification of high-risk cohorts that are generalizable

Class	Relative risk	Attributable risk
Predisposition		
Hereditary	High	Low
Familial	Moderately high	Higher
Polymorphism	Low	Highest
Epi/classical	Low	Higher
Intermediate end-point biomarker		
Preneoplasia	High	Moderate
Intermediate marker	Low	High

Optimal cohort = Relative risk (familial/metabolic polymorphism/epi) + proportionate risk (intermediate end-point marker)

ornithine to putrescine. In sensitive tissues, putrescine levels fall rapidly after administration of DFMO. Levels of the derived polyamine product spermidine also fall with time as does that of the terminal polyamine spermine, although to a much lower degree. DFMO was originally developed for therapeutic purposes, but was found to be ineffective against established malignancies (Abeloff *et al.*, 1986). A number of workers demonstrated that an increase in ornithine decarboxylase follows carcinogen exposure and inhibition of this rise with DFMO blocks cancer formation in essentially all in-vitro and animal models studied, including colon polyps and cancer (Verma, 1990; Halline *et al.*, 1990). These findings led to a renewed interest in DFMO as a potential chemopreventive agent.

Clinical studies

Two major obstacles needed to be overcome to demonstrate that DFMO was worth studying as a chemoprevention agent in humans. First, although the results in animals with DFMO were impressive, its potential as a chemopreventive compound in humans was unknown. It was soon found, however, that many human preneoplastic tissues had elevated basal ornithine decarboxylase activity compared with control tissues (Hixson *et al.*, 1993), thereby providing a rational basis for the use of DFMO as a chemoprevention agent for patients at risk in clinical settings. Second, at the high doses used in therapeutic trials, hearing loss, although reversible upon drug discontinuation, was substantial (Croghan, 1991). It was therefore necessary to establish whether a dose of DFMO could be found that would deplete polyamines in the organ of interest, but was below the threshold for producing hearing changes and other side-effects.

The colon was selected as the target organ for our studies, as animal experiments had demonstrated substantial anticarcinogenic activity of DFMO in this tissue (Nigro *et al.*, 1986; Verma, 1990) and the flat mucosa of patients with colon polyps and cancer has elevated levels of ornithine decarboxylase (Rozhin *et al.*, 1984; Hixson *et al.*, 1993). In a pilot study, we demonstrated that a modest dose of DFMO could lower polyamine levels in rectal mucosal biopsies, but not in shed oral mucosal cells (Boyle *et al.*, 1992). This was disappointing, in that these results indicated that oral mucosal cells were not a satisfactory surrogate for rectal biopsies, thereby limiting the number of studies that could be carried out in this target patient population. However, clear evidence of suppression of polyamine content in colonic mucosa by DFMO was demonstrated.

Our subsequent phase IIa trial used a unique design to determine the lowest dose at which DFMO was effective in lowering polyamines in the rectal mucosa without producing side-effects (Meyskens *et al.*, 1994). From the results of the prior therapeutic trials and pilot study in humans, we selected a moderate dose of DFMO as the first and highest dose of drug to be studied in a cohort of individuals who had prior colonic polyps removed. The flat mucosa was biopsied, before and after one month of DFMO treatment, and polyamines were measured. The dose of DFMO was progressively reduced until no effect on the polyamine content was seen. On the basis of the results from this detailed investigation, a range of doses (75–400 mg/m^2 per day) was selected for subsequent longer-term studies. Analysis of the results also indicated that the age of the participant affected baseline and changes in polyamine values in response to DFMO, parameters which were important in evaluating the overall results of the phase IIa trial and subsequent studies.

A subsequent placebo-controlled phase IIb trial of 12 months' duration measured the effect of a range of low doses of DFMO on polyamine content in rectal mucosa over time, and general side-effects (Meyskens *et al.*, 1998) and specific hearing changes, as determined by pure tone audiometry and otoacoustic emission analysis (M.J. Doyle, unpublished) were carefully assessed. The critical parameters putrescine content and spermidine/spermine ratio in the rectal biopsies were decreased by doses as low as 200 mg/m^2 per day. Although well tolerated at all doses tested, DFMO at the highest dose of 400 mg/m^2 per day produced more side-effects than the placebo arm and subtle changes in the lowest frequencies of the pure tone audiogram were evident. In contrast, doses of DFMO of 200 mg/m^2 per day produced no side-effects or audiometric changes greater than placebo and produced effects on polyamine content nearly equivalent to the higher doses of DFMO. Therefore in all subsequent chemoprevention trials, a dose of DFMO of 200 mg/m^2 per day is being used.

We next had to consider whether to undertake a phase III trial of polyp risk reduction using DFMO alone or to develop DFMO in combination with other agents. We were significantly influenced in our decision to develop combination trials by the results from animal studies, in which marked inhibition of tumour formation was achieved by using doses of individual drugs in combination that were considerably lower than the doses of single agents alone (Kelloff, 1996). Since absence of side-effects is at least as important as efficacy in developing chemoprevention agents for human beings, we accordingly elected to develop combination clinical chemoprevention regimens.

Epidemiological and experimental as well as a few clinical studies have indicated nonsteroidal anti-inflammatory agents (NSAIDs) to be compounds likely to be effective for chemoprevention (IARC, 1997). After a careful review of the mechanism of action of the NSAIDs (COX-1, COX-2, and other), the profile of putative side-effects and the experimental (as well as clinical) activity, we chose to study sulindac for use in combination with DFMO. We have recently initiated a 36-month phase IIb trial in which half of the participants receive placebo and half a combination of low doses of DFMO and sulindac. For this trial, 250 individuals with prior colonic polyps are being recruited. A number of surrogate end-point biomarkers will be measured in the flat rectal mucosa before and after 12 and 36 months of therapy. These include nuclear morphometry, uninduced apoptosis, polyamine and prostaglandin content, Ki67, and a number of preneoplastic antigens (CEA, sialy TN, p53, Bcl-2) measured by immunostaining in flat mucosa. Changes in these markers are being correlated with appearance of new incident adenomatous polyps. The presence of K-*ras* mutations in incident polyps is also being studied. The study has sufficient power to ensure that significant correlations between the appearance of colonic polyps (the primary biomarker) and changes in the spermidine/spermine ratio and occurrence of secondary biomarkers will be detectable. However, the study was not designed to have power to detect a modest difference (25%) in new incident adenomas between the two arms, although a large (75–90%) difference would be detectable.

We have also planned a large placebo-controlled chemoprevention trial of DFMO and sulindac using a 2×2 factorial design in which the rate of incident polyps after three years of therapy will be the primary end-point. Surrogate end-point biomarkers will also be measured 12 and 36 months after therapy and include those listed above. By using a Cochrane–Armitage analysis, we have been able to reduce the projected sample size from 1800 to 1000 participants without losing power; however, the design assumes that the combination is more effective in reducing the appearance of new colonic adenomas than either compound alone and that single agents will be more effective than the placebo.

Development of Bowman–Birk inhibitor as a chemoprevention agent

Experimental data

In epidemiological studies, high levels of soybean consumption have been associated with a decreased incidence of epithelial cancers (Kennedy, 1998). Four major classes of candidate chemopreventive compounds have been identified in soybeans: isoflavones, phytic acid, saponins and certain protease inhibitors. We have focused our studies on the Bowman–Birk inhibitor, a soybean-derived serine protease inhibitor with anticarcinogenic activity at doses well below those of other chemopreventive agents identified in soybeans. A series of studies in animals have demonstrated that a Bowman–Birk inhibitor concentrate is an effective inhibitor of protease activity and oral carcinogenesis (Messadi *et al.*, 1986; Kennedy *et al.*, 1993).

Clinical trials

Bowman–Birk inhibitor concentrate, which contains active Bowman–Birk inhibitor and has the same anticarcinogenic profile as the purified substance, has been developed for human trials. Proteolytic activities are elevated in the oral mucosa cells of patients with oral leukoplakia, and markedly so in those who are actively smoking (Manzone *et al.*, 1995), providing a direct rationale for the use of this agent in clinical trials at this tissue site. Overexpression of c-erbB-2 (neu) is important in human oral carcinogenesis and progression (Craven *et al.*, 1992; Hou *et al.*, 1992), and the relationship between protease activity and *neu* oncogene expression in patients with oral leukoplakia treated with the Bowman–Birk inhibitor has been established (Wan *et al.*, 1999). A phase I trial of

Bowman–Birk inhibitor concentrate showed that the compound was non-toxic up to the maximum doses that were allowed by regulatory agreement (Armstrong *et al.*, 2000).

We have recently completed a one-month phase IIa dose-escalation chemoprevention trial of Bowman–Birk inhibitor concentrate in patients with oral leukoplakia lesions (Wan *et al.*, 1999). There was a linear fit of the relationship between the dose of the agent and the decrease in total lesion area, which was confirmed by a blinded analysis of clinical impression of lesion photographs (W. Armstrong & F. Meyskens, unpublished). Pretreatment levels of cellular protease activity affected the clinical response, with lesions having lower initial levels of protease activity responding better. There were a number of important relationships confirmed in relation to administration of Bowman–Birk inhibitor concentrate:

- High pretreatment levels of protease activity were associated with greater decreases in protease activity.
- A dose-dependent increase in serum neu protein was observed.
- Higher pretreatment serum neu protein levels were associated with greater relative decreases in serum and cellular neu protein and cellular protease activity.
- Pretreatment levels of cellular neu protein correlated inversely with changes in cellular neu protein.

Overall, these results were quite encouraging, as they suggested that Bowman–Birk inhibitor concentrate had clinical activity against oral leukoplakia that was associated with a change in predefined surrogate end-point biomarkers. We have now begun a randomized phase IIb trial in which patients with oral leukoplakia receive either Bowman–Birk inhibitor concentrate or placebo for six months. Serial shed oral mucosal cells will be collected for analysis of surrogate end-point biomarkers and clinical lesions will be monitored. Patients achieving a partial or complete clinical response will continue treatment for up to 18 months.

Acknowledgement

The research described here was supported in part by the Chao Family Comprehensive Cancer Center and P30 CA62230.

References

Abeloff, M.D., Rosen, S.T., Luk, G.D., Baylin, S.B., Zeltzman M. & Sjoerdsma, A. (1986) Phase II trials of α-difluoromethylornithine, an inhibitor of polyamine synthesis, in advanced small cell lung cancer and colon cancer. *Cancer Treat. Rep.*, **70**, 843–845

Armstrong, W.B., Kennedy, A.R., Wan, X.S., Atiba, J., McLaren, C.E. & Meyskens, F.L., Jr (2000) Single dose administration of Bowman-Birk inhibitor concentrate in patients with oral leukoplakia. *Cancer Epidemiol. Biomarkers Prev.*, **9**, 43–47

Boyle, J.O., Meyskens, F.L., Jr, Garewal, H.S. & Gerner, E.W. (1992) Polyamine contents in rectal and buccal mucosae in humans treated with oral difluoromethylornithine. *Cancer Epidemiol. Biomarkers Prev.*, **1**, 131–135

Craven, J.M., Pavetic, Z.P., Stambrook, P.J., Pavelic, L., Gapany, M., Kelley, D.J., Gapany, S. & Gluckman, J.L. (1992) Expression of c-erb B-2 gene in human head and neck carcinoma. *Anticancer Res.*, **12**, 2273–2276

Croghan, M.K., Aicken, M.G. & Meyskens, F.L., Jr (1991) Dose-related α-difluoromethylornithine (DFMO) ototoxicity (reversible hearing loss). *Am. J. Clin. Oncol.* (CCT), **14**, 331–335

Fisher, B., Costantino, J.P., Wickerham, L., Redmond, C.K., Kavanah, M. & Cronin, W.M. (1998) Tamoxifen for prevention of breast cancer: report of the National Surgical Adjuvant Breast and Bowel Project P-1 Study. *J. Natl Cancer Inst.*, **90**, 1371–1388

Halline, A.G., Dudeja, P.K., Jacoby, R.F., Llor, X., Teng, B.B., Chowdhury, L.N., Davidson, N.O. & Brasitus, T.A. (1990) Effect of polyamine oxidase inhibition on the colonic malignant transformation process induced by 1,2-dimethylhydrazine. *Carcinogenesis*, **11**, 2127–2132

Hixson, L.J., Garewl, H.S., McGee, D., Sloan, D., Fennerty, M.B., Sampliner, R.E. & Gerner, E.W. (1993) Ornithine decarboxylase and poly-amines in colorectal neoplasia and adjacent mucosa. *Cancer Epidemiol. Biomarkers Prev.*, **2**, 369–374

Hou, L., Shi, D., Tu, S.M., Zhang, H.Z., Hung, M.C. & Ling, D. (1992) Oral cancer progression and c-erb-2/neu proto-oncogene expression. *Cancer Lett.*, **65**, 215–220

IARC (1997) *IARC Handbooks of Cancer Prevention*, Volume 1, *Non-steroidal Anti-inflammatory Drugs*, Lyon, IARC

Kelloff, G.J., Crowell, J.A., Hawk, E.T., Steele, V.E., Lubet, R.A., Boone, C.W., Covey, J.M., Doody, L.A., Omenn, G.S., Greenwald, P., Hong, W.K., Parkinson, D.R., Bagheri, D., Baxter, G.T., Blunden, M., Doeltz, M.K., Eisenhower, K.M., Johnson, K., Knapp, G.G., Longfellow, D.G., Malone, W.F., Nayfield, S.G., Seifried, H.E., Swall, L.M. & Sigman, C.C. (1996) Strategy and planning for chemopreventive drug development; clinical development phase II. *J. Cell Biochem.* (Suppl.) **26**, 64–71

Kennedy, A.R. (1998) Chemopreventive agents: protease inhibitors. *Pharmacol. Ther.*, **78**, 167–209

Kennedy, A.R., Billings, P.C., Maki, P.A. & Newberne, P. (1993) Effects of various protease inhibitor preparations on oral carcinogenesis in hamsters induced by 7,12-dimethylbenz(a)anthracene. *Nutr. Cancer,* **19**, 191–200

Manzone, H., Billings, P.C., Cummings, W.N., Feldman, R., Clark, L.C., Odell, C.S., Horan, A.M., Atiba, J.O., Meyskens, F.L. & Kennedy, A.R. (1995) Levels of proteolytic activities as intermediate marker endpoints in oral carcinogenesis. *Cancer Epidemiol. Biomarkers Prev.*, **4**, 521–527

Messadi, P.V., Billings, P., Schklar, G. & Kennedy, A.R. (1986) Inhibition of oral carcinogenesis by a protease inhibitor. *J. Natl Cancer Inst.*, **76**, 447–452

Meyskens, F.L., Jr (1992a) Biomarkers intermediate endpoints and cancer prevention. *J. Natl Cancer Inst. Monographs*, **13**, 177–182

Meyskens, F.L., Jr (1992b) Biology and intervention of the premalignant process. *Cancer Bulletin*, **43**, 475–480

Meyskens, F.L., Jr & Gerner, E.W. (1999) Development of difluoromethylornithine (DFMO) as a chemoprevention agent. *Clin. Cancer Res.*, **5**, 945–951

Meyskens, F.L., Jr, Surwit, E., Moon, T.E., Childers, J.M., Davis, J.R., Dorr, R., Johnson, C.S. & Alberts, D.S. (1994) Enhancement of regression of cervical intraepithelial neoplasia II (moderate dysplasia) with topically applied all-trans-retinoic acid: a randomized trial. *J. Natl Cancer Inst.,* **86**, 539–543

Meyskens, F.L., Jr, Gerner, E., Emerson, S., Pelot, D., Durbin, T., Doyle, K. & Lagerberg, W. (1998) A randomized double-blind placebo controlled phase IIb trial of difluoromethylornithine for colon cancer prevention. *J. Natl Cancer Inst.,* **90**, 1212–1218

Nigro, N.D., Bull, A.W. & Boyd, M.E. (1986) Inhibition of intestinal carcinogenesis in rats: effect of difluoromethylornithine with piroxicam or fish oil. *J. Natl Cancer Inst.*, **77**, 1309–1313

Rozhin, J., Wilson, P.S., Bull, A.W. & Nigro, N.D. (1984) Ornithine decarboxylase activity in the rat and human colon. *Cancer Res.*, **44**, 3226–3230

Verma, A.K. (1990) Inhibition of tumor promotion by DL-α-difluoromethylornithine, specific irreversible inhibitor of ornithine decarboxylase. *Basic Life Sci.*, **52**, 195–204

Wan, X.S., Meyskens, F.L., Armstrong, W.B., Taylor, T.H. & Kennedy, A.R. (1999) Relationship between protease activity and Neu oncogene expression in patients with oral leukoplakia treated with the Bowman Birk Inhibitor. *Cancer Epidemiol. Biomarkers Prev.*, **8**, 601–608

Corresponding author:

F. L. Meyskens, Jr
Chao Family Comprehensive Cancer Center,
101 The City Drive South,
Rt. 81, Bldg. 23, Rm. 406,
Orange,
CA 92868-3298,
USA

Biomarkers in Cancer Chemoprevention
Miller, A.B., Bartsch, H., Boffetta, P., Dragsted, L. and Vainio, H. eds
IARC Scientific Publications No. 154
International Agency for Research on Cancer, Lyon, 2001

Selection and validation of biomarkers for chemoprevention: the contribution of epidemiology

P. Vineis and F. Veglia

This chapter considers the epidemiological contribution of DNA adducts as an example of markers for use in chemoprevention studies, and highlights the potential biases inherent in the conduct of epidemiological studies with molecular markers. Although adducts have been interpreted mainly as biomarkers of exposure, 'bulky' DNA adducts such as those measured by ^{32}P-postlabelling or ELISA in white blood cells are more correctly interpretable as markers of cumulative unrepaired DNA damage. The latter concept can prove useful in cancer epidemiology, since it is consistent with existing knowledge on the importance of duration of exposure in the etiology of chemically-induced cancers. Increasing evidence suggests that in addition to prolonged exposure to genotoxic chemicals, inter-individual variability in carcinogen metabolism and DNA repair is predictive of cancer risk. Also from this point of view, measurements of 'bulky' DNA adducts can be useful as biomarkers for studies in populations, since they express the amount of carcinogen linked to DNA after repair, taking into account individual repair capacity. Finally, we suggest a theory of causality based on the work of Wesley Salmon and the concept of 'propagating mark', which is particularly attractive for molecular epidemiologists.

Categories of markers and their complex interplay

The usual categorization of biomarkers refers to three categories of markers, of exposure (internal dose), of intermediate effect and of individual susceptibility (see Chapter 1). We will examine a few examples that suggest that the nature of biological measurements may be more complex than is implied by such categorization.

There is increasing evidence that some types of DNA adduct can be considered both as markers of internal dose and as markers of susceptibility, because (*a*) they predict the onset of disease independently of exposure levels (see Example 1 below), and (*b*) they represent an integrated marker of exposure and of the individual's ability to metabolize carcinogens and repair DNA damage. Classical markers of susceptibility, such as genetically-based metabolic polymorphisms, can modulate the effect of exposures, but their expression can also be influenced by external determinants such as dietary habits. In this complex context of mutual interplay, the level of exposure is also likely to be important, as Example 2 demonstrates.

Validation

Validation of biomarkers can be interpreted in at least three different ways. One is the usual validation of a marker from the laboratory point of view, including an assessment of measurement error (for example by repeat measurements in the sample, and by the use of positive control samples with known values of the marker level). Measurement error is expressed by, for example, the coefficient of variation, the ratio between the standard deviation and the mean. Still in this category of validation measures, we can include measures of marker reliability (reproducibility and repeatability). A second category is represented by attempts to estimate the validity of the marker when compared with a standard (internal validity: sensitivity and specificity) and the impact of its use in a population (predictive value, depending on internal validity and the prevalence of the condition that we aim to measure). These aspects of validation are extensively dealt with in Toniolo *et al.* (1997).

A third concept of validation refers to the relevance of the marker in the context of population studies or intervention trials when it is used as a surrogate end-point biomarker. Even if a marker is reliable and valid, and its predictive value reasonably high, in the context of chemoprevention we need to be sure that the marker is a good surrogate end-point. This is the category of markers we mainly refer to below.

Examples

Example 1: DNA adducts as predictors of cancer risk and dietary effect modifiers

'Bulky' DNA adducts (adducts with large molecules, as opposed to small alkyl radicals such as methyl groups) can be considered both as markers of internal dose and as markers of susceptibility. Several studies have considered the association between cancer at various sites and the levels of bulky DNA adducts (Table 1). Most studies (Tang *et al.*, 1995; Li *et al.*, 1996; Peluso *et al.*, 2000; Vulimiri *et al.*, 2000) have found that cancer cases had higher levels of adducts than non-cancer controls, after adjustment for relevant exposures such as smoking. This suggests that we may consider bulky DNA adducts (as measured in white blood cells by the ^{32}P-postlabelling method) as markers of susceptibility, in addition to their being markers of exposure. Further evidence for such an interpretation comes from a recent case–control study nested in the Physicians Health Study cohort, which found DNA adducts to be predictive of the cancer outcome (Perera, 2000). In the latter study, adducts were measured in blood samples that were collected years before cancer onset, thus ruling out the possibility that the higher adduct levels were due to metabolic changes associated with an already existing cancer.

It appears that bulky DNA adducts express cumulative exposure to aromatic compounds after the action of metabolizing enzymes and before the intervention of DNA repair enzymes. They are, therefore, markers of cumulative unrepaired DNA damage. There is a large amount of additional evidence to support this interpretation, based on the observation that the lymphocytes of cancer patients (and of their healthy relatives) have higher levels of DNA adducts after treatment with a hydrophilic chemical, compared with those of noncancerous individuals (Berwick & Vineis, 2000).

Other types of adduct provide a better expression of external exposure than of cumulative unrepaired DNA damage. This is the case for protein adducts that are not repaired.

The level of DNA adducts, which is suggested to be predictive of cancer risk, can be modulated by personal habits such as intake of fruit and vegetables. In a study of healthy volunteers, conducted within the EPIC Italian cohort, an inverse association between consumption of several dietary items and adduct levels was found (Table 1) (Palli *et al.*, 2000). Out of about 120 food items that were investigated, only those shown in this table were associated in a statistically significant manner with the adduct levels. Some of these food items, namely leafy vegetables and fruits, were associated with an approximately 25–30% decrease in adduct levels, while meat consumption was associated with a slight and non-significant increase (Palli *et al.*, 2000). When nutrients were assessed by means of food–nutrient conversion tables, associations were found with monounsaturated fatty acids and β-carotene (Table 2). Similar relationships have been observed in the investigation on bladder cancer by Peluso *et al.* (2000).

Other types of change that are related to DNA damage can be modified by fruit or vegetable intake. For example, several functional tests have been developed to explore individual DNA repair capacity (mutagen sensitivity tests; see Berwick & Vineis, 2000). The repair capacity thus assessed seems to be modified by several exposures or personal habits. In cultured lymphocytes, antioxidants such as α-tocopherol exercised a dose-dependent protective effect in preventing bleomycin-induced chromosome damage (Trizna *et al.*, 1992, 1993). In another study, Kucuk *et al.* (1995) found strong inverse correlations between plasma nutrients and results from the mutagen sensitivity assay. Correlation coefficients were 0.76 with β-carotene and 0.72 with total carotenoids. However, in randomized double-blind trials, Hu *et al.* (1996) and Goodman *et al.* (1998) did not find any association between supplementation and DNA repair activity.

From a mechanistic point of view, the modulating effect of fruit or vegetables on adduct levels may be explained by induction of enzymes involved in carcinogen detoxification or by repression of enzymes involved in carcinogen activation. As far as 'mutagen sensitivity' is concerned, this is usually interpreted as an indirect expression of

Table 1. Adjusted means[a] of relative adduct labelling (DNA adducts per 10^9 normal nucleotides) according to tertiles of daily consumption of selected food groups or food items, as reported on a food frequency questionnaire (EPIC Italy, 1993–98)

	Tertile of consumption				*p*-value
	Adjusted mean ± SE			% change[b]	for trend*
	I	II	III		
Leafy vegetables (except cabbage)	9.07±1.22	8.60±1.13	6.34±1.21	–30.10	**0.02**
Fruiting vegetables	8.34±1.34	8.48±1.14	7.29±1.19	–12.59	0.06
Root vegetables	7.60±1.14	10.29±1.12	5.89±1.21	–22.50	0.07
Cruciferous	10.22±1.2	16.57±1.15	7.42±1.15	–27.40	0.2
Grain and pod vegetables	7.16±1.20	9.51±1.23	7.65±1.16	6.84	0.7
Stalk vegetables, sprouts	8.76±1.16	8.70±1.17	6.66±1.17	–23.97	0.2
Mixed salad, mixed vegetables	8.47±1.19	8.45±1.18	7.02±1.33*	–17.12	0.2
All vegetables	8.92±1.27	8.66±1.13	6.58±1.18	–26.23	**0.01**
Legumes	8.24±1.22	8.18±1.16	7.69±1.15	–6.67	**0.01**
Potatoes	6.98±1.19	7.93±1.19	9.19±1.20	31.66	0.7
Onion, garlic	7.89±1.23	8.03±1.23	8.11±1.13	2.79	0.8
Fresh fruit (all types)	8.56±1.23	6.88±1.20	6.47±1.24*	–24.4	**0.04**
Nuts and seeds	6.54±1.38	7.79±1.19*	9.34±1.15*	42.81	0.1
Fruit and vegetable juices	9.34±1.10	6.40±1.43	7.42±1.22	–20.56	0.5
Milk	8.81±1.19	8.27±1.15	6.98±1.20	–20.77	0.8
Yoghurt	7.54±1.00	9.23±1.35	7.83±1.28	3.85	0.3
Cheese (including fresh cheeses)	7.02±1.20	6.86±1.14	10.20±1.20	45.30	0.1
Pasta, other grain	6.82±1.26	7.37±1.13	9.92±1.27	45.45	0.09
Rice	7.08±1.30	9.00±1.16	7.85±1.13	10.88	0.9
Bread	7.76±1.28	8.03±1.14	8.29±1.34	6.83	0.9
Fish	8.60±1.25	8.77±1.18	7.02±1.14	–18.37	0.2
Seafood	8.43±1.25	9.03±1.18	6.90±1.14	–18.15	0.1
Eggs	8.99±1.19	7.46±1.14	8.69±1.21	–3.34	0.9
Processed meat	8.82±1.22	8.08±1.18	6.95±1.34	–21.20	0.8
Offal	9.09±1.13	8.27±1.19	6.40±1.21	–29.59	0.1
All red meat	7.65±1.18	8.58±1.14	7.80±1.24	1.96	0.5
All white meat	7.85±1.19	8.03±1.13	8.19±1.19	4.33	0.8
Seed oil	7.70±1.20	7.96±1.22	8.31±1.10	7.92	0.3
Olive oil	8.28±1.16	8.88±1.18	6.76±1.25	–18.36	**0.05**
Butter	6.71±1.13	7.00±1.20	10.52±1.20	56.78↑	0.1
Sugar, honey, jam	6.45±1.20	8.23±1.19	9.31±1.19	44.34	0.06
Cakes, pies, pastries, puddings	8.15±1.26	7.98±1.15	7.95±1.23	–2.45	0.9
Ice cream	8.70±1.13	8.92±1.20	6.06±1.30*	–30.34↓	**0.003**
Wine	7.89±1.29	7.55±1.12	8.68±1.35	10.01	0.1
Beer	8.04±1.23	8.55±1.23	7.55±1.20	–6.09	0.9
Coffee	8.12±1.30	8.24±1.13	7.72±1.20	–4.93	0.5

From Palli *et al.* (2000)

[a] From analysis for covariance model including terms for age, sex, centre, smoking habits (never, ex and current), period of blood drawing and total caloric intake (kcal)

[b] Percentage change from I to III tertile (↑ highest and ↓ lowest variation)

* Dunnett test for multiple comparisons significant at the 0.05 level (first level is the reference category)

Table 2. Adjusted means[a] of relative adduct labelling (DNA adducts per 10^9 normal nucleotides) according to tertiles of estimated daily intake of various nutrients (EPIC Italy, 1993–98)

Nutrient	Tertile of consumption				*p*-value for trend *
	Adjusted mean ± SE			% change[b]	
	I	II	III		
Total protein	9.73±1.60	7.13±1.12	7.33±1.66	–32.74	0.9
Animal protein	9.40±1.34	7.77±1.15	6.74±1.44	–39.47	0.9
Vegetable protein	6.92±1.57	8.25±1.15	8.77±1.50	21.09	0.7
Total fat	9.99±1.44	7.01±1.14	6.96±1.54	–30.33	0.09
Animal fat	9.01±1.41	7.56±1.14	7.58±1.44	–15.87	0.6
Vegetable fat	8.86±1.29	8.75±1.13	6.17±1.32	–30.36	**0.009**
Fatty acids					
Total saturated fatty acids	8.57±1.43	7.92±1.14	7.55±1.51	–11.90	0.9
Oleic acid	9.95±1.35	7.79±1.15	5.87±1.46	–41.01	0.03
Total monounsaturated fatty acids	10.20±1.34	7.67±1.15	6.67±1.47	–34.61	**0.008**
Linoleic acid	8.20±1.48	8.09±1.18	7.82±1.40	–4.63	0.4
Linolenic acid	11.04±1.45	6.86±1.14	6.44±1.42	–41.67↓	**0.01**
Total polyunsaturated fatty acids	8.16±1.46	7.91±1.16	8.02±1.39	–1.72	0.2
Cholesterol	8.25±1.38	7.52±1.14	8.35±1.42	1.21	0.5
Carbohydrates	8.08±1.68	6.76±1.14	9.36±1.63	15.84	0.3
Starch	6.46±1.51	7.15±1.16	10.22±1.48	58.20↑	0.1
Sugar	7.84±1.35	7.32±1.15	8.92±1.33	13.78	0.8
Fibre	8.04±1.43	8.03±1.24	8.01±1.52	–0.37	0.1
Alcohol	7.41±1.30	7.93±1.17	8.66±1.29	16.87	**0.05**
Total calories	8.26±1.78	7.84±1.17	7.91±2.04	–4.24	0.9
Minerals					
Iron	10.31±1.52	7.53±1.16	6.26±1.55	–64.70	0.4
Calcium	7.79±1.37	7.71±1.14	8.62±1.40	9.63	0.9
Sodium	9.35±1.42	6.45±1.13	8.34±1.51	–12.11	0.5
Potassium	10.19±1.43	7.11±1.16	6.89±1.39*	–47.90	**0.004**
Vitamins					
Thiamine	9.74±1.42	7.32±1.14	6.91±1.53	–40.96	0.2
Riboflavin	9.58±1.37	7.24±1.13	7.28±1.43	–31.59	0.4
Niacin	8.35±1.46	8.58±1.13	7.02±1.47	–18.95	0.3
Vitamin B6	8.26±1.42	9.11±1.14	6.65±1.40	–24.21	0.2
Folic acid	8.71±1.36	7.21±1.15	8.19±1.35	–6.35	0.4
Retinol	8.96±1.21	7.42±1.16	7.62±1.23	–17.59	0.2
β-Carotene	9.37±1.22	7.63±1.13	7.15±1.23	–31.05	**0.01**
Vitamin E	8.91±1.46	7.51±1.23	7.73±1.30	–15.27	**0.02**
Vitamin C	9.03±1.24	7.28±1.17	7.79±1.21*	–15.92	**0.01**

From Palli *et al.* (2000)

[a] From analysis for covariance model including terms for age, sex, centre, smoking habits (never, ex and current), period of blood drawing and total caloric intake (kcal)

[b] Percentage change from I to III tertile (↑ highest and ↓ lowest variation)

* Dunnett test for multiple comparisons significant at the 0.05 level (first level is the reference category)

DNA repair capacity. One can speculate that fruit and vegetables, or vitamins, may interfere with DNA repair enzymes, but there is in fact no evidence supporting this hypothesis.

In the light of such observations, a possible explanation for the higher levels of bulky adducts among cancer cases than in controls can be found in the concept of cumulative unrepaired DNA damage. What causes cancer would be the total burden of a genotoxic chemical that remains bound to DNA, after the repair processes. This burden may be higher because DNA repair is impaired, because higher levels of carcinogenic metabolites are present (due to genetic or acquired effects) or because of repeated exposures to the same agent.

We also have to consider the limitations of this model. First, the level of measurement error for bulky adducts is not certain, but seems to be high (coefficient of variation around 20–30%). However, the effect of measurement error is to attenuate a relationship, if error is evenly distributed between the compared groups (Copeland *et al.*, 1977). Table 3 shows an example: in this case, the laboratory error was measured by the intra-class correlation coefficient, which, in turn, was used to correct the observed (attenuated) relative risks so as to obtain a more realistic estimate of the true effects on the risk of cancer. Thus, measurement error is expected to blur existing associations rather than reveal false associations.

A second relevant question concerning Example 1 is whether the effect is attributable to specific agents or to fruits and vegetables as a whole. This question has been addressed by Thompson *et al.* (1999) in an experimental study that aimed to test the hypothesis that increased consumption of vegetables and fruits would reduce levels of markers of oxidative cellular damage. Twenty-eight women participated in a 14-day dietary intervention. The primary end-points assessed were levels of 8-hydroxydeoxyguanosine (8-OHdG) in DNA isolated from peripheral lymphocytes, 8-OHdG excreted in urine, and urinary 8-isoprostane F-2α (8-EPG). Overall, the levels of 8-OHdG in DNA isolated from lymphocytes and in urine and the level of 8-EPG in urine were reduced by the intervention. The reduction in lymphocyte 8-OHdG was greater (32 versus 5%) in individuals with lower average pre-intervention levels of plasma α-carotene than in those with higher levels. The results of this study indicate that consumption of a diet that significantly increased

Table 3. Intraclass correlation coeffcients (*r*) for the measurement of estrone by different laboratories and resulting observed relative risks given true relative risks of 1.5, 2.0 and 2.5

Laboratory	*r*	Observed relative risk		
		RR_t=1.5	RR_t=2.0	RR_t=2.5
Laboratory 1	0.12	1.1	1.1	1.1
Laboratory 2				
analysis 1	0.82	1.4	1.8	2.1
analysis 2	0.53	1.2	1.4	1.6
Laboratory 3				
analysis 1	0.57	1.3	1.5	1.7
analysis 2	0.65	1.3	1.6	1.8
Laboratory 4	0.90	1.4	1.9	2.3

RR_t, true relative risk
Observed RR = exp (ln $RR_t \times r$)
From Hankinson *et al.* (1994).

vegetable and fruit intake led to significant reductions in markers of oxidative cellular damage to DNA and lipids, in contrast to previous studies that were based on administration of single components of diet.

Thus, we may argue that the choice of the type of intervention can be crucial, and in some circumstances positive results can be obtained more easily with a general category (fruit and vegetable intake) than with a specific component.

Example 2: Modulation by genetic susceptibility and the effect of dose; the case of methylenetetrahydrofolate reductase polymorphism

Common genetic polymorphisms have been reported in the gene encoding methylenetetrahydrofolate reductase (MTHFR), the enzyme that produces 5-methyltetrahydrofolate (5-methyl-THF) required for the conversion of homocysteine to methionine. In individuals with the genotype corresponding to a val/val polymorphism, functional effects include elevation of plasma homocysteine levels and differences in response to folic acid supplementation. The metabolic changes associated with the genotype have been reported to modify the risk for chronic disease (e.g., vascular disease and cancer) and neural tube defects in conjunction with folate deficiency. Folate intake requirements may be different in affected individuals to those of normal or heterozygous individuals. The complex interaction between this common genetic polymorphism of MTHFR and folate intake is the focus of intense investigation (Bailey *et al.*, 1999).

In a study in the United States, an inverse association of this MTHFR gene polymorphism with colorectal cancer was found. The inverse association of methionine intake and positive association of alcohol with colorectal cancer were stronger among val/val individuals. These interactions were not seen for colorectal adenomas (Chen *et al.*, 1998, 1999). In another study, the association between the V (val) allele of the MTHFR gene and ischaemic stroke in an elderly Japanese population was examined. In 256 stroke patients and 325 control subjects, the frequencies of the V allele were 0.45 and 0.32, respectively. The odds ratios and 95% confidence intervals adjusted for the other risk factors were 1.51 (1.02–2.23) for the AV (alanine/valine) genotype and 3.35 (1.94–5.77) for the VV genotype, compared with the AA genotype (Morita *et al.*, 1998).

Moderate elevation of plasma total homocysteine (tHCY) level is a strong and independent risk factor for coronary artery disease. The polymorphism in MTHFR, plasma tHCY and folate using baseline blood levels were examined among 293 Physicians' Health Study participants who developed myocardial infarction during up to eight years of follow-up and 290 control subjects. Compared with those with genotype AA, the relative risk (RR) of myocardial infarction were 1.1 (95% CI, 0.8–1.5) among those with the AV genotype and 0.8 (0.5–1.4) for the VV genotype; none of these RRs was statistically significant. However, those with genotype VV had an increased mean tHCY level (mean ± SEM, 12.6 ± 0.5 nmol/ml), compared with those with genotype AA (10.6± 0.3) ($p < 0.01$). This difference was most marked among men with low folate levels (the lowest quartile distribution of the control subjects): those with genotype VV had tHCY levels of 16.0 ± 1.1 nmol/ml, compared with 12.3 ± 0.6 nmol/ml ($p < 0.001$) for genotype AA. Therefore, the modulating effect of MTHFR seems to be exerted through the tHCY level, especially when folate intake is low (Ma *et al.*, 1996).

Methodological issues

Gene–environment interactions and their potential role in chemoprevention trials: some calculations

To assess the potential role of gene–environment interactions in chemoprevention trials, it is essential to know the penetrance of the genetic trait and its prevalence in the target groups. Table 4 presents an attempt to address this complex issue. Let us imagine we wish to screen high-risk families for a highly penetrant gene (*BRCA1*). In this case, the cumulative risk of breast cancer is approximately 80% in the mutation carriers and the prevalence of the mutations in families is about 50%. Let us hypothetically suppose that tamoxifen reduces the risk by 50%. This means that we have to treat 2.5 family members (carriers of the mutation) to prevent one cancer, i.e., to screen five members to have the same result. However, if we aim at the general population, things change dramatically. Now the cumulative risk is 40%, with an absolute risk reduction of 20%, which means a number needed to treat (NNT) of five women among those carrying the mutations. However, since in the general population only 0.2% are mutation carriers

Table 5. Calculation of the number needed to treat in the case of a screening for a low-penetrance gene (*GSTM1* in smokers), and a high-penetrance gene (*BRCA1*), respectively in the general population or in families

	Smokers		*BRCA1*	
	GSTM1-null	*GSTM1* wild	General population	Families
Relative risk	1.34 (1.21–1.48)[a]	1.0	5	10
Cumulative risk	13%	10%	40% [b]	80%
Risk reduction	50%[c]	50%[c]	50% (tamoxifen)[d]	50%
Cumulative risk after intervention	6.5%	5%	20%	40%
Absolute risk reduction	6.5%	5%	20%	40%
NNT in mutation carriers	15	20	5	2.5
Prevalence	50%	50%	0.2%[e]	50%
NNT in whole target population	30	40	2500	5
NNT in all smokers	35			

NNT, number needed to treat

[a] From Vineis *et al.* (1999); the OR for *GSTM1* in smokers was 1.22 (0.96–1.54)
[b] From Hopper *et al.* (1999)
[c] Theoretical maximum reduction in risk of lung cancer due to chemopreventive agent
[d] Theoretical benefit
[e] Coughlin *et al.* (1999)

(1 in every 500), the number we need to screen is as large as 2500 in order to prevent one cancer, based on the rather optimistic (and theoretical) 50% benefit of tamoxifen treatment.

Considering next a low-penetrance gene, *GSTM1*, we might plan to screen smokers for the *GSTM1* genotype and to address chemoprevention only to them. What would be the advantage? In a meta-analysis (Vineis *et al.*, 1999), the risk of lung cancer associated with the *GSTM1* genotype was 1.34 (it was 1.22 if the meta-analysis was restricted to smokers). Therefore, if the cumulative risk of lung cancer in smokers is 10%, it will be 13% among the null *GSTM1* carriers. Let us suppose again that the chemopreventive intervention has a 50% efficacy. This leads to a cumulative risk of 6.5% among smokers who are *GSTM1*-null, with a number needed to treat (NNT) of 15 (1/6.5%). However, since the carriers of the null genotype are 50% in the population, we need to screen 30 individuals to prevent one cancer. Now we repeat the same calculations with the carriers of the wild genotype, ending up with an NNT of 40. Without screening the population for *GSTM1*, we would have an NNT of 35 (the average of the previous two). Clearly, there is very little advantage in screening for a low-penetrance gene if the NNT just increases from 35 to 40.

Validation of biomarkers: bias and confounding

We have considered several examples of contributions that might come from epidemiology for the selection and validation of biomarkers, and particularly surrogate end-point biomarkers. We suggest that, subject to further validation, bulky DNA adducts could become one such surrogate, particularly in view of the cohort study results mentioned above. Also, mutational spectra in cancer genes and measures of gene–environment interaction are potentially important end-points. However, epidemiological studies are prone to several types of bias (Kleinbaum *et al.*, 1982; Murphy, 1976; Hennekens & Buring, 1987).

Information bias is related to material mistakes in conducting laboratory or other analyses, or in reporting mutations; for example, a distortion arose from incorrect reporting of the *p53* gene sequence in an early paper, which influenced subsequent reports (Lamb & Crawford, 1986). Another example is the measurement of oxidative damage to DNA: routine phenol-based DNA purification procedures can increase 8-hydroxydeoxyguanosine levels 20-fold in samples that are exposed to air following removal of the phenol (Wiseman *et al.*, 1995). Such gross contamination would seriously bias an epidemiological study if subsets (batches) of samples from different subgroups in the study population (e.g., exposed vs. unexposed) were to undergo different technical procedures subject to different levels of error.

Selection bias is certainly relevant to molecular epidemiology. Consider hospital-based case–control studies of disease risk in relation to metabolic polymorphisms. We can imagine at least three mechanisms by which selection bias (and, more specifically, Berkson's bias) can occur. First, if a person is hospitalized for a specific reason, but has more than one pathological condition, it is possible that the concurrent disease is also associated with the genetic polymorphism(s) under investigation. Second, patients with a certain allele at the polymorphic locus under investigation can have adverse reactions to drugs and be hospitalized for this reason. Third, induction of an enzyme by treatment can influence the phenotypic indicator of genotype. For example, administration of methotrexate can induce hydroxyfolate-reductase by gene amplification; therefore, if in a case–control study we include cancer patients among the controls, we may have a distorted association between the disease under study and hydroxyfolate-reductase activity.

Specific characteristics of bias and confounding in studies on mutational spectra of cancer genes should also be considered. The size of the biopsies that are selected for investigation provides a clear example of the type of selection bias that can occur in studies on cancer genes. In bladder cancer, for example, it is likely that early-stage tumours are too small to allow the urologist to obtain a biopsy large enough for both research and clinical purposes. However, large biopsies tend to correspond to more advanced cases, which in turn may show a higher proportion of mutations in certain genes (Yaeger *et al.*, 1998).

Detection bias is likely to be a common problem in case–control studies in which the risk factor investigated itself leads to increased diagnostic investigations and thus increases the probability that the disease is identified in that subset of persons.

Detection bias can be considered as a form of information bias, in that the probability of identifying the diseased people is conditional on the clinical information collected, which differs between categories of the risk factor.

In molecular epidemiological studies, this may happen if molecular markers of early disease are prospectively analysed in a cohort. This will lead to easier detection of the eventual clinical disease in those who test positively, even if the marker is not necessarily intermediate in the causal chain leading from exposure to disease. For example, exposure to certain agents, such as formaldehyde, can induce micronuclei in mucosal cells of the oral cavity; these, in turn, may lead to earlier detection of oral cancer through subsequent periodic examination of the workers with positive test results. A similar phenomenon can occur with identification of mutated oncogenes or tumour-suppressor genes in exposed workers, well before the onset of clinical disease.

Publication bias is particularly difficult to characterize and quantify. Publication bias refers to the greater probability that studies with positive findings (e.g., those showing an association between *p53* mutation spectrum and exposure) get published. A way to identify publication bias is to plot the result of each study of a particular phenomenon (expressed, for example, as an odds ratio) against its size. In the absence of publication bias, the plot would be expected to show great variability with small samples and lower variability with large samples, around a central value of the true odds ratio. If publication bias occurs, negative results are not published, particularly if they arise from small investigations, and their results thus do not appear in the plot. For example, in the large database available at the International Agency for Research on Cancer on *p53* mutations (Hainaut *et al.*, 1998), the distribution of the proportion of mutations reported by different studies is skewed: all the studies with a proportion greater than 50% had less than 50 cases, while lower proportions were found in both small and large studies. This distribution does not necessarily imply that publication bias occurred; it might also suggest that large studies were based on heterogeneous populations, with a variable prevalence of mutations, while small studies refer to small subgroups with specific exposures to carcinogens and a genuinely high proportion of mutations.

Confounding occurs when a third variable creates a spurious association between the exposure at issue and the biomarker measurement. Several variables may act as confounders, for example if they modify the expression of oncogenes or tumour-suppressor genes. One such variable is chemotherapy: for example, cytostatic treatment for leukaemia induces characteristic cytogenetic abnormalities in chromosomes 5 and 7. Confounding arises if, for some reason, therapy is related to the exposure at issue. Disease stage is another potential confounder. Therefore, studies that aim to determine the expression of cancer genes in humans should be restricted to untreated patients or specific stages, or statistical analyses should be stratified according to treatment and stage.

The definition of confounder in molecular epidemiology may be more subtle than in traditional epidemiology. In the study of metabolic polymorphisms in relation to cancer risk, a confounder can be an exogenous exposure which is associated not with another exogenous exposure, but rather with gene expression or enzyme induction (Tajoli & Garte, unpublished). For example, ethanol is an inducer of various metabolic enzymes, and, in turn, excessive intake has been related to colon cancer. The observation of an association between the phenotype for CYP2E1 and colon cancer could be ambiguous, since it could be attributed to a genuine role of this metabolic polymorphism or to confounding by ethanol, which on the one hand would increase the risk of colon cancer and on the other would induce the enzyme (thus creating a spurious association between the two). A somewhat different type of example concerns the observed association of the CYP1A2 polymorphism with variation in the risk of colon cancer; this association is plausible since CYP1A2 is involved in the metabolism of heterocyclic aromatic amines. However, it has also been shown that consumption of cruciferous and other vegetables induces the activity of the CYP1A2 enzyme (Kall *et al.*, 1996), and we know that vegetables reduce the risk of colon cancer (Potter, 1996) (perhaps through their content in antioxidants, acting via pathways unrelated to the CYP1A2 enzyme). Therefore, the association between CYP1A2 and colon cancer can be confounded by dietary habits: specifically, a positive association between the CYP1A2 phenotype and colon cancer may be missed or underestimated because protective factors such as cruciferous vegetables induce CYP1A2 (an example of *negative* confounding). In general, the assessment of inducible enzymes is problematic in case–control studies.

In conclusion, the selection and validation of biomarkers for chemopreventive trials should consider the potential limitations of observational studies.

Interpretation of causal pathways

We can refer to the causal model proposed by a philosopher, Wesley Salmon (1984), to devise a framework for the interpretation of causal pathways in carcinogenesis and in the role played by different biomarkers. Salmon proposes two different models for causality, which are supposed to be complementary. The first model was suggested by Reichenbach many years earlier, and was simply based on probabilistic computation (the 'positive relevance' criterion). The basic idea is that two events (A and B) have a common cause C if we can show (1) that the joint probability of A and B is greater than the product of their separate probabilities, i.e., $p(AB)>p(A)p(B)$; (2) that the introduction of C into the equation entirely or at least partially explains the association between A and B. Condition (1) simply expresses the concept that the probability that two events occur jointly is the product of the individual probabilities if the two events are independent, while it is greater than the product if they are not independent. Condition 2 means that the joint occurrence of A and B can be explained by a common third event: for example, subjects A and B both have an angiosarcoma of the liver, and they were both exposed to vinyl chloride (VC) in the same plant. We can calculate that the probability of such joint events by simple chance is very low (condition 1). In addition, hypothesizing that the common exposure is VC, we observe that all the excess risk (observed/expected cases) is explained by such exposure (condition 2).

However, according to Salmon, this line of reasoning is not sufficient, precisely for the reason we have considered in the case of metabolizing enzymes and dietary habits. In that example, we were not able to disentangle the complex relationships between cruciferous vegetables, the *CYP1A2* metabolic polymorphism and the risk of colon cancer. Salmon suggests that a way to establish whether an event is a genuine cause, in addition to the statistical considerations above, is to include it in a process, and establish whether along the process there is what he calls the *propagation of a 'mark'*. In other words, a genuine causal process is one in which you can follow a 'mark' that propagates over the course of time, precisely because the causal events are able to induce a *structural change* that becomes a part of the effect.

This reasoning is quite relevant to the identification of good intermediate-effect markers for chemoprevention. According to a classical example, yellow fingers are a risk indicator for lung cancer: lung cancer patients have yellow fingers more frequently than population controls, thus fulfilling Reichenbach's definition of causality (positive relevance). However, clearly yellow fingers are just an indirect marker of exposure to tobacco smoke, and not a genuine cause of lung cancer. This is because they do not fulfil Salmon's second criterion of causality, the propagation of a marker along the process. In fact, whereas one can identify *p53* mutations in the lung cancers that are characteristic of tobacco carcinogens, thus following the structural changes left by tobacco into the lung, the same does not hold true for yellow fingers.

In the field of chemoprevention, we propose that certain types of DNA adduct may fulfil such requirements to be used as surrogate end-points. Mutations in cancer genes tend to be late events, and they are too rare to be detected in healthy subjects (in a study of 60 normal subjects in Lyon and Paris, no *p53* mutation in serum was found; P. Hainaut, personal communication), although clearly they may be structural 'marks' for cancer. At the other extreme, protein adducts tend to reflect exposure faithfully, but they are not repaired and are probably weaker predictors of cancer. The advantage of DNA adducts is that apparently they represent the *cumulative unrepaired DNA damage*, and thus they may be used to test several types of chemopreventive agent.

Acknowledgements

This research has been made possible by a grant of the S. Paolo Foundation (Torino) to the ISI Foundation. We are grateful to Anthony B. Miller and Jenny Chang-Claude for helpful comments.

References

Bailey, L.B. & Gregory, J.F., 3rd (1999) Polymorphisms of methylenetetrahydrofolate reductase and other enzymes: metabolic significance, risks and impact on folate requirement. *J. Nutr.*, **129**, 919–922

Berwick, M. & Vineis, P. (2000) Markers of DNA repair and susceptibility to cancer in humans: an epidemiologic review. *J. Natl Cancer Inst.*, **92**, 874–897

Chen, J., Giovannucci, E., Hankinson, S.E., Ma, J., Willett, W.C., Spiegelman, D., Kelsey, K.T. & Hunter, D.J. (1998) A prospective study of methylenetetrahydrofolate reductase and methionine synthase gene polymorphisms, and risk of colorectal adenoma. *Carcinogenesis*, **19**, 2129–2132

Chen, J., Giovannucci, E.L. & Hunter, D.J. (1999) MTHFR polymorphism, methyl-replete diets and the risk of colorectal carcinoma and adenoma among U.S. men and women: an example of gene-environment interactions in colorectal tumorigenesis. *J. Nutr.*, **129**, 560S–564S

Copeland, K.T., Checkoway, H., Holbrook, R.H. & McMichael, A.J. (1977) Bias due to misclassification in the estimate of relative risk. *Am. J. Epidemiol.*, **105**, 488–495

Coughlin, S.S., Khoury, M.J. & Steinberg, K.K. (1999) BRCA1 and BRCA2 gene mutations and risk of breast cancer. Public health perspectives. *Am. J. Prev. Med.*, **16**, 91–98

Goodman, M.T., Hernandez, S., Wilkens, L.R., Lee, J., Le Marchand, L., Liu, L.Q., Franke, A.A., Kucuk, O. & Hsu, T.C. (1998) Effects of beta-carotene and alpha-tocopherol on bleomycin-induced chromosomal damage. *Cancer Epidemiol. Biomarkers Prev.*, **7**, 113–117

Hainaut, P., Hernandez, T.M., Robison, A., Rodriguez-Tome, P., Flores, T., Hollstein, M., Harris, C.C. & Montesano R. (1998) IARC database of p53 gene mutations in human tumors and cell lines: updated compilation, revised formats and new visualization tools. *Nucl. Acid Res.*, **26**, 205–213

Hankinson, S.E., Manson, J.E., London, S.J., Willett, W.C. & Speizer, F.E. (1994) Laboratory reproducibility of endogenous hormone levels in postmenopausal women. *Cancer Epidemiol. Biomarkers Prev.*, **3**, 51–56

Hennekens, C.H. & Buring, J.E. (1987) *Epidemiology in Medicine*, Boston, Little, Brown

Hopper, J.L., Southey, M.C., Dite, G.S., Jolley, D.J., Giles, G.G., McCredie, M.R., Easton, D.F. & Venter, D.J. (1999) Population-based estimate of the average age-specific cumulative risk of breast cancer for a defined set of protein-truncating mutations in BRCA1 and BRCA2. Australian Breast Cancer Family Study. *Cancer Epidemiol. Biomarkers Prev.*, **8**, 741–747

Hu, J.J., Roush, G.C., Berwick, M., Dubin, N., Mahabir, S., Chandiramani, M. & Boorstein, R. (1996) Effects of dietary supplementation of alpha-tocopherol on plasma glutathione and DNA repair activities. *Cancer Epidemiol. Biomarkers Prev.*, **5**, 263–270

IARC (1986) *IARC Monographs on the Evaluation of the Carcinogenic Risk of Chemicals to Humans*. Vol. 38, *Tobacco Smoking*, Lyon, IARC

Kall, M.A., Vang, O. & Clausen, J. (1996) Effects of dietary broccoli on human in vivo drug metabolizing enzymes: evaluation of caffeine, oestrone and chlorzoxazone metabolism. *Carcinogenesis*, **17**, 791–799

Kleinbaum, D., Kupper, L.L. & Morgenstern, H. (1982) *Epidemiologic Research*, Belmont, CA, Lifetime Learning Publ., p. 186

Kucuk, O., Pung, A., Franke, A.A., Custer, L.J., Wilkens, L.R., Le Marchand, L., Higuchi, C.M., Cooney, R.V. & Hsu, T.C. (1995) Correlations between mutagen sensitivity and plasma nutrient levels of healthy individuals. *Cancer Epidemiol. Biomarkers Prev.*, **4**, 212–221

Lamb P. & Crawford L. (1986) Characterization of the human p53 gene. *Mol. Cell Biol.*, **6**, 1379–1385

Li, D., Wang, M., Cheng, L., Spitz, M.R., Hittelmen, W.N. & Wei, Q. (1996) In vitro induction of benzo(a)pyrene diol epoxide-DNA adducts in peripheral lymphocytes as a susceptibility marker for human lung cancer. *Cancer Res.*, **56**, 3638–3641

Ma, J., Stampfer, M.J., Hennekens, C.H., Frosst, P., Selhub, J., Horsford, J., Malinow, M.R., Willett, W.C. & Rozen, R. (1996) Methylenetetrahydrofolate reductase polymorphism, plasma folate, homocysteine, and risk of myocardial infarction in US physicians. *Circulation*, **94**, 2410–2416

Morita, H., Kurihara, H., Tsubaki, S., Sugiyama, T., Hamada, C., Kurihara, Y., Sindo, T., Oh-Hashi, Y., Kitamura, K. & Yazaki, Y. (1998) Methylenetetrahydrofolate reductase gene polymorphism and ischemic stroke in Japanese. *Arterioscler. Thromb. Vasc. Biol.*, **18**, 465–469

Murphy, E.A. (1976) *The Logic of Medicine*, Baltimore, Johns Hopkins University Press

Palli, D., Vineis, P., Russo, A., Berrino, F., Krogh, V., Masala, G., Munnia, A., Panico, S., Taioli, E., Tumino, S., Garte, S. & Peluso, M. (2000) Diet, metabolic polymorphisms and DNA adducts: The EPIC-Italy cross-sectional study *Int. J. Cancer*, **87**, 444–451

Peluso, M., Airoldi, L., Magagnotti, C., Fiorini, L., Munnia, A., Hautefeuille, A., Malaveille, C. & Vineis, P. (2000) White blood cell DNA adducts and fruit and vegetable consumption in bladder cancer. *Carcinogenesis,* **21**, 183–187

Perera, F. (2000) Clues to cancer etiology and prevention from molecular epidemiology. In: *Molecular Epidemiology: A New Tool in Cancer Prevention*. Keystone Symposia, Taos (USA), February 10–15, 2000

Potter, J. (1996) Nutrition and colorectal cancer. *Cancer Causes Control*, **7**, 127–146

Salmon, W. (1984) *Scientific Explanation and the Causal Structure of the World*, Princeton, NJ, Princeton University Press

Tang, D., Santella, R.M., Blackwood, A.M., Young, T.L., Mayer, J., Jaretzki, A., Grantham, S., Tsai, W.Y. & Perera, F.P. (1995) A molecular epidemiological case-control study of lung cancer. *Cancer Epidemiol. Biomarkers Prev.*, **4**, 341–346

Thompson, H.J., Heimendinger, J., Haegele, A., Sedlacek, S.M., Gillette, C., O'Neill, C., Wolfe, P. & Conry, C. (1999) Effect of increased vegetable and fruit consumption on markers of oxidative cellular damage. *Carcinogenesis*, **20**, 2261–2266

Toniolo, P., Boffetta, P., Shuker, D.E.G., Rothman, N., Hulka, B. & Pearce, N., eds (1997) *Application of Biomarkers in Cancer Epidemiology* (IARC Scientific Publications No. 142), Lyon, IARC

Trizna, Z., Hsu, T.C. & Schantz, S.P. (1992) Protective effects of vitamin E against bleomycin-induced genotoxicity in head and neck cancer patients in vitro. *Anticancer Res.*, **2**, 325–327

Trizna, Z., Schantz, S.P., Lee, J.J., Spitz, M.R., Goepfert, H., Hsu, T.C. & Hong, W.K. (1993) In vitro protective effects of chemopreventive agents against bleomycin-induced genotoxicity in lymphoblastoid cell lines and peripheral blood lymphocytes of head and neck cancer patients. *Cancer Detect. Prev.*, **17**, 575–583

Vineis, P. & Perera, F. (2000) DNA adducts as markers of exposure to carcinogens and risk of cancer. *Int. J. Cancer* (in press)

Vineis, P., Malats, N., Lang, M., d'Errico, A., Caporaso, N., Cuzick, J. & Boffetta, P., eds (1999) *Metabolic Polymorphisms and Susceptibility to Cancer* (IARC Scientific Publications No. 148), Lyon, IARC

Vulimiri, S.V., Wu, X., Baer-Dubowska, W., de Andrade, M., Detry, M., Spitz, M.R. & DiGiovanni, J. (2000) Analysis of aromatic DNA adducts and 7,8-dihydroxy-8-oxo-2'-deoxyguanosine in lymphocyte DNA from a case control study of lung cancer involving minority populations. *Mol. Carcinogenesis*, **27**, 34–46

Wiseman, H., Kaur, H. & Halliwell, B. (1995) DNA damage and cancer: measurement and mechanism. *Cancer Letters*, **93**, 113–120

Yaeger, T.R., de Vries, S., Jarrard, D.F., Kao, C., Nakada, S.Y., Moon, T.D., Bruskewitz, R., Stadler, W.M., Meisner, L.F., Gilchrist, K.W., Newton, M.A., Waldman, F.M. & Reznikoff, C.A. (1998) Overcoming cellular senescence in human cancer pathogenesis. *Genes Dev.*, **12**, 163–174

Corresponding author:

P. Vineis
University of Torino and CPO-Piemonte,
Via Santena 7,
Torino,
Italy

Biomarkers in Cancer Chemoprevention
Miller, A.B., Bartsch, H., Boffetta, P., Dragsted, L. and Vainio, H. eds
IARC Scientific Publications No. 154
International Agency for Research on Cancer, Lyon, 2001

Ultraviolet radiation-induced photoproducts in human skin DNA as biomarkers of damage and its repair

K. Hemminki, G. Xu and F. Le Curieux

We have developed a ^{32}P-postlabelling method for quantifying ultraviolet irradiation (UV)-induced cyclobutane dimers and 6-4 photoproducts in human skin *in situ*. We review the application of the method in studies with human volunteers, demonstrating dose–response relationships over a wide range of administered doses, repair kinetics of UV-damaged DNA among healthy individuals and melanoma patients, and modulation by sunscreens, tan and constitutive pigmentation of damage induction. A notable finding is the wide interindividual variation in DNA damage immediately after irradiation and in its repair. Moreover, the protective effects of sunscreens against erythema and DNA damage also show wide interindividual variation. These results cannot be explained by variation in the experimental methods used. The worst-case scenario is that the differences between individuals are multiplicative, resulting in 1000-fold differences in sensitivity in the population, which would be likely to translate into differences in risk of skin cancer.

Introduction

Exposure to ultraviolet (UV) light has deleterious effects on human skin including sunburn, elastoses, cancer and wrinkling. Although only 1–2% of the intensity of the solar radiation reaching the Earth is UVB (280–320 nm), this is believed to be the main cause of the deleterious effects (Sayre, 1992). Some 60% of the sun's energy reaching the Earth's surface is visible light (400–750 nm) and 25% is infrared radiation, i.e., heat. UVA (320–400 nm), accounting for some 15% of the solar energy reaching the Earth, also has deleterious effects but to a smaller extent than UVB. Thus 500–1000 times higher doses of UVA than of UVB are required to cause skin reddening (erythema) (Augustin *et al.*, 1997) and the evidence relating to skin cancer incriminates only UVB (English *et al.*, 1997; Tomatis *et al.*, 1990). Solar energy, and to some extent the emission spectrum, depends on the solar altitude, relating to the time of the day and latitude. The energy in the UVA and UVB range is over four times higher at solar altitude 70° than at 23° (Sayre, 1992).

UVB causes specific types of DNA damage, known as photoproducts, including cyclobutane pyrimidine dimers (CPDs) and 6-4 photoproducts between two pyrimidines on the same DNA strand (Bykov & Hemminki, 1995). The quantification of specific photoproducts in human skin *in situ* poses many scientific and practical problems. A number of methods are available to measure UV-induced DNA damage, including assays for DNA strand-breaks (Gao *et al.*, 1994; Kasren *et al.*, 1995), unscheduled DNA synthesis (UDS) (Mu & Sancar, 1997), host-cell reactivation (Wei *et al.*, 1993; Runger *et al.*, 1997), immunological detection (Nakagawa *et al.*, 1998) and ^{32}P-postlabelling methods (Liuzzi *et al.*, 1989). Each method has advantages and disadvantages in terms of sensitivity, specificity and practicality. We have developed a ^{32}P-postlabelling method to determine UV-induced DNA damage in human skin *in situ*. One valuable feature of this method is that it can distinguish between the different dipyrimidine combinations of CPDs and 6-4 photoproducts in a single analysis. Another is that a small amount of DNA (less than 3 μg) is needed to analyse the photoproducts, which is especially useful in molecular epidemiological studies. Using this method, we have studied DNA photodamage and repair in general population samples (Bykov *et al.*, 1998a,b, 1999, 2000) and in people with skin disease (Xu *et al.*, 2000a). Moreover, application of this method to determination of photoproducts in human urine is possible, as described below.

Formation of UV-induced photoproducts is believed to be a crucial event initiating photocarcinogenesis in human skin. However, three evolutionary mechanisms protect against such DNA damage, namely, constitutive pigmentation, DNA repair and tanning. Humans have also invented protective strategies against UV, such as sunscreens, protective glasses, shading, clothes and hats. Thus UV exposures and the induced effects vary widely across the population. Melanoma and squamous cell carcinoma of the skin are the most rapidly increasing types of cancer in Sweden and in many other countries with fair-skinned populations, probably because of changing habits of sunbathing and exposure to UV (English *et al.*, 1997; Center for Epidemiology, 1998). Several lines of evidence suggest that constitutional factors such as skin type, level and type of pigmentation, age, tanning ability and DNA repair capacity partly explain the interindividual differences in sensitivity to photocarcinogenesis (Young *et al.*, 1996; Lock-Andersen *et al.*, 1997; Rosso *et al.*, 1998; Bykov et *al.*, 2000). An extreme example of host susceptibility is xeroderma pigmentosum, a pleiotropic disease in which deficient nucleotide excision repair results in extreme sensitivity to UV and a 1000-fold increase in the incidence of skin cancer (Kraemer *et al.*, 1994; Kraemer, 1997). Even in the general population, photoproduct induction and repair vary substantially among individuals (Freeman, 1988; Bykov *et al.*, 1998b, 1999). Elucidation of the mechanisms underlying interindividual differences in photodamage and its repair will be a key to understanding photocarcinogenesis.

UV-induced DNA damage provides a valuable example in terms of the validation and utilization of biomarkers, for several reasons: (1) the doses can be well controlled and it is ethically acceptable to expose normal humans to doses of about 200–400 J/m² that inflict a minimal erythemal response (MED) in a fair-skinned population; (2) UV radiation induces specific types of DNA damage leading to specific tandem CC to TT mutations, also found in defined genes from skin tumours (Harris, 1996); (3) UV radiation is an established cause of human cancer; (4) target tissue is available for experimental studies; (5) the human target tissue is amenable to chemopreventive trials; and (6) an individual DNA repair test can be based on the measurement of removal rates of UV-induced DNA damage. In this chapter we review examples of many of these applications.

Methods

³²P-Postlabelling of skin samples

Most of the study subjects were healthy volunteers. However, in two studies cutaneous malignant melanoma patients and healthy controls were used. No significant differences in DNA damage and repair between the two groups were found, thus justifying the pooling of the study populations (Xu *et al.*, 2000a, b). The UV sources used and the doses administered have been described in the cited papers. When solar-simulating radiation was used, the spectral curves mimicked closely the spectrum of solar radiation at the Earth's surface (summer, noontime, Helsinki latitude). Biopsies were taken from buttock skin (usually 4 mm diameter) and were immediately put into ice, frozen and stored at –20 °C to await DNA isolation. DNA extraction from epidermis was performed using a chloroform–isoamyl alcohol method after separation of epidermis from dermis with a blunt scalpel.

The ³²P-postlabelling method is based on enzymatic digestion of DNA to nucleoside-3′-phosphates and 5′-labelling of adducts with ³²P of high specific activity from [γ-³²P]ATP using T4 polynucleotide kinase (Bykov & Hemminki, 1995). One distinct difference in the method for analysing photoproducts compared with other adduct measurements using ³²P-postlabelling is that the photoproducts were assayed as trinucleotides with an unmodified nucleotide on the 5′-side. The nucleotide at the 5′-side of each photoproduct can be any one of the four types of nucleotide. The labelled products were detected with a Beckman ³²P radioisotope detector and identified by coelution with external standards (Bykov *et al.*, 1999). The level of photoproducts was expressed per 10^6 nucleotides.

In what follows, T=C and T=T designate CPDs, and T–C and T–T designate 6-4 photoproducts.

^{32}P-Postlabelling of urine samples

Approach: Our search for photoproducts is based on the assumption that cyclobutane thymidine dimers (T=T) released by DNA repair are not degraded further and are excreted in urine as a dimer. This was tested by incubating cyclobutane thymidine dimer with S_9 mix enzymes (homogenate from rat liver) at 37°C. After 24 hours, the amounts of unmodified cyclobutane thymidine dimers in the reaction mixture were unchanged, suggesting that the hypothesis was correct, but validation *in vivo* is required.

The cyclobutane thymidine dimer is not a good substrate for T4 polynucleotide kinase and, consequently, cannot be labelled directly. As the parent dinucleotide TpT is easily labelled, we chose to convert the T=T dimer to TpT by UVC irradiation at 254 nm, using a Stratalinker UV Crosslinker 2400 with lamps providing almost monochromatic 254 nm light. This lamp was used both to prepare the reference T=T dimer, and to convert it back to the parent compound TpT. The reversion of T=T to TpT was not quantitative and was quantified in each experiment. Typical yields were 50%.

T=T purification: Urine samples were filtered through 0.22-μm filters and 1 μl was injected onto an HPLC system with UV detection. Preliminary analysis showed that the retention time for the T=T dimer was about 16 minutes, so the fraction eluting between 15 and 17 minutes was collected, freeze-dried and redissolved in 40 μl distilled water. The mixture was then subjected to UVC irradiation (10 kJ/m^2) for conversion of the T=T dimer to the parent dinucleotide (TpT). TpT was labelled on the 5′-side using a protocol described previously (Bykov *et al.*, 1995). The labelled samples were then analysed using an HPLC system with radioisotope detection.

HPLC analysis of TpT: UV-HPLC analyses were performed on a Beckman instrument (model 126 pump) operated with System Gold and coupled to a model 168 diode-array detector (Beckman Instruments, San Ramon, CA, USA). The urine samples were chromatographed on a 5 mm, 4.6 × 250-mm reversed-phase C18 Luna column from Phenomenex (Genetec, Kungsbacka, Sweden). A precolumn filter was positioned before the column. The column was eluted isocratically for 5 minutes with 50 mM ammonium formate buffer (pH 4.6) and then with a gradient from 0 to 30% methanol over 45 minutes at a flow rate of 0.7 ml/min. The labelled samples were analysed with the same Beckman instrument as in the human skin work. However, a different gradient was used: isocratic elution for 3 minutes with buffer (500 mM ammonium formate, 20 mM orthophosphoric acid, pH 4.6), and then a gradient from 0 to 20% methanol over 30 minutes. The identification was based on coelution with standard labelled thymidine dimer.

Results and discussion

Analysis of UV photoproducts required a modification of the postlabelling technique because it was found that cross-linked dinucleotides labelled very poorly (Bykov *et al.*, 1995). In this modification, a normal nucleotide was left on the 5′-side of the cross-linked dinucleotide, resulting in a number of labelled trinucleotides. This is at present the only way to label cross-linked products. Radioactivity was analysed by HPLC with assignment of radioactive products based on the standards used. This assay has been used to study dose–response relationships in humans. In skin biopsies from UV-irradiated skin, there was a linear relationship between dose from 50 to 400 J/m^2 and adduct levels (Bykov *et al.*, 1998a). The relationship was also linear between 150 and 2000 J/m^2 when a sunscreen was used to protect skin, reducing the level of adducts (Bykov *et al.*, 1998b). The amount of photoproducts induced by an MED dose of solar-simulating UV radiation was about 1000 TT=T dimers per 10^8 normal nucleotides (Table 1). This amount is only a quarter of all T=T dimers because equal amounts of dimer would be expected to be formed at all dithymidine sites (TTT, ATT, CTT and GTT), with approximately equal levels at all TC sites. This is a remarkably high level of DNA damage compared with any other known human carcinogen, as shown in Table 1. For example, lung DNA from smokers contains methylation products and benzo[*a*]pyrene types of adduct at levels of only 100 and 10 adducts per 10^8 normal nucleotides, respectively. This high level of UV damage in human DNA is probably the basis for UV-induced carcinogenesis and for the extreme sensitivity to

Table 1. Levels of UV-induced photoproducts in human skin DNA as compared to those of DNA adducts from other exogenous carcinogens that cause cancer in humans

Adduct	Tissue	Level (per 10^8 bases)	Comment
7-Methyl-G	Lung	100	Smoker
7-Hydroxyethyl-G	Leukocyte	100	Smoker
Benzo[*a*]pyrene-G	Lung	10	Smoker
Benzidine-G	Urothelium	3	Dye workers
PhiP-G	Colon	3	Roasted meat
Tamoxifen-DNA	Endometrium	0.3	Dose c. 40 mg/d
Cyclobutane T=T	Skin	1000	400 J/m^2 UV radiation

Modified from Bykov *et al.* (1998a).
G = guanine, T =thymine.

solar UV-induced skin cancer in xeroderma pigmentosum.

Induction of photoproducts by UV radiation

Irradiation of previously unexposed skin (from buttock) with a defined dose of UV radiation (Xu *et al.*, 2000b) generated high levels of photoproduct detected immediately after irradiation. CPDs and 6-4 photoproducts were formed with different efficiency in human skin, the 6-4 photoproduct levels being about one eighth of those of CPDs. In addition, the levels of photoproducts measured immediately after irradiation showed wide interindividual differences. The levels of each type of photoproduct were correlated with each other and each type showed interindividualvariation approximately proportionate to the absolute level. Figure 1(*a*) shows a 15-fold difference in formation of TT=C (one of the analysed CPDs) between individuals. Up to 30-fold interindividual differences have also been found in previous studies by our group (Bykov *et al.*, 1998b, 1999). The subjects in the present study were melanoma patients and matched healthy controls, but there was no difference between these groups (Xu *et al.*, 2000a). The level of interindividual variation is so large that DNA damage cannot be considered as a simple marker of dose.

Factors modulating UV-induced photoproduct formation

UV-induced DNA damage in skin includes formation of CPDs (TT=C and TT=T) and 6-4 photoproducts (TT–T and TT–C). This is not a random process. A number of factors influence the formation of photoproducts, such as UV wavelength, DNA sequence context and chromosomal proteins (Black *et al.*, 1997; Pfeifer, 1997). We have studied the effects of host factors (e.g., age, skin type, gender) on the formation of photoproducts (Xu *et al.*, 2000b). In older subjects (≥50 years, mean 62.5±9.1 years), the amount of each of the four types of photoproduct immediately after UV irradiation was higher than that in the younger age group (<50 years, mean 42.3±6.6 years) (Table 2). The difference in the level of TT=C reached statistical significance at $p<0.05$. As to skin type, the CPD levels (TT=C and TT=T) were both notably higher in subjects with skin types I/II than in those with skin types III/IV (TT=C was significantly higher, $p<0.05$). However, no clear effect of skin type was found for 6-4 photoproducts (TT–T and TT–C). There was no significant difference between males and females (Table 2).

Multivariate regression analysis showed that age had a systematic effect on the induction of all photoproducts ($0.07 > p \geq 0.05$, data not shown). One year of increased age appeared to cause an increase

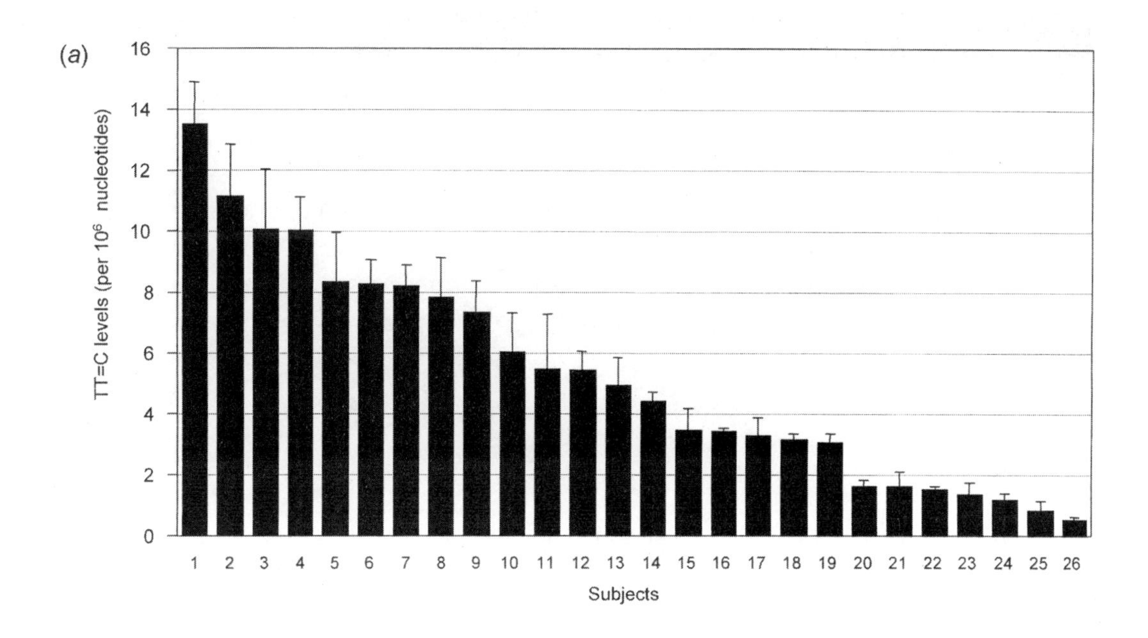

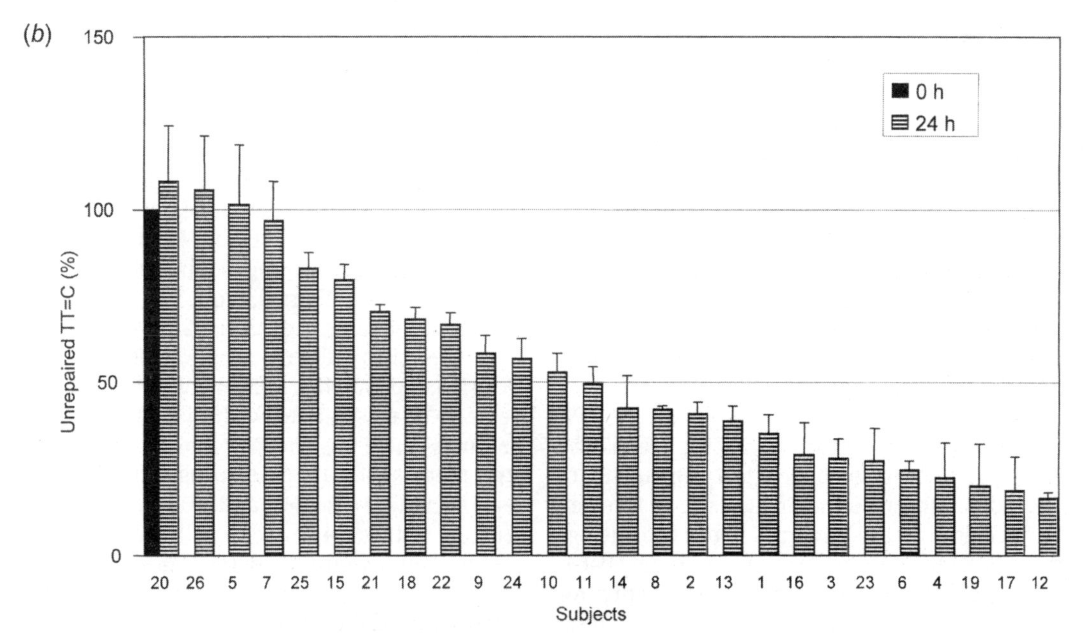

Figure 1. (a) TT=C levels at 0 h after UV irradiation. Bars indicate mean ± SD.
(b) Individual kinetics of TT=C repair in DNA after UV irradiation. The bars show the amount of photoproduct (± SD) remaining in skin DNA 24 h after irradiation. Each point represents the percentage of unrepaired pyrimidine dimers at corresponding time after irradiation.

Table. 2. Photoproduct levels[a] (per 10^6 nucleotides) at 0 hour after UV irradiation

		TT=C	TT=T	TT–T	TT–C
Age (years)	<50	4.22±2.94(15) b	5.94±3.58 (16)	0.49±0.33 (16)	0.63±0.46 (16)
	≥50	6.56±3.74* (13)	7.08±3.97 (13)	0.56±0.38 (13)	0.66±0.42 (13)
Skin type	I and II	6.11±3.13* (15)	7.26±4.02 (16)	0.51±0.32 (16)	0.60±0.47 (16)
	III and IV	4.38±3.76 (13)	5.46±3.24 (13)	0.54±0.40 (13)	0.69±0.40 (13)
Gender	Female	5.01±3.50 (7)	6.43±3.68 (7)	0.58±0.28 (7)	0.51±0.36 (7)
	Male	5.40±3.56 (21)	6.46±3.84 (22)	0.51±0.37 (22)	0.68±0.45 (22)

* Student's *t* test, $p<0.05$.
[a] Expressed as mean ± SD
[b] Number of subjects

of about 0.11 TT=C, 0.07 TT=T, 0.01 TT–T and 0.004 TT–C per 10^6 nucleotides. Skin types III and IV, analysed combined, protected against the induction of CPDs (TT=T and TT=C) and of the 6-4 photoproduct TT–T in skin compared with skin types I/II, analysed combined, but the effect was not statistically significant. Thus age was the main host factor influencing the induction of photoproducts in the present study. A plausible explanation may be the changes in skin structure with ageing. In normal skin, ageing alone causes marked changes in morphology, histology and physiology (Gilchrest, 1991). Among these age-associated changes, a decrease of some 10–20% in the number of enzymatically active melanocytes per unit surface area of the skin each decade can result in the reduction of the body's protective barrier against UV radiation (Gilchrest *et al.*, 1979). Thus, considering photoproducts as biomarkers for the risk of skin cancer, the progressively greater UV-induced DNA damage with advancing age may be associated with the steep age effect in incidence of squamous cell carcinoma of the skin. In Sweden, the incidence of this cancer is highest in the age groups over 85 years, exceeding that in the 60–64-year age group by more than 10 times (Center for Epidemiology, 1998).

Repair of photoproducts

UV-induced DNA damage and many bulky DNA lesions are repaired by means of the nucleotide excision repair (NER) enzyme system, which involves about 30 different gene products (Lehmann, 1995). NER involves two sub-pathways, global genomic repair and transcription-coupled repair. Repair of photoproducts by NER also shows heterogeneity: photoproducts are not repaired with the same efficiency within all regions of the genome (Black *et al.*, 1997; Pfeifer, 1997). In the present study, the global genomic repair of photoproducts displayed substantial interindividual differences and sequence-dependence (Xu *et al.*, 2000b). Taking the CPD TT=C as an example, the percentage of repaired photoproducts at 24 h after irradiation varied from zero to about 82% among the study population. Figure 1(b) shows the amount of TT=C remaining in skin DNA of 26 individuals 24 h after UV irradiation. The subjects were melanoma patients and matched healthy controls, but there was no difference between these groups (Xu *et al.*, 2000a). It is clear that the rate of repair bears no relationship to the initial level of photoproducts. There was evidence of photoproduct-specific repair, with TT=C showing a faster rate of repair than TT=T: 46% and 80% of TT=C was repaired within 24 h and 48 h after irradiation, respectively, compared with 25% and 54% for TT=T.

We assessed the effects of some host factors that may influence the rate of repair of photoproducts (data not shown). Age (range 32–78 years) had no consistent significant effect on the repair rate of TT=C or TT=T up to 48 h after irradiation (Xu *et al.*, 2000b). No effect of skin type or gender on DNA repair was found. In contrast, one early study showed an age-related decline in DNA repair capacity in human lymphocytes *in vitro* (Wei *et al.*, 1993). The discrepancy between the results could be due to factors such as methodological differences, repair capacities *in situ* vs *in vitro*, or differences between cell types. However, for repair of UV-induced DNA damage in human skin, only the present results are directly relevant. In all these studies, we measured the rate of removal of photoproducts, rather than direct repair. However, these photoproducts are chemically stable in DNA and require DNA repair for their removal. Cell replication would cause an increase in amounts of DNA and thus dilution of the photoproducts. We cannot rule out some minor effect of DNA replication, despite the short duration of the studies (up to two days).

Effect of sunscreen and tan

We have applied the postlabelling method to assess the effects of sunscreens and tanning on UV-induced DNA damage in human skin *in situ* (Bykov *et al.*, 1998b). Sunscreen was applied by one person, according to the manufacturer's instructions. While the protection against the erythemal response varied fivefold among nine subjects, protection against DNA damage differed by a factor of 10 and was independent of the erythemal response. On average, sunscreens protected against DNA damage in accordance with the sun-protection factor (SPF), but the degree of protection was highly individual. Figure 2 shows the individual SPF (ISPF) and the protection factor against DNA damage (PF/DNA damage); thus, for individual L, sunscreen provided an ISPF of 20 but the protection against DNA damage was only about 5. For individual C, the reverse was the case. Since the SPF is based on average erythemal response, it is no guarantee against individual DNA damage. It is likely that the main reason for the apparent interindividual difference is simply uneven spreading of the sunscreen. Figure 2 also shows that even

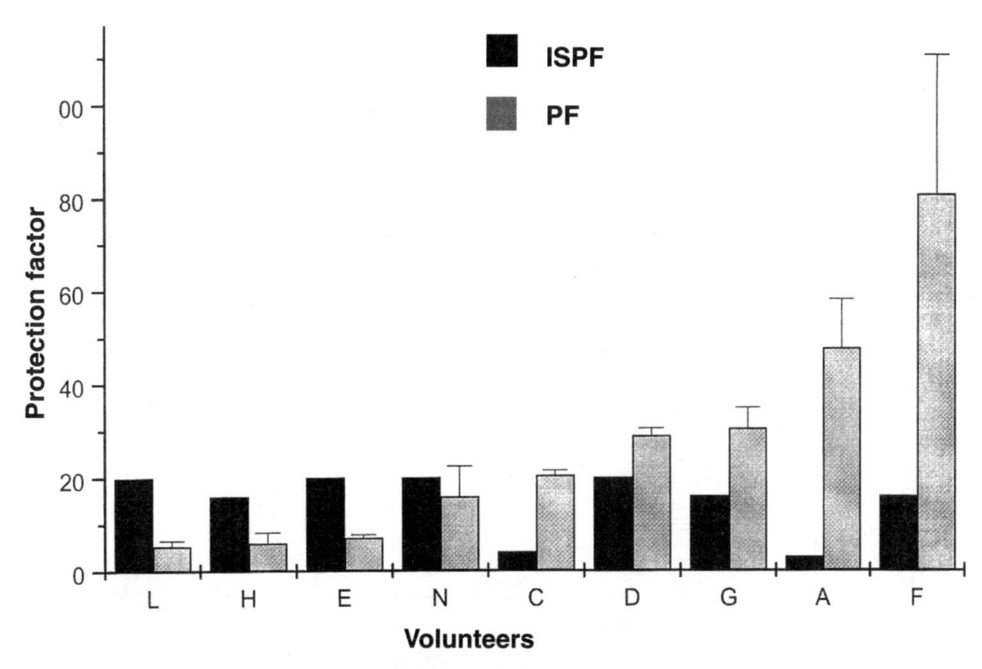

Figure 2. Individual sunscreen protection factor (ISPF), based on erythemal response, and protection factor against DNA damage (PF/DNA damage) measured in nine subjects who received 150 J/m^2 of UVB. Data from Bykov *et al.* (1998b).

The SPF of the sunscreen was 10.

protection against the erythemal response (ISPF) is highly individual.

Tanning provides an endogenous 'sunscreen' with an SPF of 3 to 5 (Gilchrest *et al.*, 1999). These measurements are based on protection against erythema, and fundamental data on how well tanning protects genetic material appear to be lacking. We have investigated the effects of tanning on DNA damage *in situ*, under conditions simulating the use of sunbeds (Hemminki *et al.*, 1999; Bykov *et al.*, 2000). We measured the protective effects of tanning by quantifying the levels of UV-induced photoproducts in skin of eight healthy, fair-skinned Caucasians. Each subject was irradiated on a 2 x 4-cm area of skin on the lower back with UVB radiation at a dose of 0.3 J/cm² and on a second 2 x 4-cm area at a dose of 0.6 J/cm². A punch biopsy was performed immediately after irradiation. An additional biopsy was performed on unirradiated skin and served as a background control. Tanning was induced by 10–13 sessions of UVA irradiation for three weeks. Tanning was observed by a clear change in skin colour towards brown and was measured with a reflectometer adjusted to record melanin pigmentation. In the course of UVA treatment, the instrumental readings indicated an increase in pigmentation of 38.8 ± 16.7 reflectometer units (mean ± SD, $n = 8$). After the last UVA dose, the challenge with UVB was repeated, and three biopsy specimens were taken, as described above, except that the control biopsy was from tanned skin. The samples were coded for blind analysis. Photoprotection was defined as the difference in photoproduct levels before and after the UVA treatment.

In subjects who received 0.6 J/cm² of UVB, the levels of CPDs were slightly lower than those in untanned skin (Figure 3). The average tanning protection factor was only 1.19 ± 0.17 (mean ± SD). Since tanning acts like a low-level sunscreen to suppress the erythemal response without the unpredictable and limited protection against DNA damage afforded by a chemical sunscreen (Bykov *et al.*, 1998b), people who have acquired a tan may prolong sun exposure, resulting in DNA damage and increased risk of skin cancer. Tanning may

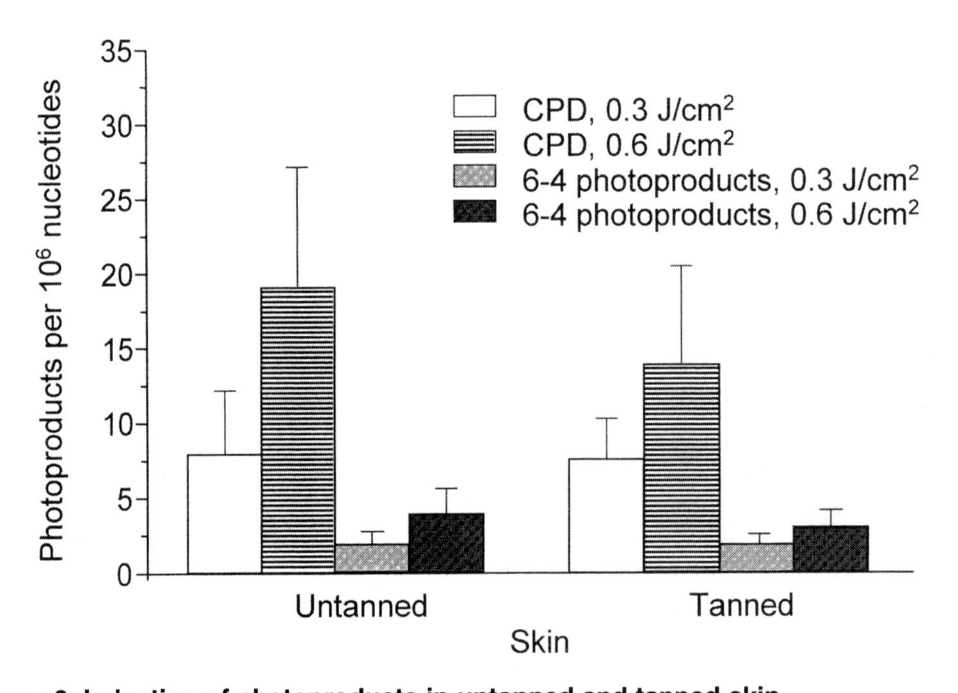

Figure 3. Induction of photoproducts in untanned and tanned skin

Levels of UV-induced DNA damage, cyclobutane pyrimidine dimers (CPD) and 6-4 photoproducts are expressed per 10^6 nucleotides (mean ± SD, $n = 8$ subjects). From Hemminki *et al.* (1999).

provide a false sense of security, leading to inadvertently lengthened recreational sun exposure like that of the high-SPF sunscreen users studied by Autier *et al.* (1999). Natural pigmentation, even in the fair-skinned Scandinavians, provides better protection against UV-induced DNA damage than UVA-induced tanning (Bykov *et al.*, 2000). Sun-induced tanning may also provide better protection against DNA damage than that induced by UVA. However, solar-simulated tan in skin types II and III afforded a protection factor of only 2 against erythemal response (Sheehan *et al.*, 1998).

Urinary photoproducts

The amount of CPDs that can theoretically be expected in urine was estimated using the following assumptions: (1) a dose of 400 J/m^2 of solar-simulating radiation (1–2 MED, about 15–30 min sun in Stockholm area in summer) induces the formation of 6 TT=T per10^6 nucleotides in skin DNA, giving a total amount of dimer (AT=T, GT=T, CT=T and TT=T) of 24 T=T per 10^6 nucleotides; (2) the amount of DNA extracted from a skin biopsy is 10 µg, correcting for yield; thus 1 m^2 of skin will contain 800 mg or 60 nmol T=T; (3) 25% of T=T is removed from skin DNA in the first 24 hours after UV irradiation (Xu *et al.*, 2000b); (4) all T=T

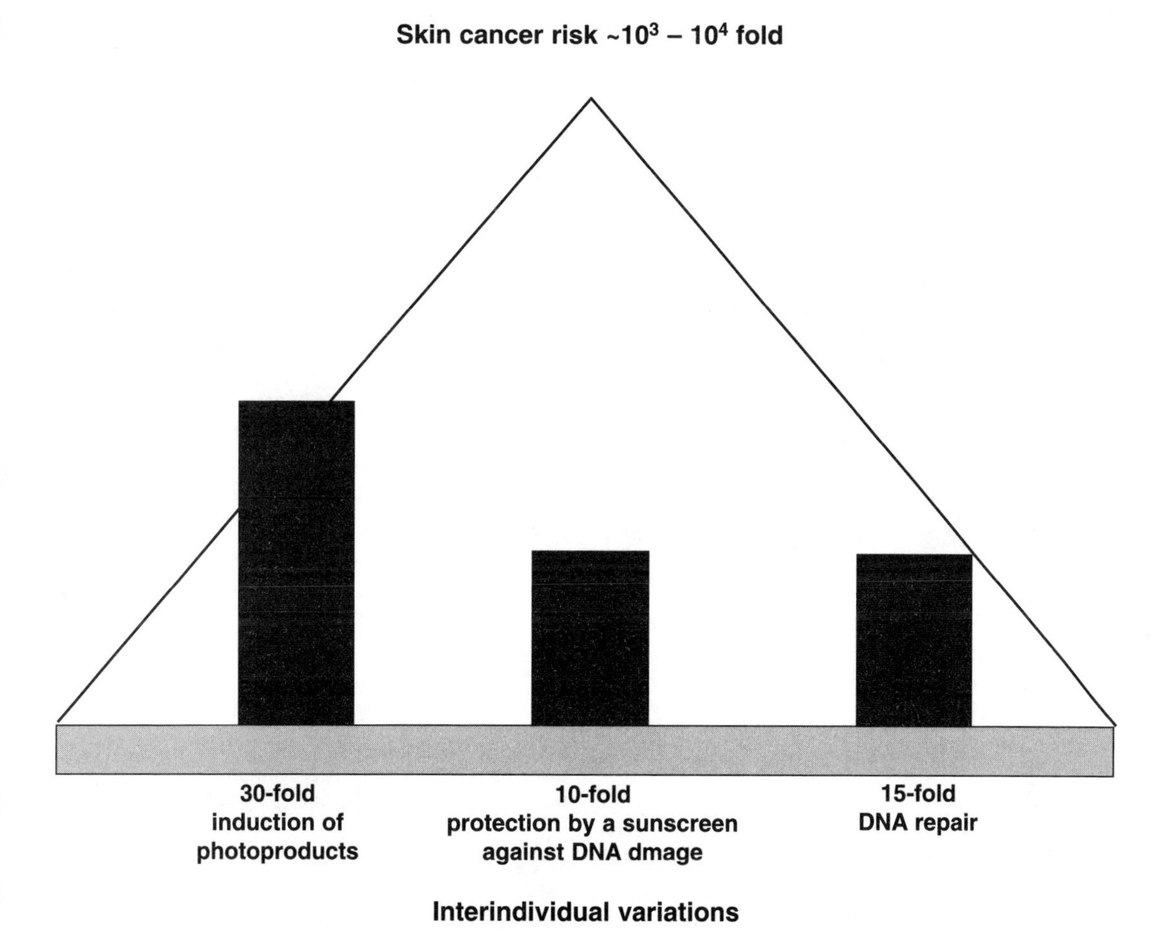

Figure 4. A scheme of the interindividual variations observed in the studies cited.

removed from skin DNA ends up in urine (1 litre per 24 h). This calculation indicated that the level of T=T to be expected in urine is 15 fmol/μl.

In chromatographic analyses of urine samples containing CPDs, the ^{32}P-labelled TpT showed a retention time of about 27 minutes. The first analyses of human urine collected after sun exposure showed T=T levels of 2 to 20 fmol per μl of urine. These amounts are consistent with the theoretical levels calculated above. If validated, this ^{32}P-postlabelling quantification of thymidine dimers in urine would provide a non-invasive method to assess the levels of photoproducts after whole-body exposure to UV light, that would offer an easy alternative to skin biopsies.

Conclusions

A surprising finding in these studies has been the large interindividual variation in the immediate DNA damage after exposure to UV and in its repair. Moreover, the protective effects of sunscreens against erythema and DNA damage also show wide interindividual variation. These results cannot be explained by variations in the experimental methods used. While the interindividual variation in the levels of immediate DNA damage can be 30-fold, our coefficient of variation among repeated analyses of the same samples was only some 30% (Bykov *et al.*, 2000). An interesting issue is the stability of individual response to UV radiation. So far we have limited data on this: there was a high correlation in the induced photoproduct levels when the same individuals were tested three weeks apart (unpublished data). Figure 4 illustrates the ranges of interindividual variations found in these relatively small studies. The worst-case scenario would be that the various differences are multiplicative, resulting in 1000-fold differences in sensitivity between individuals. These would be likely to translate into differences in risk of skin cancer.

Acknowledgements

We thank Dr V. J. Bykov for Figures 2 and 4. This work was supported by the Cancer Fund and Swedish Radiation Protection Institute.

References

Augustin, C., Collombel, C. & Damour, O. (1997) Measurements of the protective effect of topically applied sunscreens using in vitro three-dimensional dermal and skin equivalents. *Photochem. Photobiol.*, **66**, 853–859

Autier, P., Doré, J.-F., Negrier, S., Lienard, D., Panizzon, R., Lejeune, F.J., Guggisberg, D. & Eggermont, A.M. (1999) Sunscreen use and duration of sun exposure: a double-blind, randomized trial. *J. Natl Cancer Inst.*, **91**, 1304–1309

Black, H.S., deGruijl, F.R., Forbes, P.D., Cleaver, J.E., Ananthaswamy, H.N., deFabo, E.C., Ullrich, S.E. & Tyrrell, R.M. (1997) Photocarcinogenesis: an overview. *J. Photochem. Photobiol. B*, **40**, 29–47

Bykov, V.J. & Hemminki K. (1995) UV-induced photoproducts in human skin explants analysed by TLC and HPLC-radioactivity detection. *Carcinogenesis*, **16**, 3015–3019

Bykov, V.J., Janssen, C. & Hemminki, K. (1998a) High levels of dipyrimidine dimers are induced in human skin by solar-simulating UV radiation. *Cancer Epidemiol. Biomarkers Prev.*, **7**, 199–202

Bykov, V.J., Marcusson, J.A. & Hemminki, K. (1998b) Ultraviolet B-induced DNA damage in human skin and its modulation by a sunscreen. *Cancer Res.*, **58**, 2961–2964

Bykov, V.J., Sheehan, J.M., Hemminki, K. & Young, A.R. (1999) In situ repair of cyclobutane pyrimidine dimers and 6-4 photoproducts in human skin exposed to solar simulating radiation. *J. Invest. Dermatol.*, **112**, 326–331

Bykov, V.J., Marcusson, J.A. & Hemminki, K. (2000) Effect of constitutional pigmentation on ultraviolet B-induced DNA damage in fair-skinned people. *J. Invest. Dermatol.*, **114**, 40–43

Center for Epidemiology (1998) *Cancer Incidence in Sweden 1996*, pp. 1–114, Stockholm

English, D., Armstrong, B., Kricker, A. & Fleming, C. (1997) Sunlight and cancer. *Cancer Causes Control*, **8**, 271–283

Freeman, S.E. (1988) Variation in excision repair of UVB-inducd pyrimidine dimers in DNA of human skin in situ. *J. Invest. Dermatol.*, **90**, 814–817

Gao S., Drouin R. & Holmquist G.P. (1994) DNA repair rates mapped along the human PGK-1 gene at nucleotide resolution. *Science*, **263**, 1438–1440

Gilchrest, B. (1991) Physiology and pathophysiology of aging skin. In: Goldsmith, L.A., ed., *Physiology, Biochemistry and Molecular Biology of Skin*, 2nd ed., Oxford, Oxford University Press

Gilchrest, B.A., Blog, F.B. & Szabo G. (1979) Effects of ageing and chronic sun exposure on melanocytes in human skin. *J. Invest. Dermatol.*, **73**, 141–143

Gilchrest, B.A., Eller, M.S., Geller, A.C. & Yaar, M. (1999) The pathogenesis of melanoma induced by ultraviolet radiation. *New Engl. J. Med.,* **340**, 1341–1348

Harris, C.C. (1996) P53 tumor suppressor gene: at the crossroads of molecular carcinogenesis, molecular epidemiology, and cancer risk assessment. *Environ. Health Perspect.,* **104** (Suppl. 3), 435–439

Hemminki, K., Bykov, V.J. & Marcusson, J.A. (1999) Re: Sunscreen use and duration of sun exposure: a double-blind, randomized trial. *J. Natl Cancer Inst.,* **91**, 2046–2047

Kasren, U., Beyersmann, D, Dahm-Daphi, J. & Hartwig, A. (1995) Sensitive nonradioactive detection of UV-induced cyclobutane pyrimidine dimers in intact mammalian cells, *Mutat. Res.,* **336**, 143–152

Kraemer, K.H. (1997) Sunlight and skin cancer: another link revealed. *Proc. Natl Acad. Sci. USA,* **94**, 11–14

Kraemer, K.H., Lee, M.M, Andrews, A.D. & Lambert, W.C. (1994) The role of sunlight and DNA repair in melanoma and nonmelanoma skin cancer. The xeroderma pigmentosum paradigm. *Arch. Dermatol.,* **130**, 1018–1021

Lehmann, A.R. (1995) Nucleotide excision repair and the link with transcription. *Trends Biochem. Sci.,* **20**, 402–405

Liuzzi, M., Weifeld, M. & Paterson, M.C. (1989) Enzymatic analysis of isomeric trithymidylates containing ultraviolet light-induced cyclobutane pyrimidine dimers. *J. Biol. Chem.,* **264**, 6355–6363

Lock-Andersen, J., Therkildsen, P., de Fine Olivarius, F., Gniadecka, M., Dahlstrom, K., Poulsen, T. & Wulf, H.C. (1997) Epidermal thickness, skin pigmentation and constitutive photosensitivity. *Photodermatol. Photoimmunol. Photomed.,* **13**, 153–158

Mu, D. & Sancar, A. (1997) DNA excision repair assays. *Prog. Nucl. Acid Res. Mol. Biol.,* **56**, 63–81

Nakagawa, A., Kobayashi, N., Muramatsu, T., Yamashina, Y., Shirai, T., Hashimoto, M.W., Ikenaga, M. & Mori, T. (1998) Three-dimensional visualization of ultraviolet-induced DNA damage and its repair in human cell nuclei. *J. Invest. Dermatol.,* **110**, 143–148

Pfeifer, G.P. (1997) Formation and processing of UV photoproducts: effects of DNA sequence and chromatin environment. *Photochem. Photobiol.,* **65**, 270–283

Rosso, S., Zanetti, R., Mippione, M. & Sancho Garnicr, H. (1998) Parallel risk assessment of melanoma and basal cell carcinoma: skin characteristics and sun exposure. *Melanoma Res.,* **8**, 573–583

Runger, T.M., Epe, B., Moller, K., Dekant, B. & Hellfritsch, D. (1997) Repair of directly and indirectly UV-induced DNA lesions and of DNA double-strand breaks in cells from skin cancer-prone patients with the disorders dysplastic nevus syndrome or basal cell nevus syndrome. *Recent Results Cancer Res.,* **143**, 337–351

Sayre, R. (1992). Sunlight risk and how sunscreens work. *Cosmet. Toiletries,* **107**, 105–109

Sheehan, J.M., Potten, C.S. & Young, A.R. (1998) Tanning in human skin types II and III offers modest photoprotection against erythema. *Photochem. Photobiol.,* **68**, 588–592

Tomatis, L., Aitio, A., Day, N.E., Heseltine, E., Kaldor, J., Miller, A.B., Parkin, D.M. & Riboli, E. (1990) *Cancer: Causes, Occurrence and Control* (IARC Scientific Publications No. 100), Lyon, International Agency for Research on Cancer

Wei, Q., Matanoski, G.M., Farmer, E.R.,Hedayati, M.A. & Grossman, L. (1993) DNA repair and aging in basal cell carcinoma: a molecular epidemiology study. *Proc. Natl Acad. Sci. USA,* **90**, 1614–1618

Xu, G., Snellman, E., Bykov, V., Jansen, C. & Hemminki, K. (2000a) Cutaneous malignant melanoma patients have normal repair kinetics of UV-induced DNA damage in skin in situ. *J. Invest. Dermatol.,* **114**, 628–631

Xu, G., Snellman E., Bykov V., Jansen C. & Hemminki K. (2000b) Effect of age on the formation and repair of UV photoproducts in human skin in situ. *Mutat Res.,* **459**, 195–202

Young, A.R., Chadwick, C.A, Harrison, G.I, Hawk, J.L.M, Nikaido, O. & Potten C.S. (1996) The in situ repair kinetics of epidermal thymine dimers and 6-4 photoproducts in human skin types I and II. *J Invest Dermatol,* **106**, 1307–1313

Corresponding author:

K. Hemminki
Department of Biosciences at Novum,
Karolinska Institute,
141 57 Huddinge,
Sweden

Biomarkers in Cancer Chemoprevention
Miller, A.B., Bartsch, H., Boffetta, P., Dragsted, L. and Vainio, H. eds
IARC Scientific Publications No. 154
International Agency for Research on Cancer, Lyon, 2001

Intermediate-effect biomarkers in prevention of skin cancer

J.-F. Doré, R. Pedeux, M. Boniol, M.-C. Chignol and P. Autier

Skin cancers, both non-melanoma and melanoma, usually progress through sequential steps towards malignant transformation, leading to mutant clones and precancerous lesions. Prevention of skin cancers relies on reduction of exposure to solar radiation and may be evaluated by measuring induction of intermediate-effect biomarkers such as sunburn cells or *p53* mutations in the epidermis, actinic (solar) keratoses, UV-induced immunosuppression or naevi.

Sunburn cells (apoptotic keratinocytes) and *p53* mutations are indicators of UV-induced DNA lesions as early steps of malignant transformation of epidermal keratinocytes. Actinic keratoses are premalignant sun-induced skin lesions, characterized as keratinized patches with aberrant cell differentiation and proliferation; they represent risk factors for basal-cell carcinoma and melanoma and are precursors of squamous-cell carcinoma. Studies in humans have investigated UV-induced immunosuppression and its modulation by topical sunscreen application, focusing on contact hypersensitivity as measured by immunization or response to haptens, or on modulation of stimulation of allogeneic lymphocytes by epidermal cells, or local release of immunomodulatory molecules such as *cis*-urocanic acid or interleukin-10.

Naevi are focal collections of melanocytes, usually found at the junction of the epidermis and dermis or at various depths in the dermis. Common acquired naevi arise after birth both spontaneously and in response to sun exposure. Most acquired naevi are clonal, while most melanocytes in non-naeval areas are not. Although it is not yet certain whether naevi represent premalignant lesions or risk factors, many melanomas arise in acquired naevi, and the number of naevi constitutes the best predictor of individual risk of melanoma. The presence of large (i.e., >5 mm) or atypical naevi (i.e., large naevi with non-uniform colour and irregular borders) is associated with elevated melanoma risk, independently of the number of smaller naevi. Children seem particularly vulnerable to sun-induced biological events involved in the genesis of melanoma, and the greatest increase in naevus numbers per unit of skin surface occurs before adolescence. Therefore, the distribution of naevi and their development in children are relevant to understanding melanoma occurrence in adults.

Introduction

Skin cancers, both non-melanoma and melanoma, usually progress through sequential steps of malignant transformation, leading from an initiated cell to a mutant clone and to a precancerous lesion. In the absence of exogenous influences, mutant clones and precancers may remain stable or tend to regress. Only on rare occasions do these precancerous lesions finally transform into an invasive cancer (Brash & Ponten, 1998) (Table 1). Epidemiological studies, and, more recently, studies of *p53* mutations in squamous-cell carcinoma (SCC) (Ziegler *et al.*, 1994) have suggested that the main etiological agent of skin cancer in humans is exposure to solar radiation, especially early in life (Autier & Doré, 1998). Clonal expansion of an initial mutated cell may be driven by sunlight exposure either by a direct effect of UV radiation on initiated target cells or through an indirect immunosuppressive effect. Immunosuppressive drugs may exert the same effect as UV radiation in facilitating the clonal expansion of precancers. Sunlight can act twice: as tumour initiator and as tumour promoter. Predisposition to sunlight-induced precancer is a multigenic trait that involves independent factors such as skin (photo)type, eye and hair colour, individual DNA repair capacity and naevus number. However, this

Table 1. Steps in the progression of skin cancers in humans

Tumour progression steps	Squamous cell carcinoma (SCC)	Malignant melanoma
Reversible (or abortive) steps	Normal keratinocyte	Normal melanocyte
	Sun-damaged epidermis (scattered keratinocytes with p53 mutations)	Sun-damaged melanocytes
	Clonal proliferation of keratinocytes with mutant p53	Naevus (clonal melanocytic proliferation)
	Actinic keratosis	Atypical (dysplastic) naevus
Irreversible steps	Carcinoma *in situ*	Melanoma *in situ*
	Squamous cell carcinoma	Early primary melanoma (radial growth phase)
		Late primary melanoma (vertical growth phase)
	Metastasis	Metastasis

latter trait results from both individual susceptibility to UV and exposure to sunlight in childhood (Autier, 1997).

Prevention of skin cancers will largely result from reduction in exposure to solar radiation and may be evaluated by measuring induction of intermediate-effect biomarkers such as induction of sunburn cells or *p53* mutations in the epidermis, actinic (solar) keratoses, UV-induced immunosuppression or naevi (Tables 2 and 3).

Non-melanoma skin cancer

SCC of the skin results from chronic exposure to sunlight and progresses by stages (Table 1): sun-damaged epidermis with individually disordered keratinocytes; clones of keratinocytes harbouring a mutant *p53* tumour-suppressor gene within a normal epidermis; actinic (solar) keratoses, i.e., keratinized patches with aberrant cell differentiation and proliferation (these dysplastic lesions may still spontaneously regress), carcinoma *in situ*; SCC and metastasis. The set of *p53* mutations in tumours is more restricted than in precancers, suggesting the existence of additional events and selection in the progression from precancer to invasive cancer (Ziegler *et al.*, 1994; Brash & Ponten, 1998).

Basal-cell carcinoma (BCC) of the skin seems to arise without a precancer, probably from stem cells in the bulge region of the hair follicle, and contains mutations of the *Ptch* gene and, less frequently, of *p53*. Its relationship to sunlight exposure is less clear but it may, like malignant melanoma, be associated with intermittent recreational exposure rather than chronic exposure.

Sunburn cells and p53 *mutations*

Sunburn cells are identified in conventionally stained epidermal biopsies as keratinocytes with a dense, pyknotic nucleus and a homogeneously eosinophilic cytoplasm. They are keratinocytes that have sustained a lethal dose of UV radiation (apoptotic keratinocytes). It has long been observed that sunburn cells may be produced, in the absence of erythema, by doses of UV radiation below the minimum erythemal dose (Grove & Kaidbey, 1980). Hence, formation of sunburn cells in the epidermis provides a quantifiable end-point of acute damage by UV radiation.

A study conducted in healthy volunteers of skin phototypes I to III, exposed to a dose of UV radiation equivalent to the individual sun protection factor (SPF) of an SPF-15-labelled sunscreen, over areas of the middle of the back protected by

Table 2. Modulation by sunscreens of intermediate-effect biomarkers of non-melanoma skin cancer in humans

Biomarker	Comments	Sunscreen effects	Reference
Sunburn cells	Apoptotic keratinocytes identified in conventionally stained epidermal biopsies	Prevention of occurrence of sunburn cells with high-SPF sunscreen	Kaidbey (1990)
p53	Transient overexpression in response to UV radiation. Persistent nuclear accumulation in cells with *p53* mutation (diffuse pattern). Keratinocyte clones with mutated *p53* can be identified as sharply demarcated areas (compact pattern)	Decrease in overexpression of *p53* in non-sun-exposed skin following UV exposure	Ponten *et al.* (1995)
		Significant reduction in *p53*-positive keratinocytes in sun-protected skin after application of sunscreen to chronically sun-exposed skin during a summer	Berne *et al.* (1998)
Actinic keratoses	Keratinized patches with aberrant cell differentiation and proliferation induced by chronic sun exposure in a dose-dependant manner. May spontaneously regress in the absence of UV exposure	Regular application of a broad-spectrum high-SPF sunscreen prevent the development of acitinic keratoses and even enhances the regression of preexisting keratoses	Thompson *et al.* (1993) Naylor *et al.* (1995)
Immunosuppression	UV irradiation (especially UVB) induces a local and systemic depression of cell-mediated immunity/-contact hypersensitivity	Prevention of UVB-induced suppression of induction to contact hypersensitivity to DNCB	
		Reduction of UV-induced suppression of response to nickel patches in nickel-allergic subjects, or to recall antigens in normal subjects	Whitmore & Morrison (1995); Serre *et al.* (1997); Hayag *et al.* (1997) Damian *et al.*, 1997; Moyal (1998)

SPF, sun protection factor

Table 3. Studies of naevus development as a function of sun protection and sunscreen use

Study type	Subjects	Exposure	End-point / Main outcome	Comments	Reference
Case–control	418 melanoma 438 controls	Never/ever use of sunscreens	Naevus count on both arms in controls increased from no use to ever use of sunscreens. Rate ratio (RR) 1.31 (95% CI, 1.19–1.43) adjusted for age, sex, hair colour, sun exposure	Increase more pronounced when subjects had ever used psoralen sunscreen. RR 2.10 (95% CI, 182–2.43)	Autier *et al.* (1995)
Cohort	357 children (170 boys, 187 girls) 7–11 yrs (median, 9 yrs)	Regular, seldom or never use of sunscreen	Whole body naevus counts at five-year interval. RR of increase in naevi ≥1 mm: 1, 0.81 (0.65–1.01), and 0.41 (0.23–0.75) for regular, seldom or never use of sunscreen.	Univariate analysis	Luther *et al.* (1996)
Retrospective cohort	631 children (321 boys, 310 girls) 6–7 yrs	Total and average sunscreen use, wearing of clothes, sun exposure.	Whole-body naevus count. Median number of naevi increased with total and average sunscreen use, and decreased with average wearing clothes. Adjusted RR of high naevus count on the trunk 1.68 (95% CI, 1.09–2.59) for the highest level of sunscreen use and 0.59 (95% CI, 0.36–0.97) for the highest level of wearing clothes	SPF of sunscreen has no effect on naevus count. Highest risk of naevi on the trunk in children using sunscreen and who never suffered from sunburns (RR, 2.21, 95% CI 1.33–3.67)	Autier *et al.* (1998)

application of an SPF-15 or SPF-30 sunscreen, showed that the SPF-30 sunscreen more efficiently prevented the formation of sunburn cells (Kaidbey, 1990).

The p53 protein plays an important role in cellular response to DNA damage. Following exposure to genotoxic agents such as ionizing or UV radiation, wild-type p53 accumulates and becomes immunohistochemically detectable. In human skin, UV irradiation induces accumulation of p53 in the epidermis; this response is rapid and transient, being detectable as early as two hours after irradiation, peaking at 24 hours and persisting for several days (Hall *et al.*, 1993). Following UV irradiation of normally unexposed skin of healthy subjects, the pattern of p53 expression in the epidermis differs according to the UV wavelength: while UVA induces p53 expression predominantly in the basal layer, UVB induces p53 expression diffusely throughout the whole epidermis (Campbell *et al.*, 1993). In contrast, mutation of *p53* leads to persistent and strong nuclear accumulation of the protein, and in human sun-exposed skin samples from the face or the dorsal surface of the hands, epidermal areas of homogeneously stained cells sharply demarcated from their surroundings and strongly reactive with antibody to p53 (compact pattern) can be detected (Ponten *et al.*, 1995), that have been shown by microdissection and DNA sequencing to reflect clonal proliferation of keratinocytes with mutated *p53* (Ren *et al.*, 1996). Mutation of *p53* plays an important role in the onset of SCC of the skin. More than 90% of human SCCs of the skin contain UV-induced *p53* mutations that are already present in actinic keratoses (Ziegler *et al.*, 1994), but are genetically unrelated to mutations present in benign clonal keratinocyte patches with *p53* mutations in sun-damaged epidermis (Ren *et al.*, 1997).

Two studies have shown that topical application of a broad-spectrum sunscreen with a sun protection factor of 15 either decreases the overexpression of wild-type p53 (a physiological response to UV-induced DNA damage) in epidermal keratinocytes following UV exposure of normal previously non-sun-exposed skin (Ponten *et al.*, 1995) or significantly reduces the number of p53-positive keratinocytes in sun-protected skin after application of sunscreen to chronically sun-exposed skin during a one-summer period (Berne *et al.*, 1998).

Actinic keratoses

Actinic, or solar, keratoses are premalignant, sun-induced skin lesions, characterized as keratinized patches with aberrant cell differentiation and proliferation. They represent risk factors for BCC and melanoma and are precursors to SCC, although the rate of transformation is low (Green & O'Rourke, 1985; Marks *et al.*, 1988a, b). Strong clinical and experimental evidence supports their being precursors of SCC of the skin. Approximately 60% of SCCs are thought to arise in actinic keratoses (Marks *et al.*, 1988a). Similarly to SCC, they have a strong link to sun exposure and share many features with their malignant counterpart, including a dose–response relationship with cumulative sun exposure (Vitasa *et al.*, 1990) and an increased incidence in immunocompromised individuals such as transplant patients (Blohme & Larko, 1984).

Two randomized, placebo-controlled studies have investigated the prevention of new actinic keratoses and the reduction of pre-existing keratoses by sunscreens in high-risk populations with actinic keratoses or even non-melanoma skin cancer. The Australian study (Thompson *et al.*, 1993), conducted over a period of six months (one summer) in 431 subjects, as well as the North American study (Naylor *et al.*, 1995), conducted on a smaller number of subjects (50) over a two-year period, both showed that regular application of a high sun protection factor broad-spectrum sunscreen significantly reduced the total number of actinic keratoses. The Australian study further showed that this reduction in total number of actinic keratoses stemmed from a reduction in the development of new solar keratoses and an increased remission rate of keratoses present at baseline.

Immunosuppression

Experimental studies have shown that UV irradiation, and more especially UVB, induces local and systemic suppression of cell-mediated immunity that plays an important role in the control of growth of UV-induced malignant tumours (Kripke, 1994). Although there is no direct evidence that UVB-induced immunosuppression plays a role in the growth of sunlight-associated skin malignancies in humans, there is abundant circumstantial evidence; for instance, immunosuppressed patients such as transplant recipients have greatly elevated rates of non-melanoma skin cancer incidence (Granstein, 1995).

Recent studies have investigated UV induced immunosuppression and its modulation by topical sunscreen application in humans, focusing on contact hypersensitivity as measured by immunization or response to haptens, on modulation of stimulation of allogeneic lymphocytes by epidermal cells, or on release of immunomodulatory molecules such as *cis*-urocanic acid or interleukin-10.

Dinitrochlorobenzene (DNCB) is a potent contact sensitizer to which spontaneous sensitization is rarely encountered in human populations and has been widely used for evaluation of immune capacities of patients. Typically, sensitization to DNCB is induced by applying to the skin in a Finn chamber a small patch of filter paper containing 30–50 µg DNCB in acetone solution; the patch is removed after 48 hours. The sensitization induced is tested two weeks after the first contact with DNCB by application of challenge patches containing a range of concentrations of DNCB (usually from 3.125 to 12.5 µg). The challenge patches are removed after 48 hours and the contact hypersensitivity reactions are assessed 24 hours later.

Several studies have shown that in healthy volunteers an erythemal UV exposure significantly impairs the afferent arm of the contact hypersensitivity reaction, and that the application of a high sun protection factor sunscreen can at least partially prevent the UV-induced suppression of contact hypersensitivity, without itself interfering with the contact sensitization, probably by preventing the decrease in epidermal Langerhans cells at the irradiated site that usually follows exposure to UVB (Whitmore & Morison, 1995; Hayag *et al.*, 1997; Serre *et al.*, 1997). However, these tests result in the permanent sensitization of the subjects to a potent allergen, and hence may cause an allergic risk, even if the DNCB molecule is rarely encountered in everyday life.

Nickel is a frequent contact allergen in the general population. Up to 15% of women and 5% of men develop allergic contact dermatitis when exposed to nickel. UV radiation suppresses the allergic response of these individuals to patch testing with nickel, and clinical improvement of nickel allergy occurs after whole-body irradiation. This model has been developed into a technique for evaluating the immune protection afforded by sunscreens, and since neither UV irradiation nor sunscreens significantly affect erythema induced by a skin irritant, sodium lauryl sulfate, nickel patch testing appears to be a valid means of assessing UV-induced immunosuppression in humans and its modulation by sunscreens (Damian *et al.*, 1997). The fact that even with suberythemal UV irradiation, suppression of allergic response to nickel was abolished only by broad-spectrum sunscreens suggests that UVA plays an important role in UV-induced immunosuppression.

Alternatively, UV-induced immunosuppression can be assessed in normal subjects by using the delayed hypersensitivity response to common recall antigens (Moyal, 1998).

Another minimally invasive technique to explore UV-induced immunosuppression and its modulation by sunscreens is offered by the mixed epidermal cell–lymphocyte reaction (MECLR), in which epidermal cells obtained by the suction blister method after localized UV irradiation are used to stimulate allogeneic lymphocytes *in vitro*. However, conflicting reports have shown that the ability of sunscreens to interfere with UV-induced modulation of cell-mediated immune responses critically depends on the UV irradiation protocol and assay end-points that may involve different mechanisms of UV-induced immunomodulation (Van Praag *et al.*, 1991; Hurks *et al.*, 1997).

Numerous mediators are released by keratinocytes upon UV irradiation. *cis*-Urocanic acid, formed by photoisomerization in the epidermis of *trans*-urocanic acid, accumulates in the stratum corneum and is considered an important mediator of local immunosuppression resulting from exposure to UV radiation. Formation of *cis*-urocanic acid can be measured in stratum corneum strippings and has been used to evaluate photoprotection by sunscreens (Krien & Moyal, 1994; de Fine Olivarius *et al.*, 1999). More recently, interleukin-10 (IL-10) mRNA expression has been studied by reverse transcription polymerase chain reaction in epidermal cells obtained from suction blisters (Hochberg & Enk, 1999).

Melanoma

Cutaneous melanoma represents the end of a continuum starting from sun-damaged melanocytes in the epidermis and progressing through a common acquired naevus which may show cellular atypia, an atypical (dysplastic) naevus with architectural atypia, melanoma *in situ*, early (radial growth phase) primary melanoma with no competence for metastasis, late (vertical growth phase) primary melanoma, and finally to metastatic melanoma (Table 1).

Although the pathogenesis of melanoma is far from fully understood (Gilchrest *et al.*, 1999), the main exogenous etiological factor of melanoma appears to be intermittent (recreational) sun exposure, especially in childhood (Autier & Doré, 1998). It has been proposed that melanomas arising before the age of 50 years occur electively in body areas intermittently exposed to sunlight, while melanomas arising at a later age occur in body areas more likely to be chronically sun-exposed (Elwood & Gallagher, 1998).

Naevi

Naevi are focal collections of non-dendritic melanocytes, usually found at the junction of the epidermis and dermis (junctional naevi) or at various depths in the dermis (compound or dermal naevi). Common acquired naevi arise after birth, both spontaneously and in response to various factors, particularly exposure to the sun (Harrison *et al.*, 1994).

Many, if not most, cutaneous melanomas arise in acquired naevi, and it has recently been shown that most acquired naevi are clonal, while most melanocytes in non-naeval areas are not (Robinson *et al.*, 1998). Although it is not clear whether acquired naevi represent actual premalignant lesions or risk factors, the number of common naevi in any given individual, which is determined by genetic factors and sun exposure, constitutes the best predictor of individual risk of melanoma (Boyle *et al.*, 1995; Berwick & Halpern, 1997). The presence of large naevi (e.g., those with one dimension >5 mm) or of atypical naevi (e.g., large naevi with non-uniform colour and irregular borders) is also associated with higher melanoma risk, independently of the number of smaller naevi (Bataille *et al.*, 1996; Tucker *et al.*, 1997). Large naevi are not uncommon in children, but atypical naevi are rare in North American or European children before onset of puberty (Greene *et al.*, 1985).

Children seem particularly vulnerable to sun-induced biological events possibly involved in the genesis of melanoma (Holman & Armstrong, 1984; Autier & Doré, 1998). Also, the greatest increase in naevus numbers per unit of skin surface takes place before adolescence (English & Armstrong, 1994). Therefore, the distribution of naevi and their development in children is relevant to understanding melanoma occurrence in adults.

A recent study by our group (Autier *et al.*, 2000) assessed the body-distribution of naevi, with particular reference to differences in sun exposure by body-site, in 649 European children aged 6–7 years from Brussels (Belgium), Bochum (Germany), Lyon (France) and Rome (Italy). Counts of naevi of size 2–4.9 mm and ≥5 mm were performed using a standard method. The numbers of naevi 2–4.9 mm and of naevi ≥5 mm were strongly correlated, especially on the trunk. For naevi 2–4.9 mm, the highest relative densities were found on the face, back, shoulders and the external surface of the arms. Lowest relative densities were found on the hands, legs, feet and abdomen (Figure 1). The relative density of naevi ≥5 mm was higher on the trunk than on any other body site (Figure 2). Similar body distributions were observed in both sexes and at each centre. The body-site distribution of naevi 2–4.9 mm seemed to parallel the usual sun exposure patterns of young European children. It is suggested that the development of naevi ≥5 mm might be a marker of the vulnerability of melanocytes to the harmful effects of solar radiation. Such vulnerability would be maximal on the trunk, and would decrease distally, with melanocytes of the hands and feet having the lowest vulnerability. The number of naevi acquired on a specific area of skin would result from the combined effects of local vulnerability to solar radiation and local sun exposure history. The origin of acquired body-site differences in the susceptibility of melanocytes to UV radiation is unknown, although it seems to parallel the body-site density of sensory innervation.

Several studies have investigated the development of naevi as a function of sun protection or sunscreen use (Table 3). In a case–control study conducted in Europe in 418 melanoma cases and 438 controls (Autier *et al.*, 1995), it was noted that the use of sunscreen was associated with a higher density of pigmented lesions of the skin in controls. The naevus count on both arms in

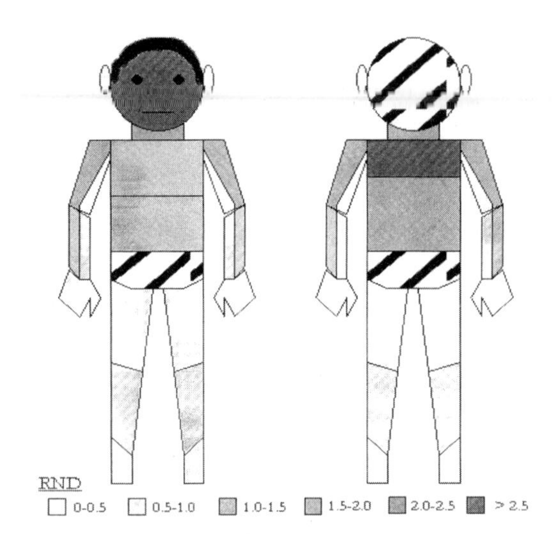

Figure 1. Relative densities of naevi of size ≥ 2 mm in 6–7-year old European children.

RND = relative naevus density.

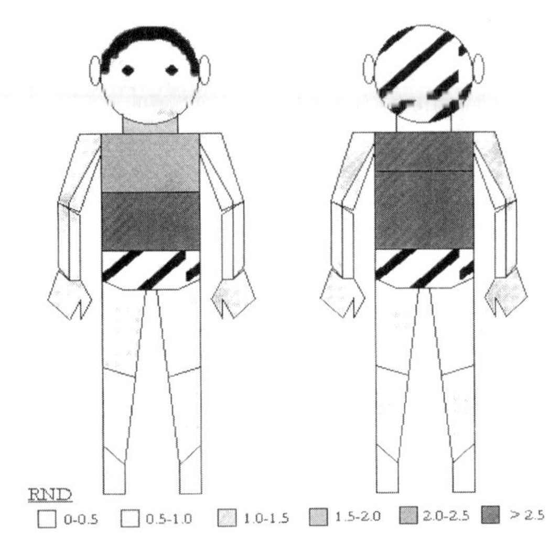

Figure 2. Relative densities of naevi of size > 5 mm in 6–7-year-old European children.

RND = relative naevus density.

control subjects significantly increased from those reporting no sunscreen use to those who ever used sunscreen.

A study of naevus development was conducted in Germany among schoolchildren (Luther *et al.*, 1996). Naevus counts were performed in 866 children between the ages of 1 and 6 years (median age, 4 years), and a second time after a five-year period in 424 children of whom the parents had agreed to participate. Univariate analysis showed a significant relationship between the regular use of sunscreen and a high increase in numbers of melanocytic naevi. This association was interpreted by the authors as resulting from a tendency of children who use sunscreen regularly to have more cumulative sun exposure.

A retrospective cohort study examined the number of naevi in 631 6–7-year-old European children in elementary schools in Brussels, Bochum, Lyon and Rome, according to their sun exposure history, physical protection, sunscreen use and sunburn history from birth to the moment of skin examination (Autier *et al.*, 1998). In all study locations, the median numbers of naevi tended to increase with total or average sunscreen use during holidays, whereas the reverse was true for average wearing of clothes in the sun. Median naevus counts increased with both increasing sun exposure and average sunscreen use. After adjustment for sun exposure and host characteristics, the relative risk for a high naevus count on the trunk was 1.68 (95% confidence interval, 1.09–2.59) for the highest level of sunscreen use and 0.59 (95% confidence interval, 0.36–0.97) for the highest level of wearing clothes while in the sun. The average sun protection factor of the sunscreen used had no demonstrable effect on naevus counts. The highest risk of naevi on the trunk associated with sunscreen use was seen in children who never suffered from sunburn (relative risk, 2.21; 95% confidence interval, 1.33–3.67).

Conclusion

Few intermediate end-points related to the biology of tumour progression can be reliably used as biomarkers to assess chemoprevention of skin cancers.

Assessment of sunburn cell formation or of *p53* mutation induction in the epidermis provides information on the early steps of non-melanoma skin cancer. However, due to their invasive measurement, these biomarkers would be difficult to use in large-scale prevention trials. Actinic keratoses, precursors to SCC, share a number of features with their malignant counterpart, including a dose-dependent link to cumulative sun exposure, and can be easily measured clinically. They have therefore been measured in large-scale trials in humans that have shown that a reduction of actinic keratoses can be obtained by use of sunscreens. The use of such an intermediate-effect biomarker in chemoprevention studies may be regarded as validated by a recent Australian trial showing a reduction in risk of SCC, but not of BCC, in a prospective controlled trial of sunscreen use in the prevention of non-melanoma skin cancer (Green *et al.*, 1999).

Naevi and melanomas share a number of common features and it is thought that long-term prospective studies on melanoma occurrence could be replaced by prospective studies of naevus development. In this respect, the maintenance of a low number of naevi should be considered in the evaluation of the efficacy of protection against long-term harmful effects of sunlight exposure. However, the prevalence of common acquired naevi exceeds by far that of melanoma, and future studies should focus on more precise identification of subjects at risk, for example by considering markers of DNA damage susceptibility or of DNA repair proficiency such as the inherited susceptibility to induction of chromatid breaks by an UV-mimetic mutagen, 4-nitroquinoline-1-oxide (Wei *et al.*, 1996; Wu *et al.*, 1996). In public health, it is increasingly evident that a reduction in the incidence of melanoma will be preceded by a reduction in naevus numbers. The evolution of naevus number at different ages could serve as an indicator of the likely trend of melanoma incidence in the coming years.

References

Autier, P. (1997) Epidémiologie des naevus. *Ann. Dermatol. Venereol.*, **124**, 735–739

Autier, P. & Doré, J.F. (1998) Influence of sun exposures during childhood and during adulthood on melanoma risk. *Int. J. Cancer*, **77**, 533–537

Autier, P., Doré, J.F., Schifflers, E., Cesarini, J.P., Bollaerts, A., Koelmel, K.F., Gefeller, O., Liabeuf, A., Lejeune, F., Liénard, D., Joarlette, M., Chemaly, P. & Kleeberg, U.R. for the EORTC Melanoma Cooperative Group (1995) Melanoma and use of sunscreens: an EORTC case-control study in Germany, Belgium and France. *Int. J. Cancer*, **61**, 749–755

Autier, P., Doré, J.F., Cattaruzza, M.S., Renard, F., Luther, H., Gentiloni-Silverj, F., Zantedeschi, E., Mezzetti, M., Monjaud, I., Andry, M., Osborn, J.F. & Grivegnée, A. (1998) Sunscreen use, wearing clothes, and number of nevi in 6- to 7-year-old European children. *J. Natl Cancer Inst.*, **90**, 1873–1880

Autier, P., Boniol, M., Severi, G., Giles, G., Cattaruzza, M.S., Luther, H., Renard, F., Pedeux, R. & Doré, J.F. (2000) The body site distribution of melanocytic nevi in 6 to 7 year old European children (submitted for publication)

Bataille, V., Newton-Bishop. J.A., Sasieni, P., Swerdlow, A.J., Pinney, E., Griffiths, K. & Cuzick, J. (1996) Risk of cutaneous melanoma in relation to the numbers, types and sites of naevi: a case-control study. *Br. J. Cancer*, **73**, 1605–1611

Berne, B., Ponten, J. & Ponten, F. (1998) Decreased p53 expression in chronically sun-exposed human skin after topical photoprotection. *Photodermatol. Photoimmunol. Photomed.*, **14**, 148–153

Berwick, M. & Halpern, A. (1997) Melanoma epidemiology. *Curr. Opin. Oncol.*, **9**, 178–182

Blohme, L. & Larko, O. (1984) Premalignant and malignant skin lesions in renal transplant patients. *Transplantation*, **37**, 165–167

Boyle, P., Maisonneuve, P. & Doré, J.F. (1995) Epidemiology of melanoma. *Br. Med. Bull.*, **51**, 523–547

Brash, D.E. & Ponten, J. (1998) Skin precancer. *Cancer Surv.*, **32**, 69–113

Campbell, C., Quinn, A.G., Angus, B., Farr, P.M. & Rees, J.L. (1993) Wavelength specific patterns of p53 induction in human skin following exposure to UV irradiation. *Cancer Res.*, **53**, 2697–2699

Damian, D.L., Halliday, G.L. & Barnetson, R.St.C. (1997) Broad-spectrum sunscreens provide greater protection against ultraviolet-radiation-induced suppression of contact hypersensitivity to a recall antigen in humans. *J. Invest. Dermatol.*, **109**, 146–151

de Fine Olivarius, F., Wulf, H.C., Crosby, J. & Norval, M. (1999) Sunscreen protection against cis-urocanic acid production in human skin. *Acta Derm. Venereol.*, **79**, 426–430

English, D.R. & Armstrong, B. (1994) Melanocytic nevi in children. I. Anatomic sites and demographic and host factors. *Am. J. Epidemiol.*, **139**, 390–401

Elwood, J.M. & Gallagher, R.P. (1998) Body site distribution of cutaneous malignant melanoma in relationship to patterns of sun exposure. *Int. J. Cancer*, **78**, 276–280

Gilchrest, B.A., Eller, M.S., Geller, A.C. & Yaar, M. (1999) The pathogenesis of melanoma induced by ultraviolet radiation. *New Engl. J. Med.*, **340**, 1341–1348

Granstein, R.D. (1995) Evidence that sunscreens prevent UV radiation-induced immunosuppression in humans. Sunscreens have their day in the sun. *Arch. Dermatol.*, **131**, 1201–1203

Green, A.C. & O'Rourke, M.G. (1985) Cutaneous malignant melanoma in association with other skin cancers. *J. Natl Cancer Inst.*, **74**, 977–980

Green, A., Williams, G., Neale, R., Hart, V., Leslie, D., Parsons, P., Marks, G.C., Gaffney, P., Battistuta, D., Frost, C., Lang, C. & Russell, A. (1999) Daily sunscreen application and betacarotene supplementation in prevention of basal-cell and squamous-cell carcinomas of the skin: a randomised controlled trial. *Lancet*, **354**, 723–729

Greene, M.H., Clark, W.H. & Tucker M.A. (1985) Acquired precursors of cutaneous malignant melanoma: the familial dysplastic naevus syndrome. *New Engl. J. Med.*, **312**, 91–97

Grove, G.L. & Kaidbey, K.H. (1980) Sunscreens prevent sunburn cells formation in human skin. *J. Invest. Dermatol.*, **75**, 363–364

Hall, P.A., McKee, P.H., Menage, H.D.P., Dover, R. & Lane, D.P. (1993) High levels of p53 protein in UV-irradiated normal human skin. *Oncogene*, **8**, 203–207

Harrison, S.L., MacLennan, R., Speare, R. & Wronski, I. (1994) Sun exposure and melanocytic naevi in young Australian children. *Lancet*, **344**, 1529–1532

Hayag, M.V., Chartier, T., DeVoursney J., Tie, C., Machler, B. & Taylor, J.R. (1997) A high SPF sunscreen's effect on UVB-induced immunosuppression of DNCB contact hypersensitivity. *J. Dermatol. Sci.*, **16**, 31–37

Hochberg, M. & Enk, C.D. (1999) Partial protection against epidermal IL-10 transcription and Langerhans cell depletion by sunscreens after exposure of human skin to UVB. *Photochem. Photobiol.*, **70**, 766–772

Holman, C.D.J. & Armstrong, B. (1984) Cutaneous malignant melanoma and indicators of total accumulated exposure to the sun: an analysis separating histogenic types. *J. Natl Cancer Inst.*, **73**, 75–82

Hurks, H.M., Van der Molen, R.G., Out-Luiting, C., Vermeer, B.-J., Claas, F.H.J. & Mommaas, A.M. (1997) Differential effects of sunscreens on UVB-induced immunomodulation in humans. *J. Invest. Dermatol.*, **109**, 699–703

Kaidbey, K.H. (1990) The photoprotective potential of the new superpotent sunscreens. *J. Am. Acad. Dermatol.*, **22**, 449–452

Krien, P.M. & Moyal, D. (1994) Sunscreens with broad-spectrum absorption decrease the trans to cis photoisomerization of urocanic acid in the human stratum corneum after multiple UV light exposures. *Photochem. Photobiol.*, **60**, 280–287

Kripke, M.L. (1994) Ultraviolet radiation and immunology: something new under the sun – Presidential address. *Cancer Res.*, **54**, 6102–6105

Luther, H., Altmeyer, P., Garbe, C., Ellwanger, U., Jahn, S., Hoffmann, K. & Segerling, M. (1996) Increase in melanocytic nevus counts in children during 5 years of follow-up and analysis of associated factors. *Arch. Dermatol.*, **132**, 1473–1478

Marks, R., Rennie G. & Selwood, T.S. (1988a) Malignant transformation of solar keratoses to squamous cell carcinoma. *Lancet*, **1**, 795–797

Marks, R., Rennie G., & Selwood, T.S. (1988b) The relationship of basal cell carcinoma and squamous cell carcinoma to solar keratoses. *Arch. Dermatol.*, **124**, 1039–1042

Moyal, D. (1998) Immunosuppression induced by chronic ultraviolet irradiation in humans and its prevention by sunscreens. *Eur. J. Dermatol.*, **8**, 209–211

Naylor, M.F., Boyd, A., Smith, D.W., Cameron, G.S., Hubbard, D. & Nelder, K.H. (1995) High sun protection factor sunscreens in the suppression of actinic neoplasia. *Arch. Dermatol.*, **131**, 170–175

Ponten, F., Berne, B., Ren, Z.P., Mister, M. & Ponten, J. (1995) Ultraviolet light induces expression of p53 and p21 in human skin: effect of sunscreen and constitutive p21 expression in skin appendages. *J. Invest. Dermatol.*, **105**, 402–406

Ren, Z.P., Ponten, F., Nister, M. & Ponten, J. (1996) Two distinct p53 immunohistochemical patterns in human squamous-cell skin cancer, precursors and normal epidermis. *Int. J. Cancer*, **69**, 174–179

Ren, Z.P., Ahmadian, A., Ponten, F., Nister, M., Berg, C., Lundeberg, J., Uhlen, M. & Ponten, J. (1997) *Am. J. Pathol.*, **150**, 1791–1803

Robinson, W.A., Lemon, M., Elefanty, A., Harrison-Smith, M., Markham, N. & Norris, D. (1998) Human acquired naevi are clonal. *Melanoma Res.*, **8**, 499–503

Serre, I., Cano, J.P., Picot, M.C., Meynadier, J. & Meunier, L. (1997) Immunosuppression induced by acute solar-simulated ultraviolet exposure in humans: prevention by a sunscreen with a sun protection factor of 15 and high UVA protection. *J. Am. Acad. Dermatol.*, **37**, 187–194

Thompson, S.C., Jolley, D. & Marks, R. (1993) Reduction of solar keratoses by regular sunscreen use. *New Engl. J. Med.*, **329**, 1147–1151

Tucker, M.A., Halpern, A., Holly, E.A., Hartge, P., Elder, D.E. & Sagebiel, R.W. (1997) Clinically recognized dysplastic nevi. A central risk factor for cutaneous melanoma. *J. Am. Med. Ass.*, **277**, 1439–1444

van Praag, M.C.G., Out-Luyting, C., Claas, F.H.J., Vermeer, B.-J. & Mommaas, A.M. (1991) Effect of topical sunscreens on the UV-radiation-induced suppression of the alloactivating capacity in human skin in vitro. *J. Invest. Dermatol.*, **97**, 629–633

Vitasa, B.C., Taylor, H.R. & Strickland, P.T. (1990) Association of nonmelanoma skin cancer and actinic keratoses with cumulative solar ultraviolet exposure in Maryland watermen. *Cancer*, **65**, 2811–2817

Wei, Q., Spitz, M.R., Gu, J., Cheng, L., Xu, X., Strom, S.S., Kripke, M.L. & Hsu, T.C. (1996) DNA repair capacity correlates with mutagen sensitivity in lymphoblastoid cell lines. *Cancer Epidemiol. Biomarkers Prev.*, **5**, 199–204

Whitmore, S.E. & Morison, W.L. (1995) Prevention of UVB-induced immunosuppression in humans by a high sun protection factor sunscreen. *Arch. Dermatol.*, **131**, 1128–1133

Wu, X., Hsu, T.C. & Spitz, M.R. (1996) Mutagen sensitivity exhibits a dose-response relationship in case-control studies. *Cancer Epidemiol. Biomarkers Prev.*, **5**, 577–578

Ziegler, A., Jonason, A.S., Lefell, D.J., Simon, J.A., Sharma, H.W., Kimmelman, J., Remington, L., Jacks, T. & Brash, D.E. (1994) Sunburn and p53 in the onset of skin cancer. *Nature*, **372**, 773–6

Corresponding author:

J.-F. Doré
INSERM U 453,
Bâtiment Cheney, Centre Léon Bérard,
28, rue Laënnec,
F-69373 Lyon Cedex 08,
France

Biomarkers in Cancer Chemoprevention
Miller, A.B., Bartsch, H., Boffetta, P., Dragsted, L. and Vainio, H. eds
IARC Scientific Publications No. 154
International Agency for Research on Cancer, Lyon, 2001

Genetically determined susceptibility markers in skin cancer and their application to chemoprevention

H. Hahn

Development of skin cancer is a result of interactions between genetic and environmental factors. Exposure to sunlight is an established cause of non-melanoma skin cancer as well as of melanoma. Other additional factors such as exposure to environmental chemicals (e.g., chimney soot, arsenic compounds), chronic irritation of the skin, viral infections and the immune status of the host may predispose to skin cancer. The high incidence of skin cancer highlights the need for development of more effective chemopreventive agents. This requires a better understanding of genetically determined host susceptibility, which is increasingly acknowledged as a major factor in the causation of skin tumours.

Introduction

Skin cancer accounts for approximately one third of all newly diagnosed cancers in the United States (Boring *et al.*, 1994). Non-melanoma skin cancers are the most common cancers in the white population, with 900 000 to 1 200 000 new cases diagnosed annually in the United States (Miller & Weinstock, 1994). The most frequently diagnosed cancer in this group is basal-cell carcinoma (BCC), followed by squamous-cell carcinoma (SCC). The incidence of non-melanoma skin cancers is rising worldwide and it has been estimated that 28% to 33% of Caucasians born after 1994 will develop a basal-cell carcinoma in their lifetime (Miller & Weinstock, 1994; Hughes *et al.*, 1995; Gloster & Brodland, 1996). In 1997, more than 40 000 cases of cutaneous melanoma were diagnosed in the United States (Gilchrest *et al.*, 1999). The incidence of no other cancer is increasing so fast, and mortality from cutaneous melanoma has doubled in the last 35–40 years (Balch *et al.*, 1997).

The reason for the increasing incidence of skin cancer is unclear. Increased recreational exposure to sunlight may play a major role, as may an increase in the ultraviolet (UV) radiation that reaches the Earth's surface (Balch *et al.*, 1997). Another factor is probably the increased use of sunscreens which induce individuals to spend more time outdoors. Sunscreens are very effective in preventing sunburn. However, there is no unequivocal evidence that they protect against melanoma formation (Ley & Reeve, 1997).

The identification of genes underlying several rare hereditary syndromes associated with increased skin cancer incidence has had an enormous impact on the understanding of cutaneous malignancies at the molecular level (Halpern & Altman, 1999). Additionally, evidence is emerging that allelic variants of genes involved in detoxification of a variety of exogenous and endogenous substrates may also contribute to the increased number of skin tumours (Lear *et al.*, 2000). Furthermore, it is likely that a number of still unknown genetic modifiers influencing individual susceptibility to skin cancer segregate within the population and their identification should allow better assessment of individual skin cancer risk (Nagase *et al.*, 1999).

High-penetrance susceptibility genes

Inherited conditions entailing increased predisposition to skin cancer and the underlying genes are listed in Table 1.

Albinism

Albinism comprises a group of genetic disorders characterized by deficient synthesis of the pigment melanin. Oculocutaneous albinism Type 1 (OCA1;

Table 1. Inherited diseases with increased skin cancer susceptibility

Disease	Gene defect or gene locus
Oculocutaneous albinism	*TYR*, *P* gene
Xeroderma pigmentosum	*XP* genes
Xeroderma pigmentosum variant	DNA polymerase eta
Naevoid basal-cell carcinoma syndrome	*PTCH*
Cutaneous malignant melanoma	*CDKN2A*, *CDK*
Inherited retinoblastoma	*pRB*
Li–Fraumeni syndrome	*p53*
Rombo syndrome	?
Bazex syndrome	Xq
Multiple self-healing epithelioma	9q31
Muir–Torre syndrome	*MSH2*, *MLH*
Cowden's syndrome	*PTEN*

autosomal recessive inheritance) is caused by mutations in the tyrosinase gene, which encodes the enzyme that catalyses at least three steps in melanin biosynthesis (Spritz, 1994). Oculocutaneous albinism Type 2 (OCA2; autosomal recessive inheritance) results from mutations in the P gene, which is probably involved in the transport of the melanin-precursor tyrosine within the melanocyte (Spritz, 1994). OCA2 is the most prevalent type of albinism worldwide. In the United States, only 1 in 36 000 inhabitants is affected with this disease (Lee *et al.*, 1994). However, in Nigeria it is one of the most common recessive genetic disorders, with a prevalence of about 1 in 1100 (Okoro, 1975). The most frequent type of skin cancer associated with albinism in African albinos is squamous-cell carcinoma, in contrast to Caucasians, in whom basal-cell carcinoma is most frequent (Yakubu & Mabogunje, 1993).

Melanin has photoprotective functions in the skin and directly absorbs both UV photons and reactive oxygen species generated by the interaction of UV radiation with membrane lipids or other cellular components (Riley, 1997). Besides mutations in the tyrosinase gene and in the P gene, mutations or polymorphisms in other genes involved in the synthesis and transport of melanin may play critical roles in susceptibility to skin cancer in the general population (Valverde *et al.*, 1995, 1996).

Xeroderma pigmentosum and xeroderma pigmentosum variant

Exposure to UVB radiation induces the formation of photoproducts which may lead to accumulation of skin cancer-inducing mutations (Brash, 1997). The main pathway by which mammalian cells remove DNA damage caused by UV radiation and other mutagens is nucleotide excision repair (NER). Genes which are defective in the inherited disorder xeroderma pigmentosum (XP) are the best characterized components of the human NER process (Araujo & Wood, 1999). XP shows autosomal recessive inheritance and predisposes the affected individual to cutaneous malignancies. The risk of developing skin cancer on sun-exposed areas is more than 1000-fold increased in XP patients and most frequently the patients develop basal-cell carcinomas and squamous-cell carcinomas (Kraemer *et al.*, 1994). An increase in skin cancer predisposition has also been described in XP heterozygotes (Swift *et al.*, 1979). In this context, it is of great interest that reduced repair of UVB-induced DNA damage in patients with basal-cell carcinoma has been described (Wei *et al.*, 1995).

The xeroderma pigmentosum variant (XPV) is an inherited disease which is associated with increased incidence of sunlight-induced skin cancer. Unlike other XP-cells, XPV cells carry out normal nucleotide excision repair but are deficient in replication of UV-damaged DNA (Cordonnier & Fuchs, 1999). Recently XPV has been shown to arise from mutations in the DNA polymerase *eta* (Masutani *et al.*, 1999; Johnson *et al.*, 1999).

Naevoid basal-cell carcinoma syndrome (NBCCS)

It is well known that inherited defects in oncogenes and tumour-suppressor genes influence skin cancer susceptibility. For example, a large proportion of sporadic basal-cell carcinomas show mutations in the tumour-suppressor gene *patched* (*PTCH*). *PTCH* was cloned from the locus for NBCCS, which is a rare autosomal dominant disorder that predisposes the affected individuals to basal-cell carcinoma, several other cancers and developmental defects (Johnson *et al.*, 1996; Hahn *et al.*, 1996). More than 50% of basal-cell carcinomas show loss of heterozygosity (LOH) at the *PTCH* locus on 9q22–31. Mutation screening revealed that 30% of basal-cell carcinomas with LOH have mutations in the remaining allele of *PTCH* (Gailani *et al.*, 1996a, b). Interestingly, missense mutations in the ligand of PTCH, *sonic hedgehog* (*SHH*) as well as in its signalling partner, *smoothened* (*SMO*) in basal-cell carcinomas have been reported (Oro *et al.*, 1997; Xie *et al.*, 1998). Overexpression of *SHH* (Oro *et al.*, 1997), *SMO* (Xie *et al.*, 1998) and of the *PTCH* pathway targets GLI1 (Nilsson *et al.*, 2000) and GLI2 (Grachtchouk *et al.*, 2000) in murine skin leads to abnormalities resembling human basal-cell carcinomas. Therefore, it is possible that carcinogenesis can be initiated by mutations in all known components of the pathway. Essentially all basal-cell carcinomas overexpress targets of an activated SHH/PTCH/SMO pathway, which include PTCH itself, GLI1 and SMO (Dahmane *et al.*, 1997; Reifenberger *et al.*, 1998; Tojo *et al.*, 1999) (Figure 1). Expression of these genes seems to be very specific for basal-cell carcinomas in comparison with other skin tumours and the level of expression correlates with progression of basal-cell carcinomas (Tojo *et al.*, 1999; Unden *et al.*, 1997; Kallassy *et al.*, 1997). Overall, these data suggest a major role for the SHH/PTCH/SMO signalling pathway in the development of basal-cell carcinoma.

Approximately 40% of *PTCH* mutations in basal-cell carcinomas are C→T or CC→TT transitions at dipyrimidine sites which are typical of UVB-induced DNA damage, thus implicating UVB in mutagenesis of *PTCH* (Aszterbaum *et al.*, 1999). A small proportion of basal-cell carcinomas have point mutations in the *Ras*-homologues H-*ras* and K-*ras* (van der Schroeff *et al.*, 1990; Ananthaswamy & Pierceall, 1990; Lieu *et al.*, 1991). *p53* mutations have been identified in 50% of basal-cell carcinomas and 65% of these are of the UVB-type (Rady *et al.*, 1992; Ziegler *et al.*, 1993; van der Riet *et al.*, 1994; Urano *et al.*, 1995). This suggests a cooperation between either *Ras* or *p53* and *PTCH* in basal-cell carcinoma formation, although no experimental evidence for a direct interaction between these genes has yet been reported.

Melanoma-prone families

Melanoma is the most aggressive of skin cancers; it affects young individuals and much effort has been made to elucidate its development. Atypical or dysplastic naevi are major risk factors in both high-risk families and in the general population. Another significant risk factor is a positive family history of melanoma. Family cases, however, constitute only a small proportion (1–2%) of all cutaneous melanomas (Kefford *et al.*, 1999). In these families, melanoma susceptibility is enhanced by mutations in the cyclin-dependent kinase inhibitor *CDKN2A* and in the cyclin-dependent kinase *CDK4*. These genes have been found to confer elevated risk in 20–40% of melanoma-prone families. The genetic basis for melanoma-predisposition in the remaining 60–80% of families is not known (Kefford *et al.*, 1999). *CDK4* mutations are assumed to generate dominant oncogenes that are resistant to normal inhibition of the cell cycle by p16 (Zuo *et al.*, 1996). The *CDKN2A* gene encodes two distinct proteins which arise through alternative splicing. The p16INK4A protein regulates G1 phase exit by inhibiting the cyclin/CDK-mediated phosphorylation of the pRB protein. The other protein, p14ARF, acts via the p53 pathway to induce cell cycle arrest or apoptosis in response to hyperproliferative signals. Consequently, mutations in *CDKN2A* impair the function of both the p53 and the pRB pathways (Chin *et al.*, 1998). The involvement of these pathways in the pathogenesis of melanoma is underlined by the fact that cutaneous melanoma is the most common second cancer in individuals with inherited retinoblastoma and occurs also in families with Li–Fraumeni syndrome (Eeles, 1995; Moll *et al.*, 1997). The penetrance of *CDKN2A*

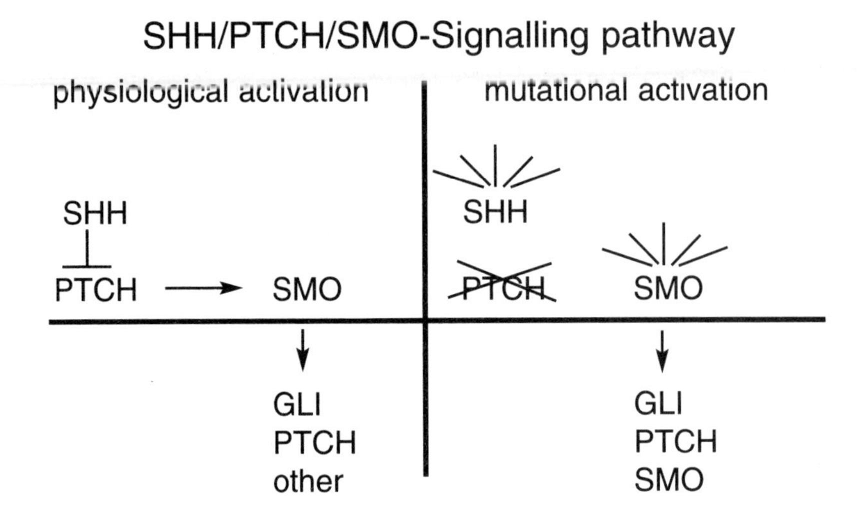

Figure 1. The SHH/PTCH/SMO signalling pathway plays a critical role in the development of basal-cell carcinoma.
Left panel: Physiological activation of the pathway occurs during embryogenesis and is triggered by the ligand of PTCH, SHH. Binding of SHH blocks PTCH, thereby activating SMO, which leads to expression of target genes. Right panel: Mutational activation of the pathway is accomplished by inactivating mutations in PTCH or activating mutations in either SHH or SMO. In basal-cell carcinomas, mutational activation of the pathway results in overexpression of GLI1, SMO and PTCH transcripts.

mutations is strongly influenced by the level of sun exposure and possibly by the action of unknown modifier genes (Goldstein *et al.,* 1998) (see also below, under *Allelic variants of tumour modifiers*). At the present time, routine clinical testing for CDKN2A mutations is not recommended (Haluska & Hodi, 1998) because of uncertainties regarding the penetrance and the correlation between *CDKN2A/CDK4* mutations and the clinical phenotype.

Low-penetrance susceptibility genes

Only 1% of cancer patients have a clearly identifiable inherited component (familial cases) (Fearon, 1997). The remaining 99% of cases are called 'sporadic'. However, there is increasing evidence that sporadic cancers also have heritable determinants that segregate within the general population and thus contribute to individual tumour susceptibility (Ponder, 1990). This appears to be true for skin cancers.

Allelic variants of effect-modulators of carcinogen exposure

UVB radiation (280–320 nm) directly damages DNA, whereas UVA radiation (320–400 nm) acts on DNA via an oxidative stress mechanism which results in the formation of reactive oxygen species in the skin. Reactive oxygen species damage DNA (as well as lipids and amino acids), leading to elevated mutation rates and tumorigenesis (Lear *et al.,* 2000). Detoxification of products of UV-induced stress is the task of numerous cellular proteins, which are also involved in the detoxification of exogenous chemicals. Many of these proteins demonstrate polymorphisms, some of which may result in reduced ability to remove potential carcinogens (Lear *et al.,* 2000). Much interest has been focused on polymorphisms in the *GSTM1* and *GSTT1* genes, which are both expressed in the skin. Homozygotes for the respective null alleles of both the *GSTM1* and *GSTT1* genes express no protein and exhibit greater skin cancer risk (Lear *et al.,* 2000) (Table 2). Further-

Table 2. Polymorphisms in detoxifying enzymes associated with increased basal-cell carcinoma number and accrual

GSTM1 null (in combination with skin type, male gender or tumour site)	Tumour number, accrual (Lear *et al.*, 1996, 1997)
GSTT1 null	Accrual (Lear *et al.*, 1996)
CYP2D6 extensive metabolizer	Tumour number, accrual (Lear *et al.*, 1996)
*CYP1A1*m1m1	Tumour number (Lear *et al.*, 1996)
*CYP1A1*Ile/Val and *CYP1A1*Val/Val	Accrual (Lear *et al.*, 1996)
NQO1 null	Tumour number (Clairmont *et al.*, 1999)

more, *GSTP*, another member of the glutathione *S*-transferase (GST) family of genes, may play an important role in skin cancer development, since homozygous GSTP knock-out mice develop up to 10 times more skin tumours (papillomas) than wild-type-mice (Henderson *et al.*, 1998).

Genes of the cytochrome P450 (*CYP*) family encode enzymes with mono-oxygenase activity. They metabolize a wide range of structurally diverse substrates and participate in the defence against oxidative stress. Many of the *CYP* genes are polymorphic and some of the polymorphisms have significant phenotypic consequences. For example, mutations in *CYP2D6* lead to extensive, intermediate or poor metabolizer phenotypes (Smith *et al.*, 1995). In patients with basal-cell carcinomas, the CYP2D6 extensive metabolizer phenotype has been associated with increased tumour number and accrual (Lear *et al.*, 1996) (Table 2). The latter phenotypic features have also been associated with specific polymorphisms in *CYP1A1* (Table 2). In addition, either *CYP2D6* extensive metabolizer phenotype, *GSTM1* null allele or male gender have been shown to significantly reduce time of presentation of further tumours in patients with truncal basal-cell carcinomas (Lear *et al.*, 1997).

Another antioxidant enzyme which protects cells against reactive oxygen species is NAD(P)H:quinone oxidoreductase (NQO1). A significant association between *NQO1* null allele and basal-cell carcinoma number has been described (Clairmont *et al.*, 1999). The elucidation of the specific roles of detoxifying enzymes and the determination of their contribution to overall susceptibility to skin cancer is becoming a hot topic in the field of skin cancer predisposition.

Allelic variants of tumour modifiers

Tumour modifiers are thought to segregate within a population and to play major functions in determination of individual tumour susceptibility. Most of the information about tumour modifiers of skin cancer has been obtained from studies using laboratory animals. In contrast to *Mus musculus*, *Mus spretus* mice are resistant to chemically induced skin cancer (Nagase *et al.*, 1995). Using a large (*NIH/Ola* × *Mus spretus*)F_1 backcross, it was possible to identify several quantitative trait loci involved in the regulation of skin cancer incidence or multiplicity. Three of these map to loci harbouring members of the family of cyclin-dependent kinase inhibitors p57Kip2, p21Waf1 and p27Kip2 (Nagase *et al.*, 1999).

Mapping of additional loci that modulate skin cancer susceptibility should be possible using the carcinogenesis-resistant (Car-R) and carcinogenesis-susceptible (Car-S) mice which have been obtained applying bi-directional selective breeding. In an initiation (7,12-dimethylbenz[*a*]anthracene)/progression (12-*O*-tetradecanoylphorbol 13-acetate) protocol for tumour induction, skin papillomas occurred in 100% of Car-S mice compared with 3.3% of Car-R mice (Saran *et al.*, 1999).

Conclusions

The agents used today for prevention of skin cancer are either sunscreens or antioxidants. Sunscreens reduce the formation of pyrimidine dimers and are undoubtedly effective in preventing sunburn.

However, epidemiological and laboratory studies indicate that sunscreens may not prevent melanoma formation (Ley *et al.*, 1997).

Only in a minority of skin cancers does the family history show a clear inherited component. For the majority of cases, the identification of the underlying genetic risk factors might be of tremendous value for preventive therapy. The assessment of the individual genetic risk factors for skin cancer requires:

(*a*) association studies using polymorphisms in genes which confer risk for skin tumours (e.g., *GST, CYP*);

(*b*) identification of predisposing polymorphisms in genes which are known to be defective in heritable diseases (e.g., *PTCH, CDKN2A*);

(*c*) identification of new skin cancer susceptibility genes using human and murine genetic studies.

References

Ananthaswamy, H.N. & Pierceall, W.E. (1990) Molecular mechanisms of ultraviolet radiation carcinogenesis. *Photochem. Photobiol.*, **52**, 1119–1136

Araujo, S.J. & Wood, R.D. (1999) Protein complexes in nucleotide excision repair. *Mutat Res.*, **435**, 23–33

Aszterbaum, M., Beech, J. & Epstein, E.H., Jr (1999) Ultraviolet radiation mutagenesis of hedgehog pathway genes in basal-cell carcinomas. *J. Investig. Dermatol. Symp. Proc.*, **4**, 41–45

Balch, C.M., Reintgen, D.S., Kirkwood, J.M., Houghton, A., Peters, L. & Kian Ang, K. (1997) Cutaneous melanoma. In: DeVita, V.T., Hellman, S. & Rosenberg, S.A., eds, *Cancer: Principles and Practice of Oncology*, 5th Edition, Philadelphia, Lippincott-Raven, pp. 1947–1994

Boring, C.C., Squires, T.S., Tong, T. & Montgomery, S. (1994) Cancer statistics, 1994. *CA Cancer J. Clin.*, **44**, 7–26

Brash, D.E. (1997) Cutaneous melanoma. In: DeVita, V.T., Hellman, S. & Rosenberg, S.A., eds, *Cancer: Principles and Practice of Oncology*, 5th Edition, Philadelphia, Lippincott-Raven, pp. 1879–1883

Chin, L., Pomerantz, J. & DePinho, R.A. (1998) The INK4a/ARF tumor suppressor: one gene — two products—two pathways. *Trends Biochem. Sci.*, **23**, 291–296

Clairmont, A., Sies, H., Ramachandran, S., Lear, J.T., Smith, A.G., Bowers, B., Jones, P.W., Fryer, A.A. & Strange, R.C. (1999) Association of NAD(P)H:quinone oxidoreductase (NQO1) null with numbers of basal-cell carcinomas: use of a multivariate model to rank the relative importance of this polymorphism and those at other relevant loci. *Carcinogenesis*, **20**, 1235–1240

Cordonnier, A.M. & Fuchs, R.P. (1999) Replication of damaged DNA: molecular defect in xeroderma pigmentosum variant cells. *Mutat. Res.*, 435, 111–119

Dahmane, N., Lee, J., Robins, P., Heller, P. & Ruiz i Altaba, A. (1997) Activation of the transcription factor Gli1 and the Sonic hedgehog signalling pathway in skin tumours. *Nature*, **389**, 876–881

Eeles, R.A. (1995) Germline mutations in the TP53 gene. *Cancer Surv.*, **25**, 101–124

Fearon, E.R. (1997) Human cancer syndromes: clues to the origin and nature of cancer. *Science*, **278**, 1043–1050

Gailani, M.R., Leffell, D.J., Ziegler, A., Gross, E.G., Brash, D.E. & Bale, A.E. (1996a) Relationship between sunlight exposure and a key genetic alteration in basal-cell carcinoma. *J. Natl Cancer Inst.*, **88**, 349–354

Gailani, M.R., Stahle-Backdahl, M., Leffell, D.J., Glynn, M., Zaphiropoulos, P.G., Pressman, C., Unden, A.B., Dean, M., Brash, D.E., Bale, A.E. & Toftgard, R. (1996b) The role of the human homologue of Drosophila patched in sporadic basal-cell carcinomas. *Nature Genet.*, **14**, 78-81

Gilchrest, B.A., Eller, M.S., Geller, A.C. & Yaar, M. (1999) The pathogenesis of melanoma induced by ultraviolet radiation. *New Engl. J. Med.*, **340**, 1341–1348

Gloster, H.M., Jr & Brodland, D.G. (1996) The epidemiology of skin cancer. *Dermatol. Surg.*, **22**, 217–226

Goldstein, A.M., Falk, R.T., Fraser, M.C., Dracopoli, N.C., Sikorski, R.S., Clark, W.H., Jr & Tucker, M.A. (1998) Sun-related risk factors in melanoma-prone families with CDKN2A mutations. *J. Natl Cancer Inst.*, **90**, 709–711

Grachtchouk, M., Mo, R., Yu, S., Zhang, X., Sasaki, H., Hui, C.C. & Dlugosz, A.A. (2000) Basal cell carcinomas in mice overexpressing Gli2 in skin. *Nature Genet.*, **24**, 216–217

Hahn, H., Wicking, C., Zaphiropoulous, P.G., Gailani, M.R., Shanley, S., Chidambaram, A., Vorechovsky, I., Holmberg, E., Unden, A.B., Gillies, S., Negus, K., Smyth, I., Pressman, C., Leffell, D.J., Gerrard, B., Goldstein, A.M., Dean, M., Toftgard, R., Chenevix-Trench, G., Wainwright, B. & Bale, A.E. (1996) Mutations of the human homolog of Drosophila patched in the nevoid basal-cell carcinoma syndrome. *Cell*, **85**, 841–851

Halpern, A.C. & Altman, J.F. (1999) Genetic predisposition to skin cancer. *Curr. Opin. Oncol.*, **11**, 132–138

Haluska, F.G. & Hodi, F.S. (1998) Molecular genetics of familial cutaneous melanoma. *J. Clin. Oncol.*, **16**, 670–682

Henderson, C.J., Smith, A.G., Ure, J., Brown, K., Bacon, E.J. & Wolf, C.R. (1998) Increased skin tumorigenesis in mice lacking pi class glutathione S-transferases. *Proc. Natl Acad. Sci. USA*, **95**, 5275–5280

Hughes, J.R., Higgins, E.M., Smith, J. & Du Vivier, A.W. (1995) Increase in non-melanoma skin cancer — the King's College Hospital experience (1970–92). *Clin. Exp. Dermatol.*, **20**, 304–307

Johnson, R.L., Rothman, A.L., Xie, J., Goodrich, L.V., Bare, J.W., Bonifas, J.M., Quinn, A.G., Myers, R.M., Cox, D.R., Epstein, E.H., Jr & Scott, M.P. (1996) Human homolog of patched, a candidate gene for the basal cell nevus syndrome. *Science*, **272**, 1668–1671

Johnson, R.E., Kondratick, C.M., Prakash, S. & Prakash, L. (1999) hRAD30 mutations in the variant form of xeroderma pigmentosum. *Science*, **285**, 263–265

Kallassy, M., Toftgard, R., Ueda, M., Nakazawa, K., Vorechovsky, I., Yamasaki, H. & Nakazawa, H. (1997) Patched (ptch)-associated preferential expression of smoothened (smoh) in human basal-cell carcinoma of the skin. *Cancer Res.*, **57**, 4731–4735

Kefford, R.F., Newton Bishop, J.A., Bergman, W. & Tucker, M.A. (1999) Counseling and DNA testing for individuals perceived to be genetically predisposed to melanoma: A consensus statement of the Melanoma Genetics Consortium. *J. Clin. Oncol.*, **17**, 3245–3251

Kraemer, K.H., Lee, M.M., Andrews, A.D. & Lambert, W.C. (1994) The role of sunlight and DNA repair in melanoma and nonmelanoma skin cancer. The xeroderma pigmentosum paradigm. *Arch. Dermatol.*, **130**, 1018–1021

Lear, J.T., Heagerty, A.H., Smith, A., Bowers, B., Payne, C.R., Smith, C.A., Jones, P.W., Gilford, J., Yengi, L., Alldersea, J., Fryer, A.A. & Strange, R.C. (1996) Multiple cutaneous basal-cell carcinomas: glutathione S-transferase (GSTM1, GSTT1) and cytochrome P450 (CYP2D6, CYP1A1) polymorphisms influence tumour numbers and accrual. *Carcinogenesis*, **17**, 1891–1896

Lear, J.T., Smith, A.G., Heagerty, A.H., Bowers, B., Jones, P.W., Gilford, J., Alldersea, J., Strange, R.C. & Fryer, A.A. (1997) Truncal site and detoxifying enzyme polymorphisms significantly reduce time to presentation of further primary cutaneous basal-cell carcinoma. *Carcinogenesis*, **18**, 1499–1503

Lear, J.T., Smith, A.G., Strange, R.C. & Fryer, A.A. (2000) Detoxifying enzyme genotypes and susceptibility to cutaneous malignancy. *Br. J. Dermatol.*, **142**, 8–15

Lee, S.T., Nicholls, R.D., Bundey, S., Laxova, R., Musarella, M. & Spritz, R.A. (1994) Mutations of the P gene in oculocutaneous albinism, ocular albinism, and Prader-Willi syndrome plus albinism. *New Engl. J. Med.*, **330**, 529–534

Ley, R.D. & Reeve, V.E. (1997) Chemoprevention of ultraviolet radiation-induced skin cancer. *Environ. Health Perspect.*, **105** Suppl. 4, 981–984

Lieu, F.M., Yamanishi, K., Konishi, K., Kishimoto, S. & Yasuno, H. (1991) Low incidence of Ha-ras oncogene mutations in human epidermal tumors. *Cancer Lett.*, **59**, 231–235

Masutani, C., Kusumoto, R., Yamada, A., Dohmae, N., Yokoi, M., Yuasa, M., Araki, M., Iwai, S., Takio, K. & Hanaoka, F. (1999) The XPV (xeroderma pigmentosum variant) gene encodes human DNA polymerase eta. *Nature*, **399**, 700–704

Miller, D.L. & Weinstock, M.A. (1994) Nonmelanoma skin cancer in the United States: incidence. *J. Am. Acad. Dermatol.*, **30**, 774–778

Moll, A.C., Imhof, S.M., Bouter, L.M. & Tan, K.E. (1997) Second primary tumors in patients with retinoblastoma. A review of the literature. *Ophthalmic Genet.*, **18**, 27–34

Nagase, H., Bryson, S., Cordell, H., Kemp, C.J., Fee, F. & Balmain, A. (1995) Distinct genetic loci control development of benign and malignant skin tumours in mice. *Nat. Genet.*, **10**, 424–429

Nagase, H., Mao, J.H. & Balmain, A. (1999) A subset of skin tumor modifier loci determines survival time of tumor-bearing mice. *Proc. Natl Acad. Sci. USA*, **96**, 15032–15037

Nilsson, M., Unden, A.B., Krause, D., Malmqwist, U., Raza, K., Zaphiropoulos, P.G. & Toftgard, R. (2000) Induction of basal-cell carcinomas and trichoepitheliomas in mice overexpressing GLI-1. *Proc. Natl Acad. Sci. USA*, **97**, 3438–3443

Okoro, A.N. (1975) Albinism in Nigeria. A clinical and social study. *Br. J. Dermatol.*, **92**, 485–492

Oro, A.E., Higgins, K.M., Hu, Z., Bonifas, J.M., Epstein, E.H., Jr & Scott, M.P. (1997) Basal cell carcinomas in mice overexpressing sonic hedgehog. *Science*, **276**, 817–821

Ponder, B.A. (1990) Inherited predisposition to cancer. *Trends Genet.*, **6**, 213–218

Rady, P., Scinicariello, F., Wagner, R.F., Jr & Tyring, S.K. (1992) p53 mutations in basal-cell carcinomas. *Cancer Res.*, **52**, 3804–3806

Reifenberger, J., Wolter, M., Weber, R.G., Megahed, M., Ruzicka, T., Lichter, P. & Reifenberger, G. (1998) Missense mutations in SMOH in sporadic basal-cell carcinomas of the skin and primitive neuroectodermal tumors of the central nervous system. *Cancer Res.*, **58**, 1798–1803

Riley, P.A. (1997) Melanin. *Int. J. Biochem. Cell Biol.*, **29**, 1235–1239

Saran, A., Pazzaglia, S., Rebessi, S., Bouthillier, Y., Pioli, C., Covelli, V., Mouton, D., Doria, G. & Biozzi, G. (1999) Skin tumorigenesis by initiators and promoters of different chemical structures in lines of mice selectively bred for resistance (Car-r) or susceptibility (Car-s) to two-stage skin carcinogenesis. *Int. J. Cancer*, **83**, 335–340

Smith, G., Stanley, L.A., Sim, E., Strange, R.C. & Wolf, C.R. (1995) Metabolic polymorphisms and cancer susceptibility. *Cancer Surv.*, **25**, 27–65

Spritz, R.A. (1994) Molecular genetics of oculocutaneous albinism. *Hum. Mol. Genet.*, **3**, 1469–1475

Swift, M. & Chase, C. (1979) Cancer in families with xeroderma pigmentosum. *J. Natl Cancer Inst.*, **62**, 1415–1421

Tojo, M., Mori, T., Kiyosawa, H., Honma, Y., Tanno, Y., Kanazawa, K.Y., Yokoya, S., Kaneko, F. & Wanaka, A. (1999) Expression of sonic hedgehog signal transducers, patched and smoothened, in human basal-cell carcinoma. *Pathol. Int.*, **49**, 687–694

Unden, A.B., Zaphiropoulos, P.G., Bruce, K., Toftgard, R. & Stahle-Backdahl, M. (1997) Human patched (PTCH) mRNA is overexpressed consistently in tumor cells of both familial and sporadic basal-cell carcinoma. *Cancer Res.*, **57**, 2336–2340

Urano, Y., Asano, T., Yoshimoto, K., Iwahana, H., Kubo, Y., Kato, S., Sasaki, S., Takeuchi, N., Uchida, N., Nakanishi, H., Arase, S. & Itakura, M. (1995) Frequent p53 accumulation in the chronically sun-exposed epidermis and clonal expansion of p53 mutant cells in the epidermis adjacent to basal-cell carcinoma. *J. Invest. Dermatol.*, **104**, 928–932

Valverde, P., Healy, E., Jackson, I., Rees, J.L. & Thody, A.J. (1995) Variants of the melanocyte-stimulating hormone receptor gene are associated with red hair and fair skin in humans. *Nature Genet.*, **11**, 328–330

Valverde, P., Healy, E., Sikkink, S., Haldane, F., Thody, A.J., Carothers, A., Jackson, I.J. & Rees, J.L. (1996) The Asp84Glu variant of the melanocortin 1 receptor (MC1R) is associated with melanoma. *Hum. Mol. Genet.*, **5**, 1663–1666

van der Riet, P., Karp, D., Farmer, E., Wei, Q., Grossman, L., Tokino, K., Ruppert, J.M. & Sidransky, D. (1994) Progression of basal-cell carcinoma through loss of chromosome 9q and inactivation of a single p53 allele. *Cancer Res.*, **54**, 25–27

van der Schroeff, J.G., Evers, L.M., Boot, A.J. & Bos, J.L. (1990) Ras oncogene mutations in basal-cell carcinomas and squamous-cell carcinomas of human skin. *J. Invest. Dermatol.*, **94**, 423–425

Wei, Q., Matanoski, G.M., Farmer, E.R., Hedayati, M.A. & Grossman, L. (1995) DNA repair capacity for ultraviolet light-induced damage is reduced in peripheral lymphocytes from patients with basal-cell carcinoma. *J. Invest. Dermatol.*, **104**, 933–936

Xie, J., Murone, M., Luoh, S.M., Ryan, A., Gu, Q., Zhang, C., Bonifas, J.M., Lam, C.W., Hynes, M., Goddard, A., Rosenthal, A., Epstein, E.H., Jr & de Sauvage, F.J. (1998) Activating Smoothened mutations in sporadic basal-cell carcinoma. *Nature*, **391**, 90–92

Yakubu, A. & Mabogunje, O.A. (1993) Skin cancer in African albinos. *Acta Oncol.*, **32**, 621–622

Ziegler, A., Leffell, D.J., Kunala, S., Sharma, H.W., Gailani, M., Simon, J.A., Halperin, A.J., Baden, H.P., Shapiro, P.E., Bale, A.E. & Brash, D.E. (1993) Mutation hotspots due to sunlight in the p53 gene of nonmelanoma skin cancers. *Proc. Natl Acad. Sci. USA*, **90**, 4216–4220

Zuo, L., Weger, J., Yang, Q., Goldstein, A.M., Tucker, M.A., Walker, G.J., Hayward, N. & Dracopoli, N.C. (1996) Germline mutations in the p16INK4a binding domain of CDK4 in familial melanoma. *Nature Genet.*, **12**, 97–99

Corresponding author:

H. Hahn
Institute of Pathology,
Technical University Munich/GSF-Research Center for Environment and Health,
Ingolstädter Landstrasse 1,
85764 Neuherberg,
Germany

Biomarkers in Cancer Chemoprevention
Miller, A.B., Bartsch, H., Boffetta, P., Dragsted, L. and Vainio, H. eds
IARC Scientific Publications No. 154
International Agency for Research on Cancer, Lyon, 2001

Biomarkers in colorectal cancer

R. W. Owen

Epidemiological studies have revealed that the major dietary constituents implicated in colorectal carcinogenesis are fat/red meat (causative) and calcium/fibre (protective). Biomarkers have been used in both animal studies and clinical trials to investigate the effect of dietary factors and chemotherapeutic agents on colon carcinogenesis. They can be used as short-term end-points when investigations based on the development of cancer are not feasible. Although they can help in elucidating dietary or pharmacological effects, important results should be confirmed with longer-term studies.

Colon cancer develops through an adenoma–carcinoma sequence. The appearance of colonic polyps in individuals at risk for colon cancer has been used as an end-point in clinical trials to assess diets and pharmacological agents for their effect on colon carcinogenesis. Normal-appearing mucosa can contain small foci of aberrant crypts, which can be dysplastic and thought of as microadenomas. The appearance and growth of such foci have been used to assess the effect of dietary factors and chemopreventive agents in experimental animals. Increased proliferation both increases the sensitivity of the colon to carcinogenesis and may represent an early step in colon carcinogenesis. Etheno-DNA adducts are an end-product of lipid peroxidation processes, and are strongly pro-mutagenic lesions. High dietary levels of *n*–6 fatty acids appear to be important here and may also increase eicosanoid or isoprostane exposure and provide a selective growth stimulus for tumour precursor cells. Low dietary calcium may lead to inhibition of apoptosis and possibly to an increase in cell proliferation. In three recently completed intervention trials, calcium moderately reduced the recurrence of adenomas, but in one study fibre increased recurrence dramatically.

Introduction

The incidence of colorectal cancer is second only to that of lung cancer in both the United States and Europe; rates are generally low in Africa, Asia and South America. Epidemiological studies provide strong support for environmental factors, especially diet, in its etiology (Armstrong & Doll, 1975; Giovannucci *et al.*, 1994; Willett *et al.*, 1990). Migrants from an area of low incidence to a region of higher incidence generally assume the colorectal cancer risk of the host population within a generation. For example, the mortality rate due to colorectal cancer in Japanese immigrants to the United States is 3–4-fold greater than that of Japanese in Japan. Similarly, colorectal cancer incidence is much higher among Puerto Ricans in New York City than in natives in Puerto Rico.

Many dietary components have been examined in relation to colorectal cancer. Dietary fat/meat, protein, alcohol and sugar as well as smoking have been linked to increased risk of colorectal cancer, whereas fibre, calcium, vitamins and selenium have been implicated as protective. Only dietary fat, fibre and calcium (Owen, 2000) have been extensively studied; there is, in general, insufficient evidence to support or refute the role of the other dietary factors.

Identification of people with susceptibility to cancer of the colorectum is an important clinical need. Five-year survival after curative surgery (Enker *et al.*, 1979) decreases markedly with severity of the disease. Early detection correspondingly offers significantly improved prognosis. A five-year survival rate of close to 100% is associated with surgical resection of Dukes A tumours, the spread of which is limited to the colonic mucosa.

Biomarkers

The American Cancer Society (2000) recommends annual digital rectal and faecal occult blood testing after the ages of 40 years and 50 years respectively for all the population. It also recommends a proctosigmoidal examination of the rectum and colon every 3–5 years after the age of 50 years. If any of these tests reveals a possible problem, more extensive tests (colonoscopy and X-ray barium enemas) are usually required. The American Cancer Society estimates that 27 000 additional lives could be saved annually in the United States alone through the use of early detection tests. However, implementation of the recommendations mentioned above would probably be too expensive. Therefore there is a need for surrogate markers (intermediate and frank biomarkers) of susceptibility to colorectal cancer; a brief review of biomarkers that have been suggested and tested (Figure 1) is presented below.

Bile acids

A role for bile acids in the causation of colorectal cancer was first proposed by Aries *et al.* (1969). Initially it was suggested that certain bile acids and neutral steroids might be transformed into carcinogens or co-carcinogens by anaerobic gut bacteria such as *Clostridia* (Aries & Hill, 1970a, b; Goddard & Hill, 1973), but due to a lack of evidence for the formation of such carcinogens *in vivo*, a different slant was proposed that perhaps constitutes the first implication of biomarkers in colorectal cancer, namely, that elevated levels of total bile acids in the stool would lead to increased susceptibility to colorectal cancer (Hill *et al.*, 1971). Many subsequent studies have shown that the major secondary bile acid metabolites (deoxycholic and lithocholic acid) formed in the bowel are either co-carcinogenic in animal model systems or co-mutagenic in mutagenicity testing systems (Bull *et al.*, 1983; Kawasumi & Shigemasa, 1988; Kelsey & Pienta, 1981; Narisawa *et al.*, 1974; Silverman *et al.*, 1977; Wilpart *et al.*, 1983). Animal and tissue-culture studies demonstrating the toxic and dysplastic effects of secondary bile acids have been supported by cohort studies correlating bile acid concentrations with adenoma size and dysplasia (Hill *et al.*, 1983) and with dysplasia in ulcerative colitis (Hill *et al.*, 1987).

However, in contrast to this extensive support from population, animal and *in-vitro* studies,

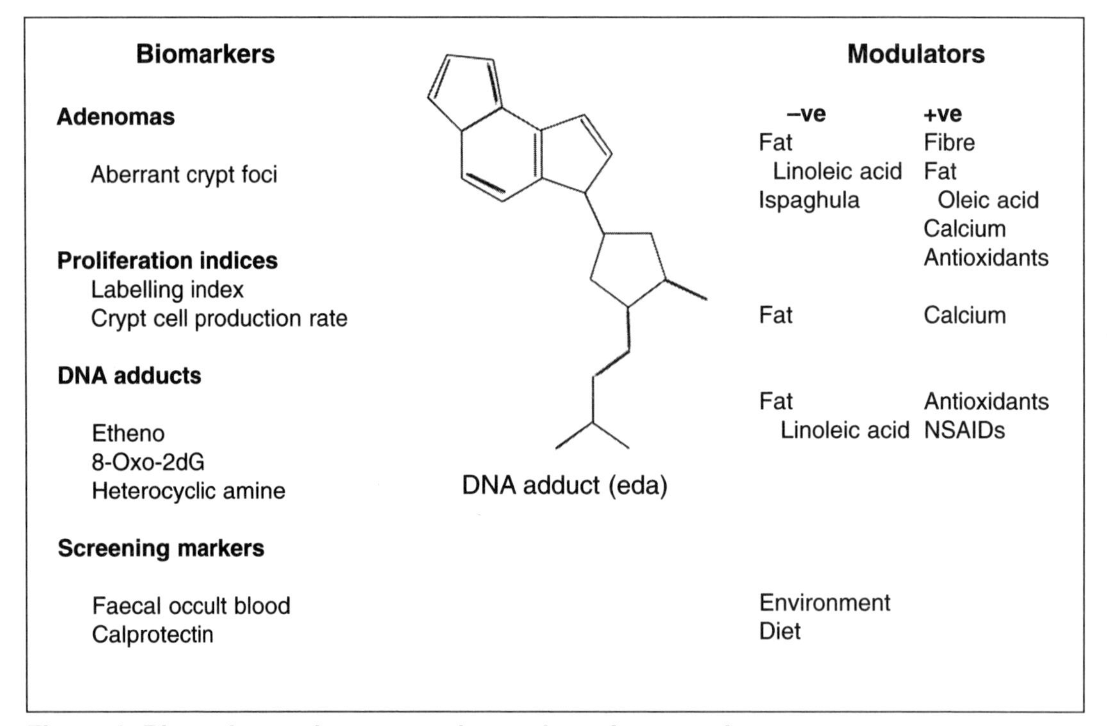

Figure 1. Biomarkers relevant to colorectal carcinogenesis

case–control studies have yielded equivocal results. Two showed positive differences between cases and controls (Hill *et al.*, 1975; Reddy & Wynder, 1977), but many other studies did not (Kaibara *et al.*, 1983; Moskowitz *et al.*, 1979; Mudd *et al.*, 1980; Murray *et al.*, 1980; Owen *et al.*, 1987; Roy *et al.*, 1999). Further studies were directed at identifying the most discriminating faecal bile acid marker. The *in-vitro* studies had highlighted the importance of the secondary bile acids lithocholic acid and deoxycholic acid, which between them account for over 90% of the total faecal bile acid concentration. Owen *et al.* (1987, 1992) and Owen (1997) showed that the ratio of lithocholic acid to deoxycholic acid was a better risk marker than deoxycholic acid, lithocholic acid or the total concentration of faecal bile acids in several case–control studies involving both adenoma and colorectal cancer patients. However, despite initial promise that this biomarker would have application in screening procedures, this was not upheld in several later case–control (Roy *et al.*, 1999) and prospective studies (Haines *et al.*, 2000).

Although studies on bile acids have tended to concentrate on faecal concentrations, some attempts have been made to correlate serum bile acid profiles with the incidence of colorectal cancer. van der Werf *et al.* (1982) and Bayerdorffer *et al.* (1993) demonstrated increased concentrations of deoxycholic acid in serum of adenoma patients compared with controls, but, as with the studies of the faecal matrix, overlap between cases and controls was too extensive to allow sensitive discrimination between high-risk individuals and healthy controls.

Cell proliferation

Hyperproliferation of the intestinal mucosa is regarded as an intermediate biomarker of colorectal cancer. It has been reported in patients with sporadic colorectal adenomas (Bleiberg *et al.*, 1985), in first-degree relatives of colorectal cancer patients (Rozen *et al.*, 1990), in patients suffering from ulcerative colitis (Biasco *et al.*, 1990), in familial adenomatous polyposis (FAP) patients (Lipkin *et al.*, 1984) as well as colorectal cancer patients (Deschner & Maskens, 1982). It appears that these events are not localized to any particular part of the colorectum, but amount to a field effect involving the whole organ. Adenomas arise with greater frequency from this hyperproliferative field, presumably because the rate of cell proliferation exceeds that of DNA repair. Risio *et al.* (1991) have reported that removal of adenomas by polypectomy restores proliferation to normal levels and this may explain why recurrence occurs in only about 40% of patients. However, the situation is very complex, involving both genetic polymorphisms and epigenetic factors, of which the relative importance is hard to determine. Despite this, hyperproliferation has been used as a surrogate end-point biomarker of colorectal neoplasia in a number of studies, especially those evaluating the effect of potential chemopreventive agents. The effect of calcium on hyperproliferation has been studied extensively and is described in more detail below in the section on Intervention Studies.

Aberrant crypt foci

Aberrant crypt foci (ACF), which are microscopic lesions (Roncucci *et al.*, 2000), are possible precursors of adenomas in the colon and therefore may be useful markers of susceptibility to neoplasia. The major advantage of ACF over adenomas as a biomarker of colorectal cancer is that animal experiments can be conducted over a very short time period (weeks rather than months). A drawback, however, is that ACF are so minute that they have to be observed by staining with methylene blue and low-power microscopic examination. In the rat model, removal of the colon for observation is of course not problematic and human resection specimens can also be obtained during surgery. Recent developments in methodology (Dolara *et al.*, 2000) also allow ACF to be visualized *in situ* in humans during colonoscopy, but since all individuals are likely to have ACF (and they cannot be removed), it is difficult to see what advantage the detection of ACF in humans would have over location of an adenoma and polypectomy. Nevertheless, they will continue to have utility in animal model systems for testing the efficacy of chemopreventive agents in short-term carcinogenicity tests (Wargovich *et al.*, 1996).

DNA adducts

Various DNA lesions such as etheno and heterocyclic amine adducts and 8-oxodeoxyguanosine may have relevance as biomarkers of colorectal cancer. However, little research in this area has been conducted in relation to the colorectal mucosa. Friesen *et al.* (1994) have shown that DNA adducts of the heterocyclic amine 2-amino-1-methyl-6-phenylimidazo[4,5-*b*]pyridine (PhIP) can be detected in colon biopsies of humans using gas chromatography/electron capture mass spectrometry. Such adducts are preneoplastic lesions in the rat and therefore may serve as biomarkers of exposure to charred red meat which is the source of heterocyclic amines in the diet.

Schmid *et al.* (2000) used an ultrasensitive and specific immunoaffinity/^{32}P-postlabelling method to detect etheno adducts in colonic tissue of familial adenomatous polyposis (FAP) and sporadic cancer patients. The levels of 1,N^6-ethenodeoxyadenosine (εdA) and 3,N^4-ethenodeoxycytidine (εdC) in polyps removed from the FAP patients were 2–3 times higher than in unaffected colon tissue. In the patients with sporadic carcinoma, no difference in the level of etheno adducts between tumours and adjacent tissue was detected. This suggests that etheno DNA adducts may act as initiators of preneoplastic lesions. The formation of etheno adducts is inherently linked to the metabolism of polyunsaturated (ω–6) long-chain fatty acids (especially arachidonic and linoleic acids) and lipid peroxidation processes and thus these adducts may represent biomarkers of exposure to fats containing high levels of such lipids (Bartsch *et al.*, 1999).

Although 8-oxodeoxyguanosine has been inferred to be a preneoplastic lesion in many cancers, no reports have yet implicated it in the etiology of colorectal cancer.

DNA adducts have potential for use as biomarkers of colorectal cancer, but as yet no convincing strategy has been formulated as to how they could fit into screening procedures. As with many biomarkers, tissue is required for analysis and therefore application would be invasive. Future research may show a specific correlation between levels of certain DNA adducts in either blood or urine and those in tissues such as the colorectum, obviating the need for biopsy material.

Screening markers

Faecal occult blood

The simplest and most dominant diagnostic aid available for colorectal screening at present is the Haemoccult test. This is a rapid colorimetric test based on a guaic–peroxide-catalysed reaction with faecal occult blood. The concept is now more than 100 years old and the test has been applied in cancer screening for over 30 years.

Faecal occult blood testing of the general population shows a rate of slide positivity from 1 to 5%, with a predictive value for neoplastic lesions of 20–50%. The percentage of patients with positive tests who are ultimately found to have adenomas or cancer increases with the age of the population, ranging from 15% in the 40–49-year age group to 60% for individuals over 69 years (Schein & Levin, 1986; Winawer & Sherlock, 1982).

However, all faecal occult blood tests suffer from a number of crucial drawbacks and over the years the test has been extensively criticized. It is not specific for human haemoglobin, also reacting with animal haem, and peroxidases and catalases ingested as part of the normal diet massively interfere with the test. Furthermore, vitamin C, by virtue of its antioxidant properties, produces false negative results with Haemoccult.

The majority of colorectal adenomas have intact epithelium and do not bleed. Thus, faecal occult blood tests are currently of little value in accurately screening the asymptomatic general population for adenomas and predisposition to colorectal cancer. In addition, faecal occult blood testing has little diagnostic value.

Calprotectin

Calprotectin is a prominent cytosol protein in neutrophil granulocytes and is regarded as a marker of inflammatory and neoplastic disease in the lower gastrointestinal tract. Kristinsson *et al.* (1998) studied the faecal concentration of calprotectin in patients with colorectal carcinoma. The median faecal calprotectin concentration in 119 colorectal cancer patients was 50 mg/L (range 2–950), significantly higher than in 125 control patients (median 5.2 mg/L) ($p < 0.0001$). In 23 patients studied after resection, excretion fell dramatically. However, a correlation was not found between plasma and faecal levels. The authors suggested that the measurement of faecal calprotectin might become a diagnostic tool for detecting colorectal carcinoma. This has gained support from a recent study by Kronborg *et al.* (2000), who studied the sensitivity and specificity of faecal calprotectin for detection of adenomas in high-risk individuals undergoing colonoscopy. Patients with confirmed adenomas had significantly higher calprotectin levels than those without (median 9.1 mg/L, 95% CI 7.5–10.1 versus 6.6 mg/L, 95% CI 5.6–7.4). Levels in patients with confirmed cancer were significantly higher than those in the adenoma and control groups (median 17.6 mg/L, 95% CI 11.5–31.0). Using a cut-off limit of 10 mg/L, the sensitivity for cancer was 74% and for adenoma 43%. The authors conclude that the measurement of faecal calprotectin as a marker of colorectal cancer would be of utility in high-risk groups but is not specific enough for use in the general population. It may be hoped that improvements in the assay of faecal calprotectin (Ton *et al.*, 2000) will lead to more generalized use, because it has the advantage over the faecal occult blood test that it does not suffer from interference by food, pharmaceuticals or nutraceuticals.

Intervention studies

Both calcium and fibre regimens have been utilized in human chemoprevention trials in an attempt to reduce recurrent preneoplasia and to verify mechanistic aspects.

In several small calcium intervention trials in humans, the results overall have been equivocal. Several studies (Lipkin *et al.*, 1989; Rozen *et al.*, 1989; Steinbach *et al.*, 1994) have shown that supplementing the human diet with calcium can reduce intestinal cell proliferation, an intermediate biomarker of colorectal cancer, but others (Gregoire *et al.*, 1989; Stern *et al.*, 1990) have shown no effect, whilst one (Kleibeuker *et al.*, 1993) has shown the opposite. In view of these equivocal clinical data, a long-term, double-blind intervention trial was undertaken by Weisgerber *et al.* (1996) in polypectomized, sporadic adenoma patients in which the putative role of calcium (2 g per day) as a protective factor in colon carcinogenesis was studied. Despite differences in stool biochemistry elicited by supplementary calcium after nine months' intervention, a similar non-significant decrease of total cell proliferation rate in sigmoidal mucosa was evident in both the calcium (13.5 down to 11.4) and placebo groups (13.7 down to 10.7). An increase in the concentration and daily excretion of total bile acids, primary bile acids, long-chain fatty acids and long-chain fatty acid soaps was observed in the calcium group, while there was no significant reduction in the concentration of the potentially toxic free bile acids and long-chain fatty acids. This indicated that even if calcium was beneficial, it did not mediate its effects via chelation of intestinal lipid. Baron *et al.* (1995) conducted a multicentre, randomized placebo-controlled, double-blinded trial in polypectomized, sporadic adenoma patients. Subjects received either 3 g of calcium carbonate (providing 1.2 g elemental calcium) or an identical-looking placebo tablet. In rectal mucosal samples obtained six to nine months later from 333 patients (intervention, $n = 173$, placebo, $n = 160$), no decrease in rectal mucosal proliferation due to calcium supplementation was detected.

A number of larger intervention studies designed to assess the efficacy of calcium and antioxidants (Hofstad *et al.*, 1998), calcium (Baron *et al.*, 1999), calcium and fibre (Bonithon-Kopp *et al.*, 2000; Faivre *et al.*, 1997) and fibre (Alberts *et al.*, 2000; Schatzkin *et al.*, 2000) on the growth and/or prevention of adenoma recurrence have been completed recently. In a placebo-controlled study in which 832 polypectomized adenoma patients received supplemental calcium (1.2 g per day) for three years, a significant reduction (RR = 0.81, 95% CI 0.67–0.99) ($p = 0.04$) in recurrence was observed in the calcium ($n = 127$) compared to the placebo arm ($n = 159$) (Baron *et al.*, 1999).

Hofstad *et al.* (1998) studied the effects of both calcium (1.6 g per day) and a cocktail of antioxidants and vitamins (β-carotene 15 mg, vitamin C 150 mg, vitamin E 75 mg, selenium 101 μg per day) in a placebo-controlled intervention trial. All adenomas up to 9 mm in diameter were left *in situ* and the effect on both adenoma growth and the appearance of new adenomas was evaluated. While intervention had no significant effect on adenoma growth, the appearance of new adenomas was significantly lower (log-rank test, $p = 0.035$) in the active group ($n = 58$) than in the placebo group.

Bonithon-Kopp *et al.* (2000), representing the European Agency for Cancer Prevention (ECP), conducted a placebo-controlled pan-European calcium/fibre intervention study in polypectomized sporadic adenoma patients ($n = 625$). In this three-arm clinical trial, 178, 176 and 198 patients were randomized (Table 1) to intervention with placebo (sucrose), calcium (2 g per day) and

Table 1. Baseline characteristics of patients in the ECP calcium fibre intervention study

	Calcium	Fibre	Placebo
Age (years)	58.8 [8.8]	59.1 [8.9]	59.3 [8.4]
Sex			
Male	116 (65.9)	128 (64.6)	107 (60.1)
Female	60 (34.1)	70 (35.4)	71 (39.9)
History of adenoma			
No	145 (82.4)	175 (88.4)	148 (83.1)
Yes	31 (16.6)	23 (11.6)	30 (16.9)
Geographical area			
S. Europe and Israel	34 (19.3)	40 (20.2)	37 (20.8)
W. Europe	72 (40.9)	86 (43.4)	70 (39.3)
N. Europe	70 (39.8)	72 (36.4)	71 (39.9)
No. of adenomas			
Single	104 (49.1)	138 (69.7)	118 (66.3)
Multiple	72 (40.9)	60 (30.3)	60 (33.7)
Adenoma size			
At least 1 > 10 mm	99 (56.3)	105 (53.0)	105 (59.0)
At least 1 > 20 mm	36 (20.5)	25 (12.6)	30 (16.8)

Mean values [SD] or number of subjects (%)
Data from Bonithon-Kopp *et al.* (2000)

Table 2. Risk of adenoma recurrence associated with fibre or calcium treatment in the ECP calcium fibre intervention study

	Calcium treatment		Fibre treatment	
	OR (95% CI)	*p* value	OR* (95% CI)	*p* value
Crude	0.75 (0.43–1.29)	0.29	1.63 (1.01–2.64)	0.042
Adjusted[a]	0.66 (0.38–1.17)	0.16	1.67 (1.01–2.76)	0.042

OR, odds ratio; CI, confidence interval
[a] Adjustment for age, sex, past adenoma history, number and location of adenomas at inclusion
Data from Bonithon-Kopp *et al.* (2000)

fibre (Ispaghula, 3.5 g per day) respectively for three years. The risk of adenoma recurrence (Table 2) associated with calcium supplementation was non-significantly reduced (OR = 0.75, 95% CI 0.43–1.29). In contrast, supplementation with Ispaghula increased the risk of adenoma recurrence significantly (OR = 1.63, 95% CI 1.01–2.64; p = 0.042) and this association was even stronger (Table 3) in those patients with > 80% compliance to the fibre supplement (OR = 1.91, 95% CI 1.08–3.35; p = 0.023). Furthermore the risk of adenoma recurrence associated with fibre treatment according to baseline dietary intake of calcium above the median (Table 4) was highly

Table 3. Risk of adenoma recurrence associated with fibre or calcium treatment according to compliance

	Calcium treatment		Fibre treatment	
	OR (95% CI)[a]	*p* value	OR (95% CI)[a]	*p* value
Compliance <80%	0.53 (0.17–1.72)	0.29	1.06 (0.34–3.33)	0.93
Compliance >80%	0.70 (0.36–1.36)	0.20	1.91(1.08–3.35)	0.023

[a] Odds ratio and 95% confidence interval adjusted for age, sex, past adenoma history, number and location of adenomas at inclusion
Data from Bonithon-Kopp *et al.* (2000)

Table 4. Risk of adenoma recurrence associated with fibre or calcium treatment according to baseline intake of calcium

Baseline dietary calcium[a]	Calcium treatment		Fibre treatment	
	OR (95% CI)[b]	*p* value	OR (95% CI)[b]	*p* value
Below the median	0.51 (0.22–1.18)	0.11	1.04 (0.49–2.18)	0.92
Above the median	0.65 (0.43–2.30)	0.99	2.81 (1.33–5.92)	0.005
Interaction test	$p = 0.12$		$p = 0.028$	

[a] Baseline dietary assessment missing for 29 patients; median value of dietary calcium intake = 918 mg/day
[b] Odds ratio and 95% confidence interval adjusted for age, sex, past adenoma history, number and location of adenomas at inclusion
Data from Bonithon-Kopp *et al.* (2000)

significantly elevated in those patients (OR = 2.81, 95% CI 1.33–5.92; $p = 0.005$).

Alberts *et al.* (2000) recently reported on a two-arm intervention study (1429 men) with high (13.5 g per day) or low (2 g per day) wheat bran supplementation for three years. Of the 1303 men who completed the study, 719 were randomly assigned to high supplementation and 584 to low supplementation. At the end of the study, at least one adenoma was detected in 47% of the high-fibre group as opposed to 51.2% in the low-fibre group, showing conclusively that intervention with high-fibre dietary regimes confers no protection against recurrent colorectal adenomas. This conclusion was supported by a similar study in which 2079 men and women were randomized either to a diet that was low in fat (20% of total calories) and high in fibre (18 g of dietary fibre per 1000 kcal) and fruits and vegetables (3.5 servings per 1000 kcal) or to a healthy eating non-supplemented usual diet for four years (Schatzkin *et al.*, 2000). Of the 1905 who completed the study (958 in the intervention group; 947 in the control group), 39.7% and 39.5% respectively had at least one recurrent adenoma. Therefore adopting a diet that is low in fat and high in fibre, fruits and vegetables does not influence the risk of adenoma recurrence.

A mechanism has been put forward by Nair *et al.* (1997) to explain the link between diet and cancer. In a dietary intervention study with either sunflower oil (high in polyunsaturated fatty acids, especially linoleic acid) or rapeseed oil (high in monounsaturated fatty acids, especially oleic acid), a dramatic increase in DNA bridged etheno adducts was detected in whole white blood cells in females on the sunflower oil diet. Etheno adducts result from lipid peroxidation processes, initiated and propagated by reactive oxygen species and the results indicate that diets which are rich in linoleic acid may be non-beneficial to health. These observations show for the first time a plausible and tangible link between an unhealthy diet and carcinogenesis.

Conclusions

It is evident that the mechanisms related to nutritional factors which lead to colorectal cancer are still somewhat of a mystery. The best and most relevant biomarker of colorectal cancer is adenoma formation and using recurrence as an endpoint is really useful in assessing the efficacy of chemopreventive agents, as exemplified by the consistent moderate beneficial effect of supplemental calcium. It should be noted that detection of adenomas in the general population in countries where screening procedures are not practised (the vast majority) is usually too late to permit hope of improved prognosis, because a large proportion of adenomas are asymptomatic and in general patients present only when it is too late due to blood in the stool, severe pain or irritable bowel syndrome. However, clinical trials are labour-intensive, time-consuming and expensive and therefore a vast range of potential health-promoting substances will never be tested by a sound method. There is a dire need for clinically relevant surrogate biomarkers which give reliable results and correlate significantly with either adenoma formation or recurrence. Preferably these should be measurable in easily accessible material such as blood, faeces or urine.

References

Alberts, D.S., Martinez, M.E., Roe, D.J., Guillen-Rodriguez, J.M., Marshall, J.R., Van Leeuwen, B., Reio, M.E., Ritenbaugh, C., Vargas, P.A., Bhattacharyya, A.B., Earnest, D.L., Sampliner, R.E. and The Phoenix Colon Cancer Prevention Physicians Network (2000) Lack of effect of a high-fiber cereal supplement on the recurrence of colorectal adenomas. *New Engl. J. Med.*, **342**, 1156–1162

American Cancer Society (1990) *Facts and Figures*, New York, American Cancer Society

Aries, V.C. & Hill, M.J. (1970a) Degradation of steroids by intestinal bacteria II. Enzymes catalysing the oxido-reduction of the 3α-, 7α- and 12α-hydroxyl groups in cholic acid and dehydroxylation of the 7-hydroxyl group. *Biochim. Biophys. Acta*, **202**, 535–543

Aries, V.C. & Hill, M.J. (1970b) The formation of unsaturated bile acids by intestinal bacteria. *Biochem. J.*, **119**, 57P

Aries, V.C., Crowther, J.S., Drasar, B.S., Hill, M.J. & Williams, R.E.O. (1969) Bacteria and the aetiology of cancer of the large bowel. *Gut*, **10**, 334-335

Armstrong, B. & Doll, R. (1975) Environmental factors and cancer incidence and mortality in different countries, with special reference to dietary practices. *Int. J. Cancer*, **15**, 617–631

Baron, J.A., Toteson, T.D., Wargowich, M.J., Sandler, R., Mandel, J., Bond, J., Hailie, R., Summers, R., van-Stolk, R. & Rothstein, R. (1995) Calcium supplementation and rectal mucosal proliferation: a randomized controlled trial. *J. Natl Cancer Inst.*, **87**, 1303–1307

Baron, J.A., Beach, M., Mandel, J.S., van-Stolk, R.U., Haile, R.W., Sandler, R.S., Rothstein, R., Summers, R.W., Snover, D.C., Beck, G.J., Bond, J.H. & Greenberg, E.R. (1999) Calcium supplements for the prevention of colorectal adenomas. Calcium Polyp Prevention Group. *New Engl. J. Med.*, **340**, 101–107

Bartsch, H., Nair, J. & Owen, R.W. (1999) Dietary polyunsaturated fatty acids and cancers of the breast and colorectum: emerging evidence for their role as risk modifiers. *Carcinogenesis*, **20**, 2209–2218

Bayerdorffer, E., Mannes, G.A., Richter, W.O., Ochsenkuhn, T., Wiebecke, B., Kopcke, W. & Paumgartner, G. (1993) Increased serum deoxycholic acid levels in men with colorectal adenomas. *Gastroenterology*, **104**, 145–151

Biasco, G., Paganelli, G.M., Miglioli, M., Brillanti, S., Di Febo, G., Gizzi, G., Ponz de Leon, M., Campieri, M. & Barbara, L. (1990) Rectal cell proliferation and colon cancer risk in ulcerative colitis. *Cancer Res.*, **50**, 1156–1159

Bleiberg, H., Buyse, M. & Galand, P. (1985) Cell kinetic indicators of premalignant stages of colorectal cancer. *Cancer*, **56**, 124–129

Bonithon-Kopp, C., Kronborg, O., Giacosa, A., Räth, U. & Faivre, J. (2000) Calcium and fibre supplementation in the prevention of colorectal adenoma recurrence: results from the European Cancer Prevention Organisation (ECP) Intervention Study. *Lancet* (in press)

Bull, A.W., Marnett, L.J., Dawe, E.J. & Nigro, N.D. (1983) Stimulation of deoxythymidine incorporation in the colon of rats treated intrarectally with bile acids and fats. *Carcinogenesis*, **4**, 207–210

Deschner, E.E. & Maskens, A.P. (1982) Significance of the labeling index and labeling distribution as kinetic parameters in colorectal mucosa of cancer patients and DMH treated animals. *Cancer*, **50**, 1136–1141

Dolara, P., Caderni, G. & Luceri, C. (2000) Surrogate endpoint biomarkers for human colon carcinogenesis. *Toxicol. Lett.*, **112–113**, 415–420

Enker, W.E., Lauffer, U.T. & Block, G.E. (1979) Enhanced survival of patients with colon and rectal cancer is based upon wide anatomic resection. *Ann. Surg.*, **190**, 350-357

Faivre, J., Couillault, C., Kronborg, O., Rath, U., Giacosa, A., De-Oliveira, H., Obrador, T., O'Morain, C., Buset, M., Crespon, B., Fenger, K., Justum, A.M., Kerr, G., Legoux, J.L., Marks, C., Matek, W., Owen, R.W., Paillot, B., Piard, F., Pienkowski, P., Pignatelli, M., Prada, A., Pujol, J., Richter, F., Seitz, J.F., Sturnolio, G.C,, Zambelli, A. & Andreatta, R. (1997) Chemoprevention of metachronous adenoma of the large bowel: design and interim results of a randomized trial of calcium and fibre. *Eur. J. Cancer Prev.*, **6**, 132–138

Friesen, M.D., Kaderlik, K., Lin, D., Garren, L., Bartsch, H., Lang, N.P. & Kadlubar, F.F. (1994) Analysis of DNA adducts of 2-amino-1-methyl-6-phenylimidazo[4,5-b]pyridine in rat and human tissues by alkaline hydrolysis and gas chromatography/electron capture mass spectrometry: validation by comparison with ^{32}P-postlabeling. *Chem. Res. Toxicol.*, **7**, 733–739

Giovannucci, E., Rimm, E.B., Stampfer, M.J., Colditz, G.A., Ascherio, A. & Willett, W.C. (1994) Intake of fat, meat and fiber in relation to risk of colon cancer in men. *Cancer Res.*, **54**, 2390–2397

Goddard, P. & Hill, M.J. (1973) The dehydrogenation of the steroid nucleus by human gut bacteria. *Biochem. Soc. Trans.*, **1**, 1113–1115

Gregoire, R.C., Stern, H.S., Yeung, K.S., Stadler, J., Langley, S., Furrer, R. & Bruce, W.R. (1989) Effect of calcium supplementation on mucosal cell proliferation in high risk patients for colon cancer. *Gut*, **30**, 376–382

Haines, A., Hill, M.J., Thompson, M.H., Owen, R.W., Williams, R.E.O., Meade, T.W. & Wilkes, H. (2000) A prospective study of faecal bile acids and colorectal cancer. *Eur. J. Cancer Prev.* (in press)

Hill, M.J., Drasar, B.S., Hawksworth, G., Aries, V., Crowther, J.S. & Williams, R.E.O. (1971) Bacteria and aetiology of cancer of large bowel. *Lancet*, **1**, 95–100

Hill, M.J., Drasar, B.S., Williams, R.E.O., Meade, T.W., Cox, A.G., Simpson, J.E.P. & Morson, B.C. (1975) Faecal bile acids and clostridia in patients with cancer of the large bowel. *Lancet*, **1**, 535–539

Hill, M.J., Morson, B.C. & Thompson, M.H. (1983) The role of faecal bile acids (FBA) in large bowel carcinogenesis. *Br. J. Cancer*, **48**, 143 (abstract)

Hill, M.J., Melville, D., Lennard-Jones, J., Neale, K. & Ritchie, J.K. (1987) Faecal bile acids, dysplasia and carcinoma in ulcerative colitis. *Lancet*, **ii**, 185–186

Hofstad, B., Almennigen, K., Vatn, M., Norheim-Anderson, S., Owen, R.W., Larsen, S. & Osnes, M. (1998) Growth of colorectal polyps: effect of antioxidants and calcium on growth and new polyp formation. *Digestion*, **59**, 148–156

Kaibara, N., Sasaki, T., Ikeguchi, M., Kog, A.S. & Yunugi, E. (1983) Faecal bile acids and neutral sterols in Japanese with large bowel carcinoma. *Oncology*, **40**, 255–258

Kawasumi, H.N.K. & Shigemasa, K. (1988) Co-carcinogenic activity of bile acids in the chemical transformation of C3H/10T1/2 fibroblasts in vitro. *Oncology*, **45**, 19–196

Kelsey, M.I. & Pienta, R.J. (1981) Transformation of hamster embryo cells by neutral sterols and bile acids. *Toxicol. Lett.*, **9**, 177–182

Kleibeuker, J.H., Welberg, J.W.M., Mulder, N.H., Van der Meer, R., Cats, A., Limburg, A.J., Kreumer, W.M.T., Hardonk, M.J. & De Vries, E.G. (1993) Epithelial cell proliferation in the sigmoid colon of patients with adenomatous polyps increases during oral calcium supplementation. *Br. J. Cancer*, **67**, 500–503

Kristinsson, J., Roseth, A., Fagerhol, M.K., Aadland, E., Schonsby, H., Bormer, O.P., Raknerud, N. & Nygaard, K. (1998) Fecal calprotectin concentration in patients with colorectal carcinoma. *Dis. Colon Rectum*, **41**, 316–321

Kronborg, O., Ugstad, M., Fuglerud, P., Johne, B., Hardcastle, J., Scholefield, J.H., Vellacott, K., Moshakis, V. & Reynolds, J.R. (2000) Faecal calprotectin levels in a high risk population for colorectal neoplasia. *Gut*, **46**, 795–800

Lipkin, M., Blattner, W.A., Gardner, E.J., Burt, R.W., Lynch, H., Deschner, E., Winawer, S. & Fraumeni, J.F. (1984) Classification and risk assessment of individuals with familial polyposis, Gardners syndrome, and familial non-polyposis colon cancer from thymidine labeling patterns in colonic epithelial cells. *Cancer Res.*, **44**, 4201–4207

Lipkin, M., Friedman, E., Winawer, S.J. & Newmark, H. (1989) Colonic epithelial cell proliferation in responders and nonresponders to supplementary dietary calcium. *Cancer Res.*, **29**, 248–254

Moskowitz, M., White, C., Barnett, R.N., Stevens, S., Russell, E., Vargo, D. & Floch, M.H. (1979) Diet, fecal bile acids and neutral sterols in carcinoma of the colon. *Dig. Dis. Sci.*, **24**, 746–751

Mudd, D.G., McKelvey, S.T.D., Norwood, W., Elmore, D.T. & Roy, A.D. (1980) Faecal bile acid concentrations of patients with carcinoma or increased risk of carcinoma in the large bowel. *Gut*, **21**, 587–590

Murray, W.R., Blackwood, A., Trotter, J.M., Calman, K.C. & Mackay, C. (1980) Faecal bile acids and clostridia in the aetiology of colorectal cancer. *Br. J. Cancer*, **41**, 923–928

Nair, J., Vaca, C.E., Velic, I., Mutanen, M., Valsta, L.M. & Bartsch, H. (1997) High dietary omega-6 polyunsaturated fatty acids drastically increase the formation of etheno-DNA base adducts in white blood cells of female subjects. *Cancer Epidemiol. Biomarkers Prev.*, **6**, 597–601

Narisawa, T., Magadia, N.E., Weisberger, J.H. & Wynder, E.L. (1974) Promoting effect of bile acids on colon carcinogenesis after intrarectal instillation of *N*-methyl-*N'*-nitro-*N*-nitrosoguanidine in rats. *J. Natl Cancer Inst.*, **53**, 1093–1097

Owen, R.W. (1997) Faecal steroids and colorectal carcinogenesis. *Scand. J. Gastroenterol.*, **32**, Suppl. 222, 76–82

Owen, R.W. (2000) Carcinogenic and anticarcinogenic factors in food. In: Eisenbrand, G., Dayan, A.D., Elias, P.S., Grunow, W. & Schlatter, J., eds, *The Role of Nutritional Factors: Colon Cancer*, Weinheim, Wiley-VCH, pp. 43–75

Owen, R.W., Dodo, M., Thompson, M.H. & Hill, M.J. (1987) Faecal steroids and colorectal cancer. *Nutr. Cancer*, **9**, 73–80

Owen, R.W., Day, D.W. & Thompson, M.H. (1992) Faecal steroids and colorectal cancer: steroid profiles in subjects with adenomatous polyps of the large bowel. *Eur. J. Cancer Prev.*, **1**, 105–112

Reddy, B.S. & Wynder, E.L. (1977) Metabolic epidemiology of colon cancer: fecal bile acids and neutral sterols in colon cancer patients and patients with adenomatous polyps. *Cancer*, **39**, 2533–2539

Risio, M., Lipkin, M., Candelaresi, G., Bertone, A., Coverlizza, S. & Rossini, F.P. (1991) Correlations between rectal mucosa cell proliferation and the clinical and pathological features of non-familial neoplasia of the large intestine. *Cancer Res.*, **51**, 1917–1921

Roncucci, L., Pedroni, M., Vaccina, F., Benatti, P., Marzona, L. & De Pol, A. (2000) Aberrant crypt foci in colorectal carcinogenesis. Cell and crypt dynamics. *Cell Prolif.*, **33**, 1–18

Roy, P., Owen, R.W., Faivre, J., Scheppach, W., Saldanha, M.H., Beckly, D.E. & Boutron, M.-C. (1999) Fecal neutral sterols and bile acids in patients with adenomas and large bowel cancers. *Eur. J. Cancer Prev.*, **8**, 409–415

Rozen, P., Fireman, Z., Fine, N., Wax, Y. & Ron, E. (1989) Oral calcium suppresses increased rectal proliferation of persons at risk of colorectal cancer. *Gut*, **30**, 650–655

Rozen, P., Fireman, Z., Fine, N., Chetrit, A. & Lubin, F. (1990) Rectal epithelial characteristics of first degree relatives of sporadic colon cancer patients. *Cancer Lett.*, **51**, 127–132

Schatzkin, A., Lanza, E., Corle, D., Lance, P., Iber, F., Caan, B., Shike, M., Weissfeld, J., Burt, R., Cooper, M.R., Kikendall, W., Cahill, J. and The Polyp Prevention Trial Study Group (2000) Lack of effect of a low-fat, high-fiber diet on the recurrence of colorectal adenomas *New Engl. J. Med.*, **342**, 1149–1155

Schein, P.S. & Levin, B. (1986) Neoplasms of the colon. In: Calabresi, P., Schein, P.S. & Rosenberg, S.A., eds, *Medical Oncology*, New York, Macmillan, pp. 884–916

Schmid, K., Nair, J., Winde, G., Velic, I. & Bartsch, H. (2000) Increased levels of promutagenic etheno-DNA adducts in colonic polyps of FAP patients. *Int. J. Cancer*, **87**, 1–4

Silverman, S.J. & Andrews, A.W. (1977) Bile acids: co-mutagenic activity in the salmonella-mammalian microsome mutagenicity test. *J. Natl Cancer Inst.*, **59**, 1557–1559

Steinbach, G., Lupton, J., Reddy, B.S., Kral, J.G. & Holt, P.R. (1994) Effect of calcium supplementation on rectal epithelial hyperproliferation in intestinal bypass subjects. *Gastroenterology*, **106**, 1162–1167

Stern, H.S., Gregoire, R.C., Kashtan, H., Stadler, J. & Bruce, W.R. (1990) Long-term effects of dietary calcium on risk markers for colon cancer in patients with familial polyposis. *Surgery*, **108**, 528–533

Ton, H., Brandsnes, O., Dale, S., Holtlund, J., Schonsby, H. & Johne, B. (2000) Improved assay for fecal calprotectin. *Clin. Chim. Acta*, **292**, 41–54

van der Werf, S.D., Nagengast, F.M., van Berge Henegouwen, G.P., Huijbregts, A.W. & van Tongeren, J.H. (1982) Colonic absorption of secondary bile acids in patients with adenomatous polyps and in matched controls. *Lancet*, **i**, 759–762

Wargovich, M.J., Chen, C.D., Jimenez, A., Steele, V.E., Velasco, M., Stephens, L.C., Price, R., Gray, K. & Kelloff, G.J. (1996) Aberrant crypts as biomarkers for colon cancer: evaluation of potential chemopreventive agents in the rat. *Cancer Epidemiol. Biomarkers Prev.*, **5**, 355–360

Weisgerber, U.M., Boeing, H., Owen, R.W., Raedsch, R. & Wahrendorf, J. (1996) Effect of long-term placebo-controlled calcium supplementation on sigmoidal cell proliferation in patients with sporadic adenomatous polyps. *Gut*, **38**, 396–402

Willett, W.C., Stampfer, M.J., Colditz, G.A., Rosner, B.A. & Speizer, F.E. (1990) Relation of meat, fat and fiber intake to the risk of colon cancer in a prospective study among women. *New Engl. J. Med.,* **323**, 1664–1672

Wilpart, M., Mainguet, P., Maskens, A. & Roberfroid, M. (1983) Mutagenicity of 1,2-dimethylhydrazine towards *Salmonella typhimurium*: co-mutagenic effect of secondary bile acids. *Carcinogenesis,* **6**, 45–48

Winawer, S.J. & Sherlock, P. (1982) Surveillance for colorectal cancer in average risk patients, familial high risk groups and in patients with adenomas. *Cancer,* **50**, 2609–2614

Corresponding author:

R. W. Owen
Division of Molecular Toxicology - C0300
Deutsches Krebsforschungszentrum
Im Neuenheimer Feld 280,
D-69120 Heidelberg,
Germany

Biomarkers in Cancer Chemoprevention
Miller, A.B., Bartsch, H., Boffetta, P., Dragsted, L. and Vainio, H. eds
IARC Scientific Publications No. 154
International Agency for Research on Cancer, Lyon, 2001

Intermediate effect markers for colorectal cancer

J.A. Baron

Recurrence or regression of adenomatous polyps is considered to be both a biomarker of risk and an intermediate (surrogate) end-point. The observational epidemiology of adenomas resembles closely that of invasive cancer, and the findings in chemoprevention trials that have been completed closely mirror that common epidemiology. Although it is possible that both the clinical trials and the epidemiology may be wrong, these common findings suggest that adenomas in general are valid end-points.

Aberrant crypt foci show promise as biomarkers, both as markers of risk and as intermediate end-points for chemoprevention trials.

If issues of cost can be overcome, assessment of *ras* mutations in stool appears to be a promising technique for screening for large bowel neoplasms. The lack of specificity of the technique limits its utility as a sole end-point in prevention studies, however.

Mucosal proliferation has been used both as a biomarker of risk and as an intermediate end-point. The utility of these measures is not clear, however, since there has been discordance between the epidemiological findings regarding proliferation and established risk factors for colorectal neoplasia. The inherent variability of the measures and the technical problems associated with their use are further impediments. However, rectal mucosal proliferation may be suitable for studies in single institutions, or in consortia with very aggressive quality control.

Adenomatous polyps

Colorectal adenomas are well demarcated tumours of the large bowel mucosa, and are composed of crypts with cells showing features of epithelial dysplasia. This histological definition implies nothing about the gross shape of the lesion, although commonly it is polypoid, protruding into the lumen of the bowel, sometimes on a stalk. Since such adenomatous polyps are raised lesions, they are easily visible by endoscopy and (unless very large) can be removed endoscopically. Adenomatous polyps have been widely applied in large bowel chemoprevention studies both as an entry criterion and as an intermediate outcome end-point (Baron, 1996).

However adenomatous polyps are used, there is measurement error associated with their identification by endoscopy. To distinguish these polyps from non-adenomatous large bowel polyps in research studies, it is important that each polyp be biopsied for histological examination. Since the microscopic identification of adenomatous tissue is generally not difficult, the net specificity of endoscopic surveillance of the bowel can be assumed to be close to 100% if pathological examination is conducted. However, the sensitivity is likely to be somewhat lower than this. For large adenomas (1 cm or greater in estimated diameter), the miss rate is small (less than 5%; Hixson *et al.*, 1991), but for smaller polyps it has been found to be 15–25% (Hixson *et al.*, 1990; Hoff & Vatn, 1985; Rex *et al.*, 1997).

The existence of 'flat adenomas'—those that are not substantially raised above the mucosal surface—somewhat complicates this assessment. Such adenomas have been noted most often in Asia, but have also been seen in North America and Europe (Jaramillo *et al.*, 1995; Lanspa *et al.*, 1992; Owen, 1996). The presence of such adenomas may decrease the sensitivity of adenoma detection by endoscopy, but will not greatly affect the predictive value of the detected lesions, at least for the raised cancers and adenomas traditionally studied in western countries. Increased awareness of these lesions and newer endoscopic techniques such as magnifying endoscopy may resolve some of the geographical differences in the apparent incidence and prevalence of these lesions.

Adenomas as biomarkers

Adenomatous polyps are clearly established markers of risk of colorectal neoplasia. Individuals with adenomas in the distal bowel (for example, within the reach of a sigmoidoscope) have an increased risk of proximal adenomas or cancer (Levin *et al.*, 1999; Lieberman & Smith, 1991). Advanced adenomas—those with large size (typically >1 cm), villous histology or advanced dysplasia—are more likely to harbour invasive cancer than simple adenomas (without those features) (Gatteschi *et al.*, 1991; Muto *et al.*, 1975; Shinya & Wolff, 1979). Advanced features in distal adenomas also are predictive of proximal adenomas (Levin *et al.*, 1999; Schoen *et al.*, 1998; Wallace *et al.*, 1998) as well as proximal advanced adenomas (Levin *et al.*, 1999; Schoen *et al.*, 1998; Wallace *et al.*, 1998; Zarchy & Ershoff, 1994). In some multivariate analyses, all of these factors have emerged as independent predictors (Gatteschi *et al.*, 1991), although others have found that while histology and number of polyps are independent predictors of risk, size is not (Atkin *et al.*, 1992; Levin *et al.*, 1999).

Patients with adenomas anywhere in the bowel also have an increased risk of later (metachronous) adenomas or cancer (Atkin *et al.*, 1992; Lotfi *et al.*, 1986; Simons *et al.*, 1992; Stryker *et al.*, 1987). Characteristics of the presenting adenomas that are predictive of subsequent neoplastic development include large size, villous histology, multiple adenomas, and higher degrees of dysplasia (Atkin *et al.*, 1992; Grossman *et al.*, 1989; Simons *et al.*, 1992; van Stolk *et al.*, 1998; Winawer *et al.*, 1993; Yang *et al.*, 1998). The same characteristics are also predictive of later adenomas with advanced pathological features (Atkin *et al.*, 1992; Yang *et al.*, 1998). In multivariate analyses, histology appears to explain the association of size with risk of metachronous tumours (Atkin *et al.*, 1992; Simons *et al.*, 1992). Unfortunately, in virtually all of the studies of later cancer risks in patients with adenomas, comparisons have been made with the general population, rather than with individuals screened intensively for colorectal cancer like the subjects with adenomas. Consequently, the excess cancer risks are almost certainly underestimated.

These data show that the association between adenomatous polyps and synchronous or metachronous neoplasia depends on the characteristics of the presenting polyp (Atkin *et al.*, 1992; Aubert *et al.*, 1982). In essence, it appears that while advanced lesions are predictive of both simple and advanced lesions, the simple lesions are associated only with an increased risk of synchronous or metachronous simple lesions. Since low-risk adenomas do not lead to high-risk adenomas (not to speak of cancer), it is possible that advanced and simple adenomas reflect different biological pathways.

Adenomas as surrogate end-points: the adenoma–carcinoma sequence

It is thought that the overwhelming majority of colorectal cancers arise from adenomatous tissue—this is the well known adenoma–carcinoma sequence. Evidence for the adenoma–carcinoma sequence is largely indirect. A substantial proportion of colorectal cancers have contiguous adenomatous tissue, suggesting that the former grew out of the latter (Eide, 1983; Peipins & Sandler, 1994). Also, in a high proportion of large adenomas (particularly large villous adenomas), there are foci of carcinoma (Muto *et al.*, 1975). The distribution of adenomas within the bowel differs from that of colorectal cancer (Peipins & Sandler, 1994), but large adenomas (carrying greater risk of developing malignancy) have a site distribution similar to that of cancer (Konishi & Morson, 1982; Matek *et al.*, 1986). Polyps (unknown histology) have been observed to grow in a location in which cancer was later found (Stryker *et al.*, 1987). Also, adenomas and carcinomas share a number of metabolic characteristics, including abnormalities in DNA ploidy, mucins, metabolic enzymes and cytoskeletal proteins (Tierney *et al.*, 1990). In each case, these changes follow what appears to be a progression of changes from normal mucosa through adenoma to carcinoma. It is important to note that this progression has been described largely in histopathological terms; it is actually a matter of dysplasia progressing to carcinoma. The term 'polyp', a gross pathology descriptor, has no particular histopathological meaning. The existence of the adenoma–carcinoma sequence does not necessarily imply that the polyp–carcinoma sequence holds all the time (Morson, 1984).

Perhaps the strongest evidence for the preneoplastic nature of adenomatous polyps is the somatic genetic changes seen in them. These somatic changes fit clearly into a progression, sitting between normal mucosa and invasive carcinoma (Kinzler & Vogelstein, 1996; Potter, 1999). There is a similarly high prevalence of mutations of the *APC* gene in colorectal cancer and in adenomas (even small adenomas) (Boland *et al.*, 1995; Kinzler & Vogelstein, 1996; Powell *et al.*, 1992); mutations in this gene seem to be important in turning individual cells onto a dysplastic path. The prevalence of Kirstin-*ras* (K-*ras*) and *p53* mutations increases as one considers progressively small adenomas (< 1 cm), large adenomas (≥ 1 cm) and invasive carcinoma (Boland *et al.*, 1995; Kinzler & Vogelstein, 1996). Various genetic changes have been seen at the transition from normal mucosa to small adenoma (*APC*), and from adenoma to severe dysplasia or carcinoma (*p53*, TGF-β in microsatellite unstable tumours) (Ahuja *et al.*, 1998; Boland *et al.*, 1995; Grady *et al.*, 1998).

Some mutations appear to be more common in early adenomas than in later tumours. β-Catenin mutations, for example, have been seen more commonly in small adenomas than in larger ones from the same patient population. It is possible that these mutations mark tumours that are unlikely to progress (at least in size) (Samowitz *et al.*, 1999), or this may have been a chance finding.

Assuming that colorectal cancers grow out of large adenomas and that large polyps were first small polyps, we can qualify small adenomas as precancerous lesions. Nonetheless, it is clear that only a small minority of these adenomas will progress to carcinoma within a human lifetime (Eide, 1986; Knoernschild, 1963; Peipins & Sandler, 1994) and some polyps (unknown histology) regress or disappear (Cole *et al.*, 1959; Hoff *et al.*, 1986; Hofstad *et al.*, 1994; Knoernschild, 1963; Nicholls *et al.*, 1988; Welin *et al.*, 1963). It is not known if large advanced adenomas grow from small advanced adenomas, or whether the advanced features are acquired with growth. It is also not clear that all large bowel adenocarcinomas have progressed through a stage of adenoma that could be distinctly recognized as such (and potentially removed before malignant conversion) (Bedenne *et al.*, 1992). Thus it is possible that an intervention that prevents only small, simple adenomatous polyps may not confer a benefit in terms of reduction of cancer incidence, if the adenomas prevented are those that would not progress. However, removing adenomas within the reach of a sigmoidoscope reduces the subsequent risk of cancer in the same region of the bowel, a finding that is difficult to explain on the basis of bias (Muller & Sonnenberg, 1995; Murakami *et al.*, 1990; Newcomb *et al.*, 1992; Selby *et al.*, 1992). Assuming that physical removal of adenomatous polyps does lower the risk of subsequent carcinoma, this is evidence that at least a substantial proportion of cancers would have grown from the small adenomas detected and removed. This also suggests that flat adenomas (presumably undetected by endoscopy) are not the principal precursors of carcinomas in the populations from which the reports derived.

Epidemiology of large bowel adenomas versus the epidemiology of colorectal cancer

In general, the epidemiology of large bowel adenomas appears to be similar to that of colorectal cancer itself (Potter, 1996; World Cancer Research Fund & American Institute for Cancer Research, 1997). Vegetable intake (and, less consistently, intake of fruits) has been inversely associated with risk, and red meat directly associated with both types of neoplasm. For both, cereal fibre has generally been unassociated with risk, while findings regarding dietary fat and other types of fibre have been variable. Protective effects of folate have been reported, particularly in association with high alcohol intake. In observational studies, calcium intake has been variably related to risk of both adenomas and cancer (Bergsma-Kadijk *et al.*, 1996; Martinez & Willett, 1998) and coffee intake may be inversely related to the risk of both (Giovannucci, 1998). Exercise seems inversely related to both colon cancer and colorectal adenomas, but findings regarding obesity have been variable for both types of tumour (Peipins & Sandler, 1994; Potter, 1996; World Cancer Research Fund & American Institute for Cancer Research, 1997). Associations with cholecystectomy also appear to be similar (Peipins & Sandler, 1994). Studies that included both colorectal cancers and colorectal adenomas have provided particularly interesting data in this regard, with broadly similar patterns of risk factors for invasive cancer and adenomas (Benito *et al.*, 1991, 1993; Boutron *et al.*, 1996; Faivre *et al.*, 1997; Fuchs *et al.*, 1999; Giovannucci *et al.*, 1993, 1994, 1995a, b, 1998; Kampman *et al.*, 1994; Kearney *et al.*, 1996; Kune *et al.*, 1987, 1991; Martinez *et al.*, 1996; Platz *et al.*, 1997), although one such study suggested a difference in the effect of folate (Boutron-Ruault *et al.*, 1996).

Adenomas in chemoprevention trials

In general, the findings from intervention trials that used adenomas as end-points have confirmed those from observational epidemiology. Recurrence of sporadic adenomatous polyps has been used as an end-point for several clinical trials of calcium supplementation. The largest of these, the Calcium Polyp Prevention Study, randomized 930 subjects to treatment with placebo or 3000 mg of calcium carbonate (1200 mg of elemental calcium). There was a modest protective effect of calcium. A subsequent trial, using 2 g of elemental calcium per day reported a similar relative risk, which failed to reach statistical significance with a smaller sample size (Faivre *et al.*, 1999). In several trials, β-carotene has had no effect on adenoma occurrence (Greenberg *et al.*, 1994; Kikendall *et al.*, 1990; MacLennan *et al.*, 1995), nor has ascorbate plus α-tocopherol (Greenberg *et al.*, 1994; McKeown-Eyssen *et al.*, 1988).

In agreement with the observational studies, cereal fibre supplements did not affect sporadic adenoma recurrence in two trials, either alone (Alberts *et al.*, 2000; MacLennan *et al.*, 1995) or with a low-fat diet (MacLennan *et al.*, 1995). Two other trials that investigated a low-fat/high-fibre diet also failed to find an effect (McKeown-Eyssen *et al.*, 1994; Schatzkin *et al.*, 2000). The earlier of these studies (McKeown-Eyssen *et al.*, 1994) relied heavily on a cereal fibre supplement, and so did not focus on the fibre moieties most strongly associated with reduced risk of colorectal neoplasia. In the more recent trial (Schatzkin *et al.*, 2000), subjects in the intervention arm were advised to increase dietary fibre by increasing intake of fruits and vegetables. No data were presented regarding changes in vegetable intake during the study (separately from changes in fruit). If vegetable intake increased substantially, these negative findings may conflict with the observational data.

In familial adenomatous polyposis (FAP), vitamin C supplementation yielded weak indications of a reduction in polyp burden (Bussey *et al.*, 1982), although in a larger study, ascorbate plus α-tocopherol had no effect (DeCosse *et al.*, 1989). There were weak suggestions that cereal fibre was similarly beneficial in conjunction with ascorbate plus α-tocopherol (DeCosse *et al.*, 1989). Several studies of sulindac have shown clear evidence that treatment with this nonsteroidal anti-inflammatory drug (NSAID) can lead to a reduction in polyps (Giardiello *et al.*, 1993; Labayle *et al.*, 1991; Nugent *et al.*, 1993).

Aberrant crypt foci

Aberrant crypt foci are groups of crypts that are morphologically altered: delineated from the surrounding glands, larger than normal, often with dilated lumina (Fenoglio-Preiser & Noffsinger, 1999). Vital stains such as methylene blue accentuate the aberrant crypt foci. The lesions are histologically heterogeneous, variably including hyperplastic and dysplastic crypts. The multiplicity of aberrant crypt foci (i.e., the number of crypts involved) can vary from one to as many as several hundred.

Aberrant crypt foci were first noted in rats treated with carcinogens, and several aspects of their occurrence suggest that they may be preneoplastic (Bird, 1995; McLellan *et al.*, 1991; Pretlow, 1995). Their numbers correlate with the dose of carcinogen used, and aberrant crypt foci have not often been seen in untreated animals, in those treated with carcinogens that target other organs or in those treated with toxic non-carcinogenic compounds such as cholic acid (Bird, 1995; Pretlow *et al.*, 1992). With increasing time after carcinogen administration, the number, size and degree of dysplasia of the aberrant crypt foci increase. Known promoters or inhibitors of colorectal carcinogenesis seem to have the same effect on the number and multiplicity of aberrant crypt foci (Bird, 1995). Clonal expansion and dysplasia may be separate processes (Bird, 1995; Jen *et al.*, 1994). In these experimental situations, aberrant crypt foci have been found to have some of the features of carcinogenesis, such as hyperproliferation and mutated *p53* (Fenoglio-Preiser & Noffsinger, 1999; Olivo & Wargovich, 1998; Pretlow, 1995; Roncucci, 1992). A natural history study found that after administration of carcinogen, areas of aberrant crypt foci were likely to later contain adenomas or carcinoma, although some areas of aberrant crypt foci clearly regressed (Shpitz *et al.*, 1996). Aberrant crypt foci have been widely used in animal studies of chemopreventive agents. In one large-scale investigation of numerous agents, the response of aberrant crypt foci to the interventions was thought to correlate reasonably well with earlier efficacy studies (Olivo & Wargovich, 1998; Wargovich *et al.*, 1996).

Aberrant crypt foci have also been observed in the human colorectal mucosa. Relatively high numbers have been found in the bowel mucosa of patients with FAP or with sporadic colorectal cancer (Fenoglio-Preiser & Noffsinger, 1999; Nascimbeni *et al.*, 1999; Roncucci *et al.*, 1991), but are also seen in patients with non-neoplastic bowel disorders (Fenoglio-Preiser & Noffsinger, 1999). The numbers of aberrant crypt foci are progressively higher in the more distal regions of the bowel (Bouzourene *et al.*, 1999; Roncucci *et al.*, 1991; Shpitz *et al.*, 1998; Yamashita *et al.*, 1995), although dysplastic aberrant crypt foci may be preferentially found more proximally (Nascimbeni *et al.*, 1999; Roncucci *et al.*, 1998).

Aberrant crypt foci may be observed *in vivo* using magnifying endoscopy and methylene blue staining (Takayama *et al.*, 1998). There appears to be an association between the appearance of the lumina of the crypts and dysplastic aberrant crypt foci (Fenoglio-Preiser & Noffsinger, 1999; Roncucci *et al.*, 1991). Although one study reported excellent concordance between endoscopic appearance and histological characteristics (Takayama *et al.*, 1998), it is likely that there is considerable measurement error in the assessment of aberrant crypt foci with magnifying endoscopy.

Some aberrant crypt foci seem to be monoclonal, and hence neoplastic (Fenoglio-Preiser & Noffsinger, 1999; Siu *et al.*, 1999). In humans, aberrant crypt foci have been found to display some of the characteristics of colorectal carcinogenesis: hyperproliferation, overexpression of carcinoembryonic antigen, MSI phenotype, and somatic *APC* and K-*ras* mutations (Augenlicht *et al.*, 1996; Fenoglio-Preiser & Noffsinger, 1999; Heinen *et al.*, 1996; Polyak *et al.*, 1996; Shpitz *et al.*, 1997; 1998). K-*ras* mutations are very common (in various studies ranging up to 100%; Fenoglio-Preiser & Noffsinger, 1999), but *APC* and *p53* mutations are relatively uncommon (Fenoglio-Preiser & Noffsinger, 1999; Jen *et al.*, 1994; Losi *et al.*, 1996; Otori *et al.*, 1998; Pretlow *et al.*, 1993; Smith *et al.*, 1994; Yamashita *et al.*, 1995). *APC* mutations in aberrant crypt foci may correlate with dysplastic histology (Otori *et al.*, 1998). Human aberrant crypt foci with carcinoma *in situ* have been observed (Konstantakos *et al.*, 1996; Siu *et al.*, 1997). In one study, a correlation of number of aberrant crypt foci with numbers of adenomas was observed (Takayama *et al.*, 1998). The histology of aberrant crypt foci is not uniform among the crypts: determination of the hyperplastic/dysplastic nature of the aberrant crypt foci may require serial sectioning, and mixed foci are common (Nascimbeni *et al.*, 1999; Pretlow, 1995; Siu *et al.*, 1997). It is possible that hyperplastic aberrant crypt foci may evolve into dysplastic aberrant crypt foci (Otori *et al.*, 1995). On the other hand, it has been suggested that aberrant crypt foci are precursors of both adenomas and hyperplastic polyps (Fenoglio-Preiser & Noffsinger, 1999). Aberrant crypt foci have been used in only one human chemoprevention study—a non-randomized parallel arm study that found sulindac to be very effective in inducing regression of aberrant crypt foci (Takayama *et al.*, 1998).

Stool K-*ras*

The mucosa of the large bowel is a rapidly proliferating tissue, turning over completely in four days (Smith-Ravin *et al.*, 1995). Despite the enzymes and debris present in the stool environment, sufficient DNA from cells shed into the lumen is passed in stool to permit amplification with the polymerase chain reaction (PCR) (Sidransky *et al.*, 1992). The amount of DNA obtained per 100 mg stool has varied from less than 1 μg to 9 μg. In patients with large bowel carcinoma, it is possible that the proportion of the DNA in stool derived from the tumour itself could be sufficient to permit amplification and detection (Sidransky *et al.*, 1992). The K-*ras* gene is particularly attractive for use in stool diagnostics because, when present, the mutations typically cluster in a few codons and are therefore relatively easy to detect.

Assay for mutated K-*ras* reveals a high proportion of carcinomas that themselves contain mutated K-*ras* (Hasegawa *et al.*, 1995; Nollau *et al.*, 1996; Ratto *et al.*, 1996; Sidransky *et al.*, 1992). Adenomas with mutated *ras* can also be detected in this way (Hasegawa *et al.*, 1995; Sidransky *et al.*, 1992). The sensitivity of the approach clearly depends on the methods used (Hasegawa *et al.*, 1995); sensitivity may be lower for lesions in the right bowel (Hasegawa *et al.*, 1995). Colonic effluent from bowel preparations has also been used as a source material (Tobi *et al.*, 1994), as have washings obtained during endoscopy (Smith-Ravin *et al.*, 1995). This technique revealed mutated *ras* in some patients with a history of colorectal neoplasia or with a family history, although colonoscopy disclosed no neoplasm (Tobi *et al.*, 1994; Villa *et al.*, 1996). Mutated *ras* has also been detected in the stool of patients with pancreatic cancer ductal hyperplasia (Caldas *et al.*, 1994).

Other molecular targets have also been described. A deletion in the *APC* gene has been detected in stool (Nollau *et al.*, 1996), but this issue has not been otherwise studied. The large size of the *APC* gene makes effective screening of this gene difficult in tumours, not to speak of stool. Similar considerations apply to *p53*. The amount of DNA on the surface of stool has also been investigated in one study: in individuals with colorectal cancer, a greater amount of DNA was detected (Loktionov *et al.*, 1998). The stool content of mRNA for CD44, a cell surface glycoprotein, has also been associated with large bowel adenocarcinoma (Yamao *et al.*, 1998).

Assessment of K-*ras* in stool has limited sensitivity for detection of large bowel neoplasia—the proportion of colorectal cancers that contain mutated *ras* is typically less than 50% (Andreyev *et al.*, 1998), although adenomas and aberrant crypt foci may have a higher prevalence of *ras* mutation (Jen *et al.*, 1994; Martinez *et al.*, 1999; McLellan *et al.*, 1993).

Mucosal proliferation

Mucosal proliferation is thought to play a prominent role in carcinogenesis, and consequently, has appeal as a biomarker. Mucosal proliferation in epithelial tissues was first measured using tissue explant culture of biopsy specimens, labelling dividing cells with tritiated thymidine, which is incorporated into DNA during S phase (Risio, 1994; Rozen, 1992). This technique suffers from the disadvantages of being time-consuming and requiring radioactive materials; and other, simpler labelling agents have superseded tritiated thymidine. One of these, bromodeoxyuridine, is also incorporated into DNA during S phase in tissue explant culture, but it can be recognized by immunohistochemistry, and so avoids the use of radioactive materials. Immunohistochemical detection of endogenous nuclear proteins, proliferating cell nuclear antigen and Ki-67 has also been used to assess proliferation, although these antigens are expressed over a wider range of the cell cycle than thymidine and bromodeoxyuridine (Biasco *et al.*, 1994; Risio, 1994). These techniques are relatively easy to use, since uptake of the label during explant culture is not required (Biasco *et al.*, 1994; Einspahr *et al.*, 1997; Risio, 1994; Rozen, 1992). The whole crypt mitotic count is another measure of proliferation for the bowel (Goodlad *et al.*, 1991; Murray *et al.*, 1995; Tosteson *et al.*, 1996). In limited studies, it did not show strong correlation with the labelling index (the proportion of cells in a crypt that are labelled) computed using proliferating cell nuclear antigen (Keku *et al.*, 1998; Murray *et al.*, 1995). Whole crypt production rate (Allan & Jewell, 1983) and flow cytometry (Nakamura *et al.*, 1995) have also been used in various studies.

There are clear impediments to the use of mucosal proliferation as a biomarker of risk or as an intermediate-effect biomarker. Application in epidemiological studies has been difficult, in part because of the demands on personnel to properly handle the biopsy specimens (Baron *et al.*, 1995a). Whatever the label, these techniques require manipulation of biopsy specimens in such a way that crypt architecture is maintained and then fixation until labelling and scoring. Scoring involves computation of the labelling index. Although measurement of proliferation indices can be reproducible (Bostick *et al.*, 1997a; Einspahr *et al.*, 1997; Lyles *et al.*, 1994), careful assessments of the variability of the measurements (Anti *et al.*, 1994a; Bostick *et al.*, 1997a; Lyles *et al.*, 1994; Macrae *et al.*, 1994; McShane *et al.*, 1998) have clearly identified the potential for measurement error and intra-subject variability.

Proliferation in the upper (luminal) parts of the crypt is a more sensitive marker of increased proliferation, and consequently the labelling index is also commonly computed in crypt compartments (typically quintiles of cells by position from bottom to top of crypt (Rozen, 1992). There is only a weak correlation between the overall labelling index and the proliferation in the upper crypts (Bostick *et al.*, 1997a; Risio *et al.*, 1991). Although there are differences in the labelling indices generated by the various labels, in general they are highly correlated (Bostick *et al.*, 1997a; Diebold *et al.*, 1992; Earnest *et al.*, 1993; Kubben *et al.*, 1994; Lacy *et al.*, 1991; Richter *et al.*, 1992; Weisgerber *et al.*, 1993). Nonetheless, some investigators have found proliferating cell nuclear antigen to be a less effective label than others (Bromley *et al.*, 1996; Risio *et al.*, 1993). No consistent differences have emerged in proliferation measurements in various segments of the large bowel (Cats *et al.*, 1991; Terpstra *et al.*, 1987).

Many studies have demonstrated that colorectal carcinomas and adenomas exhibit a higher labelling index than normal-appearing mucosa (Kanemitsu *et al.*, 1985; Risio *et al.*, 1988, 1993; Shpitz *et al.*, 1997). Moreover, the normal mucosa of individuals with colorectal cancer or colorectal adenomas seems to have higher indices than the mucosa of unaffected individuals (Bleiberg *et al.*, 1985; Lipkin *et al.*, 1987; Paganelli *et al.*, 1991; Risio *et al.*, 1991; Roncucci *et al.*, 1991; Stadler *et al.*, 1988; Terpstra *et al.*, 1987). The proliferative zone also has been observed to shift towards the lumen (Ponz de Leon *et al.*, 1988; Risio *et al.*, 1991; Terpstra *et al.*, 1987; Wilson *et al.*, 1990). A few studies, however, found no association of proliferation with the presence of adenomas or cancer (Keku *et al.*, 1998; Jass *et al.*, 1997; Kashtan *et al.*, 1993; Nakamura *et al.*, 1995; Wong *et al.*, 1995). There has been only one published investigation of the relationship between proliferative indices and subsequent neoplasia: high indices were predictive of adenoma recurrence (Anti *et al.*, 1993).

Although data are not consistent, FAP patients may have increased proliferation indices (Deschner & Lipkin, 1975; Lipkin *et al.*, 1984; Mills *et al.*, 1995; Nakamura *et al.*, 1993), and there have been indications of increased proliferation in normal-appearing mucosa of affected and unaffected subjects in other types of high-risk families (Cats *et al.*, 1991; Gerdes *et al.*, 1993; Lipkin *et al.*, 1984, 1985; Lynch *et al.*, 1985; Patchett *et al.*, 1997). One small study found no evidence of hyperproliferative epithelium in patients with hereditary nonpolyposis colon cancer (HNPCC) (Jass *et al.*, 1997). Increased proliferation has also been observed in normal mucosa of patients with hyperplastic polyps (Risio *et al.*, 1995).

In some studies, proliferation has been seen to increase with age (Paganelli *et al.*, 1990; Roncucci *et al.*, 1998), but in several large investigations, no age association was seen (Bostick *et al.*, 1997b; Caderni *et al.*, 1999; Keku *et al.*, 1998). In observational studies, mucosal proliferation has not been consistently associated with intake of any nutrient (Bostick *et al.*, 1997 b; Caderni *et al.*, 1999; Keku *et al.*, 1998). However, there have been suggestions that high intake of red meat may be associated with increased proliferation (Bostick *et al.*, 1997 b; Caderni *et al.*, 1999).

There have been numerous clinical trials of dietary supplementation that studied proliferation as an end-point. In one trial, supplementation with wheat bran had no effect on proliferation (overall, or in the upper crypt) (Alberts *et al.*, 1997), although in a smaller study (that used crypt cell production rate), decreased proliferation was reported (Rooney *et al.*, 1994). In another crossover study, a low-fat diet in association with oat bran or wheat bran had no effect on overall labelling index or proliferation in the upper crypt (Macrae *et al.*, 1997). In a five-day study, a high-fat diet (with much of the fat given as a bolus) was associated with a higher overall labelling index than a lower-fat diet or a high-fat diet without the bolus (Stadler *et al.*, 1988). Fish oil (eicosapentaenoic and docosahexaenoic acids) decreased proliferation both overall and in the upper crypt (Anti *et al.*, 1992, 1994b), and subjects given a fish oil supplement had lower proliferation than subjects given corn oil (Bartram *et al.*, 1993). In another study, only subjects with initially high proliferation had a response to fish oil (Anti *et al.*, 1994b), suggesting a component of regression to the mean.

In one investigation, a combination of vitamin A, α-tocopherol and ascorbic acid reduced proliferation in the upper crypts (Paganelli *et al.*, 1992). In another study, vitamin C lowered proliferation in all crypt compartments, while β-carotene (9 mg per day for a month) significantly reduced it only in the lower parts of the crypt, and α-tocopherol had no effect (Cahill *et al.*, 1993). Another study found that supplementation with 30 mg β-carotene did not affect proliferation in any crypt compartment (Frommel *et al.*, 1995), although an analysis, published only in an abstract, indicated a beneficial effect (lower proliferation) in the upper crypt (Macrae *et al.*, 1991).

Mucosal proliferation has been most intensively studied with regard to calcium supplementation. Here, the findings are quite mixed, as in the epidemiology of carcinoma and adenoma. Some studies have shown a decrease in proliferation with calcium supplementation (Barsoum *et al.*, 1992; Bostick *et al.*, 1995; Cats *et al.*, 1995; Lipkin & Newmark, 1985; Wargovich *et al.*, 1992), while others have shown no effect (Baron *et al.*, 1995b; Bostick *et al.*, 1993; Stern *et al.*, 1990; Weisgerber *et al.*, 1996) or even an increase (Gregoire *et al.*, 1989; Kleibeuker *et al.*, 1993). Two studies of high versus low consumption of dairy food also came to differing conclusions (Holt *et al.*, 1998; Karagas *et al.*, 1998).

Treatment with NSAIDs, consistently found to be inversely associated with risk of colorectal cancer and colorectal adenoma, appears to be unrelated to mucosal proliferation (Aoki *et al.*, 1996; Earnest *et al.*, 1993; Labayle *et al.*, 1991; Pasricha *et al.*, 1995; Spagnesi *et al.*, 1994). However, in one study of FAP, sulindac lowered the labelling index (Nugent *et al.*, 1993).

The observational epidemiology of large bowel mucosal proliferation is not well developed, but some published findings (NSAIDs, age) have diverged from clinical trial and epidemiological findings. However, in clinical trials, the mixed findings regarding calcium supplementation could be taken to reflect the weak findings in observational studies and the mixed epidemiological results.

References

Ahuja, N., Li, Q., Mohan, A.L., Baylin, S.B. & Issa, J.P. (1998) Aging and DNA methylation in colorectal mucosa and cancer. *Cancer Res.*, **58**, 5489–5494

Alberts, D., Einspahr, J., Ritenbaugh, C., Aickin, M., Rees-McGee, S., Atwood, J., Emerson, S., Mason-Liddil, N., Bettinger, L., Patel, J., Bellapravalu, S., Ramanujam, P.S., Phelps, J. & Clark, L.I. (1997) The effect of wheat bran fiber and calcium supplementation on rectal mucosal proliferation rates in patients with resected adenomatous colorectal polyps. *Cancer Epidemiol. Biomarkers Prev.*, **6**, 161–169

Alberts, D.S., Elena, M., Roe, D.J., Guillen-Rodriguez, J.M., Marshall, J.R., Leeuwen, J.B.v., Reid, M.E., Rittenbaugh, C., Vargas, P.A., Bhattacharyya, A.B., Earnest, D.L., Sampliner, R.E. and the Phenix Colon Cancer Prevention Physicians' Network. (2000) Lack of effect of a high-fiber cereal supplement on the recurrence of colorectal adenomas. *New Engl. J. Med.*, **342**, 1156–1162

Allan, A. & Jewell, D.P. (1983) In vitro model for the assessment of luminal factors on rectal mucosa. *Gut*, **24**, 812–817

Andreyev, H.J., Norman, A.R., Cunningham, D., Oates, J.R. & Clarke, P.A. (1998) Kirsten ras mutations in patients with colorectal cancer: the multicenter "RASCAL" study. *J. Natl Cancer Inst.*, **90**, 675–684

Anti, M., Marra, G., Armelao, F., Bartoli, G.M., Ficarelli, R., Percesepe, A., De Vitis, I., Maria, G., Sofo, L., Rapaccini, G.L., Gentiloni, N., Piccioni, E. & Miggiano, G. (1992) Effect of omega-3 fatty acids on rectal mucosal cell proliferation in subjects at risk for colon cancer. *Gastroenterology*, **103**, 883–891

Anti, M., Marra, G., Armelao, F., Percesepe, A., Ficarelli, R., Ricciuto, G.M., Valenti, A., Rapaccini, G.L., De Vitis, I., D'Agostino, G., Brighi, S. & Vecchio, F.M. (1993) Rectal epithelial cell proliferation patterns as predictors of adenomatous colorectal polyp recurrence. *Gut*, **34**, 525–530

Anti, M., Marra, G., Percesepe, A., Armelao, F. & Gasbarrini, G. (1994a) Reliability of rectal epithelial kinetic patterns as an intermediate biomarker of colon cancer. *J. Cell. Biochem.*, Suppl. **19**, 68–75

Anti, M., Armelao, F., Marra, G., Percesepe, A., Bartoli, G.M., Palozza, P., Parrella, P., Canetta, C., Gentiloni, N. & De Vitis, I. (1994b) Effects of different doses of fish oil on rectal cell proliferation in patients with sporadic colonic adenomas. *Gastroenterology*, **107**, 1892–1894

Aoki, T., Boland, C. & Brenner, D. (1992) Aspirin modulation of premalignant biomarkers in rectal mucosa of high-risk subjects. *Gastroenterology*, **110**, A484

Atkin, W.S., Morson, B.C. & Cuzick, J. (1992) Long-term risk of colorectal cancer after excision of rectosigmoid adenomas. *New Engl. J. Med.*, **326**, 658–662

Aubert, H., Treille, C., Faure, H., Fouret, J., Lachet, B. & Rachail, M. (1982) Interêt de la surveillance des malades polypectomisés dans la prévention du cancer réctocolique. A propos de 123 cas. *Gastroenterol. Clin. Biol.*, **6**, 183–187

Augenlicht, L.H., Richards, C., Corner, G. & Pretlow, T.P. (1996) Evidence for genomic instability in human colonic aberrant crypt foci. *Oncogene*, **12**, 1767–1772

Baron, J.A. (1996) Large bowel adenomas: markers of risk and endpoints. *J. Cell. Biochem.*, Suppl., **25**, 142–148

Baron, J.A., Wargovich, M.J., Tosteson, T.D., Sandler, R., Haile, R., Summers, R., van Stolk, R., Rothstein, R. & Weiss, J. (1995a) Epidemiological use of rectal proliferation measures. *Cancer Epidemiol. Biomarkers Prev.*, **4**, 57–61

Baron, J.A., Tosteson, T.D., Wargovich, M.J., Sandler, R., Mandel, J., Bond, J., Haile, R., Summers, R., van Stolk, R., Rothstein, R. & Weiss, J. (1995b) Calcium supplementation and rectal mucosal proliferation: a randomized controlled trial. *J. Natl Cancer Inst.*, **87**, 1303–1307

Barsoum, G.H., Hendrickse, C., Winslet, M.C., Youngs, D., Donovan, I.A., Neoptolemos, J.P. & Keighley, M.R. (1992) Reduction of mucosal crypt cell proliferation in patients with colorectal adenomatous polyps by dietary calcium supplementation. *Br. J. Surg.*, **79**, 581–583

Bartram, H.P., Gostner, A., Scheppach, W., Reddy, B.S., Rao, C.V., Dusel, G., Richter, F., Richter, A. & Kasper, H. (1993) Effects of fish oil on rectal cell proliferation, mucosal fatty acids, and prostaglandin E2 release in healthy subjects. *Gastroenterology*, **105**, 1317–1322

Bedenne, L., Faivre, J., Boutron, M.C., Piard, F., Cauvin, J.M. & Hillon, P. (1992) Adenoma-carcinoma sequence or "de novo" carcinogenesis? A study of adenomatous remnants in a population-based series of large bowel cancers. *Cancer*, **69**, 883–888

Benito, E., Stiggelbout, A., Bosch, F.X., Obrador, A., Kaldor, J., Mulet, M. & Muñoz, N. (1991) Nutritional factors in colorectal cancer risk: a case-control study in Majorca. *Int. J. Cancer*, **49**, 161–167

Benito, E., Cabeza, E., Moreno, V., Obrador, A. & Bosch, F.X. (1993) Diet and colorectal adenomas: a case-control study in Majorca. *Int. J. Cancer*, **55**, 213–219

Bergsma-Kadijk, J.A., van 't Veer, P., Kampman, E. & Burema, J. (1996) Calcium does not protect against colorectal neoplasia. *Epidemiology*, **7**, 590–597

Biasco, G., Paganelli, G.M., Santucci, R., Brandi, G. & Barbara, L. (1994) Methodological problems in the use of rectal cell proliferation as a biomarker of colorectal cancer risk. *J. Cell. Biochem.* Suppl., **19**, 55–60

Bird, R.P. (1995) Role of aberrant crypt foci in understanding the pathogenesis of colon cancer. *Cancer Letters*, **93**, 55–71

Bleiberg, H., Buyse, M. & Galand, P. (1985) Cell kinetic indicators of premalignant stages of colorectal cancer. *Cancer*, **56**, 124–129

Boland, C.R., Sato, J., Appelman, H.D., Bresalier, R.S. & Feinberg, A.P. (1995) Microallelotyping defines the sequence and tempo of allelic losses at tumour suppressor gene loci during colorectal cancer progression. *Nature Med.*, **1**, 902–909

Bostick, R.M., Potter, J.D., Fosdick, L., Grambsch, P., Lampe, J.W., Wood, J.R., Louis, T.A., Ganz, R. & Grandits, G. (1993) Calcium and colorectal epithelial cell proliferation: a preliminary randomized, double-blinded, placebo-controlled clinical trial. *J. Natl Cancer Inst.*, **85**, 132–141

Bostick, R.M., Fosdick, L., Wood, J.R., Grambsch, P., Grandits, G.A., Lillemoe, T.J., Louis, T.A. & Potter, J.D. (1995) Calcium and colorectal epithelial cell proliferation in sporadic adenoma patients: a randomized, double-blinded, placebo-controlled clinical trial. *J. Natl Cancer Inst.*, **87**, 1307–1315

Bostick, R.M., Fosdick, L., Lillemoe, T.J., Overn, P., Wood, J.R., Grambsch, P., Elmer, P. & Potter, J.D. (1997a) Methodological findings and considerations in measuring colorectal epithelial cell proliferation in humans. *Cancer Epidemiol. Biomarkers Prev.*, 6, 931–942

Bostick, R.M., Fosdick, L., Grandits, G.A., Lillemoe, T.J., Wood, J.R., Grambsch, P., Louis, T.A. & Potter, J.D. (1997b) Colorectal epithelial cell proliferative kinetics and risk factors for colon cancer in sporadic adenoma patients. *Cancer Epidemiol. Biomarkers Prev.*, **6**, 1011–1019

Boutron, M.C., Faivre, J., Marteau, P., Couillault, C., Senesse, P. & Quipourt, V. (1996) Calcium, phosphorus, vitamin D, dairy products and colorectal carcinogenesis: a French case–control study . *Br. J. Cancer*, **74**, 145–151

Boutron-Ruault, M.C., Senesse, P., Faivre, J., Couillault, C. & Belghiti, C. (1996) Folate and alcohol intakes: related or independent roles in the adenoma-carcinoma sequence? *Nutr. Cancer*, **26**, 337–346

Bouzourene, H., Chaubert, P., Seelentag, W., Bosman, F.T. & Saraga, E. (1999) Aberrant crypt foci in patients with neoplastic and nonneoplastic colonic disease. *Human Pathol.*, **30**, 66–71

Bromley, M., Rew, D., Becciolini, A., Balzi, M., Chadwick, C., Hewitt, D., Li, Y.Q. & Potten, C.S. (1996) A comparison of proliferation markers (BrdUrd, Ki-67, PCNA) determined at each cell position in the crypts of normal human colonic mucosa. *Eur. J. Histochem.*, **40**, 89–100

Bussey, H.J., DeCosse, J.J., Deschner, E.E., Eyers, A.A., Lesser, M.L., Morson, B.C., Ritchie, S.M., Thomson, J.P. & Wadsworth, J. (1982) A randomized trial of ascorbic acid in polyposis coli. *Cancer*, **50**, 1434–1439

Caderni, G., Palli, D., Lancioni, L., Russo, A., Luceri, C., Saieva, C., Trallori, G., Manneschi, L., Renai, F., Zacchi, S., Salvadori, M. & Dolara, P. (1999) Dietary determinants of colorectal proliferation in the normal mucosa of subjects with previous colon adenomas. *Cancer Epidemiol. Biomarkers Prev.*, **8**, 219–225

Cahill, R.J., O'Sullivan, K.R., Mathias, P.M., Beattie, S., Hamilton, H. & O'Morain, C. (1993) Effects of vitamin antioxidant supplementation on cell kinetics of patients with adenomatous polyps. *Gut*, **34**, 963–967

Caldas, C., Hahn, S.A., Hruban, R.H., Redston, M.S., Yeo, C.J. & Kern, S.E.I. (1994) Detection of K-ras mutations in the stool of patients with pancreatic adenocarcinoma and pancreatic ductal hyperplasia. *Cancer Res.*, **54**, 3568–3573

Cats, A., DeVries, E.G.E. & Kleibeuker, J.H. (1991) Proliferation rate in hereditary nonpolyposis colon cancer. *J. Natl Cancer Inst.*, **83**, 1687–1688

Cats, A., Kleibeuker, J.H., van der Meer, R., Kuipers, F., Sluiter, W.J., Hardonk, M.J., Oremus, E.T., Mulder, N.H. & de Vries, E.G. (1995) Randomized, double-blinded, placebo-controlled intervention study with supplemental calcium in families with hereditary nonpolyposis colorectal cancer. *J. Natl Cancer Inst.*, **87**, 598–603

Cole, J.W., Holden, W.D. & Cleveland, M.D. (1959) Postcolectomy regression of adenomatous polyps of the rectum. *Am. Med. Ass. Arch. Surg.*, **41**, 385–392

DeCosse, J.J., Miller, H.H. & Lesser, M.L. (1989) Effect of wheat fiber and vitamins C and E on rectal polyps in patients with familial adenomatous polyposis. *J. Natl Cancer Inst.*, **81**, 1290–1297

Deschner, E.E. & Lipkin, M. (1975) Proliferative patterns in colonic mucosa in familial polyposis. *Cancer*, **35**, 413–418

Diebold, J., Lai, M.D. & Lohrs, U. (1992) Analysis of proliferative activity in colorectal mucosa by immunohistochemical detection of proliferating cell nuclear antigen (PCNA) Methodological aspects and application to routine diagnostic material. *Virchows Arch. B, Cell Pathol. Molec. Pathol.*, **62**, 283–289

Earnest, D., Hixson, L., Fennerty, J., Einshpahr, J., Blackwell, G. & Alberts, D. (1993) Excellent agreement between 3 methods of measuring rectal epithelial cell proliferation in patients with resected colon cancer but lack of evidence of a suppressive effect by piroxicam treatment. *Gastroenterology*, **104**, A396

Eide, T.J. (1983) Remnants of adenomas in colorectal carcinomas. *Cancer*, **51**, 1866–1872

Eide, T.J. (1986) Risk of colorectal cancer in adenoma-bearing individuals within a defined population. *Int. J. Cancer*, **38**, 173–176

Einspahr, J.G., Alberts, D.S., Gapstur, S.M., Bostick, R.M., Emerson, S.S. & Gerner, E.W. (1997) Surrogate end-point biomarkers as measures of colon cancer risk and their use in cancer chemoprevention trials. *Cancer Epidemiol. Biomarkers Prev.*, **6**, 37–48

Faivre, J., Boutron, M.C., Senesse, P., Couillault, C., Belighiti, C. & Meny, B. (1997) Environmental and familial risk factors in relation to the colorectal adenoma–carcinoma sequence: results of a case control study in Burgundy (France) *Eur. J. Cancer Prev.*, **6**, 127–131

Faivre, J., Bonithon-Kopp, C., Kronborg, O., Rath, U. & Giacosa, A. (1999) A randomized trial of calcium and fiber supplementa0tion in the prevention of recurrence of colorectal adenomas. A European intervention study. *Gastroenterology*, **116**, A56

Fenoglio-Preiser, C. & Noffsinger, A. (1999) Aberrant crypt foci: a review. *Toxicol. Pathol.*, **27**, 632–642

Frommel, T.O., Mobarhan, S., Doria, M., Halline, A.G., Luk, G.D., Bowen, P.E., Candel, A. & Liao, Y. (1995) Effect of beta-carotene supplementation on indices of colonic cell proliferation. *J. Natl Cancer Inst.*, **87**, 1781–1787

Fuchs, C.S., Giovannucci, E.L., Colditz, G.A., Hunter, D.J., Stampfer, M.J., Rosner, B., Speizer, F.E. & Willett, W.C. (1999) Dietary fiber and the risk of colorectal cancer and adenoma in women. *New Engl. J. Med.*, **340**, 169–176

Gatteschi, B., Costantini, M., Bruzzi, P., Merlo, F., Torcoli, R. & Nicolo, G. (1991) Univariate and multivariate analyses of the relationship between adenocarcinoma and solitary and multiple adenomas in colorectal adenoma patients. *Int. J. Cancer*, **49**, 509–512

Gerdes, H., Gillin, J.S., Zimbalist, E., Urmacher, C., Lipkin, M. & Winawer, S.J. (1993) Expansion of the epithelial cell proliferative compartment and frequency of adenomatous polyps in the colon correlate with the strength of family history of colorectal cancer. *Cancer Res.*, **53**, 279–282

Giardiello, F.M., Hamilton, S.R., Krush, A.J., Piantadosi, S., Hylind, L.M., Celano, P., Booker, S.V., Robinson, C.R. & Offerhaus, G.J. (1993) Treatment of colonic and rectal adenomas with sulindac in familial adenomatous polyposis. *New Engl. J. Med.*, **328**, 1313–1316

Giovannucci, E. (1998) Meta-analysis of coffee consumption and risk of colorectal cancer. *Am. J. Epidemiol.*, **147**, 1043–1052

Giovannucci, E., Stampfer, M.J., Colditz, G.A., Rimm, E.B., Trichopoulos, D., Rosner, B.A., Speizer, F.E. & Willett, W.C. (1993) Folate, methionine, and alcohol intake and risk of colorectal adenoma. *J. Natl Cancer Inst.*, **85**, 875–884

Giovannucci, E., Rimm, E.B., Stampfer, M.J., Colditz, G.A., Ascherio, A. & Willett, W.C. (1994) Intake of fat, meat, and fiber in relation to risk of colon cancer in men. *Cancer Res.*, **54**, 2390–2397

Giovannucci, E., Ascherio, A., Rimm, E.B., Colditz, G.A., Stampfer, M.J. & Willett, W.C. (1995a) Physical activity, obesity, and risk for colon cancer and adenoma in men. *Ann. Int. Med.*, **122**, 327–334

Giovannucci, E., Rimm, E.B., Ascherio, A., Stampfer, M.J., Colditz, G.A. & Willett, W.C. (1995b) Alcohol, low-methionine-low-folate diets, and risk of colon cancer in men. *J. Natl Cancer Inst.*, **87**, 265–273

Giovannucci, E., Stampfer, M.J., Colditz, G.A., Hunter, D.J., Fuchs, C., Rosner, B.A., Speizer, F.E. & Willett, W.C. (1998) Multivitamin use, folate, and colon cancer in women in the Nurses' Health Study. *Ann. Int. Med.*, **129**, 517–524

Goodlad, R.A., Levi, S., Lee, C.Y., Mandir, N., Hodgson, H. & Wright, N.A. (1991) Morphometry and cell proliferation in endoscopic biopsies: evaluation of a technique. *Gastroenterology*, **101**, 1235–1241

Grady, W.M., Rajput, A., Myeroff, L., Liu, D.F., Kwon, K., Willis, J. & Markowitz, S. (1998) Mutation of the type II transforming growth factor-beta receptor is coincident with the transformation of human colon adenomas to malignant carcinomas. *Cancer Res.*, **58**, 3101–3104

Greenberg, E.R., Baron, J.A., Tosteson, T.D., Freeman, D.H., Jr, Beck, G.J., Bond, J.H., Colacchio, T.A., Coller, J.A., Frankl, H.D., Haile, R.W., Mandel, J.S., Nierenberg, D.W., Rothstein, R., Snover, D.C., Stevens, M.H., Summers, R.W. & van Stolk, R.W. for the Polyp Prevention Study Group (1994) A clinical trial of antioxidant vitamins to prevent colorectal adenoma. Polyp Prevention Study Group. *New Engl. J. Med.*, **331**, 141–147

Gregoire, R.C., Stern, H.S., Yeung, K.S., Stadler, J., Langley, S., Furrer, R. & Bruce, W.R. (1989) Effect of calcium supplementation on mucosal cell proliferation in high risk patients for colon cancer. *Gut*, **30**, 376–382

Grossman, S., Milos, M.L., Tekawa, I.S. & Jewell, N.P. (1989) Colonoscopic screening of persons with suspected risk factors for colon cancer: II. Past history of colorectal neoplasms. *Gastroenterology*, **96**, 299–306

Hasegawa, Y., Takeda, S., Ichii, S., Koizumi, K., Maruyama, M., Fujii, A., Ohta, H., Nakajima, T., Okuda, M., Baba, S. & Nakamura, Y. (1995) Detection of K-ras mutations in DNAs isolated from feces of patients with colorectal tumors by mutant-allele-specific amplification (MASA) *Oncogene*, **10**, 1441–1445

Heinen, C.D., Shivapurkar, N., Tang, Z., Groden, J. & Alabaster, O. (1996) Microsatellite instability in aberrant crypt foci from human colons. *Cancer Res.*, **56**, 5339–5341

Hixson, L.J., Fennerty, M.B., Sampliner, R.E., McGee, D. & Garewal, H. (1990) Prospective study of the frequency and size distribution of polyps missed by colonoscopy. *J. Natl Cancer Inst.*, **82**, 1769–1772

Hixson, L.J., Fennerty, M.B., Sampliner, R.E. & Garewal, H.S. (1991) Prospective blinded trial of the colonoscopic miss-rate of large colorectal polyps. *Gastrointest. Endosc.*, **37**, 125–127

Hoff, G. & Vatn, M. (1985) Epidemiology of polyps in the rectum and sigmoid colon. Endoscopic evaluation of size and localization of polyps. *Scand. J. Gastroenterol.*, **20**, 356–360

Hoff, G., Foerster, A., Vatn, M.H., Sauar, J. & Larsen, S. (1986) Epidemiology of polyps in the rectum and colon. Recovery and evaluation of unresected polyps 2 years after detection. *Scand. J. Gastroenterol.*, **21**, 853–862

Hofstad, B., Vatn, M., Larsen, S. & Osnes, M. (1994) Growth of colorectal polyps: recovery and evaluation of unresected polyps of less than 10 mm, 1 year after detection. *Scand. J. Gastroenterol.*, **29**, 640–645

Holt, P.R., Atillasoy, E.O., Gilman, J., Guss, J., Moss, S.F., Newmark, H., Fan, K., Yang, K. & Lipkin, M. (1998) Modulation of abnormal colonic epithelial cell proliferation and differentiation by low-fat dairy foods: a randomized controlled trial. *J. Am. Med. Ass.*, **280**, 1074–1079

Jaramillo, E., Watanabe, M., Slezak, P. & Rubio, C. (1995) Flat neoplastic lesions of the colon and rectum detected by high-resolution video endoscopy and chromoscopy. *Gastrointes. Endosc.*, **42**, 114–122

Jass, J.R., Ajioka, Y., Radojkovic, M., Allison, L.J. & Lane, M.R. (1997) Failure to detect colonic mucosal hyperproliferation in mutation positive members of a family with hereditary non-polyposis colorectal cancer. *Histopathology*, **30**, 201–207

Jen, J., Powell, S.M., Papadopoulos, N., Smith, K.J., Hamilton, S.R., Vogelstein, B. & Kinzler, K.W. (1994) Molecular determinants of dysplasia in colorectal lesions. *Cancer Res.*, **54**, 5523–5526

Kampman, E., Giovannucci, E., van 't Veer, P., Rimm, E., Stampfer, M.J., Colditz, G.A., Kok, F.J. & Willett, W.C. (1994) Calcium, vitamin D, dairy foods, and the occurrence of colorectal adenomas among men and women in two prospective studies. *Am. J. Epidemiol.*, **139**, 16–29

Kanemitsu, T., Koike, A. & Yamamoto, S. (1985) Study of the cell proliferation kinetics in ulcerative colitis, adenomatous polyps, and cancer. *Cancer*, **56**, 1094–1098

Karagas, M.R., Tosteson, T.D., Greenberg, E.R., Rothstein, R.I., Roebuck, B.D., Herrin, M. & Ahnen, D. (1998) Effects of milk and milk products on rectal mucosal cell proliferation in humans. *Cancer Epidemiol. Biomarkers Prev.*, **7**, 757–766

Kashtan, H., Gregoire, R.C., Hay, K. & Stern, H.S. (1993) Colonic epithelial proliferation indices before and after colon cancer removal. *Cancer Invest.*, **11**, 113–117

Kearney, J., Giovannucci, E., Rimm, E.B., Ascherio, A., Stampfer, M.J., Colditz, G.A., Wing, A., Kampman, E. & Willett, W.C. (1996) Calcium, vitamin D, and dairy foods and the occurrence of colon cancer in men. *Am. J. Epidemiol.*, **143**, 907–917

Keku, T.O., Galanko, J.A., Murray, S.C., Woosley, J.T. & Sandler, R.S. (1998) Rectal mucosal proliferation, dietary factors, and the risk of colorectal adenomas. *Cancer Epidemiol. Biomarkers Prev.*, **7**, 993–999

Kikendall, J.W., Burgess, M., Bowen, P.E. & Walter Reed Army Medical Center (1990) Effect of oral beta carotene on recurrence of colonic adenomas. *Gastroenterology*, **98**, A289

Kinzler, K.W. & Vogelstein, B. (1996) Lessons from hereditary colorectal cancer. *Cell*, **87**, 159–170

Kleibeuker, J.H., Welberg, J.W., Mulder, N.H., van der Meer, R., Cats, A., Limburg, A.J., Kreumer, W.M., Hardonk, M.J. & de Vries, E.G. (1993) Epithelial cell proliferation in the sigmoid colon of patients with adenomatous polyps increases during oral calcium supplementation. *Br. J. Cancer*, **67**, 500–503

Knoernschild, H. (1963) Growth rate and malignant potential of colonic polyps: early results. *Surg. Forum*, **14**, 137

Konishi, F. & Morson, B.C. (1982) Pathology of colorectal adenomas: a colonoscopic survey. *J. Clin. Pathol.*, **35**, 830–841

Konstantakos, A.K., Siu, I.M., Pretlow, T.G., Stellato, T.A. & Pretlow, T.P. (1996) Human aberrant crypt foci with carcinoma in situ from a patient with sporadic colon cancer. *Gastroenterology*, **111**, 772–777

Kubben, F.J., Peeters-Haesevoets, A., Engels, L.G., Baeten, C.G., Schutte, B., Arends, J.W., Stockbrugger, R.W. & Blijham, G.H. (1994) Proliferating cell nuclear antigen (PCNA): a new marker to study human colonic cell proliferation. *Gut*, **35**, 530–535

Kune, S., Kune, G.A. & Watson, L.F. (1987) Case-control study of dietary etiological factors: the Melbourne Colorectal Cancer Study. *Nutr. Cancer*, **9**, 21–42

Kune, G.A., Kune, S., Read, A., MacGowan, K., Penfold, C. & Watson, L.F. (1991) Colorectal polyps, diet, alcohol, and family history of colorectal cancer: a case-control study. *Nutr. Cancer*, **16**, 25–30

Labayle, D., Fischer, D., Vielh, P., Drouhin, F., Pariente, A., Bories, C., Duhamel, O., Trousset, M. & Attali, P. (1991) Sulindac causes regression of rectal polyps in familial adenomatous polyposis. *Gastroenterology*, **101**, 635–639

Lacy, E.R., Kuwayama, H., Cowart, K.S., King, J.S., Deutz, A.H. & Sistrunk, S. (1991) A rapid, accurate, immunohistochemical method to label proliferating cells in the digestive tract. A comparison with tritiated thymidine. *Gastroenterology*, **100**, 259–262

Lanspa, S.J., Rouse, J., Smyrk, T., Watson, P., Jenkins, J.X. & Lynch, H.T. (1992) Epidemiologic characteristics of the flat adenoma of Muto. A prospective study. *Dis. Colon Rectum*, **35**, 543–546

Levin, T.R., Palitz, A., Grossman, S., Conell, C., Finkler, L., Ackerson, L., Rumore, G. & Selby, J.V. (1999) Predicting advanced proximal colonic neoplasia with screening sigmoidoscopy. *J. Am. Med. Ass.*, **281**, 1611–1617

Lieberman, D.A. & Smith, F.W. (1991) Screening for colon malignancy with colonoscopy. *Am. J. Gastroenterol.*, **86**, 946–951

Lipkin, M. & Newmark, H. (1985) Effect of added dietary calcium on colonic epithelial-cell proliferation in subjects at high risk for familial colonic cancer. *New Engl. J. Med.*, **313**, 1381–1384

Lipkin, M., Blattner, W.A., Gardner, E.J., Burt, R.W., Lynch, H., Deschner, E., Winawer, S. & Fraumeni, J.F., Jr (1984) Classification and risk assessment of individuals with familial polyposis, Gardner's syndrome, and familial non-polyposis colon cancer from [^{3}H]thymidine labeling patterns in colonic epithelial cells. *Cancer Res.*, **44**, 4201–4207

Lipkin, M., Uehara, K., Winawer, S., Sanchez, A., Bauer, C., Phillips, R., Lynch, H.T., Blattner, W.A. & Fraumeni, J.F., Jr (1985) Seventh-Day Adventist vegetarians have a quiescent proliferative activity in colonic mucosa. *Cancer Lett.*, **26**, 134–144

Lipkin, M., Enker, W.E. & Winawer, S.J. (1987) Tritiated-thymidine labeling of rectal epithelial cells in 'non-prep' biopsies of individuals at increased risk for colonic neoplasia. *Cancer Lett.*, **37**, 153–161

Loktionov, A., O'Neill, I.K., Silvester, K.R., Cummings, J.H., Middleton, S.J. & Miller, R. (1998) Quantitation of DNA from exfoliated colonocytes isolated from human stool surface as a novel noninvasive screening test for colorectal cancer. *Clin. Cancer Res.*, **4**, 337–342

Losi, L., Roncucci, L., di Gregorio, C., de Leon, M.P. & Benhattar, J. (1996) K-ras and p53 mutations in human colorectal aberrant crypt foci. *J. Pathol.*, **178**, 259–263

Lotfi, A.M., Spencer, R.J., Ilstrup, D.M. & Melton, L.J.D. (1986) Colorectal polyps and the risk of subsequent carcinoma. *Mayo Clinic Proc.*, **61**, 337–343

Lyles, C.M., Sandler, R.S., Keku, T.O., Kupper, L.L., Millikan, R.C., Murray, S.C., Bangdiwala, S.I. & Ulshen, M.H. (1994) Reproducibility and variability of the rectal mucosal proliferation index using proliferating cell nuclear antigen immunohistochemistry. *Cancer Epidemiol. Biomarkers Prev.*, **3**, 597–605

Lynch, H.T., Schuelke, G.S., Kimberling, W.J., Albano, W.A., Lynch, J.F., Biscone, K.A., Lipkin, M.L., Deschner, E.E., Mikol, Y.B., Sandberg, A.A., Elston, R.C., Baily-Wilson, J.E. & Davis, B.S. (1985) Hereditary nonpolyposis colorectal cancer (Lynch syndromes I and II) II. Biomarker studies. *Cancer*, **56**, 939–951

MacLennan, R., Macrae, F., Bain, C., Battistutta, D., Chapuis, P., Gratten, H., Lambert, J., Newland, R.C., Ngu, M., Russell, A., Ward, M. & Wahlqvist, M.L., the Australian Polyp Prevention Project (1995) Randomized trial of intake of fat, fiber, and beta carotene to prevent colorectal adenomas. The Australian Polyp Prevention Project. *J. Natl Cancer Inst.*, **87**, 1760–1766

Macrae, F.A., Hughes, N.R., Bhathal, P.S., Tay, D., Selble, L. & MacLennan, R. (1991) Dietary suppression of rectal epithelial cell proliferation. *Gastroenterology*, **100**, A383

Macrae, F.A., Kilias, D., Sharpe, K., Hughes, N., Young, G.P. & MacLennan, R. (1994) Rectal epithelial cell proliferation: comparison of errors of measurement withinter-subject variance. Australian Polyp Prevention Project Investigators. *J. Cell. Biochem.*, Suppl., **19**, 84–90

Macrae, F.A., Kilias, D., Selbie, L., Abbott, M., Sharpe, K. & Young, G.P. (1997) Effect of cereal fibre source and processing on rectal epithelial cell proliferation. *Gut*, **41**, 239–244

Martinez, M.E. & Willett, W.C. (1998) Calcium, vitamin D, and colorectal cancer: a review of the epidemiologic evidence. *Cancer Epidemiol. Biomarkers Prev.*, **7**, 163–168

Martinez, M.E., Giovannucci, E.L., Colditz, G.A., Stampfer, M.J., Hunter, D.J., Speizer, F.E., Wing, A. & Willett, W.C. (1996) Calcium, vitamin D, and the occurrence of colorectal cancer among women. *J. Natl Cancer Inst.*, **88**, 1375–1382

Martinez, M.E., Maltzman, T., Marshall, J.R., Einspahr, J., Reid, M.E., Sampliner, R., Ahnen, D.J., Hamilton, S.R. & Alberts, D.S. (1999) Risk factors for Ki-ras protooncogene mutation in sporadic colorectal adenomas. *Cancer Res.*, **59**, 5181–5185

Matek, W., Hermanek, P. & Demling, L. (1986) Is the adenoma-carcinoma sequence contradicted by the differing location of colorectal adenomas and carcinomas? *Endoscopy*, **18**, 17–19

McKeown-Eyssen, G., Holloway, C., Jazmaji, V., Bright-See, E., Dion, P. & Bruce, W.R. (1988) A randomized trial of vitamins C and E in the prevention of recurrence of colorectal polyps. *Cancer Res.*, **48**, 4701–4705

McKeown-Eyssen, G.E., Bright-See, E., Bruce, W.R., Jazmaji, V., Cohen, L.B., Pappas, S.C. & Saibil, F.G. (1994) A randomized trial of a low fat high fibre diet in the recurrence of colorectal polyps. Toronto Polyp Prevention Group [erratum appears in *J. Clin. Epidemiol.*, 1995, **48**(2):i]. *J. Clin. Epidemiol.*, **47**, 525–536

McLellan, E.A., Medline, A. & Bird, R.P. (1991) Sequential analyses of the growth and morphological characteristics of aberrant crypt foci: putative preneoplastic lesions. *Cancer Res.*, **51**, 5270–5274

McLellan, E.A., Owen, R.A., Stepniewska, K.A., Sheffield, J.P. & Lemoine, N.R. (1993) High frequency of K-ras mutations in sporadic colorectal adenomas. *Gut*, **34**, 392–396

McShane, L.M., Kulldorff, M., Wargovich, M.J., Woods, C., Purewal, M., Freedman, L.S., Corle, D.K., Burt, R.W., Mateski, D.J., Lawson, M., Lanza, E., O'Brien, B., Lake, W., Jr, Moler, J. & Schatzkin, A. (1998) An evaluation of rectal mucosal proliferation measure variability sources in the polyp prevention trial: can we detect informative differences among individuals' proliferation measures amid the noise? *Cancer Epidemiol. Biomarkers Prev.*, **7**, 605–612

Mills, S.J., Shepherd, N.A., Hall, P.A., Hastings, A., Mathers, J.C. & Gunn, A. (1995) Proliferative compartment deregulation in the non-neoplastic colonic epithelium of familial adenomatous polyposis. *Gut*, **36**, 391–394

Morson, B.C. (1984) The polyp story. *Postgrad. Med. J.*, **60**, 820–824

Muller, A.D. & Sonnenberg, A. (1995) Prevention of colorectal cancer by flexible endoscopy and polypectomy. A case-control study of 32,702 veterans. *Ann. Int. Med.*, **123**, 904–910

Murakami, R., Tsukuma, H., Kanamori, S., Imanishi, K., Otani, T., Nakanishi, K., Fujimoto, I. & Oshima, A. (1990) Natural history of colorectal polyps and the effect of polypectomy on occurrence of subsequent cancer. *Int. J. Cancer*, **46**, 159–164

Murray, S.C., Sandler, R.S., Keku, T.O., Lyles, C.M., Millikan, R.C., Bangdiwala, S.I., Kupper, L.L., Jiang, W. & Ulshen, M.H. (1995) Comparison of rectal mucosal proliferation measured by proliferating cell nuclear antigen (PCNA) immunohistochemistry and whole crypt dissection. *Cancer Epidemiol. Biomarkers Prev.*, **4**, 715–720

Muto, T., Bussey, H.J. & Morson, B.C. (1975) The evolution of cancer of the colon and rectum. *Cancer*, **36**, 2251–2270

Nakamura, S., Kino, I. & Baba, S. (1993) Nuclear DNA content of isolated crypts of background colonic mucosa from patients with familial adenomatous polyposis and sporadic colorectal cancer. *Gut*, **34**, 1240–1244

Nakamura, S., Goto, J., Kitayama, Y., Sheffield, J.P. & Talbot, I.C. (1995) Flow cytometric analysis of DNA synthetic phase fraction of the normal appearing colonic mucosa in patients with colorectal neoplasms. *Gut*, **37**, 398–401

Nascimbeni, R., Villanacci, V., Mariani, P.P., Di Betta, E., Ghirardi, M., Donato, F. & Salerni, B. (1999) Aberrant crypt foci in the human colon: frequency and histologic patterns in patients with colorectal cancer or diverticular disease. *Am. J. Surg. Pathol.*, **23**, 1256–1263

Newcomb, P.A., Norfleet, R.G., Storer, B.E., Surawicz, T.S. & Marcus, P.M. (1992) Screening sigmoidoscopy and colorectal cancer mortality. *J. Natl Cancer Inst.*, **84**, 1572–1575

Nicholls, R.J., Springall, R.G. & Gallagher, P. (1988) Regression of rectal adenomas after colectomy and ileorectal anastomosis for familial adenomatous polyposis. *Br. Med. J.*, **296**, 1707–1708

Nollau, P., Moser, C., Weinland, G. & Wagener, C. (1996) Detection of K-ras mutations in stools of patients with colorectal cancer by mutant-enriched PCR. *Int. J. Cancer*, **66**, 332–336

Nugent, K.P., Farmer, K.C., Spigelman, A.D., Williams, C.B. & Phillips, R.K. (1993) Randomized controlled trial of the effect of sulindac on duodenal and rectal polyposis and cell proliferation in patients with familial adenomatous polyposis. *Br. J. Surg.*, **80**, 1618–1619

Olivo, S. & Wargovich, M.J. (1998) Inhibition of aberrant crypt foci by chemopreventive agents. *In Vivo*, **12**, 159–166

Otori, K., Sugiyama, K., Hasebe, T., Fukushima, S. & Esumi, H. (1995) Emergence of adenomatous aberrant crypt foci (ACF) from hyperplastic ACF with concomitant increase in cell proliferation. *Cancer Res.*, **55**, 4743–4746

Otori, K., Konishi, M., Sugiyama, K., Hasebe, T., Shimoda, T., Kikuchi-Yanoshita, R., Mukai, K., Fukushima, S., Miyaki, M. & Esumi, H. (1998) Infrequent somatic mutation of the adenomatous polyposis coli gene in aberrant crypt foci of human colon tissue. *Cancer*, **83**, 896–900

Owen, D.A. (1996) Flat adenoma, flat carcinoma, and de novo carcinoma of the colon. *Cancer*, **77**, 3–6

Paganelli, G.M., Santucci, R., Biasco, G., Miglioli, M. & Barbara, L. (1990) Effect of sex and age on rectal cell renewal in humans. *Cancer Lett.*, **53**, 117–121

Paganelli, G.M., Biasco, G., Santucci, R., Brandi, G., Lalli, A.A., Miglioli, M. & Barbara, L. (1991) Rectal cell proliferation and colorectal cancer risk level in patients with nonfamilial adenomatous polyps of the large bowel. *Cancer*, **68**, 2451–2454

Paganelli, G.M., Biasco, G., Brandi, G., Santucci, R., Gizzi, G., Villani, V., Cianci, M., Miglioli, M. & Barbara, L. (1992) Effect of vitamin A, C, and E supplementation on rectal cell proliferation in patients with colorectal adenomas. *J. Natl Cancer Inst.*, **84**, 47–51

Pasricha, P.J., Bedi, A., O'Connor, K., Rashid, A., Akhtar, A.J., Zahurak, M.L., Piantadosi, S., Hamilton, S.R. & Giardiello, F.M. (1995) The effects of sulindac on colorectal proliferation and apoptosis in familial adenomatous polyposis. *Gastroenterology*, **110**, 994–998

Patchett, S.E., Alstead, E.M., Saunders, B.P., Hodgson, S.V. & Farthing, M.J. (1997) Regional proliferative patterns in the colon of patients at risk for hereditary nonpolyposis colorectal cancer. *Dis. Colon Rectum*, **40**, 168–171

Peipins, L.A. & Sandler, R.S. (1994) Epidemiology of colorectal adenomas. *Epidemiol. Rev.*, **16**, 273–297

Platz, E.A., Giovannucci, E., Rimm, E.B., Rockett, H.R., Stampfer, M.J., Colditz, G.A. & Willett, W.C. (1997) Dietary fiber and distal colorectal adenoma in men. *Cancer Epidemiol. Biomarkers Prev.*, **6**, 661–670

Polyak, K., Hamilton, S.R., Vogelstein, B. & Kinzler, K.W. (1996) Early alteration of cell-cycle-regulated gene expression in colorectal neoplasia. *Am. J. Pathol.*, **149**, 381–387

Ponz de Leon, M., Roncucci, L., Di Donato, P., Tassi, L., Smerieri, O., Amorico, M.G., Malagoli, G., De Maria, D., Antonioli, A., Chahin, N.J., Perini, M., Rigo, G., Barberini, G., Manenti, A., Biasco, G. & Barbara, L. (1988) Pattern of epithelial cell proliferation in colorectal mucosa of normal subjects and of patients with adenomatous polyps or cancer of the large bowel. *Cancer Res.*, **48**, 4121–4126

Potter, J.D. (1996) Nutrition and colorectal cancer. *Cancer Causes Control*, **7**, 127–146

Potter, J.D. (1999) Colorectal cancer: molecules and populations. *J. Natl Cancer Inst.*, **91**, 916–932

Powell, S.M., Zilz, N., Beazer-Barclay, Y., Bryan, T.M., Hamilton, S.R., Thibodeau, S.N., Vogelstein, B. & Kinzler, K.W. (1992) APC mutations occur early during colorectal tumorigenesis. *Nature*, **359**, 235–237

Pretlow, T.P. (1995) Aberrant crypt foci and K-ras mutations: earliest recognized players or innocent bystanders in colon carcinogenesis? *Gastroenterology*, **108**, 600–603

Pretlow, T.P., O'Riordan, M.A., Pretlow, T.G. & Stellato, T.A. (1992) Aberrant crypts in human colonic mucosa: putative preneoplastic lesions. *J. Cell. Biochem.*, Suppl., **16G**, 55–62

Pretlow, T.P., Brasitus, T.A., Fulton, N.C., Cheyer, C. & Kaplan, E.L. (1993) K-ras mutations in putative preneoplastic lesions in human colon. *J. Natl Cancer Inst.*, **85**, 2004–2007

Ratto, C., Flamini, G., Sofo, L., Nucera, P., Ippoliti, M., Curigliano, G., Ferretti, G., Sgambato, A., Merico, M., Doglietto, G.B., Cittadini, A. & Crucitti, F. (1996) Detection of oncogene mutation from neoplastic colonic cells exfoliated in feces. *Dis. Colon Rectum*, **39**, 1238–1244

Rex, D.K., Cutler, C.S., Lemmel, G.T., Rahmani, E.Y., Clark, D.W., Helper, D.J., Lehman, G.A. & Mark, D.G. (1997) Colonoscopic miss rates of adenomas determined by back-to-back colonoscopies. *Gastroenterology*, **112**, 24–28

Richter, F., Richter, A., Yang, K. & Lipkin, M. (1992) Cell proliferation in rat colon measured with bromodeoxyuridine, proliferating cell nuclear antigen, and [^{3}H]thymidine. *Cancer Epidemiol. Biomarkers Prev.*, **1**, 561–566

Risio, M. (1994) Methodological aspects of using immunohistochemical cell proliferation biomarkers in colorectal carcinoma chemoprevention. *J. Cell. Biochem.*, Suppl., **19**, 61–67

Risio, M., Coverlizza, S., Ferrari, A., Candelaresi, G.L. & Rossini, F.P. (1988) Immunohistochemical study of epithelial cell proliferation in hyperplastic polyps, adenomas, and adenocarcinomas of the large bowel. *Gastroenterology*, **94**, 899–906

Risio, M., Lipkin, M., Candelaresi, G., Bertone, A., Coverlizza, S. & Rossini, F.P. (1991) Correlations between rectal mucosa cell proliferation and the clinical and pathological features of nonfamilial neoplasia of the large intestine. *Cancer Res.*, **51**, 1917–1921

Risio, M., Candelaresi, G. & Rossini, F.P. (1993) Bromodeoxyuridine uptake and proliferating cell nuclear antigen expression throughout the colorectal tumor sequence. *Cancer Epidemiol. Biomarkers Prev.*, **2**, 363–367

Risio, M., Arrigoni, A., Pennazio, M., Agostinucci, A., Spandre, M. & Rossini, F.P. (1995) Mucosal cell proliferation in patients with hyperplastic colorectal polyps. *Scand. J. Gastroenterol.*, **30**, 344–348

Roncucci, L. (1992) Early events in human colorectal carcinogenesis. Aberrant crypts and microadenoma. *Ital. J. Gastroenterol.*, **24**, 498–501

Roncucci, L., Medline, A. & Bruce, W.R. (1991) Classification of aberrant crypt foci and microadenomas in human colon. *Cancer Epidemiol. Biomarkers Prev.*, **1**, 57–60

Roncucci, L., Ponz de Leon, M., Scalmati, A., Malagoli, G., Pratissoli, S., Perini, M. & Chahin, N.J. (1988) The influence of age on colonic epithelial cell proliferation. *Cancer*, **62**, 2373–2377

Roncucci, L., Scalmati, A. & Ponz de Leon, M. (1991) Pattern of cell kinetics in colorectal mucosa of patients with different types of adenomatous polyps of the large bowel. *Cancer*, **68**, 873–878

Roncucci, L., Modica, S., Pedroni, M., Tamassia, M.G., Ghidoni, M., Losi, L., Fante, R., Di Gregorio, C., Manenti, A., Gafa, L. & Ponz de Leon, M. (1998) Aberrant crypt foci in patients with colorectal cancer. *Br. J. Cancer*, **77**, 2343–2348

Rooney, P.S., Hunt, L.M., Clarke, P.A., Gifford, K.A., Hardcastle, J.D. & Armitage, N.C. (1994) Wheat fibre, lactulose and rectal mucosal proliferation in individuals with a family history of colorectal cancer. *Br. J. Surg.*, **81**, 1792–1794

Rozen, P. (1992) An evaluation of rectal epithelial proliferation measurement as biomarker of risk for colorectal neoplasia and response in intervention studies. *Eur. J. Cancer Prev.*, **1**, 215–224

Samowitz, W.S., Powers, M.D., Spirio, L.N., Nollet, F., van Roy, F. & Slattery, M.L. (1999) Beta-catenin mutations are more frequent in small colorectal adenomas than in larger adenomas and invasive carcinomas. *Cancer Res.*, **59**, 1442–1444

Schatzkin, A., Lanza, E., Corle, D., Lance, P., Iber, F., Caan, B., Shike, M., Weissfeld, J., Burt, R., Cooper, M.R., Kikendall, W., Cahill, J. & Polyp Prevention Trial Study Group (2000) Lack of effect of a low-fat, high-fiber diet on the recurrence of colorectal adenomas. *New Engl. J. Med.*, **342**, 1149–1155

Schoen, R.E., Corle, D., Cranston, L., Weissfeld, J.L., Lance, P., Burt, R., Iber, F., Shike, M., Kikendall, J.W., Hasson, M., Lewin, K.J., Appelman, H.D., Paskett, E., Selby, J.V., Lanza, E. & Schatzkin, A. (1998) Is colonoscopy needed for the nonadvanced adenoma found on sigmoidoscopy? The Polyp Prevention Trial. *Gastroenterology*, **115**, 533–541

Selby, J.V., Friedman, G.D., Quesenberry, C.P., Jr & Weiss, N.S. (1992) A case-control study of screening sigmoidoscopy and mortality from colorectal cancer. *New Engl. J. Med.*, **326**, 653–657

Shike, M., Al-Sabbach, M.R., Friedman, E., Steephen, A., Bloch, A., Turtel, P. & Winawer, S.J (1991) The effect of dietary fat on human colonic cell proliferation. *Gastroenterology*, **100**, A401

Shinya, H. & Wolff, W.I. (1979) Morphology, anatomic distribution and cancer potential of colonic polyps. *Ann. Surg.*, **190**, 679–683

Shpitz, B., Hay, K., Medline, A., Bruce, W.R., Bull, S.D., Gallinger, S. & Stern, H. (1996) Natural history of aberrant crypt foci. A surgical approach. *Dis. Colon Rectum*, **39**, 763–767

Shpitz, B., Bomstein, Y., Mekori, Y., Cohen, R., Kaufman, Z., Grankin, M. & Bernheim, J. (1997) Proliferating cell nuclear antigen as a marker of cell kinetics in aberrant crypt foci, hyperplastic polyps, adenomas, and adenocarcinomas of the human colon. *Am. J. Surg.*, **174**, 425–430

Shpitz, B., Bomstein, Y., Mekori, Y., Cohen, R., Kaufman, Z., Neufeld, D., Galkin, M. & Bernheim, J. (1998) Aberrant crypt foci in human colons: distribution and histomorphologic characteristics. *Human Pathol.*, **29**, 469–475

Sidransky, D., Tokino, T., Hamilton, S.R., Kinzler, K.W., Levin, B., Frost, P. & Vogelstein, B. (1992) Identification of ras oncogene mutations in the stool of patients with curable colorectal tumors. *Science*, **256**, 102–105

Simons, B.D., Morrison, A.S., Lev, R. & Verhoek-Oftedahl, W. (1992) Relationship of polyps to cancer of the large intestine. *J. Natl Cancer Inst.*, **84**, 962–966

Siu, I.M., Pretlow, T.G., Amini, S.B. & Pretlow, T.P. (1997) Identification of dysplasia in human colonic aberrant crypt foci. *Am. J. Pathol.*, **150**, 1805–1813

Siu, I.M., Robinson, D.R., Schwartz, S., Kung, H.J., Pretlow, T.G., Petersen, R.B. & Pretlow, T.P. (1999) The identification of monoclonality in human aberrant crypt foci. *Cancer Res.*, **59**, 63–66

Smith, A.J., Stern, H.S., Penner, M., Hay, K., Mitri, A., Bapat, B.V. & Gallinger, S. (1994) Somatic APC and K-ras codon 12 mutations in aberrant crypt foci from human colons. *Cancer Res.*, **54**, 5527–5530

Smith-Ravin, J., England, J., Talbot, I.C. & Bodmer, W. (1995) Detection of c-Ki-ras mutations in faecal samples from sporadic colorectal cancer patients. *Gut*, **36**, 81–86

Spagnesi, M.T., Tonelli, F., Dolara, P., Caderni, G., Valanzano, R., Anastasi, A. & Bianchini, F. (1994) Rectal proliferation and polyp occurrence in patients with familial adenomatous polyposis after sulindac treatment. *Gastroenterology*, **106**, 362–366

Stadler, J., Stern, H.S., Yeung, K.S., McGuire, V., Furrer, R., Marcon, N. & Bruce, W.R. (1988) Effect of high fat consumption on cell proliferation activity of colorectal mucosa and on soluble faecal bile acids. *Gut*, **29**, 1326–1331

Stadler, J., Yeung, K.S., Furrer, R., Marcon, N., Himal, H.S. & Bruce, W.R. (1988) Proliferative activity of rectal mucosa and soluble fecal bile acids in patients with normal colons and in patients with colonic polyps or cancer. *Cancer Letters*, **38**, 315–320

Stern, H.S., Gregoire, R.C., Kashtan, H., Stadler, J. & Bruce, R.W. (1990) Long-term effects of dietary calcium on risk markers for colon cancer in patients with familial polyposis. *Surgery*, **108**, 528–533

Stryker, S.J., Wolff, B.G., Culp, C.E., Libbe, S.D., Ilstrup, D.M. & MacCarty, R.L. (1987) Natural history of untreated colonic polyps. *Gastroenterology*, **93**, 1009–1013

Takayama, T., Katsuki, S., Takahashi, Y., Ohi, M., Nojiri, S., Sakamaki, S., Kato, J., Kogawa, K., Miyake, H. & Niitsu, Y. (1998) Aberrant crypt foci of the colon as precursors of adenoma and cancer. *New Engl. J. Med.*, **339**, 1277–1284

Terpstra, O.T., van Blankenstein, M., Dees, J. & Eilers, G.A. (1987) Abnormal pattern of cell proliferation in the entire colonic mucosa of patients with colon adenoma or cancer. *Gastroenterology*, **92**, 704–708

Tierney, R.P., Ballantyne, G.H. & Modlin, I.M. (1990) The adenoma to carcinoma sequence. *Surg. Gynecol. Obstet.*, **171**, 81–94

Tobi, M., Luo, F.C. & Ronai, Z. (1994) Detection of K-ras mutation in colonic effluent samples from patients without evidence of colorectal carcinoma. *J. Natl Cancer Inst.*, **86**, 1007–1010

Tosteson, T.D., Karagas, M.R., Rothstein, R., Ahnen, D.J. & Greenberg, E.R. (1996) Reliability of whole crypt mitotic count as a measure of cellular proliferation in rectal biopsies. *Cancer Epidemiol. Biomarkers Prev.*, **5**, 437–439

van Stolk, R.U., Beck, G.J., Baron, J.A., Haile, R. & Summers, R. (1998) Adenoma characteristics at first colonoscopy as predictors of adenoma recurrence and characteristics at follow-up. The Polyp Prevention Study Group. *Gastroenterology*, **115**, 13–18

Villa, E., Dugani, A., Rebecchi, A.M., Vignoli, A., Grottola, A., Buttafoco, P., Losi, L., Perini, M., Trande, P., Merighi, A., Lerose, R. & Manenti, F. (1996) Identification of subjects at risk for colorectal carcinoma through a test based on K-ras determination in the stool. *Gastroenterology*, **110**, 1346–1353

Wallace, M.B., Kemp, J.A., Trnka, Y.M., Donovan, J.M. & Farraye, F.A. (1998) Is colonoscopy indicated for small adenomas found by screening flexible sigmoidoscopy? *Ann. Int. Med.*, **129**, 273–278

Wargovich, M.J., Isbell, G., Shabot, M., Winn, R., Lanza, F., Hochman, L., Larson, E., Lynch, P., Roubein, L. & Levin, B. (1992) Calcium supplementation decreases rectal epithelial cell proliferation in subjects with sporadic adenoma. *Gastroenterology*, **103**, 92–97

Wargovich, M.J., Chen, C.D., Jimenez, A., Steele, V.E., Velasco, M., Stephens, L.C., Price, R., Gray, K. & Kelloff, G.J. (1996) Aberrant crypts as a biomarker for colon cancer: evaluation of potential chemopreventive agents in the rat. *Cancer Epidemiol. Biomarkers Prev.*, **5**, 355–360

Weisgerber, U.M., Boeing, H., Nemitz, R., Raedsch, R. & Waldherr, R. (1993) Proliferation cell nuclear antigen (clone 19A2) correlates with 5-bromo-2-deoxyuridine labelling in human colonic epithelium. *Gut*, **34**, 1587–1592

Weisgerber, U.M., Boeing, H., Owen, R.W., Waldherr, R., Raedsch, R. & Wahrendorf, J. (1996) Effect of longterm placebo controlled calcium supplementation on sigmoidal cell proliferation in patients with sporadic adenomatous polyps. *Gut*, **38**, 396–402

Welin, S., Youker, J. & Spratt, J. (1963) The rates and patterns of growth of 375 tumors of the large intestine and rectum observed serially by double contrast enema study (Malmo technique) *Am. J. Roentgenol. Radium Ther. Nucl. Med.*, **90**, 673–687

Wilson, R.G., Smith, A.N. & Bird, C.C. (1990) Immunohistochemical detection of abnormal cell proliferation in colonic mucosa of subjects with polyps. *J. Clin. Pathol.*, **43**, 744–747

Winawer, S.J., Zamber, A.G., Ho, M.N., O'Brien, M.J., Gottleib, L.S., Sternberg, S.S., Waye, J.D., Schapiro, M., Bond, J.H., Panish, J.F., Ackroyd, F., Shike, M., Kurtz, R.C., Hornsby-Lewis, L., Gerdes, H., Stewart, E.T., and the National Polyp Study Workgroup (1993) Randomized comparison of surveillance intervals after colonoscopic removal of newly diagnosed adenomatous polyps. The National Polyp Study Workgroup. *New Engl. J. Med.*, **328**, 901–906

Wong, A.J., Kohn, G.J., Schwartz, H.J., Ruebner, B.H. & Lawson, M.J. (1995) Colorectal cancer and noncancer patients have similar labeling indices by microscopy and computed image analysis. *Human Pathol.*, **26**, 1329–1332

World Cancer Research Fund & American Institute for Cancer Research (1997) *Food, Nutrition and the Prevention of Cancer: a Global Perspective,* Washington, DC, American Institute for Cancer Research

Yamao, T., Matsumura, Y., Shimada, Y., Moriya, Y., Sugihara, K., Akasu, T., Fujita, S. & Kakizoe, T. (1998) Abnormal expression of CD44 variants in the exfoliated cells in the feces of patients with colorectal cancer. *Gastroenterology*, **114**, 1333–1335

Yamashita, N., Minamoto, T., Ochiai, A. & Onda, M. (1995) Frequent and characteristic K-ras activation and absence of p53 protein accumulation in aberrant crypt foci of the colon. *Gastroenterology*, **108**, 434–440

Yang, G., Zheng, W., Sun, Q.R., Shu, X.O., Li, W.D., Yu, H., Shen, G.F., Shen, Y.Z., Potter, J.D. & Zheng, S. (1998) Pathologic features of initial adenomas as predictors for metachronous adenomas of the rectum. *J. Natl. Cancer Inst.*, **90**, 1661–1665

Zarchy, T.M. & Ershoff, D. (1994) Do characteristics of adenomas on flexible sigmoidoscopy predict advanced lesions on baseline colonoscopy? *Gastroenterology*, **106**, 1501–1504

Corresponding author:

J. A. Baron
Departments of Medicine, and Community & Family Medicine,
Dartmouth-Hitchcock Medical Center,
One Medical Center Drive,
Lebanon,
NH 03767,
USA

Biomarkers in Cancer Chemoprevention
Miller, A.B., Bartsch, H., Boffetta, P., Dragsted, L. and Vainio, H. eds
IARC Scientific Publications No. 154
International Agency for Research on Cancer, Lyon, 2001

Susceptibility markers in colorectal cancer

J. Burn, P.D. Chapman, D.T. Bishop, S. Smalley, I. Mickleburgh, S. West and J.C. Mathers

Many susceptibility factors contribute to an individual's risk of developing colorectal cancer. Family history of colorectal cancer (particularly with early age of onset), maleness and increasing age are all factors associated with increasing risk. About three quarters of colorectal cancers are thought to be due to somatic mutations, and both high- and low-penetrance predisposing genes contribute to the remaining quarter of cases. Many of the highly penetrant dominant genes are known, but others remain to be identified. Describing the contribution of individual genes is likely to be very complex, as some modify the impact of other genes and other environmental factors rather than incurring a direct, easily attributable effect. The two dominant predisposing syndromes are familial adenomatous polyposis and Lynch syndrome, the first due to a mutant tumour-suppressor gene *APC*, and the second due to mutations in a number of genes responsible for mismatch repair in DNA at cell division.

Establishing genetic susceptibility for colorectal cancer will soon be possible, and could save lives by allowing targetting of screening and the encouragement of preventive behaviours. However, there will always be a risk of making healthy people "sick" through the identification of predisposing genes, and there are many potential ways by which a gene carrier may be stigmatized by society, insurance companies and employers.

Introduction

Susceptibility to colorectal cancer can be predicted on the basis of a family history of the disease, particularly when this involves early age of onset. Other factors of relevance are age and sex, increasing age and maleness being associated with increasing risk (Figure 1). Features of rare syndromes such as Gorlin's syndrome and Peutz-Jeghers syndrome are predictive of elevated risk, as is a personal history of a resected adenomatous polyp or colorectal cancer. Clinical features of familial adenomatous polyposis and Lynch syndrome (hereditary nonpolyposis colorectal cancer) are the most important predictors. In addition to clearly pathological mutations in the *APC* and mismatch repair genes, allelic variants at these loci are likely to be of importance. Interactive loci such as the 'Modifier of Min' (*MOM1*) gene discovered in the mouse are of growing interest, as are genes which interact with environmental factors to increase mutagenicity or to diminish availability of protective substances such as folic acid.

Understanding genetic susceptibility markers will bring clinical benefits and increase the possibilities for further research into the etiology of colorectal cancer *in vivo*, much as the study of genetic muscle disorders has contributed to the knowledge of the genetics, biochemistry and physiology of normal muscle function.

Family history

Population studies consistently demonstrate a two-fold increase in colorectal cancer in first-degree relatives of an individual with colorectal cancer (Brown *et al.*, 1988). The cancers are seen at a comparable age to those in the general population, and have a similar location and age of onset (Lynch & Lynch, 1998). The causes of this familial risk are largely unknown, but presumably include a contribution from partially penetrant susceptibility genes to colonic neoplasia, common environmental exposures (which are risk factors for colorectal cancer and which aggregate in families) and interactions between genetic and

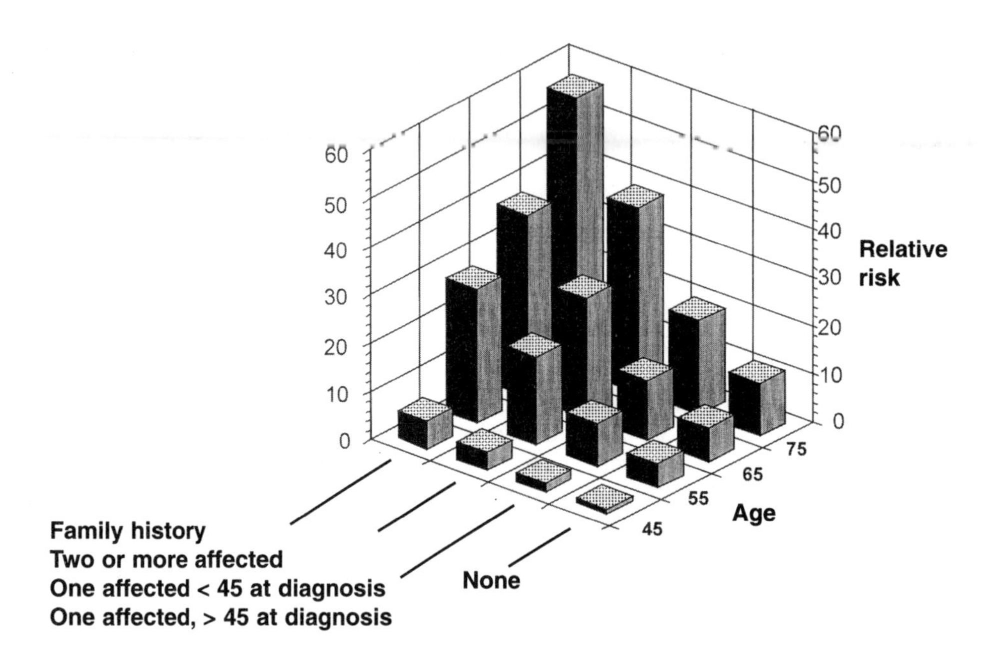

Figure 1. Risk of colorectal cancer by age and extent of family history, relative to risk of a 45-year-old with no family history
The data for this figure are taken from a study of colorectal cancer in families in Melbourne, Australia conducted by Dr J. St John (personal communication)

environmental factors (Kim, 1997). Colon cancer in these families is not linked to high-penetrance genes (described below), implying that these genes are not the major cause of familial colorectal cancer.

About three quarters of all colorectal cancers are thought to result from somatic mutations. At the present time, there is no certain way of picking out from the crowd those subjects with a predisposition who will make up the remaining quarter (NHS Executive, 1997). About 3 to 5% of colorectal cancer cases have a known, dominantly inherited predisposition and another 5% of families appear to have highly penetrant predisposing genes which have not yet been identified.

The risk of colorectal cancer in relatives of cases has been shown, in a number of studies, to be related to the age of onset in any close relative and to the number of affected relatives. Figure 1 shows the relationship between the risk of colorectal cancer by age and extent of family history. We use as a baseline the risk of colorectal cancer in a 45-year-old who has no family history. As this person gets older his or her risk increases simply on the basis of their own ageing process (at age 55 years, the risk of developing colorectal cancer in the next year is five times that at age 45, at age 65 it is seven times the risk of a 45-year-old and at age 75 it is 11 times the risk at 45 years). The risk also increases with the number of affected relatives. Thus, a 45-year-old with one affected relative diagnosed after 45 years of age has a risk 1.8 times that of a 45-year-old without a family history. The relative risk increases to 3.7 if the diagnosis was before 45 years and to 5.7 if there are two affected first-degree relatives. The risk increases across all levels of family history and at all ages, so that a 75-year-old with two affected first-degree relatives has a risk of colorectal cancer in the next year over 50 times that of a 45-year-old with no family history.

Lynch syndrome (hereditary nonpolyposis colon cancer) is due to a mutation in a highly penetrant susceptibility gene, and was referred to in the past as the cancer family syndrome. Patients are at risk of many extracolonic malignancies, including cancer of the endometrium, stomach, small bowel, hepatobiliary and urinary tract. For example, gene-carrying women have a 42% risk (to age 70 years) of developing endometrial cancer, which exceeds colorectal cancer risk in some age groups (Dunlop *et al.*, 1997). An individual with a family history including any of this spectrum of cancers, not just colorectal, could be considered to be likely to have an underlying genetic susceptibility. Lynch syndrome is most often suspected on the basis of family history. The modified Amsterdam criteria (Table 1) make use of the pattern of disease in families without access to direct mutation searching (Vasen *et al.*, 1994). Family history alone is not sufficiently sensitive for use in assessing risk for an individual. When screening strategies are being planned for someone with a Lynch syndrome family history, additional factors such as local screening availability, advice given to other family members, and the possibility of non-penetrance, early death or non-paternity should be considered. The Amsterdam criteria are useful in research to ensure inclusion of high-risk individuals in studies, whereas in clinical use they will only identify a proportion of high-risk people. An indirect application of the Amsterdam criteria is in selecting the families most likely to yield positive results when searching for mismatch repair gene defects (Wijnen *et al.*, 1998).

Where there is a family history of colorectal cancer, early age of onset in an affected relative and two or more affected generations are the most predictive factors for risk for an individual (Gaglia *et al.*, 1995). However, other independent risk factors, such as age and male sex, greatly modify an individual's risk. For example, Guillem *et al.* (1992) found that, at screening colonoscopy, those at greatest risk for harbouring an asymptomatic adenoma were males over the age of 50 years having at least one first-degree relative with colorectal cancer.

Personal history

Previous colorectal adenomas or cancer, without regular follow-up, indicate an overall increase in susceptibility to colorectal cancer. Such clinical observations can be further refined by pathological and genetic analysis of the neoplasm, and also by the age and follow-up history of the individual. Because there is variability in the normal adenoma–carcinoma sequence, some types of adenoma confer a greater risk than others. Sessile villous lesions behave differently to pedunculated adenomas, and flat adenomas are associated with an increased potential for malignant change (Jass, 1995). Microadenomata, the pathologically detectable precursors of adenomas, are very common, but only a fraction of them ever progress to malignancy, depending on the genetic status of the individual. For example, the risk is very low for an individual microadenoma in familial

Table 1. Revised criteria for recognition of Lynch syndrome (modified Amsterdam criteria)

There should be at least three relatives with a Lynch syndrome-related cancer (colorectal cancer, cancer of endometrium, small bowel, ureter or renal pelvis)

- One should be the first-degree relative of the other two.
- At least two generations should be so affected.
- At least one cancer should be diagnosed before 50 years of age.
- Familial adenomatous polyposis should be excluded.
- Tumours should be verified by pathological examination

From Vasen *et al.* (1999)

adenomatous polyposis, but high in Lynch syndrome. Similarly with adenomas; in familial adenomatous polyposis the risk of an individual adenoma becoming malignant is very small, whereas the risk is higher in Lynch syndrome.

Adenomas have much greater malignant potential than metaplastic polyps (Winawer, 1993a). Removal of polyps reduces the subsequent rate of colorectal cancer, confirming their premalignant potential (Winawer, 1993b).

Microsatellite instability in a colorectal tumour is predictive of a high risk of recurrence for an individual, regardless of family history (Brown *et al.*, 1999). Inflammatory bowel disease, and in particular ulcerative colitis, carries an increased risk for colorectal cancer. Those at highest risk have ulcerative colitis throughout the colon, rather than localized disease, are over 40 years of age, and have had ulcerative colitis for more than 10 years. Interestingly, this does not depend on continuous manifestation of the disease; those who have a short episode are at the same risk of colorectal cancer 10 years later as those who have 10 years without remission. Patients with ulcerative colitis have a small but increasing risk which is equivalent to about 0.5–1% chance of colorectal cancer per year of follow-up (Ekbom, 1998). For this reason, many gastroenterology units follow up patients for many years, despite lack of definitive evidence that this prevents colorectal cancer. Recent case–control studies have shown that the patients with ulcerative colitis who are at highest risk of colorectal cancer are those with a family history of colorectal cancer, whereas those with no family history have a risk which may not be significantly different from the average person without colitis (Nuako *et al.*, 1998). This finding implies that any inherited genetic risk is not associated with both the development of colitis and colon cancer risk, but that they are separate, discrete risk factors, and that inflammation is not a sufficient risk factor in isolation to have a clinically significant impact in most people.

Pathological susceptibility markers

Microsatellite instability is seen in about 15% of all colorectal cancers (Bodmer *et al.*, 1994), and at a much higher rate in Lynch syndrome colorectal cancers. The latter association probably explains why it is commonly seen in the younger age group. Functional assays for binding of the mismatch recognition genes can be used to test cells for a mismatch repair defect (Aquilina *et al.*, 1994).

This phenotype is likely to be recessive, reflecting total loss of function of the binding protein that recognizes DNA mismatches, for example at CA and CT repeats. This explains the observation that normal cells from Lynch syndrome gene carriers do not exhibit a microsatellite instability phenotype, as the cells in a heterozygous individual have one normal, working copy of the mutated, inherited gene.

Genetic factors predisposing to colorectal cancer

Colorectal cancer provides an excellent system for the study of genetic changes occurring during the development of a common human cancer. Most colorectal cancers arise from benign adenomas, which means that the developing carcinoma can be observed, removed and studied at all stages from an aberrant crypt focus to metastatic carcinoma. Adenomas normally arise from a single stem cell, which is demonstrated by the usually monoclonal nature of all adenomas from the smallest visible lesion, in contrast to the polyclonal composition of colorectal epithelium.

The study of the stochastic genetic events leading from early adenoma to colorectal cancer has led to the identification of three major classes of genes involved in familial risk: oncogenes which actively confer a direct growth-promoting effect, tumour-suppressor genes which normally restrain proliferation, and DNA mismatch repair (MMR) genes which identify and correct DNA replication errors. Mutations in MMR genes lead to genetic instability, as defects in somatic cell DNA replication are not corrected, resulting in mutations in other genes such as type-II TGF-β receptor (Yagi *et al.*, 1997).

Since colorectal cancer development is a multistep process, involving mutations in at least seven genes, many genotypes are likely to be involved in susceptibility to this disease. Most but not all of these genes exert a biological effect only when both alleles are mutated; in other words, they are recessive at the cellular level. However, some defects in tumour-suppressor genes such as *APC* can exert a phenotypic effect even in the heterozygous state. Many genes are known to be part of the multistep process and some have been shown to be important predictors of colorectal cancer when a germline mutation is present.

Although the genetic changes which give rise to colorectal tumours often occur in a particular sequence (Figure 2), it is the accumulation of mutations rather than the temporal sequence which determines the malignant potential of a tumour (Fearon & Vogelstein, 1990). In multipotent stem cells, the accumulation of mutations occurs in a stochastic, stepwise manner, with each mutation providing a selective advantage for subsequent cell generations, leading to an expanded population of daughter cells (Bodmer *et al.*, 1994). *APC* mutations occur early (Powell *et al.*, 1992) and are important for initiation. This explains the severe, young colorectal cancer phenotype of familial adenomatous polyposis. In contrast, *ras* oncogene mutations usually occur in larger adenomas, being present only in 10% of those smaller than 1 cm. It is thought that such mutations are responsible for the development of a small adenoma into a larger lesion. Similarly, allelic losses on chromosomes 5q, 17p and 18q and others are frequently observed in malignant tumours, but rarely in adenomas. Such mutations occurring later in the cascade of mutational events are unlikely to have major relevance in the search for cancer susceptibility, as they are largely somatic mutational events commencing in a single cell. Nevertheless, it is conceivable that variations which predispose to mutational change in any of these genes could influence the nature and rate of cancer development.

There is continuing debate about whether loss of a single *APC* allele can increase proliferation. Whereas significant changes in proliferation have not been seen in *Apc* knock-out mice, our own work in humans with familial adenomatous polyposis has shown a significant increase in the number of mitoses per crypt (Mills *et al.*, 2000). In contrast, Wasan *et al.* (1998) reported an increase in crypt fission rather than proliferation. DNA hypomethylation is probably not a pivotal event and the role of mismatch repair deficiency early in the process is also equivocal.

Pathological stage	Genetic event
Normal epithelium	
	APC mutation (in familial adenomatous polyposis)
Hyperproliferative epithelium	*APC* mutation
Early adenoma	K-*ras* mutation
Intermediate adenoma	Mismatch repair deficiency DCC (18q) loss DPC4 JV18
Late adenoma	*p53* loss
Carcinoma	
	Other changes
Metastasis ↓	

Figure 2. A genetic model for colorectal tumorigenesis
Based on Kinzer & Vogelstein (1996) and Kim (1997)

Mouse models

Mice bred to have defective copies of the major genes involved in colorectal cancer can be of great value in the evaluation of susceptibility factors (Kim *et al.*, 1993). The Min mouse, *Apc*1638N and *Apc*1638T and the $Apc^{delta\ 716}$ are the four mouse models of defective Apc function. The *Apc*1638T mouse is interesting as it involves a mutation near the end of the coding sequence, leaving the critical catenin-binding function intact. These mice do not develop significant intestinal tumorigenesis and homozygotes can survive to term.

While the phenotypes of the mouse models show important differences from the human, with predominance of small gut tumours, these models have been of great value in studies of the biology of colorectal cancer and in investigations of chemopreventive agents. Mouse mutants with defective mismatch repair genes have been less valuable, as the phenotype does not include gastrointestinal tumours.

Inherited predisposition to colorectal cancer

Mendelian syndromes and genes of major effect

Although the molecular genetics of most colorectal cancers remain unclear, some cases can be designated as having a genetic predisposition on the basis of their family history, clinical findings, pathology or molecular genetic features. Clinical overlap between the syndromes sometimes makes diagnosis difficult, but this is being clarified as genetic and functional histopathological analysis becomes available. Figure 3 shows the syndromes currently known to genetically predispose to colorectal cancer.

The two major, dominantly inherited forms of colorectal cancer are familial adenomatous polyposis and Lynch syndrome. This nomenclature describes the phenotype of these two conditions at the histological level, but the range of mutations in causative genes is much more disparate. Almost all cases of familial adenomatous polyposis result from a pathological mutation in a single gene on chromosome 5, *APC*. However, Lynch syndrome results from loss of function of one of at least five separate genes, each one of which encodes part of a protein complex which is responsible for mismatch repair during DNA replication.

Familial adenomatous polyposis

Familial adenomatous polyposis is the most genetically determined of all inherited cancer-

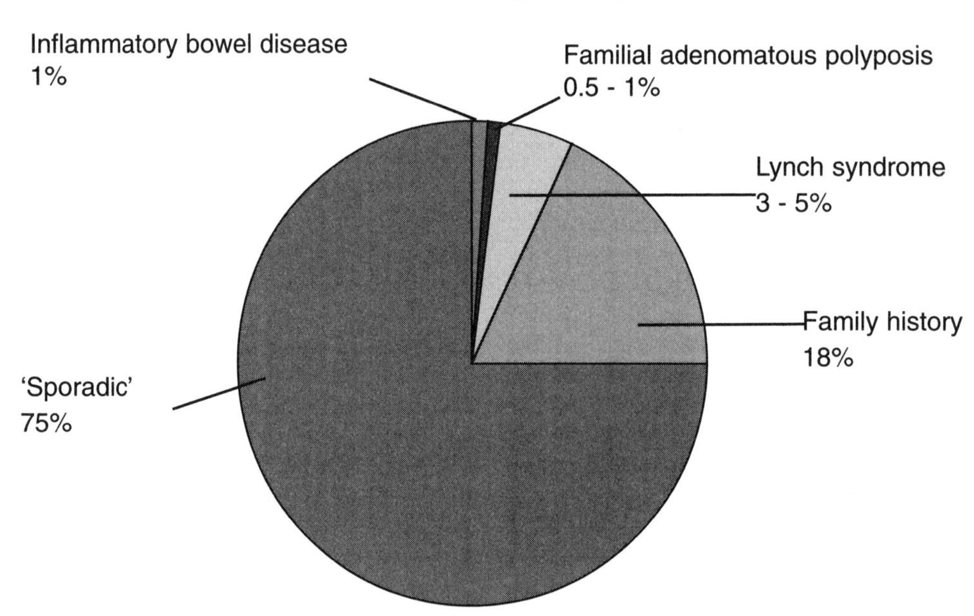

Figure 3. Predisposition to colorectal cancer

predisposing syndromes. A mutation in *APC* causes multiple colorectal adenomatous polyps to develop during the teens and early adulthood. In the absence of prophylactic colectomy, the large number of adenomas leads to the almost certain development of colorectal cancer at a young age. Much has been learned about the *APC* gene since its localization and cloning in the early 1990s (Bodmer *et al.*, 1987; Nishisho *et al.*, 1991), and the relationship of germline mutations to individual phenotype (Figure 4) has been described in more detail than for any other inherited cancer predisposition.

About 80% of familial adenomatous polyposis families have a different, distinct mutation of *APC,* although almost all of the disease-causing mutations so far found inactivate *APC* and result in protein truncation. *APC* codes for a large 2843-amino-acid protein which is involved in cell fate determination, adhesion and cytoskeleton function, and is an integral part of the Wnt signalling pathway by complex formation with glycogen synthase kinase 3 beta (GSK3 beta) (Brown *et al.*, 1999), β-catenin and other proteins including axin and conductin. There are 15 exons in *APC,* of which exons 1 to 14 are short and exon 15 encodes three quarters of the protein. A higher density of polyps occurs in families with a mutation near the centre of the gene in exon 15, whereas a sparse pattern of adenomas, known as attenuated familial adenomatous polyposis, is associated with mutations at the extreme 5′(proximal) end of the gene. It is likely that the most severe phenotype is a consequence of mutations which disrupt the mediation of β-catenin degradation in the Wnt signal transduction pathway. Events leading to oncogenic activation of β-catenin, which promotes tumour progression via interaction with a downstream target, can result from inactivation of tumour-suppressor activity of a mutated *APC* gene, from activation of Wnt receptors, or from direct mutation of the β-catenin gene itself (Polakis, 1999). In the nucleus, β-catenin upregulates the oncogene *c-myc* (He *et al.*, 1998), among other oncogenes.

Recognition of families and individuals at risk of developing familial adenomatous polyposis relies on careful pedigree analysis aided by a multidisciplinary approach including genetics, surgery, gastroenterology and pathology. This approach has been adopted in the Northern Region of England, and has been shown to be effective in reducing the burden of colorectal cancer in such families (Burn *et al.*, 1991). However, the new mutation rate for germline *APC* mutations is 20–30% and, unfortunately, new

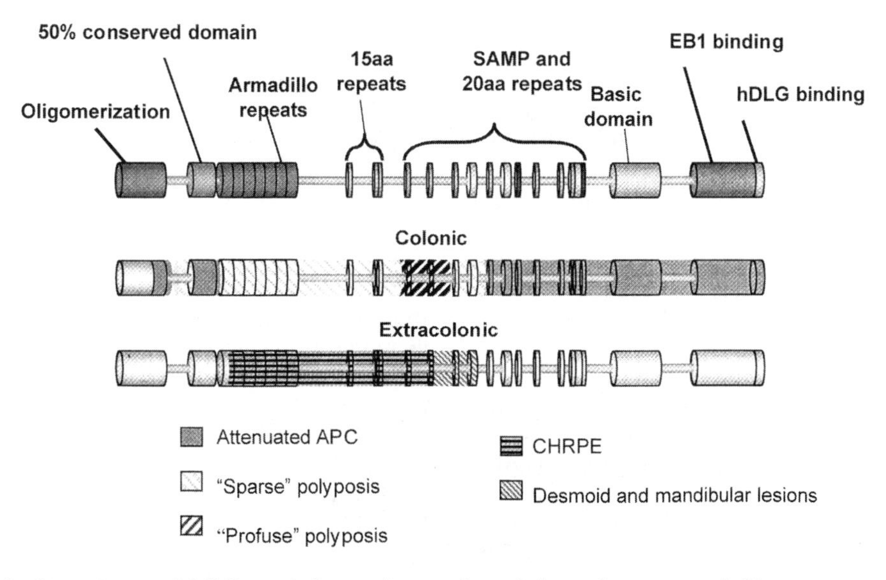

Figure 4. Structure of APC protein and genotype/phenotype correlation

mutation cases appear to exhibit a more severe phenotype than familial cases, with mutations more common at codon 1309. It is therefore possible that a high new mutation rate combined with improving survival rates into or beyond reproductive years will lead to an increase in incidence of familial adenomatous polyposis (Gayther *et al.*, 1994).

One explanation for the genotype/phenotype correlation in familial adenomatous polyposis is that some mutations, such as a truncating mutation at codon 1309, result in a dominant negative effect at the protein level. There is experimental evidence that wild-type *APC* activity is strongly inhibited by a mutant allele with this codon 1309 mutation, and this results in a severe phenotype. In contrast, a mutation associated with a mild phenotype (attenuated APC, or AAPC) produces a gene product which associates only weakly with the wild-type product (Dihlmann *et al.*, 1999). In AAPC, colorectal cancer occurs at a later age and extracolonic manifestations are less common (Lynch *et al.*, 1995). It is possible to attribute this mild phenotype to mutations in three distinct regions of the gene; at the 5´ end, within exon 9, and at the 3´ distal end of *APC*. When such a mutation is known in a family seeking genetic counselling, it is possible to modify risks according to the known genotype/phenotype relationship, and bowel examination may be less frequent than in a family with a mutation such as APC^{1309}.

Variation in the familial adenomatous polyposis phenotype even in those with identical mutations presents difficulties in counselling, both within and between families. However, in all cases of familial adenomatous polyposis, colorectal cancer susceptibility remains high, and regular screening and prophylactic surgery are essential components of care in such families (see section below on genetic modifiers).

Lynch syndrome

Lynch syndrome is the preferred term for the form of hereditary nonpolyposis colorectal cancer associated with an MMR gene defect. This relatively common syndrome is characterized by the development of neoplastic lesions in a variety of tissues (gastrointestinal, endometrial, ovarian, uroepithelial) and, most prominently, the colorectum (Aarnio *et al.*, 1995; Marra & Boland, 1995; Vasen *et al.*, 1995; Watson & Lynch, 1993). Clinically, the colorectal neoplastic process in hereditary nonpolyposis colorectal cancer appears to follow an adenoma-to-carcinoma progression similar to that described in familial adenomatous polyposis or other colorectal cancer settings, though several aspects of the clinical manifestations, as well as the molecular pathophysiologies underlying them, may be distinctive (Kinzler & Vogelstein, 1996). The disease was traditionally recognized by the familial clustering of colorectal cancers in persons without obvious polyposis.

Classical Lynch syndrome is caused by inherited mutations in one of the *Mut*-related family of MMR genes, including *hMLH1*, *hMSH2* and *hMSH6* (Fishel *et al.*, 1993; Kolodner *et al.*, 1994, 1995; Nicolaides *et al.*, 1994). Three other genes involved in the MMR complex, *PMS1*, *PMS2* and *MLH3*, are rarely or never associated with a Mendelian phenotype. This has been attributed to redundancy between them (Lipkin *et al.*, 2000) (see below). These genes encode protein products that are responsible for recognizing and correcting errors that arise when DNA is replicated (Dunlop *et al.*, 1997; Leach *et al.*, 1993). An early manifestation of this defect *in vivo* is the appearance of microsatellite instability. A second mutation is required in colorectal cells to inactivate the MMR function. Microsatellite instability contributes to the progressive accumulation of secondary mutations throughout the genome and thereby affects crucial growth-regulatory genes, ultimately leading to cancer.

More than 100 different germline mutations have been identified in the MMR genes known to be associated with the Lynch syndrome. Mutations in *hMSH2* and *hMLH1* account for roughly equal proportions of Lynch kindreds, and are together responsible for a majority of colorectal cancer in these families (Aaltonen *et al.*, 1998). However, germline disease-associated mutations are found in only about 40–70% of probable Lynch syndrome families. Other germline mutations and/or different classes of genes may be discovered which play an etiological role in susceptibility to hereditary nonpolyposis colorectal cancer. One family has been described in which a probable pathological mutation has occurred in the TGF-β receptor II gene (Yagi *et al.*, 1997), a gene known to show altered expression in colorectal cancer. This, and the attenuated form of familial adenomatous polyposis, might be regarded as falling into the broader category of hereditary nonpolyposis colorectal cancer but not Lynch syndrome.

One of the recently identified MMR genes, *MLH3*, associates with *MLH1*, *MSH2* and *MSH3* to form a complex involved with repair of insertion–deletion loops of single-stranded DNA. Its role in human cancer predisposition is uncertain but it is thought to show functional redundancy with *Pms1* and *Pms2*. This would explain why *PMS1* and *PMS2* mutations are only rarely found in Lynch syndrome families (Lipkin *et al.*, 2000).

Other syndromes

Juvenile polyposis

Juvenile polyps arise from the lamina propria, rather than the epithelium, which is the source of hyperplasia in adenomas. Solitary juvenile polyps are diagnosed in approximately 1% of children and account for the majority of gastrointestinal polyps found in childhood. Juvenile polyposis coli is a rare autosomal dominant syndrome which is characterized by multiple polyps in the colon and occasionally elsewhere in the gastrointestinal tract. This inherited condition is associated with a high risk of colorectal cancer, probably due to the development of foci of adenomatous change which progresses to dysplasia and adenocarcinoma. In some cases of juvenile polyps without a family history and of polyps arising in the juvenile polyposis syndrome, the predisposing factor is a mutation in one of two genes; *PTEN* on chromosome 10 (Jacoby *et al.*, 1997) or *SMAD4* on chromosome 18. It is likely that removal of juvenile polyps will be preventive for colorectal cancer.

There have been reports of families with atypical juvenile polyps, adenomas and colorectal cancers as well as inflammatory and metaplastic polyps. In one such family, cases of mixed polyps follows an autosomal dominant inheritance pattern, and the putative gene has been localized to the long arm of chromosome 6 (Thomas *et al.*, 1996).

Basal cell naevus syndrome (Gorlin's syndrome)

Hamartomatous intestinal polyps have been reported in Gorlin's syndrome (Schwartz, 1978), in which there is an association with broad facies, basal cell naevi, ectopic calcification of the falx and bony abnormalities of sites including the ribs, mandible and maxilla. Malignancies seen in this condition include medulloblastoma, malignant naevi and less commonly, colorectal cancer (Murday & Slack, 1989). Premature termination of the *patched* protein resulting from germline mutations in one of two tumour-suppressor genes, Patched 1 and 2, are thought to be responsible for this syndrome, but no genotype/phenotype correlation has yet been described (Wicking *et al.*, 1997).

Peutz–Jeghers syndrome

The predisposing polyps in this rare syndrome are pathologically discrete from other polyps and are called Peutz–Jeghers polyps. They exhibit some adenomatous features and have an increased malignant potential, appearing in the stomach, small bowel and colon. The distinguishing feature of this syndrome is the mucocutaneous melanin pigment seen around and inside the mouth and between the fingers. At least some cases of this disorder are due to mutations in the serine/threonine (*STK11*) tumour-suppressor gene on chromosome 19 (Jenne *et al.*, 1998).

Turcot's syndrome

In this condition, multiple colorectal adenomas are associated with central nervous system tumours, particularly of the brain. This phenotype has been seen with variants of both Lynch syndrome and familial adenomatous polyposis genotypes. With an *APC* variant, the brain tumours are cerebellar and medulloblastomas, whereas with *MLH1* or *MSH2* variants, glioblastoma multiforme is more frequently seen.

Modifiers and genes of minor effect

Allelic variants

A polymorphism at codon 1307 in *APC* is relatively common in the Ashkenazi Jewish population, and increases adenoma formation in this group (Gryfe *et al.*, 1999). This mutation causes a change of isoleucine to lysine (shown as I1307K), The underlying mutation changes a thymine to an adenine, resulting in a sequence of eight adenines which is more mutable than the wild-type. This leads to mutational susceptibility in somatic colon cells which in turn confers a higher risk of neoplastic development. Because the gene variant is dominantly inherited, but the penetrance for the phenotype is low, this form of predisposition could be described as familial rather than dominant. The lifetime risk of colorectal cancer for an individual with the polymorphism is about 10%, but because it is common (6% in New York Ashkenazim and 28% in those with a family history), it is thought to underlie 3–4% of colo-rectal cancer in this population (Laken *et al.*, 1997).

This finding has raised the issue of genetic predictive testing for the *APC* I1307K polymorphism which could be targeted towards individuals of Jewish Ashkenazi descent to identify those with an increased susceptibility as a prelude to prevention programmes. It is thought, for example, that 360 000 polymorphism carriers live in the United States, but the issue of testing is contentious, as a positive test could carry with it unfavourable psychological effects, insurance difficulties and potential sociological problems linked to selection of a population on the basis of ethnic descent. Moreover, the predictive value of such a marker in isolation is very small.

There is a strong likelihood that different mutational events might result in allelic variants at any of the "major gene" loci which increase the risk of colorectal cancer. The challenge will be to choose candidate genes for detailed sequencing. In future, high-throughput technology will make selection less critical but, at present, it is easy to spend large sums of money and achieve little. For example, a splice site mutation in the *MSH2* gene has been found in normal individuals with no history of hereditary nonpolyposis colorectal cancer. It is therefore possible that some functional effect exists which is associated with increased risk of colorectal cancer but which is not sufficiently high to show up as a positive family history.

One method of targeting is to use single nucleotide polymorphisms (SNPs) to identify haplotypes. Table 2 shows a series of five SNPs at the *MSH2* locus. Four of these are considered neutral, as they involve intronic DNA. The fifth is a coding sequence SNP in exon 6 and may have a phenotypic effect. Of the 32 possible haplotypes in our control population, only nine were identified among 99 chromosomes characterized and three (bold in the table) accounted for 85 of the 99. In other words, common ancient versions of the gene will occur in different populations. It will now be possible to examine these SNP patterns in people with colon cancer to see if the distribution of haplotypes is different from that in the local population. If, for example, there had been a mutation several hundred years ago in an *hMSH2* gene residing on a "ggtat" chromosome, this would be reflected in an overrepresentation of that haplotype in the disease population and would focus sequencing studies on *hMSH2* in affected people with the "ggtat" haplotype.

Interactive genes

In familial adenomatous polyposis, allelic heterogeneity does not appear to account for all of the observed variation in phenotype and other genes are probably involved in the phenotypic expression of this, the most "monogenic" of all cancer susceptibility genes. This is a continuing area of research, but there is evidence for a modifier gene on chromosome 1p35–36 which maps to an equivalent locus in the mouse for a known modifying gene called *Mom-1* (Dobbie *et al.*, 1997). The candidate gene for this is type 2 non-pancreatic *Pla2*, a phospholipase gene, but no mutations in the human homologue have been found to date (Spirio *et al.*, 1996). Identification of modifying genes is a powerful tool in the understanding of a gene's function, and the advent of dense genetic linkage maps has made the dissection of polygenic traits, such as colorectal cancer susceptibility, more practical. A relationship between mucin-producing genes such as *MUC2* and the pathogenesis of colorectal cancer has been suggested (Sternberg *et al.*, 1999). *MUC2* predominates within colorectal goblet cell mucin and is expressed in adenomas and mucinous carcinomas. Down-regulation of *MUC2* is seen in non-mucinous adenocarcinomas arising from adenomas, and cancers that develop *de novo* do not express *MUC2*. When more is known about colorectal cancer mucin, there will be opportunities to study different cell lineages to further explore the pathogenesis and other susceptibility factors in colorectal cancer.

CDX1 and *2* homeotic genes have the characteristics of transcriptional regulatory genes and are down-regulated in about 85% of colorectal cancers. In *CDX2*$^{+/-}$ mice, multiple intestinal polyps are seen in the proximal colon (Chawengsaksophak *et al.*, 1997). However, these polyps are not typical adenomas, and have similar histo-

Table 2. Single nucleotide polymorphisms (SNPs) at the *MSH2* locus

Haplotype	Number	Haplotype frequency
cgagt	61	0.616
ggagt	14	0.141
ggaat	10	0.101
ggtat	7	0.071
cgagc	3	0.030
ggtgt	0	0.000
ggagc	0	0.000
cgaat	2	0.020
ggaac	2	0.020
Total number of chromosomes typed	99	

(unpublished data)

logical characteristics to those seen in the *Min* mouse. These polyps occasionally contain true metaplasia and occasional large pedunculated tubulovillous colonic adenomas are seen. The fact that polyps are seen mostly in the proximal colon suggests that lowering levels of *CDX2* would induce exaggerated cell growth leading to tumour formation, and expression would stimulate cell differentiation and growth arrest (Yagi *et al.*, 1999). However, because the tumours display an unusual histological pattern and human mutations have not been identified, the role of this gene in human carcinogenesis remains unknown.

Environmentally sensitive genetic polymorphisms

A functional polymorphism in the methylene tetrahydrofolate reductase gene (*MTHFR*) makes the enzyme more thermolabile and appears to confer a 50% reduction in colorectal cancer risk in the US population. This polymorphism ([667]C-T, *alaval*), found in 10–15% of the study population, only provided protection when adequate folate was present in the diet (Ma *et al.*, 1997). Figure 5 shows the competing pathways in folate metabolism. Before the report that this polymorphism was protective against colorectal cancer in homozygotes, it might have been expected that the reverse would be true. The observation could indicate that malignancies are more susceptible to disturbance of the folate pathway and that the homozygotes for the thermolabile variant are relatively protected by the less efficient folate pathway. It is of potential importance that protection against colorectal cancer in those homozygous for the thermolabile variant was observed only in study subjects reporting zero to modest alcohol intake. With high alcohol intake, no protection was seen, indicating an important diet–gene interaction (Ma *et al.*, 1997).

The importance of gene–environment interactions in assessing cancer susceptibility was illustrated by the β-carotene trials. Intervention studies designed to test the hypothesis that supplementation with β-carotene would reduce cancer risk showed no protection. However, the trials showed increased risk of lung cancer in smokers given β-carotene (Mathers & Burn, 1999). This effect is thought to be due to increased cell proliferation and squamous metaplasia in the lung, effects that were enhanced by tobacco smoke, and is associated with suppression of retinoic acid receptor β gene expression and overexpression of *c-Jun* and *c-Fos* genes (Wang *et al.*, 1999).

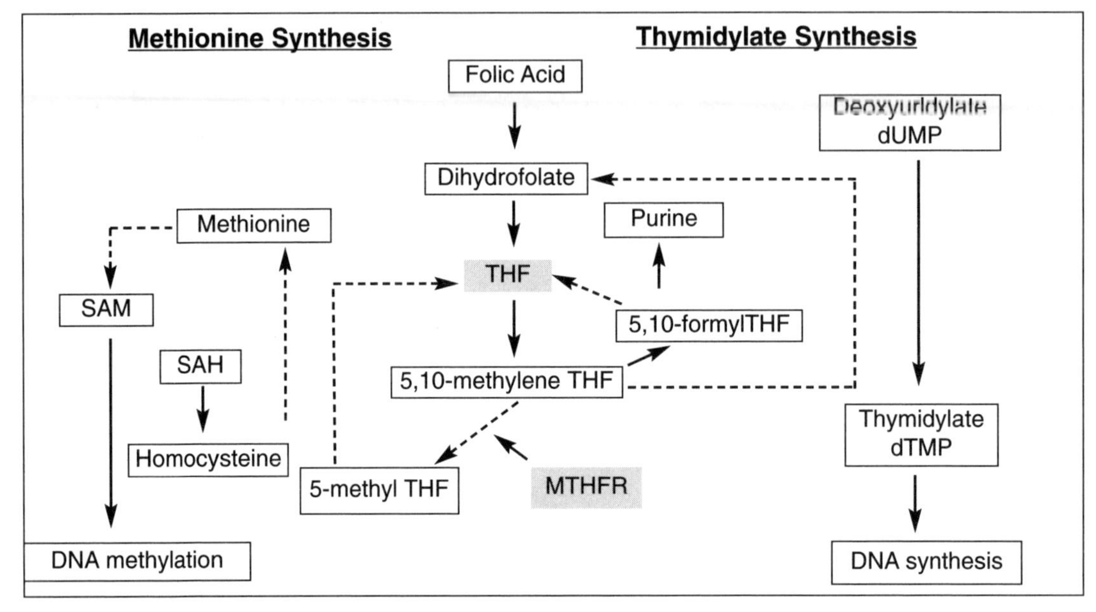

Figure 5. Competing pathways in folate metabolism
THF, tetrahydrofolate; MTHFR, methylene tetrahydrofolate reductase

Exposure to environmental carcinogens such as aromatic amines found in well cooked or preserved meat and cigarette smoke are associated with increased risk of colorectal cancer. Their metabolism is complex but central to most is activation or detoxification of amylines and heterocyclic amines. Acetylation of heterocyclic amines by *N*-acetyltransferases (NAT) is likely to be of major importance. Acetylation of heterocyclic amines by *NAT1* and *NAT2* gene products can lead to the formation of reactive carcinogenic intermediates or to detoxification. This means that the association of cancer risk and enzyme activity could go either way. Acetylation activity varies as a result of the sequence polymorphism in the *NAT2* gene, and if two non-functional alleles are inherited (slow acetylator alleles), there is no NAT2 activity. Several studies show that NAT2 rapid acetylation phenotypes are associated with an increased risk of colorectal cancer. An increased colorectal cancer risk of 1.9 results from a variation in NAT1, which is again due to a rapid acetylation genotype (Hein *et al.*, 2000). There is an association between fast acetylator status and cancer in those with high intakes of cooked meat (a source of heterocyclic amines). Conversely, slow acetylators who smoke and drink are at increased risk. These genetic polymorphisms provide a good example of how complex the interaction between genotype and environmental factors can be. It must also be remembered that "multiple slices" of a data-set might point to apparent interactions as a random event, resulting in claims and counter-claims on their predictive significance.

Prevention and colorectal cancer genotype

Familial adenomatous polyposis has been identifiable in families for over a hundred years because of the presence of multiple polyps, and more recently, by genetic testing for mutations or specific patterns in and around the *APC* gene. In the Northern Region of England, using a registry approach, and by applying clinical and genetic criteria, almost all individuals at high risk have been recruited to screening programmes (Burn *et al.*, 1991). One of the biggest advantages of a proactive approach towards offering genetic testing is the relief of uncertainty and the reduction of unnecessary colonoscopic surveillance in those not at high genetic risk. At the level of population health care, there are benefits in being able to target screening effectively to those at greatest risk.

There are many differences between making a genetic diagnosis of hereditary colorectal cancer risk and a clinical diagnosis of colorectal cancer. Firstly, hereditary cancer involves probability statements describing inheritance and penetrance. Little is known about the role of environmental influences in these families, and this contributes to the high levels of uncertainty accompanying predictive testing for late-onset disorders. Secondly, there are psychosocial implications for the individual and for the family. How people respond to the information will depend on many variables such as their personality, defence mechanisms and understanding of the disease and its consequences. The effect of a genetic diagnosis can pass through a family like ripples on a pond and extended families will become more aware of, and discuss, cancer diagnoses. Family beliefs about inheritance may be very different from accepted patterns of Mendelian inheritance. For example, some believe that only males are affected by hereditary nonpolyposis colorectal cancer.

It is easy to assume that those found not to carry a predisposing gene will be unconditionally pleased. However, some individuals react in a negative way to such information. This may be due to having to alter life plans built around an assumption that they will become ill, or to guilt about escaping the family disease. It would also be logical to assume that those found to be gene carriers would present themselves for bowel examination. This does not always happen and is a source of frustration to surgical and endoscopy staff when appointments are missed repeatedly. The reasons for this reaction by patients are complex, but may include fear of surgery, fear of the examination, or a hedonistic personality which does not easily accept hospitalization and potential illness. These reasons can differ greatly from a non-familial case where the patient is attending with symptoms (Rossi & Srivastava, 1996).

There are many ethical issues surrounding hereditary colon cancer, whether it is a clinical or a genetic diagnosis. Firstly, equitable access to services is a guiding principle for health services in the United Kingdom, but this is difficult to apply when only some regions have active registers and recall systems with individual counselling services. Bearing in mind the European history of eugenics during the early part of the 20th century, there must be caution in pursuing equity by using forms of coercion such as laws and social pressures. Where efforts are made to make contact with family members who are at risk in an attempt to offer equitable services, there is a risk of straying over the invisible line between voluntary and enforced testing. For example, even if DNA testing is undertaken, each person should be able to withdraw at any time or choose not to take any action if the result is positive. Other issues, such as those concerning confidentiality and privacy, must be discussed with family members, so together with the genetic and clinical information and exploration of psychosocial issues, much careful counselling is required. The common thread to all of this discussion is the protection of individual rights, and a multi-professional approach with specialist genetic counselling available for all is the best way to offer such a service.

Research into other risk factors and possible interventions described earlier can be undertaken with subjects at high risk as opposed to general population cohorts. This means that smaller numbers are required, as interim or end-point results such as adenomas are likely to be more frequent. Two such trials are under way using patients with familial adenomatous polyposis and mismatch repair gene mutations, in which aspirin and resistant starch are being tested in randomized controlled trials (Burn *et al.*, 1995, 1998). How much relevance the results of these trials will have to the rest of the population remains to be established, but both interventions were selected because of their favourable effect in observational studies of the general population.

Conclusions

All colorectal cancer involves multiple somatic genetic changes. Based on family history data, it is likely that in at least a half, and probably three quarters, of colorectal cancer in developed countries, these changes occur through random acquisition of mutations due to environmental influences. If and when diet improves and people take more exercise, the importance of germline defects will grow. Even now, there are very large numbers of people at significantly elevated risk of cancer who will be identifiable using molecular genetic tests for pathological mutations in *APC* and the mismatch repair genes. It is likely that there will be value in searching for less penetrant defective alleles in these genes, as this might influence treatments and screening strategies. Perhaps more importantly, knowledge of a specific personal risk factor is liable to stimulate a greater interest in chemoprevention and lifestyle factors which could lead to reduced risk.

On the negative side, this possibility also raises psychosocial issues, including the risk of making healthy people "sick" by describing them as susceptible to cancer and by identifying high-risk ethnic groups such as Ashkenazi Jews. Ethical issues relating to providing equitable screening opportunities across populations, and financial concerns when insurance risks can be stratified by susceptibility factors or medical liability of the clinician become more complex. In the United States, physicians are being pressurized into assuming the additional responsibility of establishing a family history of cancer across several generations. This in turn creates a duty for a health care worker to provide counselling to extended families, and several lawsuits have been instituted claiming negligence when a family history has not been given adequate consideration or has not been communicated to family members at risk (Nelson, 1996; Severin, 1999). In addition to legal pressures to include genealogy in colorectal cancer care, there are economic factors associated with identification and monitoring of susceptible individuals (Bolin, 1996; Brown & Kessler, 1996; Smith & DuBois, 1997). Where there are advantages to a community or society in providing a health care intervention, there will always be a risk of eugenic policies creeping into practice.

Establishing genetic susceptibility for colorectal cancer will soon become a reality, and the advantages to a member of a hereditary nonpolyposis colorectal cancer or familial adenomatous polyposis family member in finding out they are not a gene carrier can be enormous. However, the situation is more complex where the predictability of the genotype is less certain, and this is likely to be the situation for most genetic susceptibility testing (Lynch *et al.*, 1999). The needs and views of the individual must always take precedence over societal needs if maximum uptake of screening alongside freedom of choice is to be assured. Even more important is the recognition that few risk factors will be sufficiently predictive to justify extension from the realm of primary research to the clinic. In many cases, groups of individuals chosen for their susceptible genotype will be used to test which environmental changes might benefit the whole community.

Acknowledgements

Our work is supported by the Imperial Cancer Research Fund and the European Union Biomed Programme.

References

Aaltonen, L.A., Salovaara, R., Kristo, P., Canzian, F., Hemminki, A., Peltomaki, P., Chadwick, R.B., Kaariainen, H., Eskelinen, M., Jarvinen, H., Mecklin, J.P. & De la Chapelle, A. (1998) Incidence of hereditary nonpolyposis colorectal cancer and the feasibility of molecular screening for the disease. *New Engl. J. Med.*, **338**, 1481–1487

Aarnio, M., Mecklin, J.P., Aaltonen, L.A., Nystrom-Lahti, M. & Jarvinen, H.J. (1995) Life-time risk of different cancers in hereditary non-polyposis colorectal cancer (HNPCC) syndrome. *Int. J. Cancer*, **64**, 430–433

Aquilina, G., Hess, P., Branch, P., MacGeoch, C., Casciano, I., Karran, P. & Bignami, M. (1994) A mismatch recognition defect in colon carcinoma confers DNA microsatellite instability and a mutator phenotype. *Proc. Natl Acad. Sci. USA*, **91**, 8905–8909

Bodmer, W., Bishop, T. & Karran, P. (1994) Genetic steps in colorectal cancer. *Nature Genet.*, **6**, 217–219

Bodmer, W.F., Bailey, C.J., Bodmer, J., Bussey, H.J.R., Ellis, A., Gorman, P., Lucibello, F.C., Murday, V.A., Rider, S.H., Scambler, P., Sheer, D., Solomon, E. & Spurr, N.K. (1987) Localization of the gene for familial adenomatous polyposis on chromosome 5. *Nature*, **328**, 614

Bolin,T.D. (1996) Cost benefit of early diagnosis of colorectal cancer. *Scand. J. Gastroenterol.*, **31**, 142–146

Brown, M.L. & Kessler, L.G. (1996) Use of gene tests to detect hereditary predisposition to cancer: what do we know about cost effectiveness? *Int. J. Cancer*, **69**, 55–57

Brown, S.R., Finan, P.J. & Bishop, D.T. (1998) Are relatives of patients with multiple HNPCC spectrum tumours at increased risk of cancer? *Gut*, **43**, 664–668

Brown, S.R., Finan, P.J., Cawkwell, L., Quirke, P. & Bishop, D.T. (1999) Frequency of replication errors in colorectal cancer and their association with family history. *Gut*, **43**, 553–557

Burn, J., Chapman, P., Delhanty, J., Wood, C., Lalloo, F., Cachon-Gonzalez, M.B., Tsioupra, K., Church, W., Rhodes, M. & Gunn, A. (1991) The UK Northern Region genetic register for familial adenomatous polyposis coli: use of age of onset, congenital hypertrophy of the retinal pigment epithelium, and DNA markers in risk calculations. *J. Med. Genet.*, **28**, 289–296

Burn, J., Chapman, P.D., Bertario, L., Bishop, D.T., Bülow, S., Cummings, J., Mathers, J., Phillips, R. & Vasen, H. (1995) The protocol for a European double-blind trial of aspirin and resistant starch in familial adenomatous polyposis: the CAPP study. *Eur. J. Cancer*, **31A**, 1385–1386

Burn, J., Chapman, P., Bishop, T. & Mathers, J. (1998) Diet and cancer prevention: the CAPP studies. *Proc. Nutr. Soc.*, **57**, 183–186

Chawengsaksophak, K., James, R., Hammond, V.E., Kontgen, F. & Beck, F. (1997) Homeosis and intestinal tumours in Cdx2 mutant mice. *Nature*, **386**, 84–87

Dihlmann, S., Gebert, J., Siermann, A., Herfarth, C. & von Knebel Doeberitz, M. (1999) Dominant negative effect of the *APC*[1309] mutation: a possible explanation for genotype-phenotype correlations in familial adenomatous polyposis. *Cancer Res.*, **59**, 1857–1860

Dobbie, Z., Heinimann, K., Bishop, D.T., Müller, H. & Scott, R.J. (1997) Identification of a modifier gene locus on chromosome 1p35-36 in familial adenomatous polyposis. *Human Genet.*, **99**, 653–657

Dunlop, M.G., Farrington, S.M., Carothers, A.D., Wyllie, A.H., Sharp, L., Burn, J., Liu, B., Kinzler, K.W. & Vogelstein, B. (1997) Cancer risk associated with germline DNA mismatch repair gene mutations: *Human Molec. Genet.*, **6**, 105–110

Ekbom, A. (1998) Risk factors and distinguishing features of cancer in IBD. *Inflam. Bowel Dis.*, **4**, 235–243

Fearon, E.R. & Vogelstein, B. (1990) A genetic model for colorectal tumorigenesis. *Cell*, **61**, 759–767

Fishel, R., Lescoe, M.K., Rao, M.R.S., Copeland, N.G., Jenkins, N.A., Garber, J., Kane, M. & Kolodner, R. (1993) The human mutator gene homolog *MSH2* and its association with hereditary nonpolyposis colon cancer. *Cell*, **75**, 1027–1038

Gaglia, P., Atkin, W.S., Whitelaw, S., Talbot, I.C., Williams, C.B., Northover, J.M. & Hodgson, S.V. (1995) Variables associated with the risk of colorectal adenomas in asymptomatic patients with a family history of colorectal cancer. *Gut*, **36**, 385–390

Gayther, S.A., Wells, D., Sengupta, S.B., Chapman, P., Neale, K., Tsioupra, K. & Delhanty, J.D.A. (1994) Regionally clustered APC mutations are associated with a severe phenotype and occur at a high frequency in new mutation cases of adenomatous polyposis coli. *Human Molec. Genet.*, **3**, 53–56

Gryfe, R., Di Nicola, N., Lal, G., Gallinger, S. & Redston, M. (1999) Inherited colorectal polyposis and cancer risk of the *APC I1307K* polymorphism. *Am. J. Human Genet.*, **64**, 378–384

Guillem, J.G., Forde, K.A., Treat, M.R., Neugut, A.I., O'Toole, K.M. & Diamond, B.E. (1992) Colonoscopic screening for neoplasms in asymptomatic first-degree relatives of colon cancer patients. A controlled, prospective study. *Dis. Colon Rectum*, **35**, 523–529

He,T.C., Sparks, A.B., Rago, C., Hermeking, H., Zawel, L., da Costa, L.T., Morin, P.J., Vogelstein, B. & Kinzler, K.W. (1998) Identification of c-MYC as a target of the pathway. *Science*, **281**, 1509–1512

Hein, D.W., Doll, M.A., Fretland, A.J., Leff, M.A., Webb, S.J., Xiao, G.H., Devanaboyina, U.S., Nangju, N.A. & Feng, Y. (2000) Molecular genetics and epidemiology of the NAT1 and NAT2 acetylation polymorphisms. *Cancer Epidemiol. Biomarkers Prev.*, **9**, 29–42

Jacoby, R.F., Schlack, S., Cole, C.E., Skarbek, M., Harris, C. & Meisner, L.F. (1997) A juvenile polyposis tumour suppressor locus at 10q22 is deleted from nonepithelial cells in the lamina propria. *Gastroenterology*, **112**, 1398–1403

Jass, J.R. (1995) Colorectal cancer progression and genetic change: is there a link? *Ann. Med.*, **27**, 301–306

Jenne, D.E., Reimann, H., Nezu, J., Friedel, W., Loff, S., Jeschke, R., Muller, O., Back, W. & Zimmer, M. (1998) Peutz-Jeghers syndrome is caused by mutations in a novel serine threonine kinase. *Nature Genet.*, **18**, 38–43

Kim, Y.S. (1997) Molecular genetics of colorectal cancer. *Digestion*, **58**, 65–68

Kim, S.H., Roth, K.A., Moser, A.R. & Gordon, J.I. (1993) Transgenic mouse models that explore the multistep hypothesis of intestinal neoplasia. *J. Cell Biol.*, **123**, 877–893

Kinzler, K.W. & Vogelstein, B. (1996) Lessons from hereditary colorectal cancer. *Cell*, **87**, 159–170

Kolodner, R.D., Hall, N.R., Lipford, J., Kane, M.F., Rao, M.R.S., Morrison, P., Wirth, L., Finan, P.J., Burn, J., Chapman, P., Earabino, C., Merchant, E. & Bishop, D.T. (1994) Structure of the human MSH2 locus and analysis of two Muir-Torre kindreds for msh2 mutations. *Genomics*, **24**, 516–526

Kolodner, R.D., Hall, N.R., Lipford, J., Kane, M.F., Morrison, P.T., Finan, P.J., Burn, J., Chapman, P., Earabino, C., Merchant, E. & Bishop, D.T. (1995) Structure of the human *MLH1* locus and analysis of a large hereditary nonpolyposis colorectal carcinoma kindred for mLH1 mutations. *Cancer Res.*, **55**, 242–248

Laken, S.J., Peterson, G.M., Gruber, S.B., Oddoux, C., Ostrer, H., Giardiello, F.M., Hamilton, S.R., Hampel, H., Markowitz, A., Klimstra, D., Jhanwar, S., Winawar, S., Offit, K., Luce, M.C., Kinzler, K.W. & Vogelstein, B. (1997) Familial colorectal cancer in Ashkenazim due to a hypermutable tract in APC. *Nature Genet.*, **17**, 79–83

Leach, F.S., Nicolaides, N.C., Papadopoulos, N., Liu, B., Jen, J., Parsons, R., Peltomaki, P., Sistonen, P., Aaltonen L.A., Nystrom-Lahti, M., Guan, X.-Y., Zhang, J., Meltzer, P.S., Yu, J.-W., Kao, F.-T., Chen, D.J., Cerosaletti, K.M., Fournier, R.E.K., Todd, S., Lewis, T., Leach, J., Naylor, S.L., Weissenbach, J., Meklin, J.-P., Jarvinen, H., Petersen, G.M., Hamilton, S.R., Green, J., Jass, J., Watson, P., Lynch, H.T., Trent, J.M., de la Chapelle, A., Kenzler, K.W. & Vogelstein, B. (1993) Mutations of mutS homolog in hereditary nonpolyposis colorectal cancer. *Cell*, **75**, 1215–1225

Lipkin, S.M., Wang, V., Jacoby, R., Banerjee-Basu, S., Baxevanis, A.D., Lynch, H.T., Elliott, R.M. & Collins, F.S. (2000) *MLH3:* a DNA mismatch repair gene associated with mammalian microsatellite instability. *Nature Genet.*, **24**, 27–35

Lynch, H.T. & Lynch, J.F. (1998) Genetics of colonic cancer. *Digestion*, **59**, 481–492

Lynch, H.T., Smyrk, T., McGinn, T., Lanspa, S., Cavalieri, J., Lynch, J., Slominski-Caster, S., Cayouette, M.C., Priluck, I., Luce, M.C. (1995) Attenuated familial adenomatous polyposis (AFA P). *Cancer*, **76**, 2427–2433

Lynch, H.T., Watson, P., Shaw, T.G., Lynch, J.F., Harty, A.E., Franklin, B.A., Kapler, C.R., Tinley, S.T., Liu, B. & Lerman, C. (1999) Clinical impact of molecular genetic diagnosis, genetic counseling, and management of hereditary cancer. Part II Hereditary nonpolyposis colorectal carcinoma as a model. *Cancer*, **86** (11 Suppl.), 2457–2463

Ma, J., Stampfer, M.J., Giovannucci, E., Artigas, C., Hunter, D.J., Fuchs, C., Willett, W.C., Selhub, J., Hennekens, C.H. & Rozen, R. (1997) Methylenetetrahydrofolate reductase polymorphism, dietary interactions, and risk of colorectal cancer. *Cancer Res.*, **57**, 1098–1102

Marra, G. & Boland, C.R. (1995) Hereditary nonpolyposis colorectal cancer: the syndrome, the genes, and historical perspectives. *J. Natl Cancer Inst.*, **87**, 1114–1125

Mathers, J.C. & Burn, J. (1999) Nutrition in cancer prevention. *Current Opin. Oncol.*, **11**, 402–407

Mills, S., Mathers, J.C., Chapman, P.D., Burn, J. & Gunn, A. (2000) Human colonic crypt cell proliferation state as assessed by whole crypt microdissection in sporadic neoplasia and familial adenomatous polyposis. *Gut* (in press)

Murday, V. & Slack, J. (1989) Inherited disorders associated with colorectal cancer. *Cancer Surveys*, **8**, 139–157

Nelson, N.J. (1996) Caution guides genetic testing for hereditary cancer genes. *J. Natl Cancer Inst.*, **88**, 70–72

NHS Executive (1997) Guidance on commissioning cancer services. Improving outcomes in colorectal cancer. *The Manual*, London, Department of Health, pp. 1–73

Nicolaides, N.C., Papadopoulos, N., Liu, B., Wei, Y.-F., Carter, K.C., Ruben, S.M., Rosen, C.A., Haseltine, W.A., Fleischmann, R.D., Fraser, C.M., Adams, M.D., Venter, J.C., Dunlop, M.G., Hamilton, S.R., Petersen, G.M., De la Chapelle, A., Vogelstein, B. & Kinzler, K.W. (1994) Mutations of two PMS homologues in hereditary nonpolyposis colon cancer. *Nature*, **371**, 75–80

Nishisho, I., Nakamura,Y. & Miyoshi,Y. (1991) Mutations of chromosome 5q21 genes in FAP and colorectal cancer patients. *Science*, **253**, 665–669

Nuako, K.W., Ahlquist, D.A., Mahoney, D.W., Schaid, D.J., Siems, D.M. & Lindor, N.M. (1998) Familial predisposition for colorectal cancer in chronic ulcerative colitis: a case-control study. *Gastroenterology*, **115**, 1079–1083

Polakis, P. (1999) The oncogenic activation of β-catenin. *Current Opin. Genet. Devel.*, **9**, 15–21

Powell, S.M., Zilz, N., Beazer-Barclay, Y., Bryan, T.M., Hamilton, S.R., Thibodeau, S.N., Vogelstein, B. & Kinzler, K.W. (1992) APC mutations occur early during colorectal tumorigenesis. *Nature*, **359**, 235–237

Rossi, S.C. & Srivastava, S. (1996) National Cancer Institute workshop on genetic screening for colorectal cancer. *J. Natl Cancer Inst.*, **88**, 331–339

Schwartz, R.A. (1978) Basal-cell-nevus syndrome and gastrointestinal polyposis. *New Engl. J. Med.*, **299**, 49–49

Severin, M.J. (1999) Genetic susceptibility for specific cancers. *Cancer*, **86**, 1744–1749

Smith, B. & DuBois, R.N. (1997) Current concepts in colorectal cancer prevention. *Comprehen. Ther.*, **23**, 184–189

Spirio, L.N., Kutchera, W., Winstead, M.V., Pearson, B., Kaplan, C., Robertson, M., Lawrence, E., Burt, R.W., Tischfield, J.A., Leppert, M.F., Prescott, S.M. & White, R. (1996) Three secretory phospholipase A(2) genes that map to human chromosome 1P35-36 are not mutated in individuals with attentuated adenomatous polyposis coli. *Cancer Res.*, **56**, 955–958

Sternberg, L.R., Byrd, J.C., Yunker, C.K., Dudas, S., Hoon, V.K. & Bresalier, R.S. (1999) Liver colonization by human colon cancer cells is reduced by antisense inhibition of MUC2 mucin synthesis. *Gastroenterology*, **116**, 363–371

Thomas, H.J.W., Whitelaw, S.C., Cottrell, S.E., Murday, V.A., Tomlinson, I.P.M., Markkie, D., Jones, T., Bishop, D.T., Hodgson, S.V., Sheer, D., Northover, J..M.A., Talbot, I.C., Solomon, E. & Bodmer, W.F. (1996) Genetic mapping of the hereditary mixed polyposis syndrome to chromosome 6q. *Am. J. Human Genet.*, **58**, 770–776

Vasen, H.F.A., Watson, P., Mecklin, J.-P., Lynch, H.T. and ICG-HNPCC (1999) New clinical criteria for hereditary non-polyposis colorectal cancer (NHPCC, Lynch syndrome) proposed by the International Collaborative Group in HNPCC. *Gastroenterology*, **116**, 1453–1456

Vasen, H.F., Taal, B.G., Nagengast, F.M., Griffioen, G., Menko, F.H., Kleibeuker, J.H., Offerhaus, G.J. & Khan, P.M. (1995) Hereditary nonpolyposis colorectal cancer: results of long-term surveillance in 50 families. *Eur. J. Cancer*, **31A**, 1145–1148

Wang, X.D., Lin, C., Bronson, R.T., Smith, D.E., Krinsky, N.I. & Russell, M. (1999) Retinoid signaling and activator protein-1 expression in ferrets given beta-carotene supplements and exposed to tobacco smoke. *J. Natl Cancer Inst.*, **91**, 60–66

Wasan, H.S., Park, H., Liu, K.C., Mandir, N.K., Winnett, A., Sasieni, P., Bodmer, W.F., Goodlad, R.A. & Wright, N.A. (1998) APC in the regulation of intestinal crypt fission. *J. Pathol.*, **185**, 246–255

Watson, P. & Lynch, H.T. (1993) Extracolonic cancer in hereditary nonpolyposis colorectal cancer. *Cancer*, **71**, 677–685

Wicking, C., Shanley, S., Smyth, I., Gillies, S., Negus, K., Graham, S., Suthers, G., Haites, N., Edwards, M., Wainwright, B. & Chenevix-Trench, G. (1997) Most germ-line mutations in the nevoid basal cell carcinoma syndrome lead to a premature termination of the PATCHED protein, and no genotype-phenotype correlations are evident. *Am. J. Human Genet.*, **60**, 21–26

Wijnen, J.T., Vasen, H.F., Zwinderman, A.H., van der Klift, H., Mulder, A., Tops, C., Moller, P. & Fodde, R. (1998) Clinical findings with implications for genetic testing in families with clustering of colorectal cancer. *New Engl. J. Med.*, **339**, 511–518

Winawer, S.J., Zamber, A.G., Ho, M.N., O'Brien, M.J., Gottleib, L.S., Sternberg, S.S., Waye, J.D., Schapiro, M., Bond, J.H., Panish, J.F., Ackroyd, F., Shike, M., Kurtz, R.C., Hornsby-Lewis, L., Gerdes, H., Stewart, E.T. and the National Polyp Study Workgroup (1993a) Randomized comparison of surveillance intervals after colonoscopic removal of newly diagnosed adenomatous polyps. The National Polyp Study Workgroup. *New Engl. J. Med.*, **328**, 901–906

Winawer, S.J., Zamber, A.G., Ho, M.N., O'Brien, M.J., Gottleib, L.S., Sternberg, S.S., Waye, J.D., Schapiro, M., Bond, J.H., Panish, J.F., Ackroyd, F., Shike, M., Kurtz, R.C., Hornsby-Lewis, L., Gerdes, H., Stewart, E.T. and the National Polyp Study Workgroup (1993b) Prevention of colorectal cancer by colonoscopic polypectomy. The National Polyp Study Workgroup. *New Engl. J. Med.*, **329**, 1977–1981

Yagi, O.K., Akiyama, Y., Ohkura, Y., Ban, S., Endo, M., Saitoh, K. & Yuasa, Y. (1997) Analyses of the APC and TGF-beta type II receptor genes, and microsatellite instability in mucosal colorectal carcinoma. *Jap. J. Cancer Res.*, **88**, 718–724

Yagi, O.K., Akiyama, Y. & Yuasa, Y. (1999) Genomic structure and alterations of homeobox gene CDX2 in colorectal carcinomas. *Br. J. Cancer*, **79**, 440–444

Corresponding author:

J. Burn
School of Biochemistry and Genetics,
University of Newcastle,
20 Claremont Place,
Newcastle upon Tyne NE2 4AA,
UK

Biomarkers in Cancer Chemoprevention
Miller, A.B., Bartsch, H., Boffetta, P., Dragsted, L. and Vainio, H. eds
IARC Scientific Publications No. 154
International Agency for Research on Cancer, Lyon, 2001

Endogenous hormone metabolism as an exposure marker in breast cancer chemoprevention studies

R. Kaaks

There is overwhelming evidence that alterations in endogenous hormone metabolism — as a form of endogenous exposure—may be an important metabolic risk factor for the development of breast cancer. This chapter reviews current theories and major epidemiological findings that link endogenous hormones (sex steroids and their metabolites, but also insulin, and insulin-like growth factor-I (IGF-I)) to breast cancer risk. Knowledge about these metabolic risk factors can be used to identify women at increased risk of breast cancer, who might benefit most from chemoprevention. In addition, modification of high-risk endocrine profiles may itself become a central target of chemoprevention. Possible intervention strategies include improvement of insulin sensitivity, reducing concentrations of IGF-I in blood and breast tissue, reducing ovarian overproduction of androgens, inhibiting the activity of aromatase and other enzymes involved in estrogen formation within the breast, and modifying estrogen metabolism within the breast (e.g., decreasing 16α- and 4α-hydroxylation, and increasing *O*-methylation of catecholestrogens). Several of these possible strategies are illustrated with examples of chemopreventive agents currently in use or proposed for use to prevent breast cancer.

Introduction

Breast cancer is predominantly a disease of western, industrialized countries. There is overwhelming evidence from studies on migrants and from time trends in incidence rates that the high incidence rates of breast cancer in western countries are due mainly to lifestyle factors, including nutrition. Nutritional factors most strongly implicated include low levels of physical activity and an energy-dense diet rich in fats, refined carbohydrates and animal protein. These factors also appear related to an earlier average age at menarche, increasing average body height and an increasing prevalence of obesity—three well established risk factors for breast cancer that are also known to be associated with changes in hormone metabolism. An additional risk factor, which probably also affects breast cancer risk through changes in endogenous hormone metabolism, is a childbearing pattern characterized by relatively late first pregnancy and low parity. Besides endogenous hormones, the use of exogenous hormones for contraception or for treatment of menopausal symptoms has also been suggested to be related to breast cancer risk.

The identification of specific alterations in endogenous hormone metabolism as risk factors for breast cancer risk is a topic of extensive research. Knowledge of such hormonal risk factors may help understand why more traditional risk factors (e.g., obesity, lack of physical activity, late menopause, use of certain exogenous hormones) are related to breast cancer. In addition, this knowledge can be used to design and test strategies for breast cancer prevention, through changes in lifestyle (nutrition, hormone use) or by use of specific chemopreventive agents. Favourable changes in endogenous hormone profiles can then be taken as intermediate evidence for a possible preventive effect.

The first section of this chapter reviews the associations of breast cancer risk with circulating levels of sex steroids and with modifications in sex steroid metabolism locally within the breast. In the next section, basic aspects of insulin-like growth factor-I (IGF-I) metabolism are discussed, and the associations of breast cancer risk with circulating levels of IGF-I, its binding proteins (IGFBPs) and insulin are reviewed. In the third section, a number of possible strategies for breast cancer prevention through modification of endogenous hormone metabolism are presented. The final section presents concluding remarks about the possible use of markers of endogenous hormonal exposures in prevention studies.

Sex steroids

Circulating sex steroids as determinants of breast cancer risk

Ovarian sex steroids clearly play an important role in promoting the development of breast cancer. Risk is consistently increased in women with early menarche and late menopause (Key & Pike, 1988) and hence is associated with a longer lifetime production of estradiol and progesterone by the ovaries. Furthermore, breast cancer incidence rates rise more steeply with age before menopause than after, when the ovarian synthesis of estrogens and progesterone practically ceases, and when the production of androgens gradually reduces to about half the premenopausal levels. Ovariectomy at an early age reduces breast cancer incidence and ovariectomy at the time of cancer diagnosis diminishes the risk of tumour recurrence (Secreto & Zumoff, 1994).

A popular theory is that breast cancer risk is increased in women who have increased plasma levels of bioavailable estradiol unbound to sex-hormone binding globulin, and elevated estradiol concentrations within breast tissue (estrogen excess hypothesis) (Bernstein & Ross, 1993; Key & Pike, 1988). A second theory is that breast cancer risk is increased in women with elevated levels of androgenic steroid hormones, in particular testosterone and Δ4-androstenedione (ovarian androgen excess hypothesis) (Bernstein & Ross, 1993; Kaaks, 1996; Secreto & Zumoff, 1994). These two theories are complementary, as ovarian over-production of androgens and increased plasma levels of total or bioavailable estrogens often occur together, most strikingly in patients with hyperandrogenic syndromes such as the polycystic ovary syndrome (Ehrmann *et al.*, 1995), but also in postmenopausal women with upper body type (android) obesity (Kaaks, 1996). Δ4-Androstenedione and testosterone are the immediate precursors for estrogen synthesis. Especially after the menopause, when the ovaries no longer produce the enzyme aromatase needed for estrogen synthesis, the levels of bioavailable androgens unbound to sex-hormone binding globulin in plasma are a major determinant of estrogens formed within adipose tissue, including the breast.

The ovarian androgen excess hypothesis and estrogen excess hypothesis are both supported by results from several prospective cohort studies, which have shown increased breast cancer risk in postmenopausal women having elevated plasma levels of testosterone and Δ4-androstenedione, reduced levels of sex-hormone binding globulin and increased levels of total and bioavailable estradiol (Thomas *et al.*, 1997a). The increases in breast cancer risk associated with this high-androgen steroid profile may be due particularly to the rise in bioavailable plasma estrogen levels, plus increased estrogen formation within breast tissue itself. Estradiol is a strong mitogen for breast epithelial cells, and is therefore thought to have a key role in promoting tumour development (Henderson & Feigelson, 2000). However, risk may also be influenced directly by androgens, through androgen receptors within breast tissue (Adams, 1998; Birrell *et al.*, 1998).

There is considerable evidence that the combination of estrogens and progestogens increases breast cancer risk further than exposure to excess estrogens alone (Bernstein & Ross, 1993; Key & Pike, 1988). Major observations supporting this *"estrogen-plus-progestogen"* hypothesis are that: (1) breast epithelial cells have the highest mitotic activity in the luteal phase of the menstrual cycle, when progesterone production peaks; (2) premenopausal women who are obese or who have severe forms of ovarian androgen excess (as in polycystic ovary syndrome) on average experience some reduction in breast cancer risk, and this may be explained by frequent anovulatory menstrual cycles and an impaired luteal-phase progesterone production; (3) in postmenopausal women, the use of combined estrogen-plus-progestogen hormone replacement therapy increases breast cancer risk to a greater extent than replacement therapy with estrogens alone (Ross *et al.*, 2000; Schairer *et al.*, 2000). Nevertheless, several case–control studies (Secreto & Zumoff, 1994), though not one (small) prospective cohort study (Thomas *et al.*, 1997b), have shown increases in breast cancer risk in premenopausal women who have elevated androgen levels in plasma or urine. The latter suggests that only mildly hyperandrogenic women who maintain regular ovulatory cycles, but who may have somewhat elevated bioavailable estrogens, may be at increased risk, whereas the more severely hyperandrogenic women may experience a relative protection because of chronic anovulation and impaired ovarian progesterone production. Direct evaluations of associations of breast cancer risk with circulating levels of estradiol and progesterone are complicated by the wide variations of these two hormones during the menstrual cycle, and so far have not led to any definitive conclusions.

There is extensive evidence that reductions in sex-hormone binding globulin synthesis and increases in ovarian sex steroid synthesis are both related to chronic hyperinsulinaemia, which in turn is often a consequence of obesity, lack of physical activity, and hence insulin resistance (Kaaks, 1996) (Figure 1). The effects of insulin on hepatic sex-hormone binding globulin production and steroidogenesis may be mediated at least partially by an increase in IGF-I bioactivity (see below).

Estrogen synthesis and metabolism within breast tissue

Only sex steroids unbound to sex-hormone binding globulin can diffuse from the circulation into target tissues, where they may bind to estrogen and androgen receptors and exert their biological effects. Circulating levels of bioavailable sex steroids are thus a key determinant of local estrogenic or androgenic activity within the breast. Another important determinant of estrogenic activity within breast tissue, however, is the local metabolism of sex steroids (Figure 2). Indeed, studies have shown substantially higher concentrations of estrogens in breast tissue than in the circulation, especially in postmenopausal women. Furthermore, several steroid-metabolizing enzymes are present, and active, in breast tissue (Zhu & Conney, 1998). Similar enzymatic activity is found in adipose tissue elsewhere in the body.

The principal enzyme involved in producing estrogens from androgenic precursors is aromatase (CYP19), which converts Δ4-androstenedione and testosterone into estrone and estradiol, respectively. Aromatase activity is higher in adipose tissue of breast cancer patients than in tissue from women with benign breast disease (Miller & Mullen, 1993). Furthermore, in breasts containing a tumour, aromatase expression and activity are higher in quadrants bearing tumours than in the other quadrants. A second enzyme in breast tissue that may contribute to estrogen formation is 3β-hydroxysteroid dehydrogenase, which converts dehydroepiandrosterone (DHEA) and DHEA sulfate (DHEAS) into Δ4-androstenedione, the principal androgenic precursor for aromatization into estrogens (estrone).

A second important regulatory mechanism for modulation of estrogenic activity within breast tissue is the interconversion of the less potent estrogen estrone into the more potent estradiol by 17β-hydroxysteroid dehydrogenase type 1, and vice versa, by 17β-hydroxysteroid dehydrogenase type 2. In normal breast tissue and breast cancer cells, the balance in activity between the two isozymes generally appears to favour the conversion of estrone into estradiol (Zhu & Conney, 1998).

A third mechanism by which hormonally active estrogens (i.e., with receptor binding activity) can be formed is the hydrolysis of sulfate or glucuronyl groups from, respectively, estrone sulfate and estrone glucuronate or estradiol glucuronate. The sulfated or glucuronated compounds are formed in the liver and other tissues by steroid sulfotransferases and glucuronosyl transferases, respectively, to form more water-soluble compounds that can be excreted in bile and urine. These compounds have little or no estrogen receptor-binding affinity, but their large amounts in the circulation form a reservoir from which active estrogens can be formed by hydrolysis (Zhu & Conney, 1998).

A fourth mechanism by which active estrogens are formed is hydrolysis of estrogen fatty acid esters. These esters are formed by estrogen acyltransferase and, because of their high lipophilicity, accumulate in relatively high concentrations in fatty tissues. While estradiol fatty acid esters have little or no estrogen receptor-binding affinity, they form an important local reservoir for estradiol formation by the action of specific esterases.

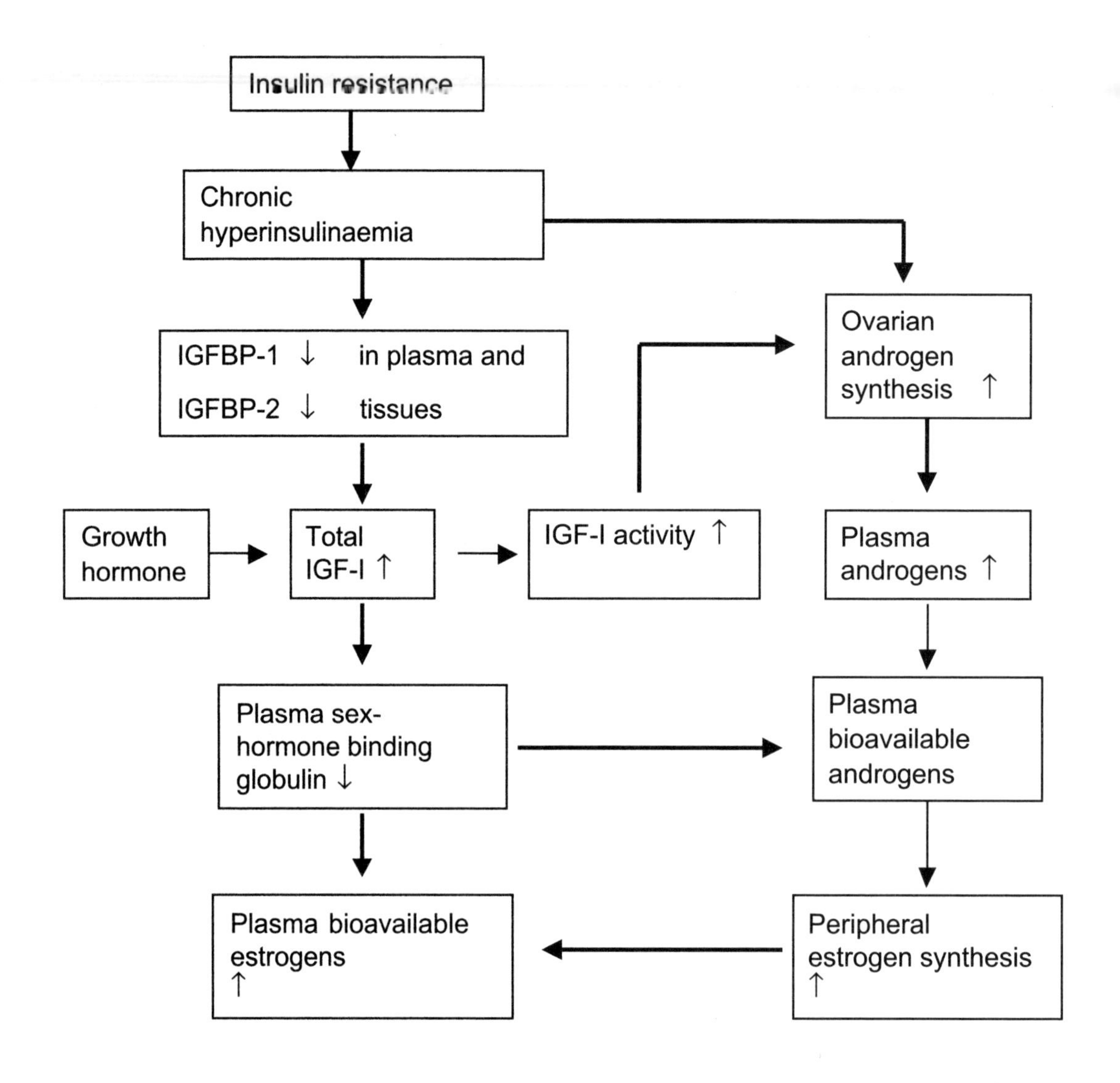

Figure 1. Role of insulin and IGF-I in regulating levels of total and bioavailable sex steroids
IGF-I, Insulin-like growth factor-I; IGFBP-1, insulin-like growth factor binding protein-1; IGFBP-2,

Besides enzymes involved in regulating levels of active versus less active or bound forms of estrogen within breast tissue, several other enzymes form estrogen metabolites that may have their own biological activities and physiological roles (Figure 2). The first major classes are cytochrome P450 enzymes (CYP1A1, CYP1A2, CYP2B1, CYP2B2, CYP2C family, CYP3A family, $P450_{C\text{-}M/F}$) that form estrogen hydroxylation products. A second type of enzyme — catechol-*O*-methyltransferase —further metabolizes hydroxy-estrogen metabolites to methoxy-estrogens.

The hydroxylating P450 enzymes all require NADPH, and are located in mitochondria. Estrogen

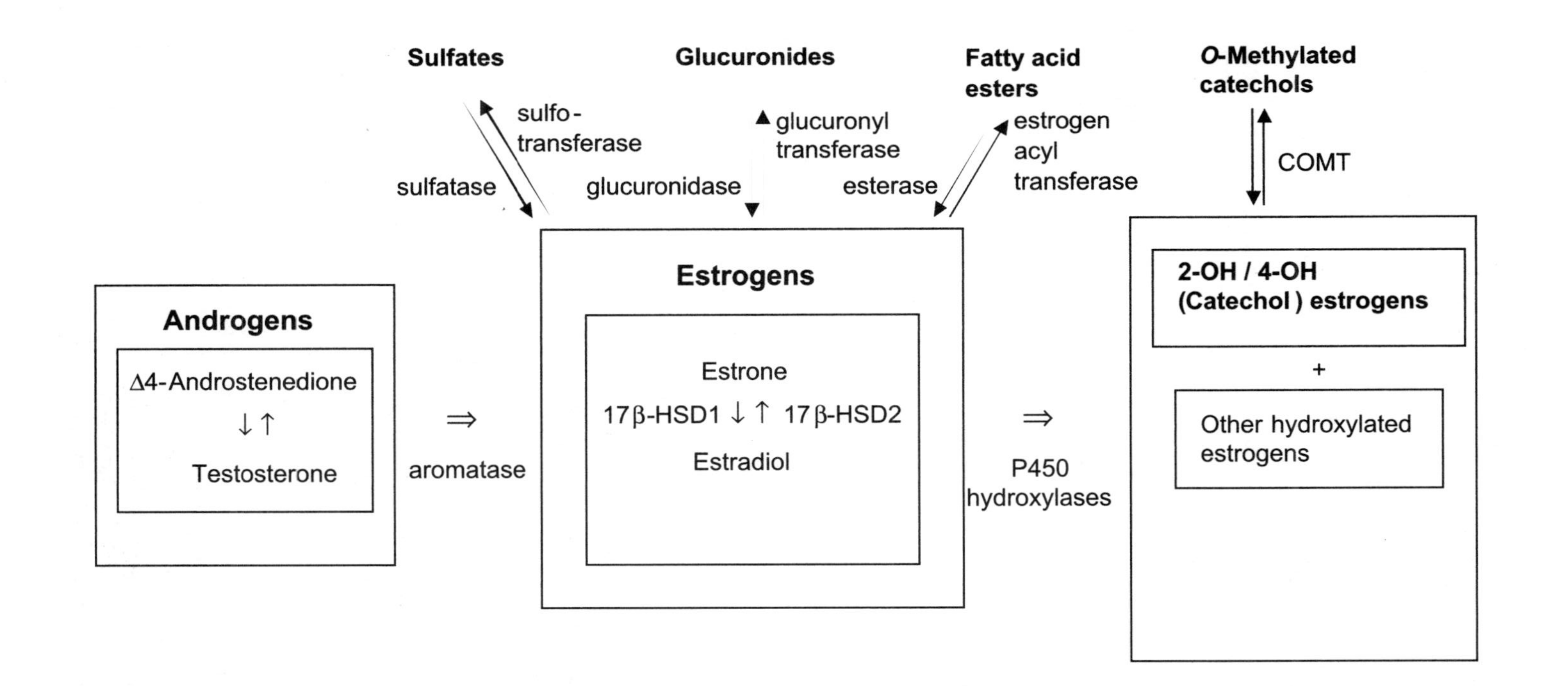

Figure 2. Synthesis and metabolism of estrogens within breast tissue
(After Zhu & Conney, 1999)

17β HSD1, 17β-hydroxysteroid dehydrogenase type 1; 17β HSD2, 17β-hydroxysteroid dehydrogenase type 2; COMT, catechol-*O*-methyltransferase metabolites with hydroxyl groups at positions 1, 2, 4, 6, 7, 14, 15, 16, 17 or 18 can be formed (Zhu & Conney, 1998). Some of these metabolites are formed predominantly in liver, but others (e.g., the 2-, 4- and 16-hydroxylation products) are also formed in relatively large amounts in breast tissue. 2-Hydroxyestradiol has markedly reduced estrogen receptor-binding activity, but 4-hydroxyestradiol and 16-hydroxyestradiol retain potent hormonal activity by binding estrogen receptors. Because of this and other lines of evidence, it has been hypothesized that an increase in 16-hydroxylation (relative to 2-hydroxylation) may be a determinant of breast cancer risk (Bradlow *et al.*, 1986). This hypothesis was confirmed in some recent epidemiological studies (Kabat *et al.*, 1997), but not by others (Ursin *et al.*, 1999), where levels of 16- and 2-hydroxylated estrogens were measured in urine (Meilahn *et al.*, 1998). Further recent studies do suggest, on the other hand, that increased formation of 4-hydroxyestrogens in target tissues, including human breast, may play an important role in estradiol-induced carcinogenesis. A key mechanism involved may be metabolic redox cycling of 4-hydroxyestradiol, catalysed by cytochrome P450 enzymes (Liehr, 2000). The potential role of increased 4-hydroxylation in breast cancer development awaits confirmation from epidemiological studies.

The 2- and 4-hydroxyestrogens have a catechol structure (i.e., with hydroxyl groups on two adjacent carbon atoms in the aromatic A ring), which means that these compounds are chemically reactive. Due to their reactivity, they may for instance bind to DNA, and thus cause DNA damage or mutations. A strong carcinogenic action has been observed only for 4-hydroxyestradiol, and not for the 2-hydroxy metabolite. The enzyme catechol-*O*-methyltransferase (COMT) further metabolizes these 'catechol estrogens' by formation of 2- and 4-methoxyestradiol compounds. *O*-Methylated estrogens are more lipophilic, have very long half-lives, and do not bind to the classical (alpha) estrogen receptor.

Earlier studies on the chemical reactivity and potential genotoxicity of catechol estrogens led to the suggestion that enzymatic *O*-methylation was primarily a detoxification pathway for these catechol intermediates. More recent results, however, suggest that the *O*-methylated products may have a number of unique biological activities apparently not mediated by the estrogen alpha receptor. In particular 2-methoxyestradiol inhibits the proliferation of several cancer cell lines *in vitro,* including human breast cancer, and *in vivo* also has strong anti-angiogenic effects (Zhu & Conney, 1998).

The distribution of catechol-*O*-methyl transferase activity appears to follow a bimodal pattern due to a polymorphism in the *COMT* gene. About 25% of Caucasians are homozygous for the low-activity allele. Some recent epidemiological studies (Huang *et al.*, 1999; Lavigne *et al.*, 1997; Thompson *et al.*, 1998) but not all (Millikan *et al.*, 1998) have shown an increase in breast cancer risk in women with this gene variant, although the results varied in association with either pre- or postmenopausal risk, and between obese or non-obese women. Taken together, there is some evidence that high catechol-*O*-methyl transferase activity may protect against breast cancer by increasing levels of 2-methoxyestradiol, which has anti-estrogenic properties, and by faster deactivation of 4-hydroxyestradiol, which is carcinogenic.

IGF-I and IGF-binding proteins

The hypothesis that risk of cancer (at various organ sites) may be related to circulating levels of total IGF-I and its binding proteins (IGFBPs), and to an increase in IGF-I bioactivity at a tissue level, has recently received much attention from epidemiologists. Reasons are that IGF-I in general inhibits apoptosis and stimulates cell proliferation (Jones & Clemmons, 1995; Werner & Le Roith, 1997) and thus stimulates tumour development. In addition, levels of IGF-I and several of its binding proteins (especially IGFBP-1, -2 and -3), as well as local IGF-I bioactivity within tissues, are intricately related to nutritional status and energy metabolism (Thissen *et al.*, 1994). Furthermore, as mentioned above, the IGF-I/IGFBP system, in interaction with insulin, appears to be central in regulating the synthesis and plasma levels of sex-hormone binding globulin and androgens.

IGF-I, IGFBPs, and IGF-I bioactivity

IGF-I bioactivity is the overall resultant of complex interactions of endocrine, paracrine and autocrine sources of IGF-I and IGFBPs with cellular receptors. Although IGF-I bioactivity is believed to increase generally when total IGF-I concentrations rise,

IGF-I bioactivity is modulated to a large extent by the IGFBPs.

A first level at which IGFBPs modulate IGF-I bioactivity is regulation of the efflux of circulating IGF-I through the capillary barrier. At least six different IGFBPs have been identified in tissues and in the circulation (Jones & Clemmons, 1995). Although IGF-I and IGFBPs are produced in practically all human tissues, most IGF-I and IGFBPs in blood plasma are produced in the liver. Over 90% of circulating IGF-I is complexed with IGFBP-3 and with another glycoprotein called acid-labile subunit (ALS). IGFBP-3 has a very high affinity for IGF-I, and the large IGF-I/IGFBP-3/ALS complex cannot pass through the capillary barrier to target tissues. There is recent evidence that IGFBP-5, which has even higher affinity for IGF-I than IGFBP-3, may form a similar IGF-I/IGFBP-5/ALS complex. Practically all of the remaining IGF-I is bound to IGFBP-1, -2, -4 and -6, which have lower binding affinities for IGF-I (compared with IGFBP-3 and IGFBP-5), do not form complexes with ALS, and are small enough to cross the endothelial barrier. Therefore, a decrease in plasma IGFBP-3 level, with a transfer of IGF-I to smaller IGFBPs not complexed with ALS, is believed to increase IGF-I availability to its tissue receptors. Reductions in plasma concentrations of the smaller IGFBPs, and particularly IGFBP-1 and -2, are also thought to increase the bioavailability of circulating IGF-I (Jones & Clemmons, 1995).

A second level at which IGFBPs modulate IGF-I bioactivity is the target tissue itself, where IGFBPs regulate the binding of IGF-I to its receptors. Although binding proteins have been mostly proposed to inhibit receptor binding by complexing IGF-I, results from *in-vitro* studies suggest that, depending on the relative concentrations of IGF-I and IGFBPs, certain IGFBPs (e.g., IGFBP-1, -2, -3 and -5) may actually enhance IGF-I binding to its receptors (Jones & Clemmons, 1995).

Epidemiological studies on insulin, IGF-I and breast cancer risk

Three case–control studies and one prospective cohort study have shown an increase in breast cancer in women with elevated plasma IGF-I or with elevated IGF-I for a given level of IGFBP-3 (Bohlke *et al.*, 1998; Bruning *et al.*, 1995; Del Giudice *et al.*, 1998; Hankinson *et al.*, 1998; Peyrat *et al.*, 1993). This relationship between risk and total plasma IGF-I was found exclusively for women who developed breast cancer before the average age at menopause. Besides increased levels of IGF-I, either as total concentration or relative to IGFBP-3 levels, two case–control studies have shown that elevated plasma insulin levels also increase breast cancer risk, in both premenopausal (Bruning *et al.*, 1995; Del Giudice *et al.*, 1998) and postmenopausal (Bruning *et al.*, 1992) women.

Possible strategies for chemoprevention

Different types of intervention aiming at reducing breast cancer risk by favourably changing endogenous hormone metabolism have been proposed, including energy restriction and body weight loss (in obese or overweight women), long-term changes in diet and physical activity level, and use of various chemopreventive agents. Strategies for prevention of breast cancer through alterations in hormone metabolism may target a number of mechanisms. These include reduction of circulating IGF-I levels and/or IGF-I bioactivity at a tissue level, reduction of ovarian (and/or adrenal) sex steroid production, increase of hepatic sex-hormone binding globulin production, reduction of local estrogen synthesis and modification of local estrogen metabolism. Each of these strategies in principle may lead to measurable alterations in levels of hormones or hormone metabolites in blood or urine. Chemopreventive agents may also be used to diminish the binding of hormones to their receptors.

Reducing IGF-I bioactivity

A class of compounds that lower IGF-I levels are growth hormone-releasing hormone (GHRH) antagonists, including the natural compound somatostatin, and a number of somatostatin analogues (e.g., octreotide, vapreotide, lanreotide, octastatin, somatuline (BIM 23014)). These compounds decrease IGF-I levels by inhibiting the pituitary secretion of growth hormone (Kath & Hoffken, 2000; Kineman, 2000). In addition to the suppression of the growth hormone/IGF-I axis, GHRH antagonists appear to inhibit tumour growth directly through a local mechanism, which may be related to reduced local synthesis of insulin-like growth factor-II (IGF-II). This direct effect might be mediated by blocking the paracrine/autocrine actions of locally produced GHRH, which may normally act through receptors for the vasoactive intestinal protein or pituitary adenylate cyclase-activating polypeptide — peptides that are structurally similar to GHRH. Somatostatin analogues have been tested as part of hormonal adjuvant therapy in breast cancer patients, but have not been evaluated for chemoprevention in healthy women.

A second class of agents that decrease especially the hepatic production and circulating levels of total IGF-I are exogenous estrogens (e.g., in oral contraceptives, postmenopausal estrogen replacement therapy, or combined estrogen–progestogen replacement therapy) (Campagnoli *et al.*, 1992, 1995; Ho *et al.*, 1996) and estrogen analogues (tamoxifen, raloxifene, droloxifene) (Pollak *et al.*, 1992). Especially weak estrogen analogues have been regularly proposed as potential chemopreventive agents against breast cancer, and this was initially motivated mainly by their anti-estrogenic effects, through competitive binding to estrogen receptors (but without receptor activation). Estrogenic compounds decrease the liver synthesis of IGF-I only if taken orally but not when administered transdermally. A likely explanation is that oral administration leads to relatively high estrogen concentrations in the portal vein and hence in the liver, and that only such high concentrations can substantially decrease hepatic IGF-I production. Oral administration of androgenic compounds, including some progestagens produced from the androgenic precursor 19-nortestosterone (e.g., norethisterone, norgestrel, levonorgestrel) and present in some types of oral contraceptive and hormone replacement therapy, leads to an increase in synthesis and plasma levels of IGF-I. So far, no clear effect of phytoestrogen intake on circulating IGF-I levels has been reported.

The epidemiological evidence for associations of breast cancer risk with oral contraceptive use, use of estrogen replacement therapy or hormone replacement therapy has been reviewed (IARC, 1999), as well as that for tamoxifen (IARC, 1996). There is a small increase in risk associated with use of combined oral contraceptives, as well as with use of estrogen replacement therapy. The association of risk with oral contraceptives based on progestogens only, and with combined estrogen plus progestogen replacement therapy was difficult to evaluate because of lack of detail in the information from published studies. Nevertheless, a large case–control study and a cohort study recently showed significantly stronger increases in risk with combined estrogen plus progestogen therapy than with estrogens alone (Ross *et al.*, 2000; Schairer *et al.*, 2000).

The use of estrogen analogues for treatment and possibly chemoprevention of breast cancer was initially motivated by the concept that such compounds would diminish estrogenic effects in breast tissue by competing with natural estradiol for estrogen receptor binding, without having a receptor-activating capacity (Goldstein, 1999). A large meta-analysis of randomized intervention trials, including more than 75 000 breast cancer patients, confirmed the efficacy of such analogues, showing dramatic improvements in ten-year recurrence-free survival in the tamoxifen treatment group (Early Breast Cancer Trialists' Collaborative Group, 1992). Another large, double-blind and placebo-controlled randomized trial also showed a strong reduction in breast cancer incidence among 13 000 initially cancer-free women (Fisher *et al.*, 1998). The latter finding was not confirmed, however, by smaller trials in about 2500 women in the United Kingdom (Powles *et al.*, 1998) and in about 5400 women in Italy (Veronesi *et al.*, 1998). A recent trial among 7700 postmenopausal women with osteoporosis showed a strong protective effect of raloxifene against breast cancer occurrence (Cummings *et al.*, 1999).

A third group of agents that have been tested for potential chemopreventive activity against breast cancer, and which also appear to decrease circulating IGF-I levels, are retinoic acid analogues, such as fenretinide (4-hydroxyphenylretinamide) and all-*trans*-retinoic acid. A randomized intervention trial among about 3000 Italian women with surgically removed stage I breast cancer showed no significant effect of fenretinide on the occurrence of contralateral or ipsilateral tumors. However, an interaction was detected between fenretinide and menopausal status, with a possible beneficial effect in premenopausal women (Veronesi *et al.*, 1999).

IGF-I bioactivity may be decreased by improving insulin sensitivity. An improvement of insulin sensitivity and decreases in hepatic glucose output lead to lower endogenous insulin secretion, and this in turn leads to higher levels of IGFBP-1 and IGFBP-2. Natural ways of improving insulin sensitivity are to lose body weight (for overweight or obese women) (Bosello *et al.*, 1997; Guzick *et al.*, 1994) or to increase physical activity. In addition, insulin-sensitizing and hypoglycaemic drugs such as metformin (Pugeat & Ducluzeau, 1999) and troglitazone (Henry, 1997; Scheen & Lefebvre, 1999), biguanides or sulfonylureas (e.g., tolbutamide and tolazamide) could be used.

Increasing plasma sex-hormone binding globulin levels

Circulating sex-hormone binding globulin is produced in the liver and, as mentioned above, is under negative control of insulin and IGF-I, which both reduce its hepatic synthesis. Approaches to increase sex-hormone binding globulin synthesis include, first of all, improvement of insulin sensitivity. Weight loss or an increase in physical activity both increase circulating sex-hormone binding globulin levels (Guzick *et al.*, 1994), as does the use of insulin-sensitizing drugs. Oral intake of exogenous estrogens and estrogen analogues (e.g., tamoxifen), which decrease the hepatic production of IGF-I, also causes a rise in plasma sex-hormone binding globulin levels (Campagnoli *et al.*, 1992). Conversely, oral intake of androgenic compounds reduces hepatic output and plasma levels of sex-hormone binding globulin (Campagnoli *et al.*, 1994). Some study results have suggested a mild increase in hepatic production (Loukovaara *et al.*, 1995) or plasma levels of sex-hormone binding globulin (Adlercreutz *et al.*, 1987) at high intakes of phytoestrogens, but this requires confirmation in larger studies.

Reducing ovarian sex steroid synthesis

Total plasma levels of testosterone are positively related to breast cancer risk. As most of the circulating testosterone comes from the ovaries, this suggests that a reduction in ovarian sex steroid output may also be an approach to prevent cancer.

A first approach to diminishing ovarian (and possibly also adrenal) androgen production is to reduce circulating insulin levels. In obese and hyperandrogenic women, energy restriction with body weight loss and use of insulin-sensitizing drugs have both been shown to decrease circulating androgen levels (Pugeat & Ducluzeau, 1999). A second approach, which has also proven effective especially in women with polycystic ovary syndrome, is the use of combination-type oral contraceptives, which reduce ovarian hyperandrogenism by diminishing the pituitary secretion of luteinizing hormone. A third, more experimental approach is to use new types of oral contraception, which contain gonadotropin-releasing hormone agonists to inhibit secretion of pituitary luteinizing hormone, as well as small amounts of exogenous estrogens to compensate for the block of endogenous (ovarian) sex steroid production (Spicer & Pike, 1994).

Reducing estrogen formation within the breast

Since steroid sulfatases may increase the local formation within the breast of estrogenic steroids with receptor-activating capacity, steroid sulfatase inhibitors could have considerable therapeutic, and possibly preventive, potential. Several such inhibitors have now been developed. The most potent to date is estrone-3-*O*-sulfamate (Purohit *et al.*, 1999). Other steroidal sulfatase inhibitors are estrone-3-methylthiophosphonate (Duncan *et al.*, 1993) and 3-*O*-methylphosphonate derivatives of dehydroepiandrosterone (DHA-3-MTP), pregnenolone or cholesterol (Purohit *et al.*, 1994). Non-steroidal estrone sulfatase inhibitors include 4-methylcoumarin 7-*O*-sulfamate and its derivatives (14, 16 and 18) (Woo *et al.*, 1998). The extent and duration of the inhibition of estrone sulfatase may be monitored by measuring the activity of this enzyme in white blood cells or by measuring the decrease in the plasma DHEA/DHEAS concentration ratio (Purohit *et al.*, 1997).

Besides steroid sulfatase inhibitors, inhibitors of the enzyme aromatase (e.g., anastrozole, letrozole, vorozole, aminoglutethimide, 4-hydroxyandrostenedione (Harvey, 1998; Singh *et al.*, 1998) and other compounds (Kelloff *et al.*, 1998)) have been tested extensively in animal models and in clinical trials for breast cancer treatment. Aromatase inhibitors are being proposed for phase III trials for breast cancer treatment. The use of aromatase inhibitors can lead to profound suppression of plasma estrogen levels (Kelloff *et al.*, 1998). Nevertheless, strategies may be developed to obtain chemopreventive effects without total suppression of aromatase and plasma estrogen levels (Kelloff *et al.*, 1998).

Several flavonoids and isoflavonoids (phytoestrogens; e.g., coumestrol, genistein) have been found to be potent inhibitors of estrone reduction by 17β-hydroxysteroid dehydrogenase type 1 in breast cancer cells (Makela *et al.*, 1995, 1998). There is little information about compounds that may be used to regulate the balance between activities of the 17β-hydroxysteroid dehydrogenase type 1 and type 2 isozymes in favour of lower estradiol synthesis.

Modification of estrogen hydroxylation

On the basis of the theory that a high ratio of 16- to 2-hydroxyestradiol would increase risk, the use of chemicals that can shift hydroxylation towards 2-hydroxylation have been proposed as chemopreventive agents. One such compound is indole-3-carbinol, a compound which occurs naturally in certain vegetables and which competes with estradiol for 16-hydroxylation (Michnovicz, 1998). The shift from 16-hydroxylation to 2-hydroxylation can be monitored by measuring the ratio of the two types of estrogen hydroxylation product in urine (Michnovicz *et al.*, 1997). It is possible, however, that inhibition of 16-hydroxylation leads to increased formation of 4-hydroxyestradiol, which is potentially carcinogenic.

Hormone measurements as markers of endogenous exposure in chemoprevention studies: conclusions and problem areas

As shown above, alterations in endogenous hormone metabolism — as a form of 'endogenous' exposure — may be an important metabolic risk factor for the development of breast cancer. Knowledge about such metabolic risk factors can be used to identify women at increased risk of breast cancer who might benefit most from chemoprevention. In addition, as discussed in the third section of this chapter, modification of high-risk endocrine profiles may become the central target for chemoprevention. In the latter case, it is assumed that chemopreventive modification of endogenous hormone levels (e.g., reduction in IGF-I levels in blood and other tissues; inhibition of aromatase activity; shifts of estradiol hydroxylation pathways towards less 16- or 4-hydroxylation) will lead to reductions in breast cancer risk.

This implies that either the hormonal parameter aimed at must be itself a direct cause of cancer or else it must be at least very closely associated with other metabolic factors that are on the causal pathway (Lippman *et al.*, 1990). Measurements of hormones or hormone metabolites may then be used to evaluate the efficacy of a given prevention strategy. There are number of important problems, however.

First, the evaluation of whether a given hormone is directly related to tumour development is generally not straightforward. One problem is that hormonal parameters can have quite strong physiological interrelationships. For example, in postmenopausal women, elevated plasma insulin levels lead to reductions in sex-hormone binding globulin, increases in plasma androgens and increases in bioavailable estrogens. Sorting out which of these interrelated hormonal factors (bioavailable plasma estrogens, androgens, insulin?) are more directly, or strongly, related to breast tumour development may not be possible using epidemiological methods alone, and may require complementary mechanistic evidence from experimental studies. When possible effects on breast tissue of physiologically interrelated hormones cannot be disentangled, a reasonable approach to prevention may be to take modification of a global endocrine profile as the target for prevention. For example, by improving insulin sensitivity, one may at the same time obtain an increase in plasma sex-hormone binding globulin levels and reductions in plasma levels of total and bioavailable androgens and estrogens.

Second, it is not always clear whether circulating levels of hormones are a main determinant, or at least a good indicator, of levels within breast tissue. For example, in postmenopausal women, who have very low plasma concentrations of total and non-sex-hormone binding globulin-bound estradiol, the main determinant of estradiol concentrations within the breast may be its local synthesis from androgen precursors or from estrone. Therefore, if estradiol is a key hormone in promoting breast tumorigenesis, increased circulating bioavailable androgens, or increased breast tissue activities of aromatase or 17β-hydroxylase, may be more important determinants of risk than plasma concentrations of bioavailable estradiol. Another example is the control of IGF-I bioactivity within breast tissue, which depends not only on circulating levels of IGF-I and IGFBPs, but also on local synthesis. Thus, it is not entirely clear, for instance, whether factors such as estrogen replacement therapy or tamoxifen that may decrease the hepatic synthesis and output of IGF-I, therefore reducing its plasma concentrations, will also substantially reduce IGF-I bioactivity within the breast. A third example is formation of hydroxy- and methoxy-metabolites of estrogens within the breast. Blood concentrations of these metabolites may not reflect very accurately their levels in breast tissue, because they can also be formed in adipose tissue elsewhere in the body.

Besides the problem that circulating levels of a given hormone may not always reflect breast tissue concentrations, there can be major technical or logistic problems in the measurement of hormones or their metabolites in blood, urine or saliva. Such problems are the lack of sensitivity and specificity of methods for the detection of specific sex steroids and their metabolites (e.g., hydroxylation products) and problems due to cyclic or pulsatile variations in hormone secretion over time (Michaud *et al.*, 1999; Xu *et al.*, 1999).

References

Adams, J.B. (1998) Adrenal androgens and human breast cancer: a new appraisal. *Breast Cancer Res. Treat.*, **51**, 183–188

Adlercreutz, H., Hockerstedt, K., Bannwart, C., Bloigu, S., Hamalainen, E., Fotsis, T. & Ollus, A. (1987) Effect of dietary components, including lignans and phytoestrogens, on enterohepatic circulation and liver metabolism of estrogens and on sex hormone binding globulin (SHBG). *J. Steroid Biochem.*, **27**, 1135–1144

Bernstein, L. & Ross, R.K. (1993) Endogenous hormones and breast cancer risk. *Epidemiol Rev.*, **15**, 48–65

Birrell, S.N., Hall, R.E. & Tilley, W.D. (1998) Role of the androgen receptor in human breast cancer. *J. Mammary Gland Biol. Neoplasia*, **3**, 95–103

Bohlke, K., Cramer, D.W., Trichopoulos, D. & Mantzoros, C.S. (1998) Insulin-like growth factor-I in relation to premenopausal ductal carcinoma in situ of the breast. *Epidemiology*, **9**, 570–573

Bosello, O., Armellini, F., Zamboni, M. & Fitchet, M. (1997) The benefits of modest weight loss in type II diabetes. *Int. J. Obes. Relat. Metab. Disord.*, **21** Suppl. 1, S10–S13

Bradlow, H.L., Hershcopf, R.E. & Fishman, J.F. (1986) Oestradiol 16 alpha-hydroxylase: a risk marker for breast cancer. *Cancer Surv.*, **5**, 573–583

Bruning, P.F., Bonfrer, J.M., Van Noord, P.A., Hart, A.A., Jong-Bakker, M. & Nooijen, W.J. (1992) Insulin resistance and breast-cancer risk. *Int. J. Cancer*, **52**, 511–516

Bruning, P.F., Van Doorn, J., Bonfrer, J.M., Van Noord, P.A., Korse, C.M., Linders, T.C. & Hart, A.A. (1995) Insulin-like growth-factor-binding protein 3 is decreased in early-stage operable pre-menopausal breast cancer. *Int. J. Cancer*, **62**, 266–270

Campagnoli, C., Biglia, N., Belforte, P., Botta, D., Pedrini, E. & Sismondi, P. (1992) Post-menopausal breast cancer risk: oral estrogen treatment and abdominal obesity induce opposite changes in possibly important biological variables. *Eur. J. Gynaecol. Oncol.*, **13**, 139–154

Campagnoli, C., Biglia, N., Lanza, M.G., Lesca, L., Peris, C. & Sismondi, P. (1994) Androgenic progestogens oppose the decrease of insulin-like growth factor I serum level induced by conjugated oestrogens in postmenopausal women. Preliminary report. *Maturitas*, **19**, 25–31

Campagnoli, C., Biglia, N., Peris, C. & Sismondi, P. (1995) Potential impact on breast cancer risk of circulating insulin-like growth factor I modifications induced by oral HRT in menopause. *Gynecol. Endocrinol.*, **9**, 67–74

Cummings, S.R., Eckert, S., Krueger, K.A., Grady, D., Powles, T.J., Cauley, J.A., Norton, L., Nickelsen, T., Bjarnason, N.H., Morrow, M., Lippman, M.E., Black, D., Glusman, J.E., Costa, A. & Jordan, V.C. (1999) The effect of raloxifene on risk of breast cancer in postmenopausal women: results from the MORE randomized trial. Multiple Outcomes of Raloxifene Evaluation. *J. Am. Med. Ass.*, **281**, 2189–2197

Del Giudice, M.E., Fantus, I.G., Ezzat, S., McKeown-Eyssen, G., Page, D. & Goodwin, P.J. (1998) Insulin and related factors in premenopausal breast cancer risk. *Breast Cancer Res. Treat.*, **47**, 111–120

Duncan, L., Purohit, A., Howarth, N.M., Potter, B.V. & Reed, M.J. (1993) Inhibition of estrone sulfatase activity by estrone-3-methylthiophosphonate: a potential therapeutic agent in breast cancer. *Cancer Res.*, **53**, 298–303

Early Breast Cancer Trialists' Collaborative Group (1992) Systemic treatment of early breast cancer by hormonal, cytotoxic, or immune therapy. 133 randomised trials involving 31,000 recurrences and 24,000 deaths among 75,000 women. *Lancet*, **339**, 1–15

Ehrmann, D.A., Barnes, R.B. & Rosenfield, R.L. (1995) Polycystic ovary syndrome as a form of functional ovarian hyperandrogenism due to dysregulation of androgen secretion. *Endocrinol. Rev.*, **16**, 322–353

Fisher, B., Costantino, J.P., Wickerham, D.L., Redmond, C.K., Kavanah, M., Cronin, W.M., Vogel, V., Robidoux, A., Dimitrov, N., Atkins, J., Daly, M., Wieand, S., Tan-Chiu, E., Ford, L. & Wolmark, N. (1998) Tamoxifen for prevention of breast cancer: report of the National Surgical Adjuvant Breast and Bowel Project P-1 Study. *J. Natl Cancer Inst.*, **90**, 1371–1388

Goldstein, S.R. (1999) Effect of SERMs on breast tissue. *J. Endocrinol. Invest.*, **22**, 636–640

Guzick, D.S., Wing, R., Smith, D., Berga, S.L. & Winters, S.J. (1994) Endocrine consequences of weight loss in obese, hyperandrogenic, anovulatory women. *Fertil. Steril.*, **61**, 598–604

Hankinson, S.E., Willett, W.C., Colditz, G.A., Hunter, D.J., Michaud, D.S., Deroo, B., Rosner, B., Speizer, F.E. & Pollak, M. (1998) Circulating concentrations of insulin-like growth factor-I and risk of breast cancer. *Lancet*, **351**, 1393–1396

Harvey, H.A. (1998) Emerging role of aromatase inhibitors in the treatment of breast cancer. *Oncology (Huntingt)*, **12**, 32–35

Henderson, B.E. & Feigelson, H.S. (2000) Hormonal carcinogenesis. *Carcinogenesis*, **21**, 427–433

Henry, R.R. (1997) Thiazolidinediones. *Endocrinol. Metab. Clin. North Am.*, **26**, 553–573

Ho, K.K., O'Sullivan, A.J., Weissberger, A.J. & Kelly, J.J. (1996) Sex steroid regulation of growth hormone secretion and action. *Hormone Res.*, **45**, 67–73

Huang, C.S., Chern, H.D., Chang, K.J., Cheng, C.W., Hsu, S.M. & Shen, C.Y. (1999) Breast cancer risk associated with genotype polymorphism of the estrogen-metabolizing genes CYP17, CYP1A1, and COMT: a multigenic study on cancer susceptibility. *Cancer Res.*, **59**, 4870–4875

IARC (1996) *IARC Monographs on the Evaluation of Carcinogenic Risks to Humans*. Vol. 66, *Some Pharmaceutical Drugs*, Lyon

IARC (1999) *IARC Monographs on the Evaluation of Carcinogenic Risks to Humans*. Vol. 72, *Hormonal Contraception and Post-menopausal Hormonal Therapy*, Lyon

Jones, J.I. & Clemmons, D.R. (1995) Insulin-like growth factors and their binding proteins: biological actions. *Endocrinol. Rev.*, **16**, 3–34

Kaaks, R. (1996) Nutrition, hormones, and breast cancer: is insulin the missing link? *Cancer Causes Control*, **7**, 605–625

Kabat, G.C., Chang, C.J., Sparano, J.A., Sepkovie, D.W., Hu, X.P., Khalil, A., Rosenblatt, R. & Bradlow, H.L. (1997) Urinary estrogen metabolites and breast cancer: a case-control study. *Cancer Epidemiol. Biomarkers Prev.*, **6**, 505–509

Kath, R. & Hoffken, K. (2000) The significance of somatostatin analogues in the antiproliferative treatment of carcinomas. *Recent Results Cancer Res.*, **153**, 23–43

Kelloff, G.J., Lubet, R.A., Lieberman, R., Eisenhauer, K., Steele, V.E., Crowell, J.A., Hawk, E.T., Boone, C.W. & Sigman, C.C. (1998) Aromatase inhibitors as potential cancer chemopreventives. *Cancer Epidemiol. Biomarkers Prev.*, **7**, 65–78

Key, T.J. & Pike, M.C. (1988) The role of oestrogens and progestagens in the epidemiology and prevention of breast cancer. *Eur. J. Cancer Clin. Oncol.*, **24**, 29–43

Kineman, R.D. (2000) Antitumorigenic actions of growth hormone-releasing hormone antagonists. *Proc. Natl Acad. Sci. USA*, **97**, 532–534

Lavigne, J.A., Helzlsouer, K.J., Huang, H.Y., Strickland, P.T., Bell, D.A., Selmin, O., Watson, M.A., Hoffman, S., Comstock, G.W. & Yager, J.D. (1997) An association between the allele coding for a low activity variant of catechol-O-methyltransferase and the risk for breast cancer. *Cancer Res.*, **57**, 5493–5497

Liehr, J.G. (2000) Is estradiol a genotoxic mutagenic carcinogen? *Endocrinol. Rev.*, **21**, 40–54

Lippman, S.M., Lee, J.S., Lotan, R., Hittelman, W., Wargovich, M.J. & Hong, W.K. (1990) Biomarkers as intermediate end points in chemoprevention trials. *J. Natl Cancer Inst.*, **82**, 555–560

Loukovaara, M., Carson, M., Palotie, A. & Adlercreutz, H. (1995) Regulation of sex hormone-binding globulin production by isoflavonoids and patterns of isoflavonoid conjugation in HepG2 cell cultures. *Steroids*, **60**, 656–661

Makela, S., Poutanen, M., Kostian, M.L., Lehtimaki, N., Strauss, L., Santti, R. & Vihko, R. (1998) Inhibition of 17beta-hydroxysteroid oxidoreductase by flavonoids in breast and prostate cancer cells. *Proc. Soc. Exp. Biol. Med.*, **217**, 310–316

Makela, S., Poutanen, M., Lehtimaki, J., Kostian, M.L., Santti, R. & Vihko, R. (1995) Estrogen-specific 17 beta-hydroxysteroid oxidoreductase type 1 (E.C. 1.1.1.62) as a possible target for the action of phytoestrogens. *Proc. Soc. Exp. Biol. Med.*, **208**, 51–59

Meilahn, E.N., De Stavola, B., Allen, D.S., Fentiman, I., Bradlow, H.L., Sepkovic, D.W. & Kuller, L.H. (1998) Do urinary oestrogen metabolites predict breast cancer? Guernsey III cohort follow-up. *Br. J. Cancer*, **78**, 1250–1255

Michaud, D.S., Manson, J.E., Spiegelman, D., Barbieri, R.L., Sepkovic, D.W., Bradlow, H.L. & Hankinson, S.E. (1999) Reproducibility of plasma and urinary sex hormone levels in premenopausal women over a one-year period. *Cancer Epidemiol. Biomarkers Prev.*, **8**, 1059–1064

Michnovicz, J.J., Adlercreutz, H. & Bradlow, H.L. (1997) Changes in levels of urinary estrogen metabolites after oral indole-3-carbinol treatment in humans. *J. Natl Cancer Inst.*, **89**, 718–723

Michnovicz, J.J. (1998) Increased estrogen 2-hydroxylation in obese women using oral indole-3-carbinol. *Int. J. Obes. Relat. Metab. Disord.*, **22**, 227–229

Miller, W.R. & Mullen, P. (1993) Factors influencing aromatase activity in the breast. *J. Steroid Biochem. Mol. Biol.*, **44**, 597–604

Millikan, R.C., Pittman, G.S., Tse, C.K., Duell, E., Newman, B., Savitz, D., Moorman, P.G., Boissy, R.J. & Bell, D.A. (1998) Catechol-O-methyltransferase and breast cancer risk. *Carcinogenesis*, **19**, 1943–1947

Peyrat, J.P., Bonneterre, J., Hecquet, B., Vennin, P., Louchez, M.M., Fournier, C., Lefebvre, J. & Demaille, A. (1993) Plasma insulin-like growth factor-1 (IGF-1) concentrations in human breast cancer. *Eur. J. Cancer*, **29A**, 492–497

Pollak, M.N., Huynh, H.T. & Lefebvre, S.P. (1992) Tamoxifen reduces serum insulin-like growth factor I (IGF-I). *Breast Cancer Res. Treat.*, **22**, 91–100

Powles, T., Eeles, R., Ashley, S., Easton, D., Chang, J., Dowsett, M., Tidy, A., Viggers, J. & Davey, J. (1998) Interim analysis of the incidence of breast cancer in the Royal Marsden Hospital tamoxifen randomised chemoprevention trial. *Lancet*, **352**, 98–101

Pugeat, M. & Ducluzeau, P.H. (1999) Insulin resistance, polycystic ovary syndrome and metformin. *Drugs*, **58** Suppl. 1, 41–46

Purohit, A., Howarth, N.M., Potter, B.V. & Reed, M.J. (1994) Inhibition of steroid sulphatase activity by steroidal methylthiophosphonates: potential therapeutic agents in breast cancer. *J. Steroid Biochem. Mol. Biol.*, **48**, 523–527

Purohit, A., Froome, V.A., Wang, D.Y., Potter, B.V. & Reed, M.J. (1997) Measurement of oestrone sulphatase activity in white blood cells to monitor *in vivo* inhibition of steroid sulphatase activity by oestrone-3-O-sulphamate. *J. Steroid Biochem. Mol. Biol.*, **62**, 45–51

Purohit, A., Hejaz, H.A., Woo, L.W., van Strien, A.E., Potter, B.V. & Reed, M.J. (1999) Recent advances in the development of steroid sulphatase inhibitors. *J. Steroid Biochem. Mol. Biol.*, **69**, 227–238

Ross, R.K., Paganini-Hill, A., Wan, P.C. & Pike, M.C. (2000) Effect of hormone replacement therapy on breast cancer risk: estrogen versus estrogen plus progestin. *J. Natl Cancer Inst.*, **92**, 328–332

Schairer, C., Lubin, J., Troisi, R., Sturgeon, S., Brinton, L. & Hoover, R. (2000) Menopausal estrogen and estrogen-progestin replacement therapy and breast cancer risk. *J. Am. Med. Ass.*, **283**, 485–491

Scheen, A.J. & Lefebvre, P.J. (1999) Troglitazone: antihyperglycemic activity and potential role in the treatment of type 2 diabetes. *Diabetes Care*, **22**, 1568–1577

Secreto, G. & Zumoff, B. (1994) Abnormal production of androgens in women with breast cancer. *Anticancer Res.*, **14**, 2113–2117

Singh, A., Purohit, A., Coldham, N.G., Ghilchik, M.W. & Reed, M.J. (1998) Biochemical control of breast aromatase. *Breast Cancer Res. Treat.*, **49** Suppl. 1, S9–14

Spicer, D.V. & Pike, M.C. (1994) Sex steroids and breast cancer prevention. *J. Natl Cancer Inst. Monogr.*, **16**, 139–147

Thissen, J.P., Ketelslegers, J.M. & Underwood, L.E. (1994) Nutritional regulation of the insulin-like growth factors. *Endocrinol. Rev.*, **15**, 80–101

Thomas, H.V., Reeves, G.K. & Key, T.J. (1997a) Endogenous estrogen and postmenopausal breast cancer: a quantitative review. *Cancer Causes Control*, **8**, 922–928

Thomas, H.V., Key, T.J., Allen, D.S., Moore, J.W., Dowsett, M., Fentiman, I.S. & Wang, D.Y. (1997b) A prospective study of endogenous serum hormone concentrations and breast cancer risk in premenopausal women on the island of Guernsey. *Br. J. Cancer*, **75**, 1075–1079

Thompson, P.A., Shields, P.G., Freudenheim, J.L., Stone, A., Vena, J.E., Marshall, J.R., Graham, S., Laughlin, R., Nemoto, T., Kadlubar, F.F. & Ambrosone, C.B. (1998) Genetic polymorphisms in catechol-O-methyltransferase, menopausal status, and breast cancer risk. *Cancer Res.*, **58**, 2107–2110

Ursin, G., London, S., Stanczyk, F.Z., Gentzschein, E., Paganini-Hill, A., Ross, R.K. & Pike, M.C. (1999) Urinary 2-hydroxyestrone/16alpha-hydroxyestrone ratio and risk of breast cancer in postmenopausal women. *J. Natl Cancer Inst.*, **91**, 1067–1072

Veronesi, U., Maisonneuve, P., Costa, A., Sacchini, V., Maltoni, C., Robertson, C., Rotmensz, N. & Boyle, P. (1998) Prevention of breast cancer with tamoxifen: preliminary findings from the Italian randomised trial among hysterectomised women. Italian Tamoxifen Prevention Study. *Lancet*, **352**, 93–97

Veronesi, U., De Palo, G., Marubini, E., Costa, A., Formelli, F., Mariani, L., Decensi, A., Camerini, T., Del Turco, M.R., Di Mauro, M.G., Muraca, M.G., Del Vecchio, M., Pinto, C., D'Aiuto, G., Boni, C., Campa, T., Magni, A., Miceli, R., Perloff, M., Malone, W.F. & Sporn, M.B. (1999) Randomized trial of fenretinide to prevent second breast malignancy in women with early breast cancer. *J. Natl Cancer Inst.*, **91**, 1847–1856

Werner, H. & Le Roith, D. (1997) The insulin-like growth factor-I receptor signaling pathways are important for tumorigenesis and inhibition of apoptosis. *Crit. Rev. Oncog.*, **8**, 71–92

Woo, L.W., Howarth, N.M., Purohit, A., Hejaz, H.A., Reed, M.J. & Potter, B.V. (1998) Steroidal and non-steroidal sulfamates as potent inhibitors of steroid sulfatase. *J. Med. Chem.*, **41**, 1068–1083

Xu, X., Duncan, A.M., Merz-Demlow, B.E., Phipps, W.R. & Kurzer, M.S. (1999) Menstrual cycle effects on urinary estrogen metabolites. *J. Clin. Endocrinol. Metab.*, **84**, 3914–3918

Zhu, B.T. & Conney, A.H. (1998) Functional role of estrogen metabolism in target cells: review and perspectives. *Carcinogenesis*, **19**, 1–27

Corresponding author:

R. Kaaks
Unit of Nutrition and Cancer,
International Agency for Research on Cancer,
150 Cours Albert Thomas,
69372 Lyon cedex 08,
France

Biomarkers in Cancer Chemoprevention
Miller, A.B., Bartsch, H., Boffetta, P., Dragsted, L. and Vainio, H. eds
IARC Scientific Publications No. 154
International Agency for Research on Cancer, Lyon, 2001

Mammographic density as a marker of susceptibility to breast cancer: a hypothesis

N.F. Boyd, G.A. Lockwood, L.J. Martin, J.W. Byng, M.J. Yaffe and D.L. Tritchler

We propose that radiological features of breast tissue provide an index of cumulative exposure to the current and past hormonal and reproductive events that influence breast cancer incidence. The changes in breast tissue that occur with ageing, and changes in the associated radiological features of the breast, are similar to the concept of "breast tissue ageing" proposed by Pike, and may explain features of the age-specific incidence of breast cancer, both within the population and between populations. These radiological features can be observed and measured, can be related directly to risk of breast cancer, and are likely to be of value in research into the etiology of breast cancer. Identification of the sources of variation in this radiological characteristic of the breast is likely to lead to a better understanding of the factors that cause breast cancer and to new approaches to prevention of the disease.

Introduction

Pike *et al.* (1983) have described a model of breast cancer incidence that incorporates the principal reproductive and endocrine risk factors for the disease. The factors included in the model are age at menarche, age at first pregnancy and age at menopause. The model has since been extended to incorporate the timing of pregnancies (Rosner & Colditz, 1996). The model is based on the concept that the rate of "breast tissue ageing", rather than chronological age, is the relevant measure for describing the incidence of breast cancer. The concept of breast tissue ageing is closely associated with exposure of breast tissue to hormones, and the effects that hormones have on the kinetics of breast cells, and is illustrated in Figure 1(*a*). The rate of breast tissue ageing is most rapid at the time of menarche, slows with each pregnancy, slows further in the perimenopausal period, and is least after the menopause. After fitting suitable numerical values for breast tissue ageing, Pike *et al.* (1983) showed that this model provided a good fit to the actual age-specific incidence curve for breast cancer. Thus, cumulative exposure to breast tissue ageing, given by the area under the curve in Figure 1(*a*), describes the age-incidence curve for breast cancer shown in Figure 1(*b*).

The hypothesis that we propose is that radiological features of breast tissue provide an index of exposure of breast tissue to the current and past hormonal and reproductive events that influence breast cancer incidence, and are a measure of susceptibility to breast cancer related to the concept of breast tissue ageing (Pike *et al.*, 1983) (by susceptibility is meant a state in which risk of disease is altered, but the presence of premalignant changes is not implied). These radiological features can be observed and measured, can be related directly to risk of breast cancer, and are likely to be of value in research into the etiology and prevention of breast cancer.

The radiographic appearance, or mammographic pattern, of the female breast varies between individuals because of differences in the relative amounts and X-ray attenuation characteristics of fat, connective and epithelial tissue (Ingleby & Gerson-Cohen, 1960). Fat is radiologically lucent and appears dark on a mammogram. Connective and epithelial tissues are radiologically dense and appear light, an appearance that we refer to in this paper as mammographic densities. These variations are illustrated in Figure 2.

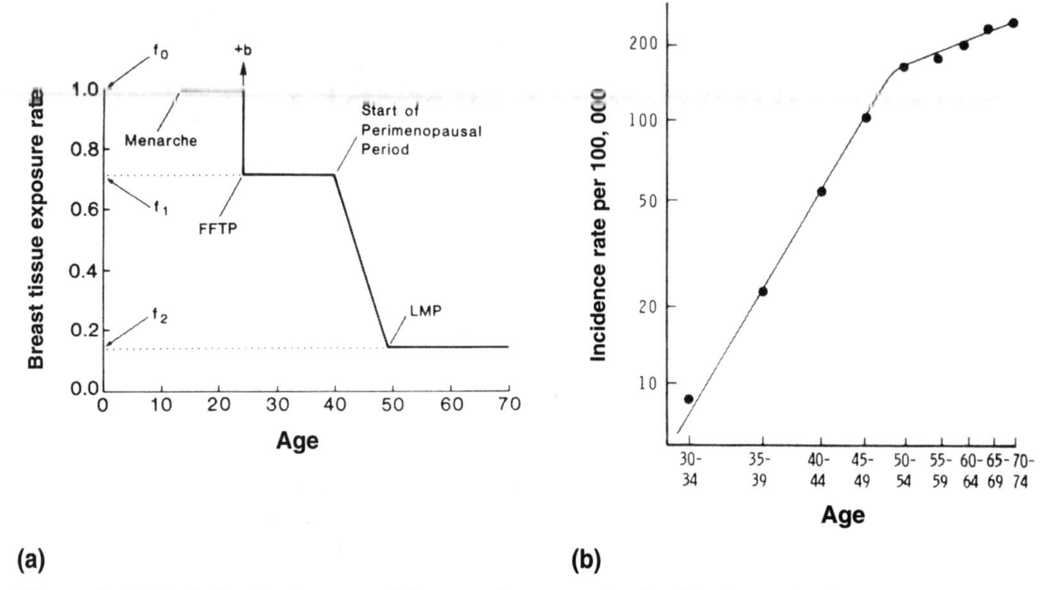

Figure 1. (a) Variation in the rate of "breast tissue ageing" with chronological age

Adapted from Pike *et al.* (1983) and Rosner & Colditz (1996). The rate of breast tissue ageing is greatest after menarche, declines with successive pregnancies and in the peri-menopausal period, and is lowest after the menopause. FFTP, first full-term pregnancy; LMP, last menstrual period

(b) Log–log plot of age-specific breast cancer incidence rates for Canada

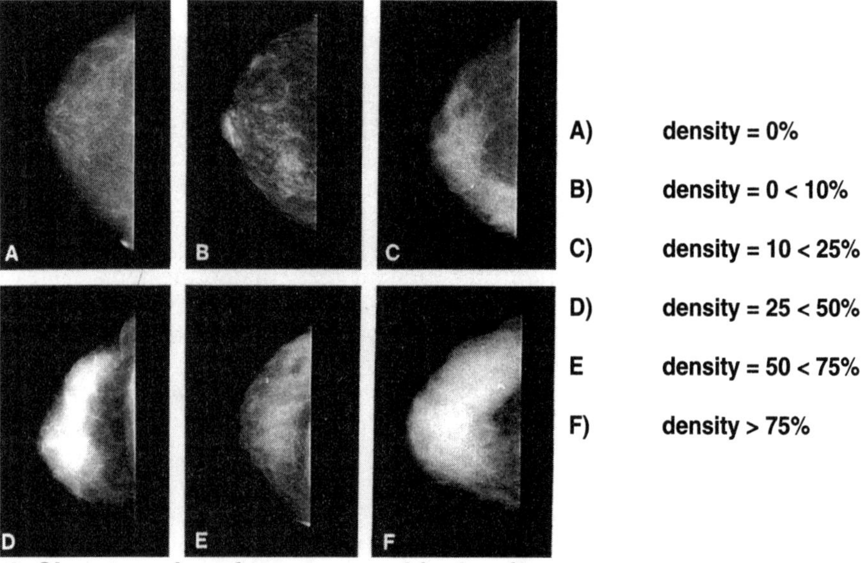

Figure 2. Six categories of mammographic density

Mammographic densities and risk of breast cancer

Wolfe (1976a,b) proposed a classification that related variations in the appearance of the mammogram to risk of breast cancer. While most well designed epidemiological studies have found that Wolfe's classification does identify individuals at different risk of breast cancer (Saftlas & Szklo, 1987; Oza & Boyd, 1993), quantitative classification of mammographic densities has given more consistent results and created larger gradients in risk. Table 1 summarizes the results of all studies published to date that have used quantitative methods to classify mammographic densities. Although definitions of categories of density vary somewhat between studies, in all, women with extensive breast tissue densities in more than 60–75% of the breast area were found to have a 4–6-fold greater risk of breast cancer than women with little or no densities. This gradient in risk is larger than is associated with any other risk factor for the disease, except age and mutations in the *BRCA1* and *BRCA2* genes (Easton *et al.*, 1995; Struewing *et al.*, 1997).

Extensive areas of mammographically dense breast tissue are common among subjects with breast cancer and estimates of attributable risk show that mammographic densities in more than 50% of the breast may account for 28% of breast cancer, a much larger fraction of disease than can be attributed to any other single risk factor (Byrne *et al.*, 1995).

Factors that influence mammographic density

Reproductive variables

Parity has been found in several studies to be related to mammographic density (Bergvist *et al.*, 1987; Brisson *et al.*, 1982b; de Waard *et al.*, 1984;

Table 1. Quantitative studies of breast density and cancer risk

Author	Design	Age	Odds ratio	95% CI	Trend
Boyd *et al.* (1982)	Case–control	40–65	a) 6.0[a,d]	2.5–14.1	Yes
			b) 2.8[a,d]	1.4–5.6	No
			c) 3.7[a,d]	1.7–4.1	Yes
Brisson *et al.* (1982a)	Case–control	20–69	a) 5.4[b]	2.5–11.4	Yes
			b) 3.8[c]	1.6–8.7	Yes
Brisson *et al.* (1984)	Case–control	–	4.4[d]	2.5–7.9	Yes
Brisson *et al.* (1989)	Case–control	40–67	a) 4.6[b]	2.4–8.5	Yes
			b) 3.2[c]	1.6–6.5	
			c) 5.5[d]	2.3–13.2	
Wolfe *et* al. (1987)	Case–control	30–85	4.3[d]	1.8–10.4	No
Saftlas *et al.* (1991)	Nested case–control	35–74	4.3[d]	2.1–8.8	Yes
Boyd *et al.* (1995)	Nested case–control	40–59	a) 6.0[d,e]	2.8–13.0	Yes
			b) 4.0[d,f]	2.1–7.7	Yes
Byrne *et al.* (1995)	Nested case–control	–	4.3[d]	3.1–6.1	Yes

[a] Odds ratios shown for each of three radiologists who estimated density
[b] Odds ratio for homogeneous density
[c] Odds ratio for nodular density
[d] Odds ratio for total density
[e] Odds ratio for estimation of area of density by radiologist
[f] Odds ratio for computer assisted measurement of area of density

Ernster *et al.*, 1980; Grove *et al.*, 1979, 1985; Kaufman *et al.*, 1991a; Whitehead *et al.*, 1985). Nulliparous women, at higher risk for breast cancer than parous women (Kelsey *et al.*, 1993), have denser breast tissue. Density decreases further with increasing number of children (de Waard *et al.*, 1984). Among parous women, density decreases further with earlier age at first birth and increasing number of pregnancies, both variables that decrease risk of breast cancer (de Waard *et al.*, 1984; Whitehead *et al.*, 1985). The effect of mammographic density on risk of breast cancer, however, remains strong after adjustment for the effect of reproductive variables.

Evidence of hormonal responsiveness

Hormone replacement therapy may increase breast densities (Bergkvist *et al.*, 1989; Berkowitz *et al.*, 1990; Cyrlack & Wong, 1993; Doyle & McLean, 1994; Kaufman *et al.*, 1991b; Laya *et al.*, 1995; Stomper *et al.*, 1990), and Spicer *et al.* (1994) have shown that the administration for one year of a hormonal contraceptive regimen that minimizes exposure of the breast epithelium to estrogen and progesterone reduces mammographic densities.

Age and the menopause

The prevalence of mammographically dense breast tissue in the population declines with increasing age (Grove *et al.*, 1979, 1985) and dense breast tissue is more common before than after the menopause (Grove *et al.*, 1979; Wolfe, 1976c). Menopausal status appears to be a stronger determinant of breast density than age. Direct evidence of a striking reduction in the proportion of the breast occupied by mammographic densities at the menopause has now been observed in a cohort of women examined by mammography before and after the cessation of menstrual activity (Boyd *et al.*, 1997).

Variations in change with age

The epidemiological data on risk presented in Table 1 show that over the range of the ages of the subjects included in the studies (between about 20 and 70 years), the mean tissue density in the mammogram of subjects who developed breast cancer was greater than in controls of the same age. Thus, the factors associated with variations in density over this range of ages may have an important influence on cancer risk. The potential sources of variation, that might give rise to differences in breast density between individuals, include differences in the quantity of breast epithelium and stroma at telarche, differences in the age at which the events of menarche, pregnancy and menopause occur, differences in the number and spacing of pregnancies, or differences in the magnitude of the reduction in mammographic density associated with pregnancy or the menopause. Diet, for example, appears to modify the magnitude of the reduction in mammographic density that occurs at the menopause (Boyd *et al.*, 1997).

Average exposure to mammographically dense breast tissue over time resembles the concept of breast tissue ageing proposed by Pike *et al.* (1983). Cumulative exposure to mammographic densities may thus also be represented by the area under the curve of Figure 1(*a*), reflecting cumulative exposure to hormonal stimuli to breast cell division, and be an important determinant of breast cancer incidence. We expect, therefore, that events that increase or decrease cumulative exposure to factors that influence the extent of mammographic densities will have corresponding effects on the incidence of the disease. Thus, late menarche, early first pregnancy, multiple pregnancies and early menopause, which are all known to decrease risk of breast cancer, also decrease cumulative exposure to mammographic densities. Conversely, early menarche, late age at first pregnancy, nulliparity, and late menopause increase cumulative exposure to dense tissue. By considering cumulative exposure to mammographically dense tissue in the population, given by the area under the curve of age-specific average densities, we may be able to account for key features of the age-specific incidence curve for breast cancer. Thus, the steeper pre-menopausal component of the age-incidence curve is associated with a higher prevalence of mammographic densities, the less steep post-menopausal component with a lower prevalence of densities. A reduction in mammographic densities at the time of the inflection in the age-specific incidence curve associated with the menopause has now been observed (Boyd *et al.*, 1997).

Biological plausibility of the hypothesis

The X-ray attenuation characteristics of epithelial and stromal tissues in the breast are responsible for mammographically dense breast tissue (Ingleby & Gerson-Cohen, 1960). Variations between individuals in the extent of mammographically dense breast tissue are thus likely to indicate variations in the number of epithelial and stromal cells in the breast, arising because of differences in stimulation by hormones and growth factors. Sex hormones in the blood, particularly estrogen, progesterone and prolactin, are likely to play a role in the etiology of mammographic densities. Blood levels of prolactin, for example, are reduced by pregnancy and the menopause (Reyes *et al.*, 1977; Wang *et al.*, 1988) and increased by hormone replacement therapy (Andersen *et al.*, 1980), all variables with effects on mammographic densities as described above.

Differences in risk of breast cancer associated with mammographic densities of different extent may thus arise, at least in part, because of differences in the number of cells in the breast that are susceptible to mutagens. Risk factors for breast cancer, such as parity, that are associated with a reduction in the extent of mammographic densities, may influence risk by reducing the size of the population of cells that are potential targets for events that give rise to mutations.

Predictions arising from the hypothesis

The testable predictions that arise from this hypothesis include the following:

(1) Age-specific estimates of the prevalence of radiologically dense breast tissue in the population will allow reconstruction of age-specific rates of breast cancer. As noted above, the principal features of the age-specific incidence curve for breast cancer are in general consistent with what is already known about the distribution of mammographic densities in the population. Little information is available about the radiological characteristics of breast tissue at earlier ages, particularly at the time of menarche, or the factors associated with them. Such information would be have to be acquired using methods of imaging the breast that do not involve exposure to radiation, such as ultrasound and magnetic resonance (Graham *et al.*, 1996; Kaizer *et al.*, 1988). The variations in the radiological appearance of the breast that are seen in middle life may in part be the result of differences that occur at the time of breast development, but events later in life, such as pregnancy and menopause, appear to be influential as well.

(2) Differences in the distribution of mammographic densities over age may be found in populations with different age-specific incidence rates of breast cancer. Studies that have compared Asians and Caucasians, Japanese and English, and North American Indians with Caucasian and Hispanic women, using Wolfe's (1976a) classification of mammographic patterns, have all shown that the population with the lower risk of breast cancer has a lower prevalence of dense patterns (Gravelle *et al.*, 1991; Hart *et al.*, 1989; Turnbull *et al.*, 1993).

(3) Age-related changes in the factors that control mammographic densities, such as hormones and growth factors, will account for the reduction in mammographic densities that occurs with increasing age.

Acknowledgements

Supported by grants from the National Cancer Institute of Canada and by a Terry Fox Programme Project Grant.

References

Andersen, J.R., Schroeder, E. & Lebech, P.E. (1980) The effect in post-menopausal women of natural human and artificial oestrogens on the concentration in serum of prolactin. *Acta Endocrinol.*, **95**, 433–437

Bergkvist, L., Tabar, L., Bergstrom, R. & Adami, H.O. (1987) Epidemiologic determinants of the mammographic parenchymal pattern. A population-based study within a mammographic screening program. *Am. J. Epidemiol.*, **126**, 1075–1081

Bergkvist, L., Tabar, L., Adami, H.O., Persson, I. & Bergstrom, R. (1989) Mammographic parenchymal patterns in women receiving noncontraceptive estrogen treatment. *Am. J. Epidemiol.*, **130**, 503–510

Berkowitz, J.E., Gatewood, O.M.B., Goldblum, L.E. & Gayler, B.W. (1990) Hormonal replacement therapy: mammographic manifestations. *Radiology*, **174**, 199–201

Boyd, N.F., O'Sullivan, B., Campbell, J.E., Fishell, E., Simor, I., Cooke G. & Germanson, T. (1982) Mammographic signs as risk factors for breast cancer. *Br. J. Cancer*, **45**, 185–193

Boyd, N.F., Byng, J.W., Jong, R.A., Fishell, E.K., Little, L.E., Miller, A.B., Lockwood, G.A., Trichler, D.L. & Yaffe, M.J. (1995) Quantitative classification of mammographic densities and breast cancer risk: results from the Canadian National Breast Screening Study. *J. Natl. Cancer Inst.*, **87**, 670–675

Boyd, N.F., Greenberg, C., Lockwood, G., Little, L., Martin. L., Byng, J., Yaffe, M. & Trichler, D. (1997) Effects at two years of a low-fat, high-carbohydrate diet on radiologic features of the breast: results from a randomized trial. *J. Natl. Cancer Inst.*, **89**, 488–496

Brisson, J., Merletti, F. & Sadowsky, N.L. (1982a) Mammographic features of the breast and breast cancer risk. *Am. J. Epidemiol.*, **115**, 428–437

Brisson, J., Sadowski, N.L., Twaddle, J.A., Morrison, A.S., Cole, P. & Merletti, F. (1982b) The relation of mammographic features of the breast to breast cancer risk factors. *Am. J. Epidemiol.*, **115**, 438–443

Brisson, J., Morrison, A.S. & Kopans, D.B. (1984) Height and weight, mammographic features of breast tissue, and breast cancer risk. *Am. J. Epidemiol.*, **119**, 371–381

Brisson, J., Verreault, R., Morrison, A., Tennina, S. & Meyer, F. (1989) Diet, mammographic features of breast tissue, and breast cancer risk. *Am. J. Epidemiol.*, **130**, 14–24

Byrne, C., Schairer, C., Wolfe, J., Parikh, N., Salane, M., Brinton, L.A., Hoover, R. & Haile, R. (1995) Mammographic features and breast cancer risk: effects with time, age, and menopause status. *J. Natl Cancer Inst.*, **87**, 1622–1629

Cyrlak, D. & Wong, C.H. (1993) Mammographic changes in postmenopausal women undergoing hormonal replacement therapy. *Am. J. Roentgenol.*, **161**, 1177–1183

de Waard, F., Rombach, J.J., Collette, H.J.A. & Slotboom, B. (1984) Breast cancer risk associated with reproductive factors and breast parenchymal patterns. *J. Natl. Cancer Inst.*, **72**, 1277–1282

Doyle, G.J. & McLean, L. (1994) Unilateral increase in mammographic density with hormone replacement therapy. *Clin. Radiol.*, **49**, 50–51

Easton, D.F., Ford, D. & Bishop, D.T. (1995) Breast cancer consortium. Breast and ovarian cancer incidence in BRCA1-mutation carriers. *Am. J. Hum. Genet.*, **56**, 265–271

Ernster, V.L., Sacks, S.T. & Peterson, C.A. (1980) Mammographic parenchymal patterns and risk factors for breast cancer. *Radiology*, **134**, 617–620

Graham, S.J., Bronskill, M.J., Byng, J.W., Yaffe, M.J. & Boyd, N.F. (1996) Quantitative correlation of breast tissue parameters using magnetic resonance and X-ray mammography. *Br. J. Cancer*, **73**, 162–168

Gravelle, I.H., Bulbrook, R.D., Wang, D.Y., Allen, D., Hayward, J.L., Bulstrode, J.C. & Takatani, O. (1991) A comparison of mammographic parenchymal patterns in premenopausal Japanese and British women. *Breast Cancer Res. Treat.*, **18** (Suppl. 1), S93–S95

Grove, J.S., Goodman, M.J., Gilbert, F.I. & Clyde, D. (1979) Factors associated with breast structures in breast cancer patients. *Cancer*, **43**, 1895–1899

Grove, J.S., Goodman, M.J. & Gilbert, F. (1985) Factors associated with mammographic pattern. *Br. J. Radiol.*, **58**, 21–25

Hart, B.L., Steinbock, R.T., Mettler, F.A., Jr, Pathak, D.R. & Bartow, S.A. (1989) Age and race related changes in mammographic parenchymal patterns. *Cancer*, **63**, 2537–2539

Ingleby, H. & Gerson-Cohen, J. (1960) *Comparative Anatomy, Pathology and Roentgenology of the Breast*, Philadelphia, University of Philadelphia Press

Kaizer, L., Fishell, E.K., Hunt, J.W., Foster, S. & Boyd, N.F. (1988) Ultrasonographically defined parenchymal patterns of the breast: relationship to mammographic patterns and other risk factors for breast cancer. *Br. J. Radiol.*, **61**, 118–124

Kaufman, Z., Garstin, W.I.H., Hayes, R., Mitchell, M.J. & Baum, M. (1991a) The mammographic parenchymal patterns of nulliparous women and women with a family history of breast cancer. *Clin. Radiol.*, **43**, 385–388

Kaufman, Z., Garstin, W.I. & Hayes, R. (1991b) The mammographic parenchymal patterns of women on hormonal replacement therapy. *Clin. Radiol.*, **43**, 389–392

Kelsey, J.L., Gammon, M.D. & John, E.S. (1993) Reproductive factors and breast cancer. *Epidemiol. Rev.*, **15**, 36–47

Laya, M.B., Gallagher, J.C. & Schreiman, J.S. (1995) Effect of postmenopausal hormonal replacement therapy on mammographic density and parenchymal pattern. *Radiology*, **196**, 433–437

Oza, A.M. & Boyd, N.F. (1993) Mammographic parenchymal patterns: a marker of breast cancer risk. *Epidemiol.Rev.*, **15**, 196–208

Pike, M.C., Krailo, M.D., Henderson, B.E., Casagrande, J.T. & Hoel, D.G. (1983) "Hormonal" risk factors, "breast tissue age" and the age-incidence of breast cancer. *Nature*, **303**, 767–770

Reyes, F.I., Winter, J.S.D. & Faiman, C. (1977) Pituitary-ovarian relationships preceding the menopause. *Am. J. Obstet. Gynecol.*, **129**, 557–564

Rosner, B. & Colditz, G. (1996) Nurses' Health Study: Log-incidence mathematical model of breast cancer incidence. *J. Natl Cancer Inst.*, **88**, 359–364

Saftlas, A.F. & Szklo, M. (1987) Mammographic parenchymal patterns and breast cancer risk. *Epidemiol.Rev.*, **9**, 146–174

Saftlas, A.F., Hoover, R.N., Brinton, L.A., Szklo, M., Olson, D.R., Salane, M. & Wolfe, J.M. (1991) Mammographic densities and risk of breast cancer. *Cancer*, **67**, 2833–2838

Spicer, D.V., Ursin, G., Parisky, Y.R., Pearce, J.G., Shoupe, D., Pike, A. & Pike, M.C. (1994) Changes in mammographic densities induced by a hormonal contraceptive designed to reduce breast cancer risk. *J. Natl. Cancer Inst.*, **86**, 431–436

Stomper, P.C., VanVoorhis, B.J., Ravnikar, V.A. & Meyer, J.E. (1990) Mammographic changes associated with postmenopausal hormone replacement therapy: a longitudinal study. *Radiology*, **174**, 487–490

Struewing, J.P., Hartge, P., Wacholder, S., Baker, S.M., Berlin, M., McAdams, M., Timmerman, M.M., Brody, L.C. & Tucker, M.A. (1997) The risk of cancer associated with specific mutations of BRCA1 and BRCA2 among Ashkenazi Jews. *New Engl. J. Med.*, **336**, 1401–1408

Turnbull, A.E., Kapera, L. & Cohen, M.E.L. (1993) Mammographic parenchymal patterns in Asian and Caucasian women attending for screening. *Clin.Radiol.*, **48**, 38–40

Wang, D.Y., De Stavola, B.L., Bulbrook, R.D., Allen, D.S., Kwa, H.G., Verstraeten, A.A., Moore, J.W., Fentiman, I.S., Hayward, J.L. & Gravelle, I.H. (1988) The permanent effect of reproductive events on blood prolactin levels and its relation to breast cancer risk: a population study of postmenopausal women. *Eur. J. Cancer Clin. Oncol.*, **24**, 1225–1231

Whitehead, J., Carlile, T., Kopecky, K.J., Thompson, D.J., Gilbert, F.I., Present, A.J., Threatt, B.A., Krook, P. & Hadaway, E. (1985) The relationship between Wolfe's classification of mammograms, accepted breast cancer risk factors, and the incidence of breast cancer. *Am. J. Epidemiol.*, **122**, 994–1006

Wolfe, J.N. (1976a) Breast patterns as an index of risk for developing breast cancer. *Am. J. Roentgenol.*, **126**, 1130–1139

Wolfe, J.N. (1976b) Risk for breast cancer development determined by mammographic parenchymal pattern. *Cancer*, **37**, 2486–2492

Wolfe, J.N. (1976c) Breast parenchymal patterns and their changes with age. *Radiology*, **121**, 545–552

Wolfe, J.N., Saftlas, A.F. & Salane, M. (1987) Mammographic parenchymal patterns and quantitative evaluation of mammographic densities: a case-control study. *Am. J. Roentgenol.*, **148**, 1087–1092

Corresponding author:

N.F. Boyd
Division of Epidemiology and Statistics,
Ontario Cancer Institute,
610 University Avenue,
Toronto, Ontario,
Canada M5G 2M7

Biomarkers in Cancer Chemoprevention
Miller, A.B., Bartsch, H., Boffetta, P., Dragsted, L. and Vainio, H. eds
IARC Scientific Publications No. 154
International Agency for Research on Cancer, Lyon, 2001

Intermediate histological effect markers for breast cancer

A.B. Miller and A.M. Borges

The evidence that ductal carcinoma of the breast (DCIS) is an obligate precursor of invasive breast cancer and thus qualifies as an intermediate effect marker in chemoprevention research is reviewed. Much of the evidence on the natural history of DCIS has been derived from the era before the introduction of mammographic screening. Thus it may not be applicable in the present situation when women are likely to be under mammographic surveillance in chemoprevention trials. Further, the data that are becoming available from breast screening trials suggest that at least over follow-up periods now exceeding a decade, detection and treatment of DCIS has no impact on subsequent incidence of breast cancer. Although there are some indications that other biomarkers of malignancy are expressed similarly in DCIS and invasive cancer of similar grade, this evidence may not be sufficient to allow exclusive reliance on DCIS as an intermediate-effect marker in chemoprevention research.

Introduction

It is generally believed that ductal carcinoma *in situ* (DCIS) is a pre-invasive neoplastic lesion in the breast, and that possibly atypical hyperplasia of the breast is a preneoplastic lesion. Indeed, chemoprevention trials are already being based on women with DCIS, in the belief that this is a valid surrogate end-point biomarker (Kelloff *et al.*, 2000). However, the evidence that women with biopsies indicating atypical hyperplasia or DCIS have a greater risk of breast cancer has largely come from long-term follow-up of cohorts of women whose biopsies were originally considered to represent benign (fibrocystic) disease (Page *et al.*, 1982, 1985). It is not clear from these studies whether the histological abnormality indicated a marker of increased risk of breast cancer, or whether it was truly a precursor lesion that if left untreated could eventually progress to invasive cancer. Part of the difficulty is that the histological abnormality was found in the original biopsies, and there was no information on whatever abnormality (if any) — either precursor lesion or early invasive cancer — had been left behind. Suspicion that the subsequent occurrence of invasive cancer in patients who developed DCIS came from a residual abnormality is derived from the fact that the risk of invasive cancer seemed to be restricted to the breast from which the original biopsy was obtained, not the contralateral breast. This is quite different from the related lesion, lobular carcinoma *in situ*, or lobular neoplasia, where the risk of subsequent breast cancer is in both breasts (Lagios & Page, 1998). Hence lobular carcinoma *in situ* seems to represent a marker of risk, not a precursor lesion, and will not be considered further here. Similarly, as atypical hyperplasia should probably be regarded as a preneoplastic lesion rather than as a defined precursor on the pathway to cancer, it will also not be considered in detail in this chapter.

Nomenclature of DCIS

Although different nomenclatures have been in vogue, DCIS has been classified in Europe as poorly differentiated, intermediately differentiated or well differentiated (Holland *et al.*, 1994). This classification is based on the degree of nuclear differentiation and on cellular polarization around intercellular spaces or towards a duct lumen. In North America, in the Van Nuys classification, DCIS is initially subdivided into those lesions with high nuclear grade or non-high nuclear grade, and the latter are then subdivided into those with or without necrosis (Silverstein *et al.*, 1995). This results in three prognostically dissimilar groups. Alternatively, DCIS has been classified as comedo, intermediate and non-comedo (solid, cribriform or papillary) (Lagios & Page, 1998). In comedo-type DCIS necrosis is extensive and the nuclear grade is high; in intermediate-type DCIS necrosis is focal or absent and nuclear grade is intermediate; in non-comedo-type DCIS, necrosis is absent and nuclear grade is low. The risk of subsequent invasive cancer is believed to be highest for the comedo type or high-grade (poorly differentiated) DCIS.

Markers of carcinogenesis, and relevance to natural history of DCIS

The use of digital image analysis of the DNA content of DCIS and invasive intraductal carcinomas associated with them showed a close concordance of DNA-ploidy and S-phase content (Fisher & Siderits, 1992). However, whether or not DCIS is directly on the pathway to cancer, and is in practice an obligate precursor to invasion, can only be inferred indirectly, given the process of biopsy that is required for its diagnosis. One approach to resolving this uncertainty is to assess whether biomarkers of neoplasia which can be identified in invasive breast cancers (logically ductal type in the majority) are also expressed in DCIS. Weinstat-Saslow *et al.* (1995) have shown that overexpression of cyclin D mRNA occurs equally in DCIS and invasive cancers, and can be used to distinguish both from non-malignant breast lesions. They found overexpression of cyclin D mRNA in 76% of 37 non-comedo DCIS, in 87% of 23 comedo DCIS and in 83% of 12 invasive breast carcinomas. These percentages are not statistically significantly different from each other. However, they found cyclin D mRNA overexpression in only 18% of 11 lesions of atypical hyperplasia and 18% of 11 benign lesions interpreted as showing no increase in risk. These percentages were significantly different from those of DCIS and invasive cancer.

Buerger *et al.* (1999) analysed specific chromosomal alterations of 38 cases of DCIS and six associated invasive breast cancers by means of comparative genomic hybridization. They found losses of 16q material almost exclusively in well and intermediately differentiated DCIS. A higher frequency of amplifications (17q12, 11q13) was found in poorly differentiated DCIS. When DCIS was adjacent to invasive carcinoma, a similar genetic pattern was seen in the DCIS and invasive component.

A slightly different approach to the same issue was taken by Dublin *et al.* (1999). These authors re-evaluated material from 98 cases originally diagnosed as minimal (< 10 mm) invasive breast cancers. Of these, 28 were found to be predominantly invasive, 48 predominantly DCIS and 22 DCIS without evidence of invasion. In the predominantly invasive group, the infiltrative component was usually > 5mm, was low-grade, and associated with well differentiated DCIS. Expression of the markers Ki-67, c-erbB2 and p53 was generally low, and that of ER, PR and Bcl-2 high. In contrast, the predominantly DCIS group had a much smaller high-grade invasive component, usually with poorly differentiated DCIS. In these, expression of Ki-67, c-erbB2 and p53 was generally high, and ER, PR and Bcl-2 low. The actuarial survival of the predominantly invasive group was 100% beyond 10 years, of the predominantly DCIS group over 80%; the difference was not statistically significant. In this study it seemed that the high-grade lesions, whether invasive or DCIS, had a similar spectrum of markers indicating, but not demonstrating in practice, poor prognosis, while the low-grade invasive cancer and DCIS also were similar. However, this study also could be interpreted as suggesting that 'invasion', in what would otherwise be regarded as a predominantly DCIS lesion, is of little or no prognostic importance. This somewhat startling conclusion suggests that for lesions that are predominantly DCIS, whether or not cells have penetrated the basement membrane is unimportant. The absence of significance of this generally regarded criterion of progressive malignancy would be explained if somehow the cells forming the outer levels of DCIS manufactured their own basement membrane. Indeed, there are other features of some cases of DCIS, especially those with associated periductal fibrosis, that suggest that DCIS is really a form of good-prognosis invasive cancer (Naresh & Borges, 1996). Whatever the true explanation, it does seem that many cases with minimal 'invasion' by standard criteria are qualitatively different from usual invasive cancer, thus adding to the doubts that DCIS is a precursor of true invasive cancer.

Whether or not recurrences of DCIS occur that are truly new disease or simply recurrent slowly growing lesions has recently been evaluated. Comparative genomic hybridization was used by

Waldman *et al.* (2000) to compare chromosomal alterations in 18 initial DCIS lesions (occurring in the absence of invasive cancer) and subsequent recurrences of DCIS in the same breast. They found a high degree of concordance of chromosomal alterations in all but one of the initial/recurrence pairs, suggesting that the large majority of the recurrences were caused by residual DCIS left at the time of the primary surgery. The most common chromosomal alterations noted were gains involving 17q and losses involving 8p and 17p.

In commenting on this study, Fisher & Fisher (2000) reported that in their experience with about 10 000 cases of DCIS associated with invasive cancer, unequivocal microscopic extension of DCIS through its basement membrane into the surrounding stroma had rarely been observed. They also noted that although about 40% of recurrences of DCIS in the same breast were invasive, the survival of such patients was 98%, again suggesting a qualitatively different natural history from classic invasive ductal carcinoma. This seems to be an example of length bias.

DCIS detected by screening

Recently, screening programmes for breast cancer have led to the diagnosis of much larger numbers of DCIS, as the calcification associated especially with comedo types of DCIS leads to their detection on mammography, biopsy and excision. The detection of DCIS is believed by many to be one of the benefits derived from breast cancer screening. Indeed, aggressive screening for what was then called minimal breast cancer used to be strongly advocated in the belief that only by the detection of such lesions would breast cancer mortality be reduced (Moskowitz *et al.*, 1976). Minimal breast cancer as then defined consisted of two components, invasive breast cancers < 10 mm in size and DCIS. Data allowing assessment of their contribution to the reduction in breast cancer potentially achieved by screening have recently become available from the long-term follow-up of the Canadian National Breast Screening Study among women aged 50–59 years on entry (CNBSS 2) (Miller *et al.*, 2000). This study was designed to assess whether annual two-view mammographic screening together with annual screening by breast physical examinations and the teaching of breast self-examination (the MP group) resulted in a greater breast cancer mortality reduction than screening by breast physical examinations and the teaching of breast self-examination alone (the PO group). Each group comrised a total of just under 20 000 women, with identical distribution of risk factors for breast cancer (CNBSS was an individually randomized trial, randomization being conducted within five-year age strata and centre). Of the 267 invasive breast cancers detected on screening in the MP group, 48 were < 10 mm in size, compared with only six of 148 in the PO group. Further, in addition, 71 *in situ* breast cancers were detected in the MP group but only 16 in the PO group. However, no reduction in breast cancer mortality was found by the addition of mammography (a cumulative rate ratio of 1.02, 95% CI 0.78–1.33). Thus the greater detection of 'minimal' breast cancers in the MP group (an excess of 97) had no impact on breast cancer mortality. Further, there was no evidence that the detection of the *in situ* cancers resulted in a reduction in breast cancer incidence, the cumulative numbers of invasive breast cancers (including those ascertained after the end of the 4–5-year screening period) were 622 in the MP group and 610 in the PO. The data from the 50 000 women aged 40–49 years on entry to CNBSS 1 are currently under analysis. Once again, there was an excess of *in situ* cancers diagnosed in the mammographic screening group compared with the usual care group, but no indication of a reduction in breast cancer incidence over the 11-year follow-up period.

Similar data have not been published from the other breast screening trials, but at present there seems to be no evidence that detection of breast cancer precursors is of any value in screening.

Boyd *et al.* (this volume) postulate that mammographic density may be an exposure marker for breast cancer, but cannot be regarded as an intermediate marker. However, Boyd *et al.* (1992, 2000) derived evidence that mammographic density is associated with atypical hyperplasia and DCIS. In a special study of women aged 40–49 years who had enrolled in CNBSS 1, it was found that women with the most extensive densities had a 9.7 times greater risk of being detected with carcinoma *in situ* or atypical hyperplasia than women with no mammary density.

Discussion

Part of the difficulty in determining the role of DCIS in chemoprevention may be that with mammographic screening, a new spectrum of disease has come to light that would have been largely undiagnosed in the absence of screening. This has contributed to the confusion as to what these lesions truly represent, and even whether it is appropriate to include the term 'carcinoma' within a terminology where no precise guidance can be provided on eventual prognosis (Foucar, 1996). It cannot therefore be assumed that the natural history data derived from the follow-up of women diagnosed in a pre-screening era applies to those now detected by screening. As already indicated, some of the presumed precursors may not be precursors at all, but simply markers of increased risk. It is quite possible that the true invasive cancer precursors, with atypical epithelial hyperplasia and incipient invasion, are largely not detectable with current screening methods. Indeed, insufficient cases of DCIS are detected by mammography to account for the numbers of invasive cancers that occur. This may be because some high-grade cancers have a transitory *in situ* phase with rapid progression to invasion, thus not allowing time for their detection as DCIS (Barnes *et al.*, 1992). The DCIS that is detectable may either be a marker of atypical epithelial hyperplasia or invasive cancer elsewhere in the same breast, an indolent lesion that progresses only slowly, or the end stage of a process well recognized for cancer of the cervix, namely the regression of the majority of the precursors that are detectable.

That detection of precursors will result in overtreatment of many patients not destined to develop invasive cancer has been well demonstrated for cancer of the cervix, and seems likely to be true for the breast also. This makes it essential that there should not be overtreatment of these lesions, as appears to have occurred in the CNBSS, for example (Miller, 1994), and that great caution be exercised in utilizing them as end-points for the evaluation of chemoprevention. It is quite possible that studies that are conducted of chemoprevention with DCIS diagnosed by mammography as an end-point may be largely irrelevant to the prevention of invasive cancer. An indication that the wrong lesions could be prevented from occurring even when invasive cancer is used as the end-point comes from the NSABP trial of tamoxifen for chemoprevention of breast cancer (Fisher *et al*, 1998). The spectrum of estrogen-positive, relatively small breast cancers that failed to occur in the tamoxifen-treated group compared with the control group suggests that these were lesions with good prognosis, with a low probability of resulting in death. This is precisely the spectrum of the lesions that were detected early in CNBSS 2, but whose earlier detection did not result in any indication of reduction in breast cancer mortality (Miller *et al.*, 2000). It is relevant that the participants in the NSABP trial were monitored by annual mammography; the follow-up of CNBSS 2 suggests that such lesions represent a classic example of length-biased sampling.

An alternative approach to using DCIS in chemoprevention research would be to select people with high-grade lesions for study of a potential chemopreventive agent, with invasive cancer as the outcome measure. Thus DCIS would be being used as an exposure or acquired susceptibility marker. However, it seems probable that the majority of women destined to develop and die of invasive cancer may be missed by such an approach, and thus any inferences derived from such a study would be subject to considerable caution.

Conclusions

DCIS detected by mammography, and currently being considered for use as an end-point for chemoprevention drug development trials, may not be an obligate precursor of invasive breast cancer. It may even be a subsegment of invasive breast cancer with very low progression potential and therefore with a very good (and non-typical) prognosis. As such, any conclusions derived from the use of DCIS as an end-point in chemoprevention trials may be in error. Research is needed into the natural history of DCIS diagnosed by mammography. Methods to detect the true obligate precursors of invasive breast cancer are urgently needed.

References

Barnes, D.M., Bartkova, J., Camplejohn, R.S., Gullick, W.J., Smith, P.J. & Millis, R.R. (1992) Overexpression of the c-*erb*B-2 oncoprotein: why does this occur more frequently in ductal carcinoma in situ than in invasive mammary carcinoma and is this of prognostic significance? *Eur. J. Cancer*, **28**, 644–648

Boyd, N.F., Jensen, H.M., Cooke, G. & Lee Han, H. (1992) Relationship between mammographic and histological risk factors for breast cancer. *J. Natl Cancer Inst.*, **84**, 1170–1179

Boyd, N.F., Jensen, H.M., Cooke, G., Lee Han, H., Lockwood, G.A., in collaboration with the reference pathologists of the Canadian National Breast Screening Study & Miller, A.B. (2000) Mammographic densities and the prevalence and incidence of histological types of benign breast disease. *Eur. J. Cancer Prevention*, **9**, 15–24

Buerger, H., Otterbach, F., Simon, R., Poremba, C., Diallo, R., Decker, T., Riethdorf, L., Brinkschmidt, C., Dockhorn-Dworniczak, B. & Boecker, W. (1999) Comparative genomic hybridization of ductal carcinoma *in situ* of the breast—evidence of multiple genetic pathways. *J. Pathol.*, **187**, 396–402

Dublin, E.A., Millis, R.R., Smith, P. & Bobrow, L.G. (1999) Minimal breast cancer: evaluation of histology and biological marker expression. *Br. J. Cancer*, **80**, 1608–1616

Fisher, B., Costantino, J.P., Wickerman, D.L., Redmond, C.K., Kavanah, M., Cronin, W.M., Vogel, V., Robidoux, A., Dimitrov, N., Atkins, J., Daly, M., Wieand, S., Tan-Chiu, E., Ford, L., Wolmark, N., and other National Surgical Adjuvant Breast and Bowel Project Investigators (1998) Tamoxifen for prevention of breast cancer. Report of the National Surgical Adjuvant Breast and Bowel Project P-1 study. *J. Natl Cancer Inst.*, **90**,1371–1388

Fisher, E.R & Fisher, B. (2000) Relation of a recurrent intraductal carcinoma (ductal carcinoma *in situ*) to the primary tumor. *J. Natl Cancer Inst.*, **92**, 288–289

Fisher, E.R. & Siderits, R. (1992) Value of cytometric analysis for distinction of intraductal carcinoma of the breast. *Breast Cancer Res. Treat.*, **21**, 165–172

Foucar, E. (1996) Carcinoma *in situ* of the breast: have pathologists run amok? *Lancet*, **347**, 707–708

Holland, R., Peterse, J.L., Millis, R.R., Eusebi, V., Faverly, D., van de Vijver, M.J. & Zafrani, B. (1994) Ductal carcinoma in situ, a proposal for a new classification. *Semin. Diagn. Pathol.*, **11**,167–180

Kelloff, G.J., Sigman, C.C., Johnson, K.M., Boone, C.W., Greenwald, P., Crowell, J.A., Hawk, E.T. & Doody, L.A. (2000) Perspectives on surrogate end points in the development of drugs that reduce the risk of cancer. *Cancer Epidemiol. Biomarkers Prev.*, **9**, 127–137

Lagios, M.D. & Page, D.L. (1998) *In situ* carcinomas of the breast: ductal carcinoma *in situ*, Paget's disease, lobular carcinoma *in situ*. In: Bland, K.L. & Copeland, E.M., eds, *The Breast. Comprehensive Management of Benign and Malignant Diseases,* Second edition, Philadelphia, W.B. Saunders, pp. 261–283

Miller, A.B. (1994) May we agree to disagree, or how do we develop guidelines for breast cancer screening in women? *J. Natl Cancer Inst.*, **86**, 1729–1731

Miller, A.B., To, T., Baines, C.J. & Wall, C. (2000) Canadian National Breast Screening Study-2: 13-year results of a randomized trial in women aged 50–59 years. *J. Natl Cancer Inst.*, **92**, 1490–1499

Moskowitz, M., Pemmaraju, S., Fidler, J.A., Sutorius, D.J., Russel, P., Scheinok, P. & Holle, J. (1976) On the diagnosis of minimal breast cancer in a screenee population. *Cancer*, **37**, 2543–2552

Naresh, K.N. & Borges, A.M. (1996) Ductal carcinoma in situ with periductal fibrosis—is it in reality an invasive carcinoma? *Human Pathol.*, **27**, 744

Page, D.L., Dupont, W.D., Rogers, L.W., & Landenberger, M. (1982) Intraductal carcinoma of the breast: follow-up after biopsy only. *Cancer*, **49**, 751–758

Page, D.L., Dupont, W.D., Rogers, L.W. & Rados, M.S. (1985) Atypical hyperplastic lesions of the female breast. A long-term follow-up study. *Cancer*, **55**, 2698–2708

Silverstein, M.J., Poller, D.N., Waisman, J.R., Colburn, W.J., Barth, A., Gierson, E.D., Lewinsky, B., Gamagami, P. & Slamon, D.J. (1995) Prognostic classification of breast ductal carcinoma-in-situ. *Lancet*, **345**, 1154–1157

Waldman, F.M., DeVries, S., Chew, K.L., Moore, D.H., Kerlikowske, K. & Ljung, B.-M. (2000) Chromosomal alterations in ductal carcinoma *in situ* and their *in situ* recurrences. *J. Natl Cancer Inst.*, **92**, 313–320

Weinstat-Saslow, D., Merino, M.J., Manrow, R.E., Lawrence, J.A., Bluth, R.F., Wittenbel, K.D., Simpson, J.F., Page, D.L. & Steeg, P.S. (1995) Overexpression of cyclin D mRNA distinguishes invasive and *in situ* breast carcinomas from non-malignant lesions. *Nature Med.*, **1**, 1257–1260

Corresponding author:

A. B. Miller
Division of Clinical Epidemiology,
Deutsches Krebsforschungszentrum,
Im Neuenheimer Feld 280,
D-69120 Heidelberg,
Germany

Biomarkers in Cancer Chemoprevention
Miller, A.B., Bartsch, H., Boffetta, P., Dragsted, L. and Vainio, H. eds
IARC Scientific Publications No. 154
International Agency for Research on Cancer, Lyon, 2001

Inherited genetic susceptibility to breast cancer

J. Chang-Claude

Inherited genetic susceptibility to breast cancer can be due both to genes which confer a high degree of risk and to polygenes which have a smaller effect on disease risk. An estimated 5–10% of breast cancer is considered to be due to mutations in genes conferring high risk which results in hereditary patterns of disease. Two major breast cancer susceptibility genes, *BRCA1* and *BRCA2*, which were identified using linkage analysis in large extended breast and/or ovarian cancer pedigrees, are estimated to account for the majority of large families with breast/ovarian cancer predisposition and about two-thirds of large breast cancer families. The associated lifetime risk for breast cancer in mutation carriers ranges from 40% to 90%, depending on the extent of family history and the population. Other genetic factors, such as HRAS or CAG repeats of AR, as well as reproductive and hormonal factors may therefore modify cancer risk.

Women at particularly high risk of developing breast cancer represent a group in whom expensive and rigorous screening programmes are cost-effective and who may benefit from trials of chemoprevention. There are only preliminary data on the efficacy of increased surveillance and on risk reduction due to prophylactic surgery. However, for chemoprevention to be equivalent to prophylactic mastectomy, it will be necessary to strive for an equivalent reduction. The efficacy of chemoprevention in this high-risk population is unknown. Existing and new agents for chemoprevention need to be carefully assessed in properly designed clinical trials among such women. In the process, other factors modifying the penetrance in mutation carriers need to be taken into account in order to evaluate the true effect of the chemopreventive agents.

Polygenes confer much lower levels of risk and may be relevant for risk assessment when the effects of multiple loci, possibly in conjunction with environmental factors, are understood and quantified. At present, it seems unlikely that the genetic information at single polygenes will be clinically relevant for risk assessment and management.

Introduction

The occurrence of a complex disease such as breast cancer can be attributed to joint effects of genetic and environmental factors. The genetic factors can include inherited genetic susceptibility due to major genes and polygenes as well as acquired somatic genetic aberrations. Major genes usually have a strong effect on disease risk and although disease-associated (mutated) alleles occur at relatively low frequency, they confer a high risk of disease to the individual. These 'highly penetrant' alleles are therefore associated with multiple occurrences of disease in a family, often in a Mendelian pattern of an autosomal dominant trait. Hereditary breast cancer is characterized by early age at onset and multifocal disease. An estimated 5–10% of breast cancer is considered to be due to the presence of a mutation in an autosomal dominant susceptibility gene. The majority of breast cancer which occurs in the absence of a strong family history and with later age at onset is likely to be due to the joint effects of several genes and environmental factors. Polygenes usually have a smaller effect on disease risk to the individual. However, the disease-associated alleles are relatively common in the population and may account for a larger attributable proportion of breast cancer than major genes in the general population.

Major susceptibility genes for breast cancer

Several major genes for breast cancer susceptibility have been identified, as shown in Table 1. They have been cloned as predisposing genes for different cancer syndromes in which breast cancer is a constituent tumour. Mutation analysis has confirmed that the cancer occurrence in these syndromes can be largely explained by the genes identified. However, there remains a proportion of such families in which the disease-associated allele has not been identified. This can be attributed to both the imperfect sensitivity of the mutation detection method and the existence of additional as yet unidentified susceptibility genes.

Mutations in several of the highly penetrant genes of the cancer syndromes associated with hereditary breast cancer, such as *TP53* (Li–Fraumeni syndrome), *MSH2* (Muir–Torre syndrome), *PTEN* (Cowden syndrome), *STK11* (Peutz–Jeghers syndrome), are extremely rare in the general population (Li *et al.*, 1988; Malkin *et al.*, 1990; Birch *et al.*, 1994; Li *et al.*, 1997; Spigelman *et al.*, 1989; Jenne *et al.*, 1998) (Table 1). They are unlikely to contribute to inherited breast cancer susceptibility manifested in families with elevated incidence of breast cancer and are considered to account for less than 1% of all breast cancer.

Families with hereditary breast cancer have been clinically recognized as multiple cases of early-onset cancer over several generations of close relatives with ovarian cancer as well as other cancer sites also involved. Two breast cancer susceptibility genes, *BRCA1* and *BRCA2*, have been identified using linkage analysis in large extended breast and/or ovarian cancer pedigrees and subsequent molecular cloning (Miki *et al.*, 1994; Wooster *et al.*, 1995; Tavtigian *et al.*, 1996). Mutations in *BRCA1* or *BRCA2* account for the majority of high-risk families in which multiple cases of breast and/or ovarian cancer occur as an autosomal dominant trait. Based on data from 237 families with four or more breast cancer cases occurring before 60 years of age (regardless of ovarian cancer) collected by the international Breast Cancer Linkage Consortium, Ford *et al.* (1998) estimated that overall 52% of hereditary breast cancer in such families was explained by mutations in the *BRCA1* gene, 32% by mutations in the *BRCA2* gene, and the remainder by one or more as yet unidentified genes. Genetic heterogeneity appears to be greater among families with breast cancer only, since the

Table 1. Genes conferring high risk for breast cancer

Associated syndrome	Clinical manifestation	Gene
Hereditary breast/ovarian cancer	Breast cancer, ovarian cancer	*BRCA1, BRCA2*
Li–Fraumeni syndrome	Sarcoma, brain and breast cancer cancer of breast and thyroid	*TP53*
Cowden's disease	Multiple hamartomatous lesions of skin, mucous membrane, cancer of breast and thyroid	*PTEN/NMAC1*
Peutz–Jeghers syndrome	Melanocytic macules of lips, multiple polyps, tumours of intestinal tract, breast, ovaries etc.	*STK11*
Hereditary nonpolyposis colorectal cancer	Colorectal cancer, predominantly also tumours of endometrium, ovaries, intestinal tract and breast	*MSH2, MLH1, PMS1, PMS2*
Ataxia telangiectasia	Progressive cerebral ataxia, hypersensitivity to radiation, increased cancer risk	*ATM*

majority of breast/ovarian cancer families have mutations in *BRCA1* (81%) and *BRCA2* (14%), compared with only two thirds of families with breast cancer only. On the other hand, familial breast cancer occurring in both males and females is mainly attributed to *BRCA2* (76%). More recent studies, using direct mutation screening, have included a broader spectrum of families seen at genetic clinics, with a larger proportion of families having only two or three breast and/or ovarian cancer cases. These studies suggest that *BRCA1* mutations are responsible for only 10–20% of breast cancer in such families, with *BRCA2* mutations accounting for about half of this fraction (Gayther *et al.*, 1997; Hakansson *et al.*, 1997; Serova Sinilnikova *et al.*, 1997; Struewing *et al.*, 1997). However, up to 45% of families with both breast and ovarian cancer may be associated with mutations in the *BRCA1* gene (Gayther *et al.*, 1995; Serova Sinilnikova *et al.*, 1997; Shattuck-Eidens *et al.*, 1995; Couch *et al.*, 1997; Stoppa-Lyonnet *et al.*, 1997; Dong *et al.*, 1998; Frank *et al.*, 1998).

Recent population-based studies of early-onset breast cancer suggest that *BRCA1* and *BRCA2* may be responsible for equal fractions of early-onset breast cancer and that these two genes together account for only about 6–10% of early-onset breast cancer in the general population (Hopper *et al.*, 1999; Peto *et al.*, 1999). Furthermore, only about 10% of patients with a first-degree family history of breast or ovarian cancer in the general population harbour a germline mutation in *BRCA1* or *BRCA2*. Based on mutation screening in ovarian cancer families, the allele frequencies of mutant *BRCA1* and *BRCA2* alleles have been estimated to be 0.0013 and 0.0017 (Antoniou *et al.*, 2000). This is compatible with the estimate of allele frequency of 0.0033 for dominant breast cancer-predisposing genes from segregation analysis, indicating that these disease-causing alleles are relatively rare in the general population (Claus *et al.*, 1991). On the other hand, in founder populations with recurrent mutations, such as among Ashkenazi Jews and in the Icelandic population, disease allele frequencies can be as high as 2.5% (Struewing *et al.*, 1995; Tonin *et al.*, 1995, 1996; Johannesdottir *et al.*, 1996; Roa *et al.*, 1996; Fodor *et al.*, 1998; Thorlacius *et al.*, 1997).

Detection of *BRCA1* and *BRCA2* mutations is considered to be of clinical importance because of the associated high lifetime risk of disease. Earlier estimates based on large extended pedigrees used for linkage analysis indicated that *BRCA1* mutations confer an 87% risk of developing breast cancer by the age of 70 years and a 40–60% risk of developing ovarian cancer (Easton *et al.*, 1994). Affected mutation carriers also have an estimated lifetime risk of 65% of developing a second breast cancer (Ford *et al.*, 1994). The risk for breast cancer was thought to be similar for *BRCA2* mutations, whereas the lifetime ovarian cancer risk was estimated to be lower at 27% (Ford *et al.*, 1998). Risk of breast cancer among males is also highly elevated in *BRCA2* families. In addition, mutations in both genes confer increased risks for cancers at other sites, such as prostate, pancreas and colon (Ford *et al.*, 1994; Easton *et al.*, 1997).

Population-based studies in Australia as well as in founder populations such as Ashkenazi Jews and the Icelandic population, however, have estimated much lower risks to mutation carriers (Struewing *et al.*, 1997; Levy-Lahad *et al.*, 1997; Thorlacius *et al.*, 1998; Hopper *et al.*, 1999). Lifetime breast cancer risk for female carriers was estimated to be 37% for the *BRCA2* 999del5 mutation in Iceland and 56% for breast cancer risk and 16.5% for ovarian cancer associated with any of the three common *BRCA1* and *BRCA2* mutations among Ashkenazi Jews.

There is substantial variation in the age of onset and the site of cancer occurrence in carriers of *BRCA1* and *BRCA2* mutations between as well as within families. This variability is also observed for founder mutations (Levy-Lahad *et al.*, 1997; Thorlacius *et al.*, 1996). Therefore, suggestions that different variants may be associated with different disease severity or may predispose differentially by cancer site cannot be the sole explanations for the variability. *BRCA1* and *BRCA2* are considered to be tumour-suppressor genes and, therefore, changes in both alleles are required for complete loss of normal gene function. Even in individuals with an inherited mutation in one gene copy, loss or aberration of the normal gene copy later in life will be required for the development of disease. Other factors, both genetic and environmental, may therefore modify cancer risk.

Several other gene loci have been reported to modify the penetrance of *BRCA1* mutations. Mutation carriers harbouring rare 'variable number of tandem repeats' (VNTR) alleles of the *HRAS* proto-oncogene have been found to have a 2.1-fold higher risk for ovarian cancer than carriers with only common alleles, but breast cancer risk was not increased (Phelan *et al.*, 1996). Rebbeck *et al.* (1999) reported that genotypes with long CAG repeats at the androgen receptor gene were associated with earlier age at diagnosis of breast cancer in *BRCA1* mutation carriers. A few studies have addressed the question of modifying effects of known reproductive risk factors on the *BRCA1*-associated risk. Some indicate similar effects of reproductive and hormonal risk factors on breast or ovarian cancer risk and others suggest that effects in mutation carriers may be different from those seen in breast and ovarian cancer in the general population (Narod *et al.*, 1995; Chang-Claude *et al.*, 1997; Ursin *et al.*, 1997; Jernstrom *et al.*, 1999). Most of these studies suffer from very small sample size and/or survival bias due to the fact that prevalent cases were used. To appropriately address these questions, an international prospective cohort study among *BRCA1/2* mutation carriers is being carried out under the coordination of the International Agency for Research on Cancer. Knowledge gained about risk modifiers in mutation carriers may be useful for refining individual risk estimation and may provide further insight into the pathways of breast cancer tumorigenesis.

Cancer risk, especially breast cancer risk, has been reported in epidemiological studies to be highly increased (4- to 8-fold) in blood relatives (thus obligate heterozygous gene carriers) of patients with the recessive disease ataxia telangiectasia (Swift *et al.*, 1987; Easton *et al.*, 1994). After the cloning of the *ATM* gene, studies based on mutation screening and haplotype analysis yielded lower estimates of about a threefold increase in the risk for breast cancer among heterozygous carriers (Athma *et al.*, 1997; Janin *et al.*, 1999). In contrast to previous estimates, heterozygosity for germline *ATM* mutations appears to be rarely observed in unilateral breast cancer in the general population and in families, but may be more prevalent among non-familial early-onset bilateral breast cancer (Vorechovsky *et al.*, 1996; FitzGerald *et al.*, 1997; Chen *et al.*, 1998; Broeks *et al.*, 2000). Most studies have screened for truncating *ATM* mutations, because missense mutations are technically more difficult to identify, and they may have therefore underestimated the prevalence. Further studies aimed at identifying germline missense mutations and rare allelic variants of the *ATM* gene may provide better estimates of the contribution of *ATM* variants to early-onset breast cancer (Izatt *et al.*, 1999; Vorechovsky *et al.*, 1999).

Polygenes in the etiology of breast cancer

Genetic variants with low penetrance are unlikely to cause extensive familial aggregation. The association between specific genetic variants and breast cancer risk is studied by comparing the allele frequency or the distribution of genotypes among unrelated cases and unrelated controls with the same genetic background. When using the traditional epidemiological study designs of case–control and cohort studies, particular attention should be paid to the genetic composition of the comparison groups, since a positive association can arise as an artefact of population admixture (Lander & Schork, 1994). To prevent spurious associations, studies could be carried out within relatively homogeneous populations or should use family-based ('internal') controls, such as parents of the affected individuals (Terwilliger & Ott, 1992). Although association studies can be performed for any random DNA polymorphism, the search for etiologically relevant genes is more likely to be rewarding if directed at functionally significant variants in genes known or assumed to be biologically related to the disease of interest.

An increasing number of studies have investigated the relationship between common allele variants and breast cancer risk. Some 20 different genes have been examined, but none of the common alleles studied was clearly shown to modify the risk of breast cancer. Such genes include those involved in steroid hormone metabolism, which may modulate the levels of bioavailable steroid hormones, such as the catechol-*O*-methyltransferase (*COMT*), aromatase cytochrome P450 genes *CYP19*, *CYP17* (steroid 17α-hydroxylase/17,20 lyase)and *CYP1B1* and the steroid hormone receptor genes, such as those for the estrogen receptor (*ER*), progesterone receptor (*PR*) and androgen receptor (*AR*) as well as the vitamin D receptor (*VDR*). Genes involved in carcinogen metabolism may also modify breast cancer risk. These include genes coding for phase I enzymes such as *CYP1A1*,

CYP1A2 and *CYP2D6*, which act on tobacco smoke-associated carcinogens, *CYP2E1* which metabolizes ethanol, as well as genes for phase II enzymes such as the glutathione *S*-transferases μ (*GSTM1*), π (*GSTP1*) and θ (*GSTT1*), and the *N*-acetyltransferases *NAT1* and *NAT2*. In addition, common alleles of high-penetrance genes such as *TP53*, *BRCA1* and *ATM* can affect the integrity of cell-cycle checkpoint and DNA repair and thus modify cancer risk. Table 2 shows examples of genetic polymorphisms studied for association with breast cancer risk.

The results of the majority of the studies have been summarized recently by Dunning *et al.* (1999). Statistically significant associations were reported in about a quarter of the 46 studies reviewed, but none was seen in more than one study. This can be explained largely by the lack of power in the majority of the studies. Less than a quarter of the studies had sufficient power to detect a relative risk of 2.5 for a rare allele homozygote if the rare allele frequency was 0.2. Dunning *et al.* (1999) performed a meta-analysis of 17 allele variants in 13 genes in order to obtain more precise estimates. Statistically significant differences in genotype frequencies were found for

Table 2. Examples of genetic polymorphisms studied for association with breast cancer risk[a]

Gene	Polymorphism	Functional effect	Frequency of risk-associated allele (controls)	Range of risk estimates
Steroid hormone metabolism genes				
COMT	Val158Met	Reduced activity	0.35–0.52	0.80–2.20
CYP17	Promoter T→C	May increase transcription	0.38–0.42	0.81–2.52
CYP19	$(TTTA)_{10}$	Unlikely	0.008–0.02	1.07–4.84
PR	306 bp insertion in intron 7	Unlikely	0.13–0.18	0.77–1.39
Carcinogen metabolism genes				
CYP1A1	Ile462Val	Possible increase in enzyme activity	0.04–0.09	0.88
CYP2D6	A, B, C alleles	Nonfunctioning enzyme	0.04–0.12	0.66–2.09
CYP2E1	Intron 6	Unlikely	0.08	1.01–1.04
GSTM1	Gene deletion	No enzyme	0.40–0.51	0.77–2.50
GSTP1	Ile105Val	Reduced activity	0.28	1.58–1.97
GSTT1	Gene deletion	No enzyme	0.21–0.28	0.63–1.50
NAT1	A1088T	Possible increased activity	0.17–0.46	1.00–1.20
NAT2	(several)	Low activity	0.51–0.62	0.70–2.08
Other genes				
TP53	Arg72Pro	Unknown	0.26–0.35	1.07–1.47
	Intron 6 G→A	Unlikely	0.10–0.13	0.28–1.08
	16 bp insertion in intron 3	Unlikely	0.12–0.16	0.51–2.08
BRCA1	Pro871Leu	Unknown	0.37	1.17
	Gln356Arg	Unknown	0.07	0.88
ATM[b]	Pro → Arg	Unknown	0.016	3.0

[a] Dunning *et al.* (1999)
[b] Larson *et al.* (1999)

CYP19 $(TTTA)_n$ polymorphism ($(TTTA)_{10}$ carrier odds ratio (OR) = 2.33, 95% confidence interval (CI) = 1.36–4.17), the *GSTP1* Ile105Val polymorphism (Val carrier OR = 1.60, 95% CI = 1.08–2.39) and the *TP53* Arg72Pro (Pro carrier OR = 1.27, 95% CI = 1.02–1.59). There was evidence for heterogeneity with the *GSTM1* gene deletion, which showed a significant association only for postmenopausal breast cancer (null homozygote OR = 1.33, 95% CI = 1.01–1.76). In addition, some evidence of protection against breast cancer was found for homozygotes of the *PR* PROGINS allele (OR = 0.41, 95% CI = 0.15–0.95). Table 3 presents some low-penetrance alleles showing significant associations with breast cancer risk in joint analyses.

Some positive associations of allele variants have been found in conjunction with environmental exposures, for example, cigarette smoking and variant alleles of genes encoding carcinogen-metabolizing enzymes, *CYP1A1, NAT2, CYP2E1* and *GSTM1*. Although major effects of variant genotypes were not always found, some interactions with smoking were reported but withinconsistent results across studies (Ambrosone *et al.*, 1996; Shields *et al.*, 1996; Hunter *et al.*, 1997; Kelsey *et al.*, 1997; Ishibe *et al.*, 1998; Millikan *et al.*, 1998). A few other studies have investigated the possibility of gene–gene interaction involving *CYP1A1, GSTM1, GSTT1, COMT* and *CYP17* (Lavigne *et al.*, 1997; Bailey *et al.*, 1998; Helzlsouer *et al.*, 1998; Huang *et al.*, 1999). The interactions reported have yet to be confirmed by other studies (Lavigne *et al.*, 1997; Huang *et al.*, 1999).

Many of the earlier association studies suffer from both small and/or poorly designed samples. Careful consideration for selection of controls was often neglected. While the meta-analysis of published studies by Dunning *et al.* (1999) provides an apparently precise estimate, the inclusion of all studies without consideration for their quality does not necessarily produce a summary estimate which is accurate and relevant, thus limiting the inferences that can be drawn about the role of these gene variants in the etiology of breast cancer. Much larger studies are required to elucidate the complex interplay between many genes and environmental factors.

Implications for breast cancer risk assessment, management and prevention

With the identification of major genes for inherited breast cancer, it is now possible to screen for germline mutations in high-risk families. When a test result is positive, family members can be identified as carriers or non-carriers based on the germ-line mutation specific to the family. The genetic diagnosis improves risk assessment, since carriers will suffer from a high lifetime cancer risk, while non-carriers have only the population level of risk. Except for certain founder populations, it may at present be appropriate to cite a range of risk estimates for carriers until more data are available to differentiate risk to carriers dependent upon the extent of family history and other modifying factors. In the case of a negative test result, it may be difficult to interpret the finding for an affected woman with high *a priori* probability of carrying a susceptibility gene. All routinely available tests fail to detect a minimum of 10% of the mutations in both *BRCA1* and *BRCA2* (Stoppa-Lyonnet *et al.*, 1999). The identification of women at particularly high risk of developing breast cancer will provide a group among whom expensive and rigorous screening programmes are likely to be cost-effective, as well as a cohort of women who may benefit from chemoprevention.

At present, there is still uncertainty about recommendations for management of women with inherited susceptibility to breast cancer, although guidelines have been proposed (Eisinger *et al.*, 1998; Møller *et al.*, 1999a). There are only preliminary data suggesting that increased surveillance will reduce breast cancer mortality through detection of early tumours in high-risk women (Møller *et al.*, 1999b; Macmillan, 2000; Tilanus-Linthorst *et al.*, 2000). The appropriateness of breast and ovarian cancer screening schedules as well as the effectiveness of magnetic resonance imaging (MRI) of the breast need to be evaluated. The question of breast-conserving therapy is also not resolved, in view of the high risk of second primary breast cancer, as well as a possibly higher risk of ipsilateral breast tumour recurrence and increased radiation sensitivity in *BRCA1* and *BRCA2* mutation carriers (Robson *et al.*, 1999; Turner *et al.*, 1999).

Table 3. Some low-penetrance alleles showing significant associations with breast cancer rist in joint analyses

Gene	Polymorphism	Frequency of risk-associated allele in controls		Risk estimates for carrier	Study size No. cases/No. controls	Reference
CYP19	$(TTTA)H_{10}$	0.018		1.07 (0.35–3.91)	348/145	Siegelmann-Danieli *et al.*, 1999
		0.008		1.56 (0.59–4.57)	599/433	Healey *et al.*, 2000
		0.005		4.84 (1.87–14.8)	464/619	Haiman *et al.*, 2000
				2.33 (1.36–4.17)	Combined	Dunning *et al.*, 1999
GSTP1	Ile105Val	0.29	heterozygous	1.48 (0.81–2.73)	110/113	Helzlsouer *et al.*, 1998
			homozygous	1.97 (0.77–5.02)		
		0.28	heterozygous	1.53 (0.83–2.84)	62/155	Harries *et al.*, 1997
			homozygous	1.58 (0.49–5.06)		
			heterozygous	1.61 (1.10–2.34)	Combined	Dunning *et al.*, 1999
			homozygous	1.83 (0.95–4.48)	Combined	
TP53	Arg72Pro	0.76		1.10 (0.69–1.75)	107/305	Wang-Gohrke *et al.*, 1998
		0.37		1.47 (1.08–2.00)	212/689	Sjalander *et al.*, 1996
		0.35		1.07 (0.66–1.76)	93/347	Kawajiri *et al.*, 1993
				1.27 (1.02–1.59)	Combined	Dunning *et al.*, 1999
GSTM1	Deletion	0.46	postmen	2.50 (1.34–4.65)	110/113	Helzlsouer *et al.*, 1998
		0.50	postmen	1.10 (0.73–1.64)	177/233	Ambrosone *et al.*, 1995
		0.51	≥ 50yr	1.99 (1.19–3.37)	361/437	Charrier *et al.*, 1999
				1.33 (1.01–1.76)	Combined	Dunning *et al.*, 1999
PR	Progins	0.13		0.90 (0.38–2.09)	68/101	Lancaster *et al.*, 1998
		0.14		0.77 (0.50–1.18)	292/220	Manolitsas *et al.*, 1997
		0.18		1.39 (0.79–2.45)	187/90	Garrett *et al.*, 1995
			homozygous	0.41 (0.15–0.95)	Combined	Dunning *et al.*, 1999

Preventive options being discussed for high-risk women are prophylactic surgery and chemoprevention. Recent reports from retrospective follow-up and emerging data from prospective follow-up suggest that prophylactic mastectomy and prophylactic oophorectomy can reduce the risk of breast cancer in high-risk women (Evans *et al.*, 1999; Hartmann *et al.*, 1999; Rebbeck *et al.*, 1999). A cohort of high-risk women who underwent prophylactic mastectomy at the Mayo Clinic experienced an 81% reduction in incident breast cancer compared with their sisters (Hartmann *et al.*, 1999). However, long-term prospective follow-up on an extended group of women will be necessary to fully address the risk of subsequent breast cancer and the psychological sequelae. The risk reduction conferred by oophorectomy reported in 122 *BRCA1* mutation carriers was 47% and increased with longer duration of follow-up (Rebbeck *et al.*, 1999). The use of compounds that reduce the production of ovarian hormones may provide a non-surgical alternative to prophylactic surgery.

Effective chemoprevention would reduce the need for prophylactic surgery. If mammographic screening is not effective in this high-risk group, chemoprevention will become a higher priority. Tabar *et al.* (1999) reported that there was no reduction in mortality in women aged 40–49 years with grade 3 ductal carcinoma in the Swedish two-county study. Women with *BRCA1* and *BRCA2* germline mutations typically present with early-onset tumours of higher grade (grade 3) (Eisinger *et al.*, 1996; Lakhani *et al.*, 1998). However, if chemoprevention is to be equivalent to (the gold standard of) prophylactic mastectomy, it will be necessary to strive for an 80% reduction. The approximately 50% reduction in the risk of developing breast cancer due to tamoxifen treatment in a trial in the United States (Fisher *et al.*, 1998) was not reproduced in two smaller European trials (Powles *et al.*, 1998; Veronesi *et al.*, 1998). Raloxifene, another selective estrogen receptor modulator (SERM), decreased the risk of estrogen receptor-positive breast cancer by 90%, but not estrogen receptor-negative invasive breast cancer (Cummings *et al.*, 1999). However, the issues relating to the use of SERMs as chemopreventive agents will have to include the estrogen receptor-negativity of *BRCA1* tumours (Lidereau *et al.*, 2000). The synthetic retinoids are another class of compounds being explored for chemoprevention (IARC, 1999; Lippman & Lotan, 2000). Fenretinide, a synthetic retinoic acid derivative, may decrease the occurrence of a second breast malignancy in premenopausal women (Veronesi *et al.*, 1999).

Chemoprevention is still in its infancy and the efficacy of chemoprevention in this high-risk population is unknown. The efficacy of new as well as existing agents for chemoprevention needs to be carefully assessed in properly designed clinical trials of women with increased risk of breast cancer. In the process, other modifying factors of penetrance in mutation carriers should be accounted for in order to evaluate the true effect of the chemopreventive agents.

Polygenes confer much lower levels of risk and may be relevant for risk assessment when the effects of multiple loci, possibly in conjunction with environmental factors, are understood and quantified. At present, it is unlikely that the genetic information about single polygenes will be clinically relevant for risk assessment and management. However, the knowledge gained on the role of different single polygenes and their interaction with environmental factors may help to direct research efforts towards identifying different pathways in breast cancer carcinogenesis which may be amenable for preventive measures, including chemoprevention.

References

Ambrosone, C.B., Freudenheim, J.L., Graham, S., Marshall, J.R., Vena, J.E., Brasure, J.R., Laughlin, R., Nemoto, T., Michalek, A.M., Harrington, A., Ford, T.D. & Shields, P.G. (1995) Cytochrome P4501A1 and glutathione S-transferase (M1) genetic polymorphisms and postmenopausal breast cancer risk. *Cancer Res.*, **55**, 3483–3485

Ambrosone, C.B., Freudenheim, J.L., Graham, S., Marshall, J.R., Vena, J.E., Brasure, J.R., Michalek, A.M., Laughlin, R., Nemoto, T., Gillenwater, K.A., Harrington, A.M. & Shields, P.G. (1996) Cigarette smoking, N-acetyltransferase 2 genetic polymorphisms, and breast cancer risk. *J. Am. Med. Ass.*, **276**, 1494–1501

Antoniou, A.C., Gayther, S.A., Stratton, J.F., Ponder, B.A. & Easton, D.F. (2000) Risk models for familial ovarian and breast cancer. *Genet. Epidemiol.*, **18**, 173–190

Athma, P., Rappaport, R. & Swift, M. (1996) Molecular genotyping shows that ataxia-telangiectasia heterozygotes are predisposed to breast cancer. *Cancer Genet. Cytogenet.*, **92**, 130–134

Bailey, L., Roodi, N., Verrier, C.S., Yee, C.J., Dupont, W.D. & Parl, F.F. (1998) Breast cancer and CYPIA1, GSTM1 and GSTT1 polymorphisms: evidence of a lack of association in Caucasians and African Americans. *Cancer Res.*, **58**, 65–70

Birch, J.M., Hartley, A.L., Tricker, K.J., Prosser, J., Condie, A., Kelsey, A.M., Harris, M., Jones, P.H., Binchy, A. & Crowther, D. (1994) Prevalence and diversity of constitutional mutations in the p53 gene among 21 Li-Fraumeni families. *Cancer Res.*, **54**, 1298–1304

Broeks, A., Urbanus, J.H., Floore, A.N., Dahler, E.C., Klijn, J.G., Rutgers, E.J., Devilee, P., Russell, N.S., van Leeuwen, F.E. & van't Veer, L.J. (2000) ATM-heterozygous germline mutations contribute to breast cancer-susceptibility. *Am. J. Hum. Genet.*, **66**, 494–500

Chang-Claude, J., Becher, H., Eby, N., Bastert, G., Wahrendorf, J. & Hamann, U. (1997) Modifying effect of reproductive risk factors on the age at onset of breast cancer for German BRCA1 mutation carriers. *J. Cancer. Res. Clin. Oncol.*, **123**, 272–279

Charrier, J., Maugard, C.M., Le Mevel, B. & Bignon, Y.J. (1999) Allelotype influence at gluatathione-S-transferase M1 locus on breast cancer susceptibility. *Br. J. Cancer*, **79**, 346–353

Chen, J., Birkholtz, G.G., Lindblom, P., Rubio, C. & Lindblom, A. (1998) The role of ataxia-telangiectasia heterozygotes in familial breast cancer. *Cancer Res.*, **58**, 1376–1379

Claus, E.B., Risch, N. & Thompson, W.D. (1991) Genetic analysis of breast cancer in the cancer and steroid hormone study. *Am. J. Hum. Genet.*, **48**, 232–242

Couch, F.J., Deshano, M.L., Blackwood, M.A., Calzone, K., Stopfer, J., Campeau, L., Ganguly, A., Rebbeck, T. & Weber, B.L. (1997) BRCA1 mutations in women attending clinics that evaluate the risk of breast cancer. *New Engl. J. Med.*, **336**,1409–1415

Cummings, S.R., Eckert, S., Krueger, K.A., Grady, D., Powles, T.J., Cauley, J.A., Norton, L., Nickelsen, T., Bjarnason, N.H., Morrow, M., Lippman, M.E., Black, D., Glusman, J.E., Costa, A. & Jordan, V.C. (1999) The effect of raloxifene on risk of breast cancer in postmenopausal women: results from the MORE randomized trial. Multiple outcomes of raloxifene evaluation. *J. Am. Med. Assoc.*, **281**, 2189–2197

Dong, J., Chang-Claude, J., Wu, Y., Schumacher, V., Debatin, I., Tonin, P. & Royer-Pokora, B. (1998) A high proportion of mutations in the BRCA1 gene in German breast/ovarian cancer families with clustering of mutations in the 3′ third of the gene. *Hum. Genet.*, **103**, 154–161

Dunning, A.M., Healy, C.S., Pharoah, P.D.P., Teare, M.D., Ponder, B.A. & Teare, M.D. (1999) A systematic review of genetic polymorphisms and breast cancer risk. *Cancer Epidemiol. Biomarkers Prev.*, **8**, 843–854

Easton, D.F. (1994) Cancer risks in A-T heterozygotes. *Int. J. Radiat. Biol.*, **66**, 177–182

Easton, D.F., Steele, L., Fields, P., Ormiston, W., Averill, D., Daly, P.A., McManus, R., Neuhausen, S.L., Ford, D., Wooster, R., Cannon-Albright, L.A., Stratton, M.R. & Goldgar D.E. (1997) Cancer risks in two large breast cancer families linked to BRCA2 on chromosome 13q12-13. *Am. J. Hum. Genet.*, **61**, 120–128

Eisinger, F., Stoppa-Lyonnet, D., Longy, M., Kerangueven, F., Noguchi, T., Bailly, C., Vincent-Salomon, A., Jacquemier, J., Birnbaum, D. & Sobol, H. (1996) Germ line mutation at BRCA1 affects the histoprognostic grade in hereditary breast cancer. *Cancer Res.*, **56**, 471–474

Eisinger, F., Alby, N., Bremond, A., Dauplat, J., Espie, M., Janiaud, P., Kuttenn, F., Lebrun, J.P., Lefranc, J.P., Pierret, J., Sobol, H., Stoppa-Lyonnet, D., Thouvenin, D., Tristant, H. & Feingold, J. (1998) Recommendations for medical management of hereditary breast and ovarian cancer: the French National Ad Hoc Committee. *Ann. Oncol.*, **9**, 939–950

Evans, D.G., Anderson, E., Lalloo, F., Vasen, H., Beckmann, M., Eccles, D., Hodgson, S., Moller, P., Chang-Claude, J., Morrison, P., Stoppa-Lyonnet, D., Steel, M. & Haites, N. (1999) Utilisation of prophylactic mastectomy in 10 European centres. *Dis. Markers*, **15**, 148–151

Fisher, B., Dignam, J., Wolmark, N., DeCillis, A., Emir, B., Wickerham, D.L., Bryant, J., Dimitrov, N.V., Abramson, N., Atkins, J.N., Shibata, H., Deschenes, L. & Margolese, R.G. (1997) Tamoxifen and chemotherapy for lymph node-negative, estrogen receptor-positive breast cancer. *J. Natl Cancer Inst.*, **89**, 1673–1682

Fitzgerald, M.G., Bean, J.M., Hegde, S.R., Unsal, H., MacDonald, D.J., Harkin, D.P., Finkelstein, D.M., Isselbacher, K.J. & Haber, D.A. (1997) Heterozygous ATM mutations do not contribute to early onset of breast cancer. *Naure Genet.*, **15**, 307–310

Fodor, F.H., Weston, A., Bleiweiss, I.J., McCurdy, L.D., Walsh, M.M., Tartter, P.I., Brower, S.T. & Eng, C.M. (1998) Frequency and carrier risk associated with common BRCA1 and BRCA2 mutations in Ashkenazi Jewish breast cancer patients. *Am. J. Hum. Genet.*, **63**, 45–51

Ford, D., Easton, D.F., Bishop, D.T., Narod, S.A. & Goldgar, D.E. (1994) Risks of cancer in BRCA1-mutation carriers. Breast Cancer Linkage Consortium. *Lancet*, **343**, 692–695

Ford, D., Easton, D.F., Stratton, M., Narod, S., Goldgar, D., Devilee, P., Bishop, D.T., Weber, B., Lenoir, G., Chang-Claude, J., Sobol, H., Teare, M.D., Struewing, J., Arason, A., Scherneck, S., Peto, J., Rebbeck, T.R., Tonin, P., Neuhausen, S., Barkardottir, R., Eyfjord, J., Lynch, H., Ponder, B.A., Gayther, S.A., Birch, J.M., Lindblom, A., Stoppa-Lyonnet, D., Bignon, Y., Borg, A., Hamann, U., Haites, N., Scott, R.J., Maugard, C.M., Vasen, H., Seitz, S., Cannon-Albright, L.A., Schofield, A., Zelada-Hedman, M., & Breast Cancer Linkage Consortium (1998) Genetic heterogeneity and penetrance analysis of the BRCA1 and BRCA2 genes in breast cancer families. *Am. J. Hum. Genet.*, **62**, 676–689

Frank, T.S., Manley, S.A., Olopade, O.I., Cummings, S., Garber, J.E., Bernhardt, B., Antman, K., Russo, D., Wood, M.E., Mullineau, L., Isaacs, C., Peshkin, B., Buys, S., Venne, V., Rowley, P.T., Loader, S., Offit, K., Robson, M., Hampel, H., Brener, D., Winer, E.P., Clark, S., Weber, B., Strong, L.C., Rieger, P., McClure, M., Ward, B.E., Shattuck-Eidens, D., Oliphant, A., Skolnick, M.H. & Thomas, A. (1998) Sequence analysis of BRCA1 and BRCA2: correlation of mutations with family history and ovarian cancer risk. *J. Clin. Oncol.*, **16**, 2417–2425

Garrett, E., Rowe, S.M., Coughlan, S.J., Horan, R., McLinden, J., Carney, D.N., Fanning, M., Kieback, D.G. & Headon, D.R. (1995) Mendelian inheritance of a Taq I restriction fragment length polymorphism due to an insertion in the human progesterone receptor gene and its allelic imbalance in breast cancer. *Cancer Res.Ther. Control*, **4**, 217–222

Gayther, S.A., Warren, W., Mazoyer, S., Russell, P.A., Harrington, P.A., Chiano, M., Seal, S., Hamoudi, R., van Rensburg, E.J., Dunning, A.M., Love R., Evans, G., Easton, D., Clayton, D., Stratton, M.R. & Ponder, B.A.J. (1995) Germline mutations of the BRCA1 gene in breast and ovarian cancer families provide evidence for a genotype-phenotype correlation. *Nature Genet.*, **11**, 428–433

Gayther, S.A., Harrington, P., Russell, P., Kharkevich, G., Garkavtseva, R.F. & Ponder, B.A. (1997) Frequently occurring germ-line mutations of the BRCA1 gene in ovarian cancer families from Russia. *Am. J. Hum. Genet.*, **60**, 1239–1242

Haiman, C.A., Hankinson, S.E., Spiegelman, D., De Vivo I., Colditz, G.A., Willett, W.C., Speizer, F.E. & Hunter, D.J. (2000) A tetranucleotide repeat polymorphism in CYP19 and breast cancer risk. *Int. J. Cancer*, **87**, 204–210

Hakansson, S., Johannsson, O., Johansson, U., Sellberg, G., Loman, N., Gerdes, A.M., Holmberg, E., Dahl, N., Pandis, N., Kristoffersson, U., Olsson, H. & Borg, A. (1997) Moderate frequency of BRCA1 and BRCA2 germ-line mutations in Scandinavian familial breast cancer. *Am. J. Hum. Genet.*, **60**, 1068–1078

Hartmann, L.C., Schaid, D.J., Woods, J.E., Crotty, T.P., Myers, J.L., Arnold, P.G., Petty, P.M., Sellers, T.A., Johnson, J.L., McDonnell, S.K., Frost, M.H. & Jenkins, R.B. (1999) Efficacy of bilateral prophylactic mastectomy in women with a family history of breast cancer. *New Engl. J. Med.*, **340**, 77–84

Harries, L.W., Stubbins, M., Forman, D., Howard, C.W. & Wolf, C.R. (1997) Identification of genetic polymorphisms at the glutathione S-transferase Pi locus and association with susceptibility to bladder, testicular and prostate cancer. *Carcinogenesis*, **18**, 641–644

Healey, C.S., Dunning, A.M., Durocher, F., Teare, D., Pharoah, P.D., Luben, R.N., Easton, D.F. & Ponder, B.A. (2000) Polymorphisms in the human aromatase cytochrome P450 gene (CYP19) and breast cancer risk. *Carcinogenesis*, **21**, 189–193

Helzlsouer, K.J., Selmin, O., Huang, H.-Y., Strickland, P.T., Hoffmann, S., Alberg, A.J. Watson, M., Comstock, G.W. & Bell, D. (1998) Association between glutathione S-transferase M1, P1, and genetic polymorphisms and development of breast cancer. *J. Natl Cancer Inst.*, **90**, 512–518

Hopper, J.L., Southey, M.C., Dite, G.S., Giles, G.G., McCredie, M., Easton, D.F. & Venter, D.J. (1999) Population-based estimate of the average age-specific cumulative risk of breast cancer for a defined set of protein-truncating mutations in BRCA1 and BRCA2. *Cancer Epidemiol. Biomarkers Prev.*, **8**, 741–747

Huang, C.S., Chern, H.D., Chang, K.J., Cheng, C.W., Hsu, S.M. & Shen, C.Y. (1999) Breast cancer risk associated with genotype polymorphism of the estrogen-metabolizing genes CYP17, CYP1A1, and COMT: a multigenic study on cancer susceptibility. *Cancer Res.*, **59**, 4870–4875

Hunter, D.J., Hankinson, S.E., Hough, H., Gertig, D.M., Garcia-Closas, M., Spiegelman, D., Manson, J.E., Colditz, G.A., Willett, W.C., Speizer, F.E. & Kelsey, K. (1997) A prospective study of NAT2 acetylation genotype, cigarette smoking, and risk of breast cancer. *Carcinogenesis*, **18**, 2127–2132

IARC (1999) *IARC Handbooks of Cancer Prevention.* Volume 4, *Retinoids*, Lyon, IARC

Ishibe, N., Hankinson, S.E., Colditz, G.A., Spiegelman, D., Willett, W.C., Speizer, F.E., Kelsey, K.T. & Hunter, D.J. (1998) Cigarette smoking cytochrome P450 1A1 polymorphisms, and breast cancer risk in the Nurses' Health Study. *Cancer Res.*, **58**, 667–671

Izatt, L., Greenman, J., Hodgson, S., Ellis, D., Watts, S., Scott, G., Jacobs, R.L., Zvelebil, M.J., Mathew, C. & Solomon, E. (1999) Identification of germline missense mutations and rare allelic variants in the ATM gene in early-onset breast cancer. *Genes Chromos. Cancer*, **26**, 286–294

Janin, N., Andrieu, N., Ossian, K., Laugé, A., Croquette, M.-F., Griscelli, C., Debré, M., Bressac-de Paillerets, B., Aurias, A. & Stoppa-Lyonnet, D. (1999) Breast cancer risk in ataxia telangiectasia (AT) heterozygotes: haplotype study in French AT families. *Br. J. Cancer*, **80**, 1042–1045

Jenne, D.E., Reimann, H., Nezu, J., Friedel, W., Loff, S., Jeschke, R., Müller, O., Back, W. & Zimmer, M. (1998) Peutz-Jeghers syndrome is caused by mutations in a novel serine threonine kinase. *Nature Genet.*, **18**, 38–43

Jernstrom, H., Lerman, C., Ghadirian, P., Lynch, H.T., Weber, B., Garber, J., Daly, M., Olopade, O.I., Foulkes, W.D., Warner, E., Brunet, J.S. & Narod, S.A. (1999) Pregnancy and risk of early breast cancer in carriers of BRCA1 and BRCA2. *Lancet*, **354**, 1846–1850

Johannesdottir, G., Gudmundsson, J., Bergthorsson, J.T., Arason, A., Agnarsson, B.A., Eiriksdottir, G., Johannsson, O.T., Borg, A., Ingvarsson, S., Easton, D.F., Egilsson, V. & Barkardottir, R.B. (1996) High prevalence of the 999del5 mutation in Icelandic breast and ovarian cancer patients. *Cancer Res.*, **56**, 3663–3665

Kelsey, K.T., Hankinson, S.E., Colditz, G.A., Springer, K., Garcia-Closas, M., Spiegelman, D., Manson, J.E., Garland, M., Stampfer, M.J., Willett, W.C., Speizer, F.E. & Hunter, D.J. (1997) Glutathione S-transferase class mu deletion polymorphism and breast cancer: results from prevalent versus incident cases. *Cancer Epidemiol. Biomarkers Prev.*, **6**, 511–515

Lakhani, S.R., Jacquemier, J., Sloane, J.P., Gusterson, B.A., Anderson, T.J., van-de-Vijver, M.J., Farid, L.M., Venter, D., Antoniou, A., Storfer-Isser, A., Smyth, E., Steel, C.M., Haites, N., Scott, R.J., Goldgar, D., Neuhausen, S., Daly, P.A., Ormiston, W., McManus, R., Scherneck, S., Ponder, B.A., Ford, D., Peto, J., Stoppa-Lyonnet, D., Bignon, Y.J., Streuwing, J., Spurr, N.K., Bishop, D., Klijn, J.G.M., Devilee, P., Chang-Claude, J., Sobol, H., Weber, B., Stratton, M. & Easton, D.F. (1998) Multifactorial analysis of differences between sporadic breast cancers and cancers involving BRCA1 and BRCA2 mutations. *J. Natl Cancer Inst.*, **90**, 1138–1145

Lancaster, J.M., Berchuck, A., Carney, M.E., Wiseman, R. & Taylor, J.A. (1998) Progesterone receptor gene polymorphism and risk for breast and ovarian cancer (Letter to the Editor). *Br. J. Cancer*, **78**, 277

Lander, E.S. & Schork, N.J. (1994) Genetic dissection of complex traits. *Science*, **265**, 2037–2048

Larson, G.P., Zhang, G., Ding, S., Foldenauer, K., Udar, N., Gatti, R.A., Neuberg, D., Lunetta, K.L., Ruckdeschel, J.C., Longmate, J., Flanagan, S., Krontiris, T.G. (1997-98) An allelic variant at the ATM locus is implicated in breast cancer susceptibility. *Genet. Test.*, **1**, 165–170

Lavigne, J., Helzlsouer, J., Huang, H.-Y., Stickland, T., Bell, D., Selmin, O., Watson, M., Hoffmann, S., Comstock, G.W. & Yager, J.D. (1997) An association between the allele coding for a low activity variant of catechol-O-methyltransferase and the risk for breast cancer. *Cancer Res.*, **57**, 5493–5497

Levy Lahad, E., Catane, R., Eisenberg, S., Kaufman, B., Hornreich, G., Lishinsky, E., Shohat, M., Weber, B.L., Beller, U., Lahad, A. & Halle, D. (1997) Founder BRCA1 and BRCA2 mutations in Ashkenazi Jews in Israel: frequency and differential penetrance in ovarian cancer and in breast-ovarian cancer families. *Am. J. Hum. Genet.*, **60**, 1059–1067

Li, F.P., Fraumeni, J.F., Jr, Mulvihill, J.J., Blattner, W.A., Dreyfus, M.G., Tucker, M.A. & Miller, R.W. (1988) A cancer family syndrome in twenty-four kindreds. *Cancer Res.*, **48**, 5358–5362

Li, J., Yen, C., Liaw, D., Podsymanina, K., Bose, S., Wang, S.I., Puc, J., Miliaresis, C. Rodgers, L., McCombie, R., Bigner, S.H., Giovanella B.C., Ittmann, M., Tycko, B., Hibshoosh, H., Wigler, M.H. & Parsons R. (1997) PTEN, a putative protein typrosine phosphatase gene mutated in human brain, breast and prostate cancer. *Science*, **275**, 1943–1947

Lidereau, R., Eisinger, F., Champeme, M.H., Nogues, C., Bieche, I., Birnbaum, D., Pallud, C., Jacquemier, J. & Sobol, H. (2000) Major improvement in the efficacy of BRCA1 mutation screening using morphoclinical features of breast cancer. *Cancer Res.*, **60**, 1206–1210

Lippman, S.M. & Lotan, R. (2000) Advances in the development of retinoids as chemopreventive agents. *J. Nutr.*, **130** (2S Suppl.), 479S–482S

Macmillan, R.D. (2000) Screening women with a family history of breast cancer – results from the British Familial Breast Cancer Group. *Eur. J. Surg. Oncol.*, **2**, 149–152

Malkin, D., Li, F.P., Strong, L.C., Fraumeni, J.F., Jr, Nelson, C.E., Kim, D.H., Kassel, J., Gryka, M.A., Bischoff, F.Z. & Tainsky, M.A. (1990) Germ line p53 mutations in a familial syndrome of breast cancer, sarcomas, and other neoplasms. *Science*, **250**, 1233–1238

Manolitsas, T.P., Englefield, P., Eccles, D.M. & Campbell, I.G. (1997) No association of a 306-bp insertion polymorphism in the progesterone receptor gene with ovarian and breast cancer. *Br. J. Cancer*, **75**, 1398–1399

Miki, Y., Swensen, J., Shattuck-Eidens, D., Futreal, A., Harshman, K., Tavtigian, S. Liu, Q., Cochran, C., Bennett, L.M., Ding, W., Bell, R., Rosenthal, J., Hussey, C. Tran, T., McClure, M., Frye, C., Hattier, T., Phelps, R., Haugen-Strano, A., Katcher, H., Yakumo, K., Gholami, Z., Shaffer, D., Stone, S., Bayer, S., Wray, C., Bogden, R., Dayananth, P., Ward, J., Tonin, P., Narod, S., Bristow, P.K., Norris, F.H., Helvering, L., Morrison, P., Rosteck, P., Lai, M., Barrett, J.C., Lewis, C., Neuhausen, S., Cannon-Albright, L., Goldgar, D., Wiseman, R., Kamb, A. & Skolnick, M.H. (1994) A strong candidate gene for the breast and ovarian cancer susceptibility gene BRCA1. *Science*, **266**, 66–71

Millikan, R.C., Pittman, G.S., Tse, C.K., Duell, E., Newman, B., Saitz, D., Moorman, P.G., Boissy, R.J. & Bell, D.A. (1998) Catechol-O-methyltransferase and breast cancer risk. *Carcinogenesis*, **19**, 1943–1947

Møller, P., Evans, G., Haites, N., Vasen, H., Reis, M., Anderson, E., Apold, J., Hodgson, S., Eccles, D., Olsson, H., Stoppa-Lyonnet, D., Chang-Claude, J., Morrison, P.J., Bevilacqua, G., Heimdal, K., Maehle, L., Lalloo, F., Gregory, H. Preece, P., Borg, A., Nevin, N.C., Caligo, M. & Steel, C.M. (1999a) Guidelines for follow-up of women at high risk for inherited breast cancer: consensus statement from the Biomed 2 Demonstration Programme on Inherited Breast Cancer. *Dis. Markers*, **15**, 207–211

Møller, P., Reis, M.M., Evans, G., Vasen, H., Haites, N., Anderson, E., Steel, C.M., Apold, J., Lalloo, F., Maehle, L., Preece, P., Gregory, H. & Heimdal, K. (1999b) Efficacy of early diagnosis and treatment in women with a family history of breast cancer. European Familial Breast Cancer Collaborative Group. *Dis. Markers*, **15**, 179–186

Narod, S.A., Ford, D., Devilee, P., Barkadottir, R.B., Lynch, H.T., Smith, S.A., Ponder, B.A.J., Weber, B.L., Garber, J.E., Birch, J.M., Cornelis, R.S., Kelsell, D.P., Spurr, N.K., Smyth, E., Haites, N., Sobol, H., Bignon, Y.-J., Chang-Claude, J., Hamann, U., Lindblom, A., Borg, A., Piver, M.S., Gallion, H.H., Struewing, J.P., Whittemore, A., Tonin, P., Goldgar, D.E., Easton, D.F. & Breast Cancer Linkage Consortium (1995) An evaluation of genetic heterogeneity in 145 breast-ovarian cancer families. *Am. J. Hum. Genet.*, **56**, 254–264

Peto, J., Collins, N., Barfoot, R., Seal, S., Warren, W., Rahman, N., Easton, D.F., Evans, C., Deacon, J. & Stratton, M.R. (1999) Prevalence of BRCA1 and BRCA2 gene mutations in patients with early-onset breast cancer. *J. Natl Cancer Inst.*, **11**, 943–949

Phelan, C.M., Rebbeck, T.R., Weber, B.L., Devilee, P., Ruttledge, M.H., Lynch, H.T., Lenoir, G.M., Stratton, M.R., Easton, D.F., Ponder, B.A., Cannon Albright, L., Larsson, C., Goldgar, D.E. & Narod, S.A. (1996) Ovarian cancer risk in BRCA1 carriers is modified by the HRAS1 variable number of tandem repeat (VNTR) locus. *Nature Genet.*, **3**, 309–311

Powles, T., Eeles, R., Ashley, S., Easton, D., Chang, J., Dowsett, M., Tidy, A., Viggers, J. & Davey, J. (1998) Interim analysis of the incidence of breast cancer in the Royal Marsden Hospital tamoxifen randomised chemoprevention trial. *Lancet*, **352**, 98–101

Rebbeck, T.R., Levin, A.M., Eisen, A., Snyder, C., Watson, P., Cannon-Albright, L., Isaacs, C., Olopade, O., Garber, J.E., Godwin, A.K., Daly, M.B., Narod, S.A., Neuhausen, S.L., Lynch, H.T. & Weber, B.L. (1999) Breast cancer risk after bilateral prophylactic oophorectomy in BRCA1 mutation carriers. *J. Natl Cancer Inst.*, **17**, 1475–1479

Rebbeck, T.R., Kantoff, P.W., Krithivas, K., Neuhausen, S., Blackwood, A.M., Godwin, A., Daly, M.B., Narod, S.A., Garber, J.E., Lynch, H.A.T., Weber, B.L. & Brown, M. (1999) Modification of BRCA1-associated breast cancer risk by the polymorphic androgen-receptor CAG repeat. *Am. J. Hum. Genet.*, **64**, 1371–1377

Roa, B.B., Boyd, A.A., Volcik, K. & Richards, C.S. (1996) Ashkenazi Jewish population frequencies for common mutations in BRCA1 and BRCA2. *Nature Genet.*, **2**, 185–187

Robson, M., Levin, D., Federici, M., Satagopan, J., Bogolminy, F., Heerdt, A., Borgen, P., McCormick, B., Hudis, C., Norton, L., Boyd, J. & Offit K. (1999) Breast conservation therapy for invasive breast cancer in Ashkenazi women with BRCA gene founder mutations. *J. Natl Cancer Inst.*, **91**, 2112–2117

Serova Sinilnikova, O.M., Boutrand, L., Stoppa Lyonnet, D., Bressac de Paillerets, B., Dubois, V., Lasset, C., Janin, N., Bignon, Y.J., Longy, M., Maugard, C., Lidereau, R., Leroux, D., Frebourg, T., Mazoyer, S. & Lenoir, G.M. (1997) BRCA2 mutations in hereditary breast and ovarian cancer in France [letter]. *Am. J. Hum. Genet.*, **5**, 1236–1239

Schmidt, S., Becher, H. & Chang-Claude, J. (1998) Breast cancer risk assessment: Use of complete pedigree information and the effect of misspecified ages at diagnosis of affected relatives. *Hum. Genet.*, 102, 348–356

Shattuck-Eidens, D., McClure, M., Simard, J., Labrie, F., Narod, S., Couch, F., Hoskins, K. Weber, B., Castilla, L., Erdos, M., Brody, L., Friedman, L., Ostermeyer, E., Szabo, C., King, M.-C., Jhanwar, S., Offit, K., Norton, L., Gilewski, T., Lubin, M., Osborne, M., Black, D., Boyd, M., Steel, M., Ingles, S., Haile, R., Lindblom, A., Olsson, H., Borg, A., Bishop, D.T., Solomon, E., Radice, P., Spatti, G., Gayther, S., Ponder, B., Warren, W., Stratton, M., Liu, Q., Fujimura, F., Lewis, C., Skolnick, M.H. & Goldgar, D.E. (1995) A collaborative survey of 80 mutations in the BRCA1 breast and ovarian cancer suscptibility gene. *J. Am. Med. Assoc.*, **273**, 535–541

Shields, P.G., Ambrosone, C.B., Saxon, G., Bowman, Elise, D., Harrington, A.M., Gillenwater, K.A., Marshall, J.P., Vena, J.E., Laughlin, R., Nemoto, T. & Feudenheim, J.L. (1996) A cytochrome P4502E1 genetic polymorphism and tobacco smoking in breast cancer. *Mol. Carcinog.*, **17**, 144–150

Siegelmann-Danieli, N. & Buetow, K.H. (1999) Constitutional genetic variation at the human aromatase gene (Cyp19) and breast cancer risk. *Br. J. Cancer*, **79**, 456–463

Sjalander, A., Birgander, R., Hallmans, G., Cajander, S., Lenner, P., Athlin, L., Beckman, G. & Beckman, L. (1996) p53 polymorphisms and haplotypes in breast cancer. *Carcinogenesis*, **17**, 1313–1316

Spigelman, A.D., Murday, V. & Phillips, R.K. (1989) Cancer and the Peutz-Jeghers syndrome. *Gut*, **11**, 1588–1590

Stoppa-Lyonnet, D., Laurent-Puig, P., Essioux, L., Pages, S., Ithier, G., Ligot, L., Fourquet, A., Salmon, R.J., Clough, K.B., Pouillart, P., Bonaiti Pellie, C. & Thomas, G. (1997) BRCA1 sequence variations in 160 individuals referred to a breast/ovarian family cancer clinic. Institut Curie Breast Cancer Group. *Am. J. Hum. Genet.*, **60**, 1021–1030

Stoppa-Lyonnet, D., Caligo, M., Eccles, D., Evans, D.G., Haites, N.E., Hodgson, N.S., Moller, P., Morrison, P.J., Steel, C.M., Vasen, H.F. & Chang-Claude, J. (1999) Genetic testing for breast cancer predisposition in 1999: which molecular strategy and which family criteria? *Disease Markers*, **15**, 67–68

Struewing, J.P., Abeliovich, D., Peretz, T., Avishai, N., Kaback, M.M., Collins, F.S. & Brody L.C. (1995) The carrier frequency of the BRCA1 185delAG mutation is approximately 1 percent in Ashkenazi Jewish individuals. *Nature Genet.*, **11**, 198–200

Struewing, J.P., Hartge, P., Wacholder, S., Baker, S.M., Berlin, M., McAdams, M., Timmerman, M.M., Brody, L.C. & Tucker, M.A. (1997) The risk of cancer associated with specific mutations of BRCA1 and BRCA2 among Ashkenazi Jews. *New Engl. J. Med.*, **336**, 1401–1408

Swift, M., Reitnauer, P.J., Morrell, D. & Chase, C.L. (1987) Breast and other cancers in families with ataxia-telangiectasia. *New Engl. J. Med.*, **316**, 1289–1294

Tabar, L., Duffy, S.W., Vitak, B., Chen, H.H. & Prevost, T.C. (1999) The natural history of breast carcinoma: what have we learned from screening? *Cancer*, **86**, 449–462

Tavtigian, S.V., Simard, J., Rommens, J., Couch, F., Shattuck-Eidens, D., Neuhausen, S., Merajver, S., Thorlacius, S., Offit, K., Stoppa-Lyonnet, D., Belanger, C., Bell, R., Berry, S., Bogden, R., Chen, Q., Davis, T., Dumont, M., Frye, C., Hattier, T., Jammulapati, S., Janecki, T., Jiang, P., Kehrer, R., Leblanc, J.F., Mitchell, J.T., McArthur-Morrison, J., Nguyen, K., Peng, Y., Samson, C., Schroeder, M., Snyder, S.C., Steele, L., Stringfellow, M., Stroup, C., Swedlund, B., Swensen, J., Ten, D., Thomas, A., Tran, T., Tranchant, M., Weaver-Feldhaus, J., Wong, A.K.C., Shizuya, H., Eyfjord, J.E., Cannon-Albright, L., Labrie, F., Skolnick, M.H., Weber, B., Kamb, A. & Goldgar, D.E. (1996) The complete BRCA2 gene and mutations in chromosome 13q-linked kindreds. *Nature Genet.*, **12**, 333–337

Terwilliger, J.D. & Ott, J. (1992) A haplotype-based 'haplotype relative risk' approach to detecting allelic associations. *Hum. Hered.*, **42**, 337–346

Thorlacius, S., Olafsdottir, G., Tryggvadottir, L., Neuhausen, S., Jonasson, J.G., Tavtigian, S.V., Tulinius, H., Ogmundsdottir, H.M. & Eyfjord, J.E. (1996) A single BRCA2 mutation in male and female breast cancer families from Iceland with varied cancer phenotypes. *Nature Genet.*, 13, 117–119

Thorlacius, S., Sigurdsson, S., Bjarnadottir, H., Olafsdottir, G., Johasson, J.G., Tryggvadottir, L., Tulinius, H. & Eyfjord, J.E. (1997) Study of a single BRCA2 mutation with high carrier frequency in a small population. *Am. J. Hum. Genet*, **60**, 1079–1084

Thorlacius, S., Struewing, J.P., Hartge, P., Olafsdottir, G.H., Sigvaldason, H., Tryggvadottir, L., Wacholder, S., Tulinius, H. & Eyfjord, J.E. (1998) Population-based study of risk of breast cancer in carriers of BRCA2 mutation. *Lancet*, **352**, 1337–1339

Tilanus-Linthorst, M.M., Bartels, C.C., Obdeijn, A.I. & Oudkerk, M. (2000) Earlier detection of breast cancer by surveillance of women at familial risk. *Eur. J. Cancer*, **36**, 514–519

Tonin, P., Serova, O., Lenoir, G., Lynch, H., Durocher, F., Simard, J., Morgan, K. & Narod, S. (1995) BRCA1 mutations in Ashkenazi Jewish women. *Am. J. Hum. Genet.*, **57**, 189

Tonin, P., Weber, B., Offit, K., Couch, F., Rebbeck, T.R., Neuhausen, S., Godwin, A.K., Daly, M., Wagner-Costalos, J., Berman, D., Grana, G., Fox, E., Kane, M.F., Kolodner, R.D., Krainer, M., Haber, D.A., Struewing, J.P., Warner, E., Rosen, B., Lerman, C., Peshkin, B., Norton, L., Serova, O., Foulkes, W.D., Lynch, H.T., Lenoir, G.M., Narod, S.A. & Garber, J.E. (1996) Frequency of recurrent BRCA1 and BRCA2 mutations in Ashkenazi Jewish breast cancer families. *Nature Med.*, **11**, 1179–1183

Turner, B.C., Harrold, E., Matloff, E., Smith, T., Gumbs, A.A., Beinfield, M., Ward, B., Skolnick, M., Glazer, P.M., Thomas, A. & Haffty, B.G. (1999) BRCA1/BRCA2 germline mutations in locally recurrent breast cancer patients after lumpectomy and radiation therapy:

implications for breast-conserving management in patients with BRCA1/BRCA2 mutations. *J. Clin Oncol.*, **17**, 3017–3024

Ursin, G., Henderson, B.E., Haile, R.W., Pike, M.C., Zhou, N., Diep, A. & Bernstein, L. (1997) Does oral contraceptive use increase the risk of breast cancer in women with BRCA1/BRCA2 mutations more than in other women? *Cancer Res.*, **57**, 3678–3681

Veronesi, U., Maisonneuve, P., Costa, A., Sacchini, V., Maltoni, C., Robertson, C., Rotmensz, N. & Boyle, P. (1998) Prevention of breast cancer with tamoxifen: preliminary findings from the Italian randomised trial among hysterectomised women. Italian Tamoxifen Prevention Study. *Lancet*, **352**, 93–97

Veronesi, U., De Palo, G., Marubini, E., Costa, A., Formelli, F., Mariani, L., Decensi, A., Camerini, T., Del Turco, M.R., Di Mauro, M.G., Muraca, M.G., Del Vecchio, M., Pinto, C., D'Aiuto, G., Boni, C., Campa, T., Magni, A., Miceli, R., Perloff, M., Malone, W.F. & Sporn, M.B. (1999) Randomized trial of fenretinide to prevent second breast malignancy in women with early breast cancer. *J. Natl Cancer Inst.*, **91**, 1847–1856

Vorechovsky, I., Rasio, D., Luo, L., Monaco, C., Hammarstrom, L., Webster, A.D., Zaloudik, J., Barbanti-Brodani, G., James, M., Russo, G., Croce, C.M. & Negrini, M. (1996) The ATM gene and susceptibility to breast cancer: analysis of 38 breast tumors reveals no evidence for mutation. *Cancer Res.*, **56**, 2726–2732

Vorechovsky, I., Luo, L., Ortmann, E., Steinmann, D. & Dork, T. (1999) Missense mutations at ATM gene and cancer risk. *Lancet*, **353**, 1276

Wang-Gohrke, S., Rebbeck, T.R., Besenfelder, W., Kreienberg, R. & Runnebaum, I.B. (1998) P53 germline polymorphisms are associated with an increased risk for breast cancer in German women. *Anticancer Res.*, **18**, 2095–2099

Wooster, R., Bignell, G., Lancaster, J., Swift, S., Seal, S., Mangion, J., Collins, N., Gregory, S., Gumbs, C., Micklen, G., Barfoot, R., Hamoudi, R., Patel, S., Rice, C., Biggs, P., Hashin, Y., Smith, A., Connor, F., Ararson, A., Gudmundsson, J., Ficenec, D., Kelsell, D., Ford, D., Tonin, P., Bishop, D.T., Spurr, N.K., Ponder, B.A.J., Eeles, R., Peto, J., Devilee, P., Cornelisse, C., Lynch, H., Narod, S., Lenoir, G., Eglisson, V., Barkadottir, R.B., Easton, D.F., Bentley, D.R., Futreal, P.A., Ashworth, A. & Stratton, M.R. (1995) Identification of the breast cancer susceptibility gene BRCA2. *Nature*, **378**, 789–792

Corresponding author:

J. Chang-Claude
Division of Clinical Epidemiology,
Deutsches Krebsforschungszentrum,
Im Neuenheimer Feld 280,
69120 Heidelberg,
Germany

Biomarkers in Cancer Chemoprevention
Miller, A.B., Bartsch, H., Boffetta, P., Dragsted, L. and Vainio, H. eds
IARC Scientific Publications No. 154
International Agency for Research on Cancer, Lyon, 2001

High-grade prostatic intraepithelial neoplasia as an exposure biomarker for prostate cancer chemoprevention research

J.R. Marshall

There is a tremendous need for exposure biomarkers, which need to function as intermediate end-points in cancer chemoprevention studies. Likely candidates include process biomarkers, atypical adenomatous hyperplasia, and a newly identified atrophic state. High-grade prostatic intraepithelial neoplasia (HGPIN) has the potential to be a useful exposure biomarker, having substantial predictive value for prostate cancer in chemoprevention trials. A limitation of the use of HGPIN as a biomarker is accessibility, since it requires the use of a highly invasive procedure that would not normally be applied unless malignancy is suspected to be present. However, most other biomarkers of prostatic tissue are similarly invasive. The HGPIN lesion appears to be highly measurable; however, problems of sampling coupled with the heterogeneity of the prostate raise questions about the degree to which the presence of HGPIN can be seen to characterize a given person's prostate gland. HGPIN has the advantage that it appears to be quite highly proximal to the development of cancer and to be modifiable. It remains less clear to what degree it reflects the exposures that are believed to alter prostate cancer risk. HGPIN has been identified as a clinical entity only recently and much additional research on the utility of this marker is needed.

Need for biomarkers

Epidemiological research on prostate and other cancers has been plagued by imprecision in the measurement of diet, occupation, physical activity and other exposures (Marshall *et al.*, 1999). Occupational studies have suggested that some industrial exposures, such as cadmium, may be relevant to risk (Ross & Schottenfeld, 1996). Dietary studies have suggested that obesity or perhaps consumption of a diet high in fat or in vegetable fat may increase risk (Giovannucci *et al.*, 1993). It has also been suggested that a diet low in some plant-based antioxidants, especially lycopene, may be a risk factor for prostate cancer (Giovannucci *et al.*, 1995; Clinton & Giovannucci, 1998). For both etiological and epidemiological research on diet and prostate cancer, biological markers of exposure would be useful, if it can be shown that they provide greater accuracy in exposure assessment. It would be preferable to link these markers to long-term exposure. Prostate cancer is believed to involve an etiology that may span several decades. The most likely premalignant lesion for prostate cancer — high-grade prostatic intraepithelial neoplasia (HGPIN) — begins to appear in men in the third or fourth decade of life, three decades before prostate cancer becomes highly incident (Sakr *et al.*, 1996; Sakr, 1998). Research on smoking as a risk factor for prostate cancer also suggests effects that only appear after several decades of exposure (Giovannucci *et al.*, 1999). Clearly, we need biomarkers of exposure that span long periods of time.

Also needed are markers of premalignant change that might be related to risk-enhancing exposures. As an indicator of premalignant change, HGPIN could be useful as a marker of carcinogenic exposure. There is evidence that HGPIN is related to substantially increased risk of prostate cancer (Bostwick & Brawer, 1987; Davidson *et al.*, 1995). Studies which might link this premalignant lesion to exposure would be extremely valuable.

HGPIN could serve as a highly useful exposure biomarker, perhaps as an intermediate biomarker of prostate cancer risk.

Possible candidates

A number of process indicators proposed as possible exposure biomarkers are associated with increased risk of prostate cancer (Kelloff *et al.*, 1997). Among these are markers of change related to excess cellular proliferation. A deficiency of apoptosis (programmed cell death) has been prominently mentioned. Another likely exposure biomarker would be an indicator of oxidative damage or of excessive oxidation. Markers of differentiation or of mutagenesis might be useful. As markers of advanced carcinogenesis or of tumour progression, angiogenesis indicators could be informative (Kelloff *et al.*, 1997).

In addition, several nonmalignant disease states could provide important information. Adenomatous atypical hyperplasia (AAH) has been proposed (Sakr & Grignon, 1998). The limitation of AAH as an exposure marker is that there is, at present, only weak evidence that AAH is causally linked to increased risk of the majority of prostate cancer. More recently, a form of prostatic inflammatory atrophy has been proposed as related to increased risk (De Marzo *et al.*, 1999), although research on this condition is not yet fully developed.

HGPIN has been proposed as a biomarker. The evidence linking HGPIN to elevated risk of prostate cancer is strong enough that researchers have claimed that HGPIN is almost certainly the premalignant lesion out of which prostate cancer arises (Brawer, 1992; Davidson *et al.*, 1995). There are some prospective data supporting this assertion. On the other hand, a recent study by O'Dowd *et al.* (2000) suggests that the importance of HGPIN as a risk factor for prostate cancer among high-risk populations may have been overestimated.

Access

Given the importance of HGPIN as a risk factor for prostate cancer, it holds tremendous promise for research. However, one of the major limitations to its use is that it can be evaluated only by means of prostatic biopsy. The probability of complications, especially of infection due to prostatic biopsy, is non-trivial and this rules out using evaluation of HGPIN except as part of clinical care. Metabolites isolated from blood or seminal fluid may permit more accessible and valid indirect examination of the prostate. Environmental factors do not have direct access to the prostate, so that any tissue-based marker may be subject to substantial metabolic modification.

There is also some debate about the prevalence of HGPIN. Early reports suggested that the prevalence of this condition in populations at elevated risk could be as high as 15–16% (Bostwick, 1996a). More recent data suggest that it is only a third of that level and that many of the high-risk individuals with HGPIN have synchronous cancer (Weinstein & Epstein, 1993; and van der Kwast, this volume). Research based on large numbers of patients from a range of clinical practices throughout the United States indicates that the probability of detecting HGPIN and no cancer is in the vicinity of 4% of all prostatic biopsies performed (O'Dowd *et al.*, 2000; Weinstein & Epstein, 1993). These figures could change as the treatment of prostatic disease evolves over time. They are also subject to the vagaries of pathology: HGPIN may be occasionally overlooked, or it may be diagnosed as cancer. The incidence of asymptomatic clinical prostate cancer is less than 10%, and only 3% of men in western industrial societies die as a result of prostate cancer. The frequency of the joint presence of HGPIN and prostate cancer is substantially higher than that of HGPIN alone, in keeping with the suspected etiological significance of HGPIN.

Measurability

Whether HGPIN will be useful as a biological marker depends in large part on our ability to identify the extent of its presence in men who are at substantial risk of prostate cancer (Bostwick & Brawer, 1987; Bostwick, 1996b). HGPIN is usually characterized as present in a gland or in a series of glands (Epstein *et al.*, 1995). However, while there is good agreement on the presence or absence of HGPIN in a gland or a limited region of prostatic tissue, there is less agreement about the characterization of an entire prostate or patient in terms of the extent to which HGPIN is present. Clearly, one could consider the number of HGPIN lesions present, the percentage of area within a biopsy occupied by an HGPIN lesion, the extent to which HGPIN lesions are dysplastic, the percentage of non-stromal tissue taken up by HGPIN, or perhaps the percentage of ductal tissue taken up by HGPIN (Montironi *et al.*, 1995). The extent to which any of these characteristics of HGPIN can be used to characterize the prostate of an individual has not been well defined or studied. The lesion itself can be quantified; one of the marks of HGPIN is degradation of the basal cell layer within the prostatic ducts and glands (Bartels *et al.*, 1998a; Sakr & Grignon, 1998), so that the percentage of the circumference of the basal cell layer that is degraded can be quantified. Grouping this information for a larger region of prostatic tissue, though, is not straightforward. Another potential marker of the extent of HGPIN is nuclear karyometry reflecting dysplasia in different cells (Bartels *et al.*, 1998b, c). There is substantial variation in the extent of dysplasia or cancer that is present. Means to evaluate sampling variability and incorporate this information into an attempt to characterize the tissue are required. A marker of the extent of HGPIN in a gland is the extent to which the circumference of the basal cell layer is eroded. However, the implications of a duct that is badly eroded, as opposed to several that are mildly so, for cancer risk is not clear.

Proximity to the causal pathway

As Schatzkin *et al.* (1990, 1996) and Kelloff *et al.* (1997) have made clear, the usefulness of a biological marker for chemoprevention depends on three characteristics of that biomarker. First is the degree to which it is dependent on a factor that increases risk of the disease. Ideally, the biomarker would be perfectly correlated with (a perfect marker of) exposure to the risk factor. The second characteristic is that the biomarker is predictive of disease. As has already been mentioned, HGPIN is an excellent predictor of prostate cancer risk (Brawer, 1992; Bostwick, 1996a; Davidson *et al.*, 1995), although O'Dowd *et al.* (2000) provide somewhat more equivocal evidence. There is a need for additional analyses of the differences in the results obtained by these studies as to the relevance of HGPIN to prostate cancer risk. The third criterion of biomarker usefulness is the degree to which the biomarker explains the association between exposure and risk. As Schatzkin *et al.* (1990, 1996) have pointed out, controlling statistically for a biomarker which explains the association of exposure and disease will eliminate the association between the exposure and disease. In other words, if the only path from exposure to disease is the one that is transmitted by the biomarker, statistical control for the biomarker would eliminate that pathway or association.

Modifiability

An important characteristic of a biological marker of exposure as relates to the development of chemoprevention strategies is that the marker be modifiable. Removal of the exposure which increases risk, or application of a chemopreventive agent, should decrease the extent of the marker. Thus, for evaluation of HGPIN as an exposure biomarker, it is of interest to determine whether the presence of HGPIN, as a neoplastic structure, is modified by removal of the risk-enhancing exposure or by the application of a chemopreventive agent. There is good evidence that HGPIN is modifiable. Androgen blockade or irradiation administered to individuals with prostate cancer appears to decrease the extent of HGPIN lesions (Ferguson *et al.*, 1994; Montironi *et al.*, 1994; Bostwick, 1996a). However, the extent to which the treatment of HGPIN lesions actually 'normalizes' the tissue, beyond shrinking the volume of lesions, has not been established. Whether the genotypic or phenotypic structure of neoplastic growth is decreased by application of either androgen blockade or irradiation requires additional study. The cells themselves may have undergone repair or the severely dysplastic cells may have undergone apoptosis and replacement by normal cells.

Limitations

The use of HGPIN as an exposure biomarker for use in studies of chemoprevention of prostate cancer appears to be highly promising. Prostate cancer is in part environmentally mediated. It is likely that HGPIN, an important predictor of prostate cancer risk, is similarly environmentally mediated. Nonetheless, it must be recognized that there is no easily accessible path to the prostate from environmental exposures except through a series of extensive metabolic pathways. Therefore, any exposure biomarker that is to be extracted from prostatic tissue has undergone substantial metabolic processing. Since metabolic factors and processes have multiple opportunities to alter any effect of environmental exposures on tissue characteristics, the prostate is highly unlikely to passively record the effects of exogenous exposures. This may substantially limit the degree to which HGPIN can be used as an exposure biomarker. At present, the necessary epidemiological data on HGPIN that would allow linking of exposure to the presence or absence of HGPIN are lacking.

As noted, the high predictive value of HGPIN in relation to prostate cancer indicates that it may represent a very late post-initiation phase of carcinogenesis. Thus, while HGPIN may function as an important exposure biomarker, it may not develop until several decades after exposure. In addition, coming as it does before completion but after initiation of carcinogenesis, it may be difficult to alter its course by chemopreventive agents. As noted, androgen deprivation seems to suppress it. Whether the application of other agents will have the same effect remains to be seen. It may be necessary to identify earlier biomarkers of exposure that can be altered by chemopreventive agents. A low-grade prostatic intraepithelial neoplasia has been recognized, and it has been assumed that progression of this condition leads to HGPIN. It would thus make sense that this could be related to prostate cancer risk (Bostwick, 1996a). However, there is little evidence, to date, that this lower-grade lesion is indeed related to increased prostate cancer risk. Additional research is needed to clarify the link between low-grade intraepithelial neoplasia and prostate cancer.

It is important to consider the way in which HGPIN as an exposure biomarker uniquely identifies and quantifies a subject with respect to the risk of prostate cancer. The case of another supposedly important biomarker, the index of crypt proliferation in colonic mucosa, is instructive. Crypt proliferation has been used for over two decades as a supposed marker of increased colon cancer risk and as a marker of exposure to carcinogens. Recently, however, it was shown that the degree to which crypt proliferation actually characterizes an individual is so low as to render markers of crypt proliferation virtually useless for most epidemiological and chemopreventive evaluation purposes (McShane *et al.*, 1998). Thus, the degree to which an individual can be characterized with respect to HGPIN and thus prostate cancer risk needs to be established, if measurement of HGPIN as a biomarker is to be interpretable.

An example of the use of HGPIN in chemoprevention of prostate cancer

In a chemoprevention clinical trial, prostate cancer risk was unexpectedly decreased by over 60% following the administration of 200 μg per day of selenium to men in a region with low soil levels of selenium (Clark *et al.*, 1996). All of these men had been diagnosed with basal- or squamous-cell cancer of the skin. These findings led Colditz (1996) to recommend that replication studies be undertaken. This appears to be a situation in which a protective agent has been identified, but the biological mechanisms by which that agent acts are not well understood (Ip, 1998). Combs and Gray (1998) suggested that none of the common forms of selenium (selenite, selenate, selenomethionine or selenocysteine) is likely to be the most biologically active protective form. Among possible protective mechanisms, selenium could contribute to antioxidant protection, immune enhancement or cell-cycle or apoptosis regulation (Combs & Gray, 1998) and it might also block the angiogenesis that is critical to neoplastic growth.

Clearly, another clinical trial of selenium and prostate cancer is needed to replicate the findings of Clark *et al.* (1996). In addition, it will be necessary to identify processes within the prostate that are affected by elevated intake of selenium; indeed, the biology of selenium needs to be delineated in a great deal more detail (Ip, 1998). For these purposes, a population of high-risk individuals would be useful; it would be preferable that, for these individuals, sampling of prostatic tissue should be a part of standard clinical care.

We have initiated a two-armed study, comparing three-year prostate cancer rates among HGPIN patients treated with 200 μg per day of selenium with the rates among HGPIN patients treated with placebo. Each patient will be randomized to one of the two treatment arms within strata of race, baseline selenium, baseline α-tocopherol and baseline prostate-specific antigen (PSA) levels. A total of 470 patients, 235 in each arm, will be randomized. Before randomization, there will be an enrolment period of three months, during which the patient will undergo a second biopsy in order to rule out prostate cancer. Approximately 1100 patients may need to be enrolled, chiefly to compensate for pre-randomization withdrawals due to prevalent prostate cancer not discovered until the second biopsy. The enrolment period will not include a placebo run-in, since very few noncompliant patients are identified by such a run-in (Feigl *et al.*, 1995); most will be identified by their failure to obtain the second biopsy or to forward the necessary pre-registration materials and information.

There will be two registrations for this study. The first will be at official study enrolment and the second at randomization. Central pathological verification of the presence of HGPIN and the absence of prostate cancer will be documented before study enrolment. Blood will be drawn. At the second registration, the patient's continued willingness to participate will be assessed. Between the first and second registrations, a second biopsy will be conducted; the absence of prostate cancer in the second biopsy will have to be confirmed before the second registration. The patient will then be randomized.

Treatment will be for three years, unless the patient is taken off the study due to toxicity, withdrawal from treatment or diagnosis of prostate cancer. Patients taken off treatment due to toxicity or for other reasons will remain on study for regular follow-up including the biopsy three years after randomization. The study goal is to follow and biopsy every participant at three years (whether on treatment or off) for ascertainment of end-point data. The only exceptions will be men who die, are diagnosed in the interim with prostate cancer, or in the interim request no more contact for study purposes.

At the first visit, the patient will be informed of the slight risk of mild toxicity associated with long-term use of selenium at the study dose (200 μg selenium). He will also be informed that participation in this study requires two additional biopsies, one before randomization and the other three years later, at the end of the study. He will be informed that the purpose of the first additional biopsy will be to confirm that he does not in fact have cancer, and that the second will be to provide a definitive evaluation of whether cancer is present after three years of treatment. He will be informed that additional risk accompanies these additional biopsies.

Slides from each patient who is willing to participate, and who remains eligible after first contact, will be forwarded to the study pathologist for confirmation that HGPIN is present and that cancer is not. The first visit will be scheduled so that, before the date of the visit, confirmation can be received that the patient is eligible. At the second visit, the patient's continued interest will be evaluated. If his second biopsy confirms that he does not have cancer, he will be randomized and provided with his first supply of pills. The study pathologist will review the second biopsy to ensure that the patient does not have cancer. Both the patient and the clinic will be blinded to the treatment assignment.

Each patient will be monitored, in the clinic at which he was initially identified, every six months, from 6 to 36 months after randomization. At each six-month visit, he will be evaluated by a blood PSA test, and he will receive a digital rectal examination (DRE). Compliance will be monitored by pill counts; the patient will be asked to bring his pill packets with him to the clinic. At each visit, blood samples will be drawn for evaluation of selenium. Between the six-month visits, each patient will be contacted by telephone about pill-taking, toxicity and health status. At each visit, the patient will be physically examined and evaluated for evidence of selenium toxicity, including an examination for lens opacity, as well as for the development of serious illness. Any patient whose PSA level increases by 50% or more above his previous level will have a new blood sample drawn and evaluated. If the second sample confirms a rise in PSA of over 50%, or if the DRE is abnormal, the patient will be scheduled for a new sextant biopsy.

Circumstances that will cause the patient to leave the study include a biopsy-based diagnosis of prostate cancer and request by the patient for no further contact. Other circumstances may cause a patient to discontinue study medication but remain on study and in active follow-up. Every effort will be made to obtain three-year outcome data, including biopsy, from these patients. Off-treatment but on-study conditions include, for example, unacceptable toxicity (i.e., drug-related > grade 3) and failure to pick up study drugs. If a patient does not comply with the schedule for clinic visits, blood samples etc., or is persistently non-compliant with respect to pill taking (< 75%

or > 125% of assigned dose), he will be counselled appropriately and kept on study. Of course, a patient may withdraw from chemopreventive treatment or the study at any time of his own volition.

At the 36-month evaluation at the end of the study, the patient will be examined by sextant biopsy, informed of the results, and provided the opportunity to ask questions. The data collection and treatment schedule is summarized in Figure 1.

Drop out rates will be assessed and noted at the time of randomization and during chemopreventive treatment. The proportion of initially enrolled patients who are not randomized and the proportion of randomized patients who do not complete the study will be noted. The proportions of randomized patients who either leave the study or are dropped will be calculated at each one-year interval.

Prostate cancer is the primary outcome to be evaluated in this trial. Data analysis will be based upon intention to treat: the focus of analysis will be prostate cancer among those assigned to treatment compared with that among subjects assigned to placebo. Thus, the three-year risk of prostate cancer among experimental subjects, relative to that among control subjects, will be the central conclusion of the study. The randomization and stratification of the trial will tend to ensure that experimental and control subjects are alike in terms of demographic and disease characteristics: age, race, PSA, and amount of HGPIN in biopsy material. However, in the event of unexpected imbalance, it will be possible to adjust estimates of the treatment effect using multiple covariate control procedures.

This trial will provide the opportunity to study several surrogate end-point biomarkers of prostate

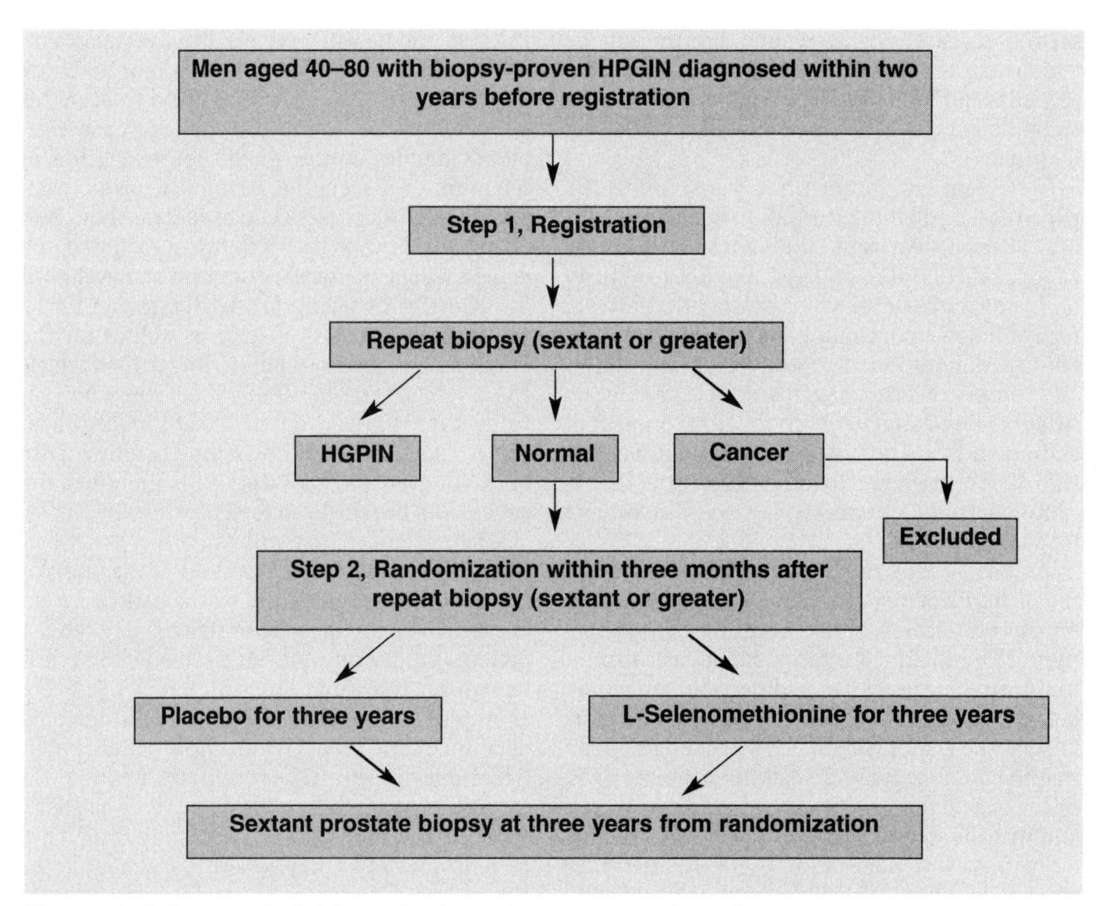

Figure 1. Scheme of trial for selenium chemoprevention of prostate cancer

cancer risk: proliferation as measured by Ki-67 (Raymond *et al.*, 1988); apoptosis as measured by TUNEL (Gavrieli *et al.*, 1992); and karyometry as measured by an automated machine vision system (Bartels *et al.*, 1995; 1996; 1998a,b,c). Use of these biomarkers will add significantly to understanding the means by which selenium exerts any effect on the risk of prostate cancer. It will also help to describe the nature of HGPIN as an exposure and risk biomarker. The first analytical task for examination of the biomarkers will be to evaluate the extent to which each of these biomarkers distinctly characterizes the subject. The model for this approach will be the proliferation biomarker analysis published by McShane *et al.* (1998). With the biomarkers indicating altered proliferation and apoptosis, as well as distorted nuclear chromatin patterns, it will be of interest to assess whether, within categories of treatment or placebo, these biomarkers predict progression to prostate cancer. Given their possible relevance to the risk of progression, it will be important to assess whether treatment has any impact upon changes in the biomarkers.

References

Bartels, P.H., Thompson, D., Bartels, H.G., Montironi, R., Scarpelli, M. & Hamilton, P.W. (1995) Machine vision-based histometry of premalignant and malignant prostatic lesions. *Pathol. Res. Pract.*, **191**, 935–944

Bartels, P.H., Thompson, D. & Montironi, R. (1996) Knowledge-based image analysis in the precursors of prostatic adenocarcinoma. *Eur. Urol.*, **30**, 234–242

Bartels, P.H., da Silva, V.D., Montironi, R., Hamilton, P.W., Thompson, D., Vaught, L. & Bartels, H.G. (1998a) Chromatin texture signatures in nuclei from prostate lesions. *Anal. Quant. Cytol. Histol.*, **20**, 407–416

Bartels, P.H., Montironi, R., Hamilton, P.W., Thompson, D., Vaught, L. & Bartels H.G. (1998b) Nuclear chromatin texture in prostatic lesions. I. PIN and adenocarcinoma. *Anal. Quant. Cytol. Histol.*, **20**, 389–396

Bartels, P.H., Montironi, R., Hamilton, P.W., Thompson, D., Vaught, L. & Bartels H.G. (1998c) Nuclear chromatin texture in prostatic lesions. II. PIN and malignancy associated changes. *Anal. Quant. Cytol. Histol.*, **20**, 397–406

Bostwick, D.G. (1996a) Prostatic intraepithelial neoplasia. In: Vogelzang N.J., Scardino, P.T., Shipley, W.V. & Coffey, D.S., eds, *Comprehensive Textbook of Genitourinary Oncology*, Baltimore, Williams and Wilkins, pp. 639–645

Bostwick, D.G. (1996b) Prospective origins of prostate carcinoma. Prostatic intraepithelial neoplasia and atypical adenomatous hyperplasia. *Cancer*, **78**, 330–336

Bostwick, D.G. & Brawer, M.K. (1987) Prostatic intraepithelial neoplasia and early invasion in prostate cancer. *Cancer*, **59**, 788–794

Brawer, M.K. (1992) Prostatic intraepithelial neoplasia: a premalignant lesion. *J. Cell. Biochem.*, Suppl., 171–174

Clark, L.C., Combs, G.F., Jr, Turnbull, B.W., Slate, E.H., Chalker, D.K., Chow, J., Davis, L.S., Glover, R.A., Graham, G.F., Gross, E.G., Krongard, A., Lesher, J.L., Park, H.K., Sanders, B.B., Smith, C.L. & Taylor, J.R. for the Nutritional Prevention of Cancer Study Group (1996) Effects of selenium supplementation for cancer prevention in patients with carcinoma of the skin. A randomized controlled trial [published erratum appears in *J. Am. Med. Ass.*, 1997, **277**, 1520]. *J. Am. Med. Ass.*, **276**, 1957–1963

Clinton, S.K. & Giovannucci, E. (1998) Diet, nutrition, and prostate cancer. *Ann. Rev. Nutr.*, **18**, 413–440

Colditz, G.A. (1996) Selenium and cancer prevention. Promising results indicate further trials required. *J. Am. Med. Ass.*, **276**, 1984–1985

Combs, G.F., Jr & Gray, W.P. (1998) Chemopreventive agents: selenium. *Pharmacol.Ther.*, **79**, 179–192

Davidson, D., Bostwick, D.G., Qian, J., Wollan, P.C., Oesterling, J.E., Rudders, R.A., Siroky, M. & Stilmant, M. (1995) Prostatic intraepithelial neoplasia is a risk factor for adenocarcinoma: predictive accuracy in needle biopsies. *J. Urol.*, **154**, 1295–1299

De Marzo, A.M., Marchi, V.L., Epstein, J.I. & Nelson, W.G. (1999) Proliferative inflammatory atrophy of the prostate: implications for prostatic carcinogenesis. *Am. J. Pathol.*, **155**, 1985–1992

Epstein, J.I., Grignon, D.J., Humphrey, P.A., McNeal, J.E., Sesterhenn, I.A., Troncoso, P. & Wheeler, T.M. (1995) Interobserver reproducibility in the diagnosis of prostatic intraepithelial neoplasia. *Am. J. Surg. Pathol.*, **19**, 873–886

Feigl, P., Blumenstein, B., Thompson, I., Crowley, J., Wolf, M., Kramer, B.S., Coltman, C.A., Jr, Brawley, O.S. & Ford, L.G. (1995) Design of the prostate cancer prevention trial (PCPT). *Controlled Clin. Trials*, **16**, 150–163

Ferguson, J., Zincke, H., Ellison, E., Bergstrahl, E. & Bostwick, D.G. (1994) Decrease of prostatic intraepithelial neoplasia following androgen deprivation therapy in patients with stage T3 carcinoma treated by radical prostatectomy. *Urology*, **44**, 91–95

Gavrieli, Y., Sherman, Y. & Ben-Sasson, S.A. (1992) Identification of programmed cell death in situ via specific labeling of nuclear DNA fragmentation. *J. Cell Biol.*, **119**, 493–501

Giovannucci, E., Rimm, E.B., Colditz, G.A., Stampfer, M.J., Ascherio, A., Chute, C.C. & Willett, W.C. (1993) A prospective study of dietary fat and risk of prostate cancer. *J. Natl Cancer Inst.*, **85**, 1571–1579

Giovannucci, E., Ascherio, A., Rimm, E.B., Stampfer, M.J., Colditz, G.A. & Willett W.C. (1995) Intake of carotenoids and retinol in relation to risk of prostate cancer. *J. Natl, Cancer Inst.*, **87**, 1767–1776

Giovannucci, E., Rimm, E.B., Ascherio, A., Colditz, G.A., Spiegelman, D., Stampfer, M.J. & Willett, W.C. (1999) Smoking and risk of total and fatal prostate cancer in United States health professionals. *Cancer Epidemiol. Biomarkers Prev.*, **8**, 277–282

Ip, C. (1998) Lessons from basic research in selenium and cancer prevention. *J Nutr.*, **128**, 1845–1854

Kelloff, G.J., Hawk, E.T., Karp, J.E., Crowell, J.A., Boone, C.W., Steele, V.E., Lubet, R.A. & Sigman, C.C. (1997) Progress in clinical chemoprevention. *Semin Oncol.*, **24**, 241–252

Marshall, J.R., Hastrup, J.L. & Ross, J.S. (1999) Mismeasurement and the resonance of strong confounders: correlated errors. *Am. J. Epidemiol.*, **150**, 88–96

McShane, L.M., Kulldorff, M., Wargovich, M.J., Woods, C., Purewal, M., Freedman, L.S., Corle, D.K., Burt, R.W., Mateski, D.J., Lawson, M., Lanza, E., O'Brien, B., Lake, W., Jr, Moler, J. & Schatzkin, A. (1998) An evaluation of rectal mucosal proliferation measure variability sources in the polyp prevention trial: can we detect informative differences among individuals' proliferation measures amid the noise? *Cancer Epidemiol. Biomarkers Prev.*, **7**, 605–612

Montironi, R., Magi-Galluzzi, C., Muzzonigro, G., Prete, E., Polito, M. & Fabris, G. (1994) Effects of combination endocrine treatment on normal prostate, prostatic intraepithelial neoplasia, and prostatic adenocarcinoma. *J. Clin. Pathol.*, **47**, 906–913

Montironi, R., Bartels, P.H., Thompson, D., Bartels, H.G. & Scarpelli, M. (1995) Prostatic intraepithelial neoplasia. Quantitation of the basal cell layer with machine vision system. *Pathol. Res. Pract.*, **191**, 917–923

O'Dowd, G.J., Miller, M.C., Orozco, R. & Veltri, R.W. (2000) Analysis of repeated biopsy results within 1 year after a noncancer diagnosis. *Urology*, **55**, 553–558

Raymond, W.A., Leong, A.S., Bolt, J.W., Milios, J. & Jose, J.S. (1988) Growth fractions in human prostatic carcinoma determined by Ki-67 immunostaining. *J. Pathol.*, **156**, 161–167

Ross, R. & Schottenfeld, D. (1996) Prostate cancer. In: Schottenfeld, D. & Fraumeni, J., Jr, eds, *Cancer Epidemiology and Prevention*, 2nd ed., New York, Oxford University Press, pp. 1180–1206

Sakr, W.A. (1998) High-grade prostatic intraepithelial neoplasia: additional links to a potentially more aggressive prostate cancer? *J. Natl Cancer Inst.*, **90**, 486–487

Sakr, W.A. & Grignon, D.J. (1998) Prostatic intraepithelial neoplasia and atypical adenomatous hyperplasia. Relationship to pathologic parameters, volume and spatial distribution of carcinoma of the prostate. *Anal. Quant. Cytol. Histol.*, **20**, 417–423

Sakr, W.A., Grignon, D.J., Haas, G.P., Heilbrun, L.K., Pontes, J.E. & Crissman, J.D. (1996) Age and racial distribution of prostatic intraepithelial neoplasia. *Eur. Urol.*, **30**, 138–144

Schatzkin, A., Freedman, L.S., Schiffman, M.H. & Dawsey, S.M. (1990) Validation of intermediate end points in cancer research. *J. Natl Cancer Inst.*, **82**, 1746–1752

Schatzkin, A., Freedman, L.S., Dorgan, J., McShane, L.M., Schiffman, M.H. & Dawsey, S.M. (1996) Surrogate end points in cancer research: a critique [editorial]. *Cancer Epidemiol. Biomarkers Prev.*, **5**, 947–953

Weinstein, M.H. & Epstein, J.I. (1993) Significance of high-grade prostatic intraepithelial neoplasia on needle biopsy. *Hum. Pathol.*, **24**, 624–629

Corresponding author:

J.R. Marshall,
Professor of Public Health,
Arizona Cancer Center College of Medicine,
University of Arizona,
Tucson, Arizona, 85724,
USA

Biomarkers in Cancer Chemoprevention
Miller, A.B., Bartsch, H., Boffetta, P., Dragsted, L. and Vainio, H. eds
IARC Scientific Publications No. 154
International Agency for Research on Cancer, Lyon, 2001

Intermediate biomarkers for chemoprevention of prostate cancer

T. H. van der Kwast

Use of high-grade prostatic intraepithelial neoplasia (PIN) as an intermediate biomarker for prostate cancer requires additional data concerning its natural biological behaviour. Moreover, it should be recognized that a proportion of PIN lesions may represent intraductal spread of an accompanying prostate cancer rather than a precancerous lesion. The detection rate of isolated PIN in the general population is low, and its clinical significance in the short term may be limited. Additional long-term studies on the significance of isolated PIN detected during population screening are required. Due to inadequate tissue sampling by current biopsy procedures, the presence of an accompanying prostate cancer is difficult to rule out.

Endocrine therapy changes the morphology of PIN, hampering its identification by making it more closely resemble the normal benign glands. In addition, endocrine therapy may lead to molecular changes in PIN, with a potential risk of induction of resistance to endocrine therapy. Prolonged androgen deprivation (for six months) does not generally lead to eradication of PIN. Cessation of endocrine therapy is likely to lead to renewed expansion of PIN, since PIN continues to express androgen receptors and the cell-cycle protein MIB-1 under conditions of low androgen levels.

Recent findings indicate that most high-grade prostate cancers seem to develop from low-grade cancers. The development of a high-grade focus of prostate cancer within a clinically latent low-grade tumour might be a suitable target for future intervention studies, provided that appropriate monitoring for development of high-grade cancer can be achieved in individual patients.

Introduction

Prostate cancer has now become the second most frequent cause of death in the ageing male population of western society (Landis *et al.*, 1999). Its incidence has risen dramatically since the introduction of methods for early detection of prostate cancer (Mettlin *et al.*, 1998). Attempts have been made to identify the mechanisms that underlie the development of prostate cancer, but a major obstacle has been the rarity of this disease in other species, hampering the development of relevant and easily accessible animal models (Riverson & Silverman, 1979). Much research effort has been devoted to the identification of precursor lesions of human prostate cancer. Since the development of prostate cancer is strongly dependent upon the presence of androgens, the effects of androgen deprivation on the potential precursor lesions are of interest.

Prostate cancer has a highly variable biological behaviour. Some prostate cancers are highly aggressive, while others remain indolent for a long period (Chodak *et al.*, 1994). Therefore the ethical acceptability of early detection of prostate cancer by population screening remains a controversial issue, as it is likely that some men diagnosed with prostate cancer will not benefit from treatment. Only randomized studies on prostate screening, measuring both mortality reduction and quality of life, will resolve this issue (Schröder, 1995). Unfortunately, it is not yet possible to predict prostate cancer behaviour preoperatively due to lack of sufficiently reliable serum or tissue markers (Murphy, 1998). Tumour heterogeneity and multifocality of screening-detected prostate cancer has further confused this issue (Hoedemaeker *et al.*, 2000). This chapter discusses current knowledge of the epidemiology and biology of precursor lesions of prostate cancer and provides some data on early-detected prostate cancers which may serve as a potential target for future chemoprevention studies.

Prostatic intraepithelial neoplasia

Potential precursor lesions of the prostate

In the past, certain lesions have been proposed to be likely precursors of prostate cancer. These include atrophy, atypical adenomatous hyperplasia or adenosis and dysplastic lesions of the prostatic ducts. The latter is now generally referred to as prostatic intraepithelial neoplasia (PIN) (Bostwick, 1995). It has been suggested that such lesions may not account for all prostatic adenocarcinomas and that the human prostate may harbour other hitherto unrecognized premalignant lesions. Some authors have considered atrophy and atypical adenomatous hyperplasia as potential precursor lesions of prostate cancer (Cheng *et al.*, 1998; De Marzo *et al.*, 1999), but this view has not been generally accepted. Evidence for PIN as a premalignant lesion is based on morphological, molecular and epidemiological data. The similarity in morphology of the dysplastic cells of PIN to that of peripheral zone prostatic adenocarcinoma, including nuclear features, and the preferential localization of PIN in the peripheral zone of the prostate, as well as the similarity in molecular changes, strongly support the hypothesis that PIN is a precursor of adenocarcinoma (Myers & Grizzle, 1996). PIN was initially graded in three classes, and subsequently low and high grades were distinguished, but particularly because of great inter-observer variation among pathologists and its lack of clinical relevance, low-grade PIN is now not reported to clinicians (Epstein *et al.*, 1995). Therefore, in the subsequent discussion, high-grade PIN is referred to as PIN.

The histopathology of PIN

Microscopically, at least five architectural patterns of PIN can be distinguished, based on the arrangement of the dysplastic cells within the pre-existing duct or gland. The most common variants of PIN are the tufted and micropapillary patterns. Less common are the flat PIN and the cribriform PIN (Bostwick *et al.*, 1993). In addition, occasional PIN lesions may contain a lumen filled with necrotic debris, resembling comedocarcinoma of the breast. In particular, cribriform PIN and comedocarcinoma-like PIN are associated with concurrent prostatic adenocarcinoma. It has been hypothesized that these two lesions may actually represent intraductal spread of the associated carcinoma rather than precursor lesions (Cohen *et al.*, 2000). Likewise, it cannot be entirely excluded that a proportion of the other, more common, variants of PIN also represent spread of adenocarcinoma within pre-existing ducts or glands rather than a precursor lesion. The similarity of molecular changes in PIN and associated cancer could also be explained in this way. Thus, it is conceivable that morphologically similar PIN lesions represent the extremes of the spectrum from precancerous lesion to spread of an overt cancer.

Prostate-specific antigen and PIN

Since an elevated prostate-specific antigen (PSA) level is now a standard criterion to determine if a man should undergo additional diagnostic procedures for prostate cancer, it has been suggested that a PSA window might be defined to detect specifically men with isolated PIN lesions. Some studies have reported that men with PIN have a PSA level intermediate between those with cancer and those with benign tissue (Brawer & Lange, 1989). More recent studies have not confirmed this and it is now accepted that no PSA window can be defined for isolated PIN (Bostwick, 1999). Immunohistochemical staining of PIN lesions reveals a lower level of PSA expression than in benign prostatic glands. Furthermore, PSA is secreted in the lumina of PIN-containing glands and drained via the prostatic ducts, largely preventing its leakage into the blood circulation. The latter would offer an explanation why PIN is not associated with raised serum PSA levels.

Epidemiology of PIN lesions

Autopsy studies

In 80–90% of radical prostatectomy specimens, PIN can be observed in association with prostate cancer (Kovi *et al.*, 1988; Qian *et al.*, 1997). This high percentage supports the presumed relationship between PIN and prostate cancer. A few (forensic) autopsy studies have compared the frequency of PIN with that of adenocarcinoma. One report suggested that PIN precedes prostate cancer by almost a decade (Bostwick, 1992), confirming an earlier autopsy study reported by Kovi *et al.*

(1988) that showed that the median age of men with atypical acinar hyperplasia (including low- and high-grade PIN) was 56.2 years, while that of men with cancer and PIN was 63.8 years, a lag time of 7.6 years. Sakr *et al.* (1995), however, suggested a much shorter interval between the presence of PIN and adenocarcinoma. In their autopsy series, they noted PIN in 26% and latent cancer in 31% of the prostates of American Caucasian men aged between 30 and 40 years. They also reported that the extent of PIN in prostates of African Americans was much greater than in Caucasians of corresponding ages. This makes it likely that the extent of PIN determines the risk of development of an overt carcinoma.

Isolated PIN in prostatic needle biopsies

Initial studies on prostate needle biopsies of men referred to urological clinics demonstrated a surprisingly high percentage of isolated PIN (without accompanying prostate cancer), with figures reaching 16% (Bostwick *et al.*, 1995). In a retrospective case–control study, repeat biopsies were performed within a period of two years in men with a previous biopsy diagnosis of isolated PIN. This led to the detection of prostate cancer in about 35% of cases, compared with 13% in men without a previous diagnosis of PIN or carcinoma in a previous prostate needle biopsy (Davidson *et al.*, 1995). The outcome of this study was considered to provide evidence for the clinical relevance of isolated PIN in the early detection of prostate cancer.

Later studies on screened populations reported much lower frequencies, in the range 1–2.5% (Hoedemaeker *et al.*, 1999). The differences in reported frequency of PIN can most likely be attributed to differences between the populations selected and in biopsy procedure. In the earlier studies, systematic sextant needle biopsies were not yet a common practice and probably only suspect lesions were biopsied. In these cases, the prostate cancer may have been missed but the frequently occurring adjacent PIN lesion was seen in the biopsy. It was recently suggested that five-region biopsy procedures might increase the yield of isolated PIN in screened men (Rosser *et al.*, 1999). Data on isolated PIN from the Rotterdam section of the European Randomized Screening Program of Prostate Cancer (ERSPC, coordinated by Professor F. Schröder) revealed an incidence of 1% in men aged between 55 and 75 years, using a PSA cut-off value of 3.0 ng/ml for systematic sextant needle biopsies. More importantly, repeat biopsies in these men within six months led to the discovery of prostate cancers in about 10% of the cases (Van der Kwast, unpublished). This is within the background levels of prostate cancers detected in men with a previous benign outcome of their biopsies (Davidson *et al.*, 1995; Epstein *et al.*, 1999), and casts doubt on the value of the current practice of repeat biopsy in this subset of men. In the ERSPC study, the vast majority of isolated PIN lesions were tufted and micropapillary lesions, while cribriform PIN was very rare and comedocarcinoma-like PIN has not yet been found (Van der Kwast, unpublished). Longer follow-up studies of isolated PIN detected in a screened population may give additional insight into the natural behaviour of PIN.

Androgen sensitivity of PIN

Controversies in the literature

Since early studies demonstrated that prostate cancer cannot develop in the absence of adequate testosterone levels, it has been suggested that reduction of androgens may be a useful approach to prevention. Furthermore, it is well known that most (organ-confined or metastasized) prostate cancers initially regress during androgen deprivation. Similarly, several studies have demonstrated that benign prostatic glands show regressive features manifested by apoptosis, vacuolization of cytoplasm and shrinkage of nuclei of the secretory epithelial cells (Armas *et al.*, 1994; Vaillancourt *et al.*, 1996). Some studies on radical prostatectomy specimens from men pretreated for various periods of time for organ-confined prostate cancer (neoadjuvant androgen blockade) have also shown a decline in frequency and extent of PIN (Ferguson *et al.*, 1994). However, published data differ widely, some claiming no decrease and others even a decrease to about 6% of cases (Table 1). Criteria to define PIN in prostates of men pretreated by androgen blockade have not been clearly established, potentially leading to inter-observer variation between pathologists. In particular, one study in which the presence of prominent nucleoli was used as a prerequisite for diagnosis of PIN suggested a strong effect of androgen deprivation on frequency of residual PIN (Vaillancourt *et al.*, 1996).

Table 1. Persistence of PIN during androgen deprivation

Authors	Treatment	% residual PIN
Armas *et al.*, 1994	3 months CAB	77
Montironi *et al.*, 1995	3 months CAB	83
Vaillancourt *et al.*, 1996	3 months CAB	6
Civantos *et al.*, 1995	≥ 3 months CAB	35
Ferguson *et al.*, 1994	(4–53 weeks) Variable	50
Van der Kwast *et al.*, 1999	3 months CAB	72
	6 months CAB	59

CAB, combined androgen blockade

Effects of androgen deprivation on morphology of PIN

It is now well established that in prostatic adenocarcinoma that persists during androgen deprivation therapy mediated by combined androgen blockade, several morphological changes occur, including loss of prominent nucleoli, and it is very likely that this also occurs in PIN lesions. Thus, androgen deprivation could lead to a metamorphosis of dysplastic cells constituting PIN lesions into cells with a histopathologically less apparent phenotype. Since at present only few markers exist that can be employed to distinguish PIN from benign glands, detection of residual PIN in androgen-deprived prostatectomy specimens would depend on histomorphological features of which the definition would be adapted. This is a good illustration of the phenomenon that chemoprevention may lead to morphological changes in such a way that the lesion, though still present, can no longer be identified with certainty. On the other hand, evaluation of the effect on PIN of the mild anti-androgen agent 5α reductase inhibitor (finasteride) did not reveal any changes in PIN lesions (Cote *et al.*, 1998). According to these authors, it is questionable whether finasteride can serve as an effective chemopreventive agent.

We have studied PIN in radical prostatectomy specimens from a series of 40 men with a clinically organ-confined prostate cancer, randomized to three or six months of combined androgen blockade before surgery (Van der Kwast *et al.*, 1999). In radical prostatectomy specimens, foci of PIN were detected in 72% of specimens from men pretreated for their prostate cancer for three months and in 59% of those from men treated for six months. The number of glands involved by PIN decreased from a median number of 19 (± 21 glands) to 7 (± 12) glands with the longer treatment. These differences were, however, not significant. In contrast, the volume of prostate cancer after six months' treatment was significantly reduced by 60% compared with the volume after three months' treatment, while the number of PIN lesions within or adjacent to residual cancer increased (Van der Kwast *et al.*, 1999). The latter observations suggest that prostate cancer may be more susceptible to androgen deprivation than PIN lesions.

Recovery of PIN after cessation of androgen deprivation therapy

Since in residual PIN, nuclear androgen receptors might be detected as well as occasional cells with expression of the cell-cycle molecule MIB-1, the data strongly suggest that PIN may recover and even further expand after cessation of androgen deprivation therapy. Importantly, even a severe regimen of androgen deprivation (so-called combined androgen blockade using flutamide and LHRH agonists) during a period of six months seems not to be sufficient to eradicate all PIN lesions, although a tendency towards further reduction of the extent of PIN was noted after six months of androgen blockade. Nevertheless, prolonged treatment with this combination therapy may lead to a greater reduction of PIN. In a few patients randomized to six months' combined androgen blockade therapy, the therapy was stopped before surgery. In three out of five of these cases, PIN with the classical features, including prominent nucleoli, was found. This confirms our view that PIN lesions that persist for several months during androgen deprivation can recover rapidly.

Another purpose of chemoprevention of prostate cancer could be to stop progression of PIN to prostatic adenocarcinoma, if eradication of the precursor lesion is not possible. In the latter case, life-long administration of anti-androgens should be envisaged. A study on the prolonged administration of finasteride to men with benign prostatic hyperplasia has suggested that such chemopreventive measures may not be without risk, since in a proportion of men who developed a carcinoma during this treatment, amplification of the androgen receptor gene in the prostate cancer was observed (Koivisto *et al.*, 1999). This androgen receptor amplification was previously shown to mediate resistance to endocrine therapy in metastasized prostate cancers. Androgen receptor gene amplifications have not been documented in cancers not exposed to endocrine therapy (Koivisto *et al.*, 1995).

Prevention of progression of clinically latent low-grade adenocarcinoma to intermediate-grade cancer

Features of early-detected prostate cancers

In radical prostatectomy specimens from men with a prostate cancer detected by first-round screening on the basis of elevated PSA levels, but not clinically, some interesting features can be observed (Hoedemaeker *et al.*, 2000). In about 50% of such specimens, the cancer appears to be multifocal, with a maximum of five different tumours detected in a single prostatectomy specimen. This heterogeneity is also reflected in tumour grade, which may differ considerably. On the other hand, 50% of the detected cancers represent intermediate-grade cancers (Gleason score 7), with variable proportions of high-grade (Gleason grade 4 or 5) tumour. Careful examination of the screen-detected radical prostatectomy specimens obtained in the Rotterdam section of the ERSPC study revealed that about 15% of the early-detected prostate cancers satisfied the criteria of minimal cancer, a category of men who might not benefit from radical prostatectomy or radiotherapy (Hoedemaeker *et al.*, 1997). Tumour volume and grade were well correlated, although a wide scatter of values existed. High-grade tumour was present particularly in the larger tumours. High-grade tumour areas were seen at the centre of tumour areas. This relationship between tumour volume and presence of a high-grade tumour component suggests that most prostate cancers initially are low-grade (Gleason growth pattern 3), while in due course Gleason growth pattern 4 or 5 develops within these low-grade tumour areas. Thus, the development of high-grade cancer within such low-grade tumour areas may represent a potential target for measures aimed at the prevention of tumour progression. PSA velocity could prove a suitable parameter to monitor tumour volume or progression during therapeutic intervention in individual patients with a low-grade latent prostate cancer. This strategy would, however, require a sensitive technique to detect specifically high-grade cancer areas in an otherwise low-grade prostate cancer.

Clinical assessment of prostate cancer grade and stage

Serum PSA level and tumour size do correlate well, but the correlation coefficient is low. Serum PSA levels are influenced by prostatic gland volume, inflammation and obstruction of prostatic glands, in addition to prostate cancer volume. Therefore, in individual patients with prostate cancer, the PSA level cannot be used to predict prostate cancer volume or pathological stage. In larger groups of patients, preoperative parameters such as proportion of involvement of needle biopsies by cancer and tumour grade determined in needle biopsies show a good correlation with tumour volume and grade in prostatectomy specimens. Probably due to sampling problems, these preoperative parameters cannot give a prediction of the biological behaviour of a tumour on an individual basis. Given the multifocality and heterogeneity of early prostate cancer, the problem of adequate sampling seems insurmountable for the application of tissue-based prognostic markers, whether conventional pathological markers or molecular ones. It may be more practical to focus research efforts upon molecules shed by prostate cancer cells into the blood circulation. Although PSA is such a molecule, its lack of specificity for (poorly differentiated) prostate cancer restricts its use as an intermediate end-point marker.

References

Armas, O.A., Aprikian, A.G., Melamed, J., Cordon-Cardo, C., Cohen, D.W., Erlandson, R., Fair, W.R. & Reuter, V.E. (1994) Clinical and pathobiological effects of neoadjuvant total androgen ablation therapy on clinically localized prostatic adenocarcinoma. *Am. J. Surg. Pathol.*, **18**, 979–991

Bostwick, D (1992). Prostatic intraepithelial neoplasia: current concepts. *J. Cell Biochem.*,**16** (S), 10–19

Bostwick, D.G. (1995) High grade prostatic intra-epithelial neoplasia: the most likely precursor of prostate cancer. *Cancer*, **75**, 1823–1836

Bostwick, D.G. (1999) Prostatic intraepithelial neoplasia is a risk factor for cancer. *Semin. Urol. Oncol.*, **17**, 187–198

Bostwick, D.G., Amin, M.B., Dundore, P., Marsh, W. & Schultz, D.S. (1993) Architectural patterns of high-grade prostatic intraepithelial neoplasia. *Human Pathol.*, **24**, 298–310

Bostwick, D.G., Qian, J. & Frankel, K. (1995) The incidence of high grade prostatic intraepithelial neoplasia in needle biopsies. *J. Urol.*, **154**, 1791–1794

Brawer, M.K. & Lange, P.H. (1989) Prostate-specific antigen and premalignant change: implications for early detection. *CA Cancer J. Clin.*, **39**, 361–375

Cheng, L., Shan, A., Cheville, J.C., Qian, J. & Bostwick, D.G. (1998) Atypical adenomatous hyperplasia of the prostate: a premalignant lesion? *Cancer Res.*, **58**, 389–391

Chodak, G.W., Thisted, R.A., Gerber, G.S., Johansson, J.E., Adolfsson, J., Jones, G.W., Chisholm, G.D., Moskovitz, B., Livne, P.M. & Warner, J. (1994) Results of conservative management of clinically localized prostate cancer. *New Engl. J. Med.*, **330**, 242–248

Civantos, F., Marcial, M.A., Banks, E.R., Ho, C.K., Speights, V.O., Drew, P.A., Murphy, W.M. & Soloway, M.S. (1995) Pathology of androgen deprivation therapy in prostate carcinoma. A comparative study of 173 patients. *Cancer*, **75**, 1634–1641

Cohen, R.J., McNeal, J.E. & Baillie, T. (2000) Patterns of differentiation and proliferation in intraductal carcinoma of the prostate: significance for cancer progression *Prostate*, **43**, 11–19

Cote, R.J., Skinner, E.C., Salem, C.E., Mertes, S.J, Stanczyk, F.Z., Henderson, B.E., Pike, M.C. & Ross, R.K. (1998) The effect of finasteride on the prostate gland in men with elevated serum prostate-specific antigen levels. *Br. J. Cancer*, **78**, 413–418

Davidson, D., Bostwick, D.G., Qian, J., Wollan, P.C., Oesterling, J.E., Rudders, R.A., Siroky, M. & Stilmant, M. (1995) Prostatic intraepithelial neoplasia is a risk factor for adenocarcinoma: predictive accuracy in needle biopsies. *J. Urol.*, **154**, 1295–1299

De Marzo, A.M., Marchi, V.L., Epstein, J.I. & Nelson, W.G. (1999) Proliferative inflammatory atrophy of the prostate: implications for prostatic carcinogenesis. *Am. J. Pathol.*, **155**, 1985–1992

Epstein, J.I., Grignon, D.J., Humphrey, P.A., McNeal, J.E., Sesterhenn, I.A., Troncoso, P. & Wheeler, T.M. (1995) Interobserver reproducibility in the diagnosis of prostatic intraepithelial neoplasia. *Am. J. Surg. Pathol.*, **19**, 873–886

Epstein, J.I., Walsh, P.C., Akingba, G. & Carter, H. (1999) The significance of prior benign needle biopsies in men subsequently diagnosed with prostate cancer. *J. Urol.*, **162**, 1649–1652

Ferguson, J., Zincke, H., Ellison, E., Bergstrahl, E. & Bostwick, D.G. (1994) Decrease of prostatic intra-epithelial neoplasia following androgen deprivation therapy in patients with stage T3 carcinoma treated by radical prostatectomy. *Urology*, **44**, 91–95

Hoedemaeker, R.F., Rietbergen, J.B., Kranse, R., Van der Kwast, T.H. & Schroder, F.H. (1997) Comparison of pathologic characteristics of T1c and non-T1c cancers detected in a population-based screening study, the European Randomized Study of Screening for Prostate Cancer. *World J. Urol.*, **15**, 339–345

Hoedemaeker, R.F., Kranse, R., Rietbergen, J.B., Kruger, A.E., Schroder, F.H. & Van der Kwast, T.H. (1999) Evaluation of prostate needle biopsies in a population-based screening study: the impact of borderline lesions. *Cancer*, **85**, 145–152

Hoedemaeker, R.F., Rietbergen, J.B.W., Kranse, R., Schröder, F.H. & Van der Kwast, T.H. (2000) Histopathological prostate cancer characteristics at radical prostatectomy after population-based screening. *J. Urol.*, **164**, 411–415

Koivisto, P., Hyytinen, E., Palmberg, C., Tammela, T., Visakorpi, T., Isola, J. & Kallioniemi, O.P. (1995) Analysis of genetic changes underlying local recurrence of prostate carcinoma during androgen deprivation therapy. *Am. J. Pathol.*, **147**, 1608–1614

Koivisto, P.A., Schleutker J, Helin, H., Ehren-van Eekelen, C., Kallioniemi, O.P. & Trapman, J. (1999) Androgen receptor gene alterations and chromosomal gains and losses in prostate carcinomas appearing during finasteride treatment for benign prostatic hyperplasia. *Clin. Cancer Res.*, **5**, 3578–3582

Kovi, J., Mostofi, F.K., Heshmat, M.Y. & Enterline, J.P. (1988) Large acinar atypical hyperplasia and carcinoma of the prostate. *Cancer*, **61**, 551–561

Landis, S.H., Murray, T., Bolden, S. & Wingo, P.A. (1999) Cancer statistics, 1999. *CA Cancer J. Clin.*, **49**, 8–31

Mettlin, C.J., Murphy, G.P., Rosenthal, D.S. & Menck, H.R. (1998) The National Cancer Data Base report on prostate carcinoma after the peak in incidence rates in the U.S. The American College of Surgeons Commission on Cancer and the American Cancer Society. *Cancer*, **83**, 1679–1684

Montironi, R., Magi-Galuzzi, C. & Fabris G. (1995) Apoptotic bodies in prostatic intraepithelial neoplasia and prostatic adenocarcinoma following total androgen ablation. *Pathol. Res. Pract.*, **191**, 873–880

Murphy, W.M. (1998) Prognostic factors in the pathological assessment of prostate cancer [editorial]. *Human Pathol.*, **29**, 427–430

Myers, R.B. & Grizzle, W.E. (1996) Biomarker expression in prostatic intra-epithelial neoplasia. *Eur. Urol.*, **30**, 153–166

Qian, J., Wollan, P. & Bostwick, D.G. (1997) The extent and multicentricity of high-grade prostatic intraepithelial neoplasia in clinically localized prostatic adenocarcinoma. *Human Pathol.*, **28**, 143–148

Riverson, A. & Silverman, J. (1979) The prostatic carcinoma in laboratory animals: a bibliographic survey from 1900 to 1977. *Invest. Urol.*, **16**, 468–472

Corresponding author:

Th. H. van der Kwast
Department of Pathology,
Josephine Nefkens Institute,
Erasmus University,
Postbox 1738,
3000 DR Rotterdam,
The Netherlands

Rosser, C.J., Broberg, J., Case, D., Eskew, L.A. & McCullough, D. (1999) Detection of high-grade prostatic intraepithelial neoplasia with the five-region biopsy technique. *Urology*, **54**, 853–856

Sakr, W.A., Grignon, D.J., Haas, G.P., Schomer, K.L., Heilbrun, L.K., Cassin, B.J., Powell, J., Montie, J.A., Pontes, J.E. & Crissman, J.D. (1995) Epidemiology of high grade prostatic intraepithelial neoplasia. *Pathol. Res. Pract.*, **191**, 838–841

Schröder, F.H. (1995) Screening, early detection, and treatment of prostate cancer: a European view. *Urology*, **46** (Suppl A), 62–70

Vaillancourt, L., Têtu, B., Fradet, Y., Dupont, A., Gomez, J., Cusan, L., Suburu, E.R., Diamond, P., Candas, B. & Labrie, F. (1996) Effect of neoadjuvant endocrine therapy (combined androgen blockade) on normal prostate and prostatic carcinoma: a randomized study. *Am. J. Surg. Pathol.*, **20**, 86–93

Van der Kwast, T.H., Labrie, F. & Têtu, B. (1999) Persistence of high-grade prostatic intra-epithelial neoplasia under combined androgen blockade therapy. *Human Pathol.*, **30**, 1503–1507

Biomarkers in Cancer Chemoprevention
Miller, A.B., Bartsch, H., Boffetta, P., Dragsted, L. and Vainio, H. eds
IARC Scientific Publications No. 154
International Agency for Research on Cancer, Lyon, 2001

The role of molecular genetics in chemoprevention studies of prostate cancer

R.K. Ross

Research into the molecular genetics of prostate cancer to date has largely focused on the possible existence of one or several single-locus high-penetrance susceptibility genes and several candidate regions have been identified, but confirmatory studies of these regions have been inconclusive. Increasingly, attention has turned to identification of candidate genes which may increase prostate cancer risk because their products play an important role in possible etiological pathways for prostate cancer. Of various such pathways which have been suggested for prostate cancer, the best studied in terms of molecular genetics is the androgen signalling pathway. Two genes in this pathway, the androgen receptor (*AR*) gene and the steroid 5-alpha reductase type II (*SRD5A2*) gene, have been under particular scrutiny and polymorphic markers in each of these genes that reproducibly predict prostate cancer risk have been identified. Such studies may have important implications for prostate cancer chemoprevention trials. As etiological pathways become better understood at the molecular level, piecing together multiple genetic variants in a pathway will allow identification of high-risk individuals and potential targets for chemopreventive interventions. Moreover, understanding the role of these genes in prostate cancer etiology may help in defining heterogeneity in response to such interventions. Finally, these genes or their products may themselves be legitimate targets for building a chemoprevention strategy.

Introduction

The molecular genetic epidemiology of prostate cancer is an evolving field (Ross *et al.*, 1998, 2000). Much of the work in this area has until recently focused on the identification of one or several single-locus high-penetrance susceptibility genes that might carry with them, for individuals with mutated forms, very high lifetime risk of developing prostate cancer. Interest in this type of susceptibility has been stimulated by the highly reproducible finding that prostate cancer is a strongly familial disease. Men with a first-degree relative with prostate cancer have a 2–3-fold increased risk relative to the population as a whole (Monroe *et al.*, 1995). This strong familial risk has been found in populations with both a high and low risk of the disease. Having several such relatives and/or a relative with prostate cancer at a relatively young age are associated with further increases in risk (Carter *et al.*, 1992). Prostate cancer has an unusual family risk pattern in that if an individual has a brother with prostate cancer, risk is roughly twice as high as if the father had the disease (Monroe *et al.*, 1995). This pattern of risk contrasts with risk associated with other familial cancers, such as breast cancer, for which risk is roughly the same whether a sister or a mother is affected. This pattern of risk has provided leads as to the mode of transmission of the purported susceptibility gene(s). Linkage analyses in multiplex families have led to the identification of several candidate regions for susceptibility loci, but so far, confirmatory studies have generally been inconclusive for each of these regions and final identification and cloning of a major locus gene for prostate cancer is unlikely to be forthcoming (Smith *et al.*, 1996; Xu *et al.*, 1998).

Increasingly, molecular genetic epidemiological research on prostate cancer has focused on the identification of candidate genes which increase prostate cancer risk because their protein product plays a role in an etiological pathway of disease development. Although many etiological/prevention pathways have been suggested for human prostate cancer, currently the strongest epidemiological and experimental evidence supports four: androgen signalling; antioxidation; vitamin D signalling; and insulin-like growth factor (IGF) signalling. Although susceptibility genes may modify individual risk through any of these pathways, evaluation of such genetic influences is not well understood for any of them and has not even begun for some. Thus although there is substantial epidemiological evidence that the antioxidant carotenoid lycopene may lower risk of prostate cancer (Giovannucci *et al.*, 1995), both epidemiological and randomized clinical trial evidence that selenium lowers risk (either as an antioxidant or through other anticarcinogenic effects) (Clark *et al.*, 1998; Yoshizawa *et al.*, 1998) and experimental evidence that the antioxidant vitamin tocopherol lowers risk (Heinonen *et al.*, 1998), no attempt has yet been made to determine if any genetic factors might modify any chemopreventive efficacy. Similarly, there is prospective epidemiological evidence that IGF-I levels are predictive of prostate cancer occurrence (Chan *et al.*, 1998). As polymorphic markers in genes for IGF and its binding proteins have been identified (Rosen *et al.*, 1998), these are potential future targets for determining individual susceptibility related to IGF-induced carcinogenesis, but no research in this area has yet been reported and validation of genotype/phenotype relationships has not been completed for all such markers.

However, there is a steadily increasing number of reports on polymorphic variants of low-penetrance genes in the androgen signalling pathway and in the vitamin D signalling pathway in relation to prostate cancer risk. The remainder of this chapter reviews the current state of knowledge in these areas, summarizes how current knowledge can affect existing or planned chemopreventive activities and provides some general thoughts about the future of this field as it pertains to chemoprevention of prostate cancer. Various strategies already available to assess chemopreventive efficacy in human studies in terms of biochemical and histological parameters or of prostate cancer risk *per se* are described.

Androgen-related susceptibility genes

Although a number of genes in the androgen signalling pathway have been suggested as candidates for investigation (e.g., *CYP17* and *HSD17B3* as genes involved in testosterone biosynthesis; *SHBG* as a gene involved in testosterone transport and *HSD3a* and *HSD3b* as genes involved in androgen degradation in the prostate), for only three genes has there been any direct investigation of a particular marker in relation to prostate cancer risk (the androgen receptor (*AR*) gene, the steroid 5-alpha reductase type II (*SRD5A2*) gene and the *CYP3A4* genes) and confirmatory findings have been obtained for only two of these markers, the CAG trinucleotide repeat polymorphic marker in the androgen receptor gene and the A49T polymorphic marker in the *SRD5A2* gene (Ross *et al.*, 1998).

The majority of testosterone biosynthesis in males occurs in the testis under regulation of luteinizing hormone stimulation (Coffey, 1979). Testosterone is transported to target cells in the circulation either as its free form, weakly bound to albumin or more tightly bound to sex-hormone-binding globulin, the latter thought to be non-bioavailable to target tissues. Free testosterone diffuses freely into prostate cells, where it is irreversibly converted to its reduced, more bioactive form dihydrotestosterone. Dihydrotestosterone, and also testosterone with lower binding affinity, bind to the androgen receptor and this complex of ligand and receptor translocates to the nucleus for DNA binding and transactivation of genes with androgen response elements in their promoter regions (Ross *et al.*, 1998). Most of these "downstream" genes transactivated by the androgen receptor have not yet been characterized, but they are thought to include the major genes regulating cell division.

Cell division is thought to be a prerequisite for the development of much if not most human cancer, as cell division is thought to be necessary for cells to accumulate the genetic changes required for transformation to a malignant phenotype (Preston-Martin *et al.*, 1990). As cell division is largely controlled by androgen activity, any gene in the transactivation pathway for androgens becomes a legitimate candidate gene in relation to prostate carcinogenesis. However, there are several additional lines of evidence that androgens and, hence, the genes which regulate androgen activity, are involved in prostate carcinogenesis. These have been previously reviewed (Ross *et al.*, 1998), but include the importance of androgen in induction or progression of prostate cancer in the few experimental models that mimic the human disease (Noble, 1977), the apparent absence of prostate cancer in men with constitutional underdeveloped prostate glands due to androgen deficiency (Ross *et al.*, 1998), the importance of androgen deprivation as an effective initial therapy in men with early advanced prostate cancer (Ross & Schottenfeld, 1996), the predictive value of circulating testosterone levels for subsequent prostate cancer development (Gann *et al.*, 1996a), and the differences in the hormonal environment in men of different racial/ethnic backgrounds with markedly different patterns of prostate cancer incidence (Ross *et al.*, 1986, 1992).

The androgen receptor (AR) is a transcription factor encoded by the *AR* gene on the X chromosome (Coetzee & Ross, 1994). The *AR* gene encodes three distinct regions of the AR molecule: a hormone (androgen)-binding domain, a DNA-binding domain and a transcription modulatory domain. The latter is completely encoded by a very large exon 1 which contains two well characterized polymorphic trinucleotide repeat sequences. The length of one of these sequences, a CAG repeat encoding a polyglutamine tract, was hypothesized by Coetzee and Ross (1994) to be related to androgen transactivation activity and to prostate cancer risk. The hypothesis was based initially on the observations that an expansion of this repeat is the cause of an X-linked adult-onset motor neuron disease, spinal and bulbar muscular atrophy or Kennedy's disease (La Spada *et al.*, 1991) and that men with this disorder transactivate androgens suboptimally and have evidence of hypoandrogenicity (Arbizu *et al.*, 1983). The hypothesis stated that men with longer CAG repeats within the normal range (9–33) will have progressively lower androgen transactivation activity (despite normal DNA-binding by the AR) and correspondingly lower prostate cancer risk (Coetzee & Ross, 1994). This hypothesis has received some support from *in vitro* studies demonstrating that there is a linear inverse relationship between CAG repeat length and transactivation activity as measured by reporter genes in transfection assays (Chamberlain *et al.*, 1994). Indirect support has also come from observations that average CAG length varies by race/ethnicity, with African Americans having shorter repeats on average, followed by Caucasians, with Oriental populations having, on average, the longest, as predicted from their respective prostate cancer incidence rates (Coetzee & Ross, 1994). In a population-based case–control study in Los Angeles, white men with less than the average number of CAG repeats in the control population had twice the risk of prostate cancer compared with men having more than the average number (2.5 times the risk of advanced disease) (Ingles *et al.*, 1997). This relationship has been confirmed by other studies (Stanford *et al.*, 1997; Giovannucci *et al.*, 1997).

The other gene in the androgen signalling pathway that has been a subject of fairly detailed study as a candidate gene for prostate cancer is the steroid 5-alpha reductase type II (*SRD5A2*) gene. *SRD5A2* is one of two 5-alpha reductase isozymes, but is the most active in prostate tissue (Thigpen *et al.*, 1992). The *SRD5A2* gene is located on chromosome 2. In addition to silent single nucleotide polymorphisms, seven substitution single nucleotide polymorphisms (SNPs) have been described by Reichardt *et al.* (1995) and a TA repeat polymorphism has been described in the 3´-untranslated region (3´UTR) region of the gene. Although unique allelic variants have been described for both African Americans and Asians for the TA repeat, no clear functional relevance has yet been ascribed to this marker (Ross *et al.*, 1995).

For the seven substitution polymorphisms, Makridakis *et al.* (1999) have conducted transfection assays in which they compared the pharmacokinetic properties of the mutant enzymes *in vitro* with those of the wild-type enzyme. These assays suggested that some of these substitution changes represented true polymorphisms, in that despite an amino acid change, the mutant enzyme kinetic properties were identical to those of the wild-type enzyme. However, others resulted in substantial increases in enzyme activity (in particular an alanine to threonine substitution at codon 49 (A49T)), while others resulted in decreased enzyme activity (e.g., a valine to leucine substitution at codon 89 (V89L)). The results in this artificial system have been validated by the demonstration that the A49T mutation, despite being quite uncommon in the general population, with a variant allele frequency of < 1.0%, was strongly associated with prostate cancer, especially advanced disease, in two populations in Los Angeles, Latinos (RR for advanced disease in men with at least one T allele = 3.6, $p = 0.04$) and African Americans (RR for advanced disease in men with at least one T allele = 7.1, $p = 0.001$) (Makridakis *et al.*, 1999). Although the V89L allele has not yet been reported to be inversely associated with prostate cancer risk, as predicted by the *in vitro* assays, it has been demonstrated to be correlated with low circulating androstanediol glucuronide levels, an index of whole-body 5-alpha reductase activity (Reichardt *et al.*, 1995).

A third gene in the androgen signalling pathway that has undergone preliminary epidemiological evaluation is the *CYP3A4* gene, whose product is involved in oxidation of testosterone in the prostate to a series of biologically inert metabolites. Rebbeck *et al.* (1998) compared the allele distribution of an SNP in a series of high-grade advanced versus low-grade localized prostate cancer. Among men in the advanced prostate cancer group, 46% carried a variant allele compared with only 5% of men in the low-grade/early-stage group (OR = 9.5, $p < 0.001$).

Vitamin D signalling pathways

Vitamin D (or, more accurately, 1,25-dihydroxyvitamin D, the bioactive vitamin D metabolite) is a potent antiproliferative agent in the prostate as well as a prodifferentiation agent for prostate cells *in vitro* (Peehl *et al.*, 1994). As prostate cells themselves metabolize vitamin D precursor compounds to 1,25-dihydroxyvitamin D, vitamin D stimulation of the prostate is under both local and systemic control. There are some, although not totally consistent, epidemiological data indicating that circulating levels of 1,25-dihydroxyvitamin D are inversely associated with prostate cancer development (Corder *et al.*, 1995; Gann *et al.*, 1996b). Vitamin D has inhibitory effects on prostate growth in experimental models, further supporting a possible chemopreventive role in prostate cancer. Vitamin D signalling is mediated by the vitamin D receptor. A number of polymorphisms in the vitamin D receptor (*VDR*) gene with common allele variants have been identified (Morrison *et al.*, 1992). Several of these have been shown to have biological correlates in relation to bone mineral metabolism, in which vitamin D plays an important role, and have been extensively studied in relation to fractures or other health outcomes caused by altered bone mineral metabolism. A few of these polymorphic markers have also been studied in relation to prostate cancer risk. Ingles *et al.* (1997) reported that a polyA microsatellite with a bimodal polymorphic distribution was strongly related to prostate cancer risk in whites; men with at least one 'long' A allele of this marker, which is located in the 3′UTR of the gene, had a 4.6-fold higher prostate cancer risk than men homozygous for 'short' polyA alleles. Similar results have been obtained for other markers in linkage disequilibrium with the polyA microsatellite in whites (Taylor *et al.*, 1996). Studies in African Americans have also supported the notion that polymorphic markers in and around the 3′UTR of the *VDR* gene are associated with increased risk of prostate cancer (Ingles *et al.*, 1998).

Strategies to evaluate chemopreventive agents for prostate cancer in human populations

There are three strategies currently in use in human populations to evaluate chemopreventive efficacy either against prostate cancer *per se* or in the context of biomarkers of prostate cancer risk or histological precursors of prostate cancer. Each of these strategies has its own particular strengths and limitations.

One strategy is to give a potential chemopreventive agent to a patient with biopsy-proven prostate cancer who is awaiting definitive prostate surgery. This strategy has the advantage of pre- and post-intervention tissue availability and large quantities of tissue to evaluate post-intervention. As the individual undergoing therapy has prostate cancer, there is a low likelihood of causing harm through administration of the agent. Disadvantages of this strategy include the short duration of the intervention (usually a matter of weeks), our current general lack of knowledge regarding changes in biomarkers suggesting possible efficacy, and concern that the response of patients with prostate cancer to a chemopreventive intervention might be different to that of healthy individuals. [*This strategy would also fall outside the definition of chemoprevention adopted by the participants in the workshop – Ed.*]

The second strategy takes advantage of the fact that men with elevated prostate-specific antigen

(PSA) who are sextant biopsy-negative for prostate cancer require a second biopsy as part of routine clinical management, typically one year later. This creates a one-year window for evaluating a chemopreventive intervention with small amounts of tissue available pre- and post-treatment. This strategy has the advantage of allowing a longer-term evaluation of the agent but, like the pre-prostatectomy approach, it is unclear what biomarkers should be evaluated to provide evidence of efficacy as a chemopreventive agent. As many of these men have high-grade prostatic intraepithelial neoplasia (PIN) lesions, a probable histological precursor of prostate cancer, alteration in these lesions has been suggested as a possible target of efficacy. Availability of tissue after the intervention as part of the routine clinical management of these patients is an enormous advantage of this approach. This second strategy has been tested in a small randomized study of the 5-alpha reductase inhibitor finasteride (Cote *et al.*, 1998).

The final strategy is a full-fledged prevention trial in healthy individuals. This is a long-term strategy and involves large numbers of participants. The duration and size of such trials yield advantages of contributing to our understanding of the efficacy of an intervention over a long period and permitting evaluation of interesting subgroups. On the other hand, these same factors contribute to the enormous cost of such studies and to their logistic complexity. Unlike the other two strategies, prostate cancer is, by definition, the outcome being assessed. However, to assess the development of histological, as opposed to clinical prostate cancer, using this design requires an invasive procedure (i.e., a prostate biopsy) outside the course of routine clinical care.

Implications of prostate cancer molecular genetics for chemoprevention trials

While chemoprevention trials are typically designed to have adequate statistical power to measure the overall impact of an intervention against a placebo or non-intervention group (or one intervention against another), one should not expect homogeneity in response. A strategy to help identify those individuals who will derive the greatest (or least) benefit from the intervention is highly desirable. A full understanding of etiological pathways at the molecular level would undoubtedly help achieve these goals. Recent work on the *SRD5A2* gene illustrates the potential importance of such an understanding in a chemoprevention setting. Studies of polymorphic variants of the *SRD5A2* gene *in vitro* have not only demonstrated huge variability in the pharmacogenetic properties of the mutant enzymes versus the wild type (as much as 150-fold variability for some parameters), but also substantially variability in response to the 5-alpha reductase inhibitor finasteride (Makridakis & Reichardt, 2000). Thus, in the current national chemoprevention trial of finasteride intervention in the United States, it is highly probable that any chemopreventive efficacy, or lack of it, will be modified by underlying genetic susceptibility related to the *SRD5A2* gene. Determining *SRD5A2* genotype should become an important component of that large study.

Another potential value of including candidate gene studies in chemoprevention trials is the possibility of identifying appropriate target groups for the intervention. It is generally considered that, although polymorphic variants of candidate genes which alter cancer risk may be relatively common in the population, these variants have only a modest influence on risk, but there are notable exceptions (Makridakis *et al.*, 1999). Full elucidation of molecular etiological pathways, as is being achieved in the area of androgen signalling in the prostate, can potentially allow the development of a polygenic etiological model of the disease (Ross *et al.*, 1998). Such a model, in turn, can allow identification of groups of individuals at very high risk of prostate cancer development by virtue of multiple high-risk allelic variants in candidate genes involved in these pathways. These individuals can then become targets for chemoprevention interventions, hopefully increasing the efficiency in terms of number of patients if not time for the completion of such a study.

Finally, understanding the relationship between candidate genes and disease risk opens up the possibility that a particular gene or its product may itself become a target for chemopreventive interventions. As understanding these relationships may also allow a fuller understanding of complex molecular etiological pathways, this approach can also potentially contribute to the development of new chemopreventive agents (Ross *et al.*, 1998).

References

Arbizu, T., Santamaria, J., Gomex, J.M., Quilez, A. & Serra, J. P. (1983) A family with adult spinal and bulbar muscular atrophy X-linked inheritance and associated with testicular failure. *J. Neurol. Sci.*, **59**, 371–382

Carter, B.S., Beaty, T.H., Steinberg, G.D., Childs, B. & Walsh, P.C. (1992) Mendelian inheritance of familial prostate cancer. *Proc. Natl. Acad. Sci. USA*, **89**, 3367–3371

Chamberlain, N.L., Driver, E.D. & Miesfeld, R.L.(1994) The length and location of CAG trinucleotide repeats in the androgen receptor N-terminal domain affect transactivation function. *Nucl. Acids Res.*, **22**, 3181–3186

Chan, J.M., Stampfer, M.J., Giovannucci, E., Gann, P.H., Ma, J., Wilkinson, P., Hennekens, C.H. & Pollak, M. (1998) Plasma insulin-like growth factor-I and prostate cancer risk: a prospective study. *Science*, **279**, 563–566

Clark, L.C., Dalkin, B., Krongrad, A., Combs, G.F.J., Turnbull, B.W., Slate, E.H., Witherington, R., Herlong, J.H., Janosko, E., Carpenter, D., Borosso, C., Falk, S. & Rounder, J. (1998) Decreased incidence of prostate cancer with selenium supplementation: results of a double-blind cancer prevention trial. *Br. J. Urol.*, **81**, 730–734

Coetzee, G.A. & Ross, R.K. (1994) Prostate cancer and the androgen receptor. *J. Natl Cancer Inst.*, **86**, 872–873

Coffey, D.S. (1979) Physiological control of prostatic growth: an overview. In: Coffey, D.S. & Isaacs, J.T., eds, *Prostate Cancer* (UICC Technical Report Series Vol. 48), Geneva, International Union Against Cancer, pp. 4–23

Corder, E.H., Friedman, G.D., Vogelman, J.H. & Orentreich, N. (1995) Seasonal variation in vitamin D, vitamin D-binding protein, and dehydroepiandrosterone: risk of prostate cancer in black and white men. *Cancer Epidemiol. Biomarkers Prev.*, **4**, 655–659

Cote, R.J., Skinner, E.C., Salem, C.E., Mertes, S.J., Stanczyk, F.Z., Henderson, B.E., Pike, M.C. & Ross, R.K. (1998) The effect of finasteride on the prostate gland in men with elevated serum prostate-specific antigen level. *Br. J. Cancer*, **78**, 413–418

Gann, P.H., Hennekens, C.H., Ma, J., Longcope, C. & Stampfer, M.J. (1996a) Prospective study of sex hormone levels and risk of prostate cancer. *J. Natl Cancer Inst.*, **88**, 1118–1126

Gann, P.H., Ma, J., Hennekens, C.H., Hollis, B.W., Haddad, J.G. & Stampfer, M.J. (1996b) Circulating vitamin D metabolites in relation to subsequent development of prostate cancer. *Cancer Epidemiol. Biomarkers Prev.*, **5**, 121–126

Giovannucci, E., Ascherio, A., Rimm, E.B., Stampfer, M.J., Colditz, G.A. & Willett, W.C. (1995) Intake of carotenoids and retinol in relation to risk of prostate cancer. *J. Natl Cancer Inst.*, **87**, 1767–1776

Giovannucci, E., Stampfer, M.J., Krithivas, K., Brown, M., Brufsky, A., Talcott, J., Hennekens, C.H. & Kantoff, P.W. (1997) The CAG repeat within the androgen receptor gene and its relationship to prostate cancer. *Proc. Natl Acad. Sci. USA*, **94**, 3320–3323

Heinonen, O.P., Albanes, D., Virtamo, J., Taylor, P.R., Huttunen, J.K., Hartman, A.M., Haapakoski, J., Malila, N., Rautalahti, M., Ripatti, S., Mäenpää, H., Teerenhovi, L., Koss, L., Virolainen, M. & Edwards, B.K. (1998) Prostate cancer and supplementation with alpha-tocopherol and beta-carotene: incidence and mortality in a controlled trial. *J. Natl Cancer Inst.*, **90**, 440–446

Ingles, S.A., Ross, R.K., Yu, M.C., Irvine, R.A., La Pera, G., Haile, R.W. & Coetzee, G.A. (1997) Association of prostate cancer risk with genetic polymorphisms in vitamin D receptor and androgen receptor. *J. Natl Cancer Inst.*, **89**, 166–170

Ingles, S.A., Coetzee, G.A., Ross, R.K., Henderson, B.E., Kolonel, L.N., Crocitto, L., Wang, W. & Haile, R.W. (1998) Association of prostate cancer with vitamin D receptor haplotypes in African-Americans. *Cancer Res.*, **58**, 1620–1623

La Spada, A.R., Wilson, E.M., Lubahn, D.B., Harding, A.E. & Fischback, K.H. (1991) Androgen receptor gene mutations in X-linked spinal and bulbar muscular atrophy. *Nature*, **352**, 77–79

Makridakis, N.M. & Reichardt, J.K. (2000) Biochemical and pharmacogenetic dissection of human steroid 5α-reductase type II. *Pharmacogenetics* (in press)

Makridakis, N.M., Ross, R.K., Pike, M.C., Crocitto, L.E., Kolonel, L.N., Pearce, C.L., Henderson, B.E. & Reichardt, J.K. (1999) Association of mis-sense substitution in SRD5A2 gene with prostate cancer in African-American and Hispanic men in Los Angeles, USA. *Lancet*, **354**, 975–978

Monroe, K.R., Yu, M.C., Kolonel, L.N., Coetzee, G.A., Wilkens, L.R., Ross, R.K. & Henderson, B.E. (1995) Evidence of an X-linked genetic component to prostate cancer risk. *Nature Med.*, **1**, 827–829

Morrison, N.A., Yeoman, R., Kelly, P.J. & Eisman, J.A. (1992) Contribution of trans-acting factor alleles to normal physiological variability: vitamin D receptor gene polymorphism and circulating osteocalcin. *Proc. Natl Acad. Sci. USA*, **89**, 6665–6669

Noble, R.L. (1977) The development of prostatic adenocarcinoma in Nb rats following prolonged sex hormone administration. *Cancer Res.*, **37**, 1929–1933

Peehl, D.M., Skowronski, R.J., Leung, G.K., Wong, S.T., Stamey, T.A. & Feldman, D. (1994) Antiproliferative effects of 1,25-dihydroxyvitamin D3 on primary cultures of human prostatic cells. *Cancer Res.*, **54**, 805–810

Preston-Martin, S., Pike, M.C., Ross, R.K., Jones, P.A. & Henderson, B.E. (1990) Increased cell division as a cause of human cancer. *Cancer Res.*, **50**, 7415–7421

Rebbeck, T.R., Jaffe, J.M., Walker, A.H., Wein, A.J. & Malkowicz, S.B. (1998) Modification of clinical presentation of prostate tumors by a novel genetic variant in CYP3A4. *J. Natl Cancer Inst.*, **90**, 1225–1229

Reichardt, J.K.V., Makridakis, N., Henderson, B.E., Yu, M.C., Pike, M.C. & Ross, R.K. (1995) Genetic variability of the human SRD5A2 gene: implications for prostate cancer risk. *Cancer Res.*, **55**, 3973–3975

Rosen, C.J., Kurland, E.S., Vereault, D., Adler, R.A., Rackoff, P.J., Craig, W.Y., Witte, S., Rogers, J. & Bilezikian, J.P. (1998) Association between serum insulin growth factor-I (IGF-I) and a simple sequence repeat in IGF-I gene: implications for genetic studies of bone mineral density. *J. Clin. Endocrinol. Metab.*, **83**, 2286–2290

Ross, R.K. & Schottenfeld, D. (1996) Prostate cancer. In: Schottenfeld, D. & Fraumeni, J.F., eds, *Cancer Epidemiology and Prevention*, 2nd Edition, New York, Oxford University Press, pp. 1180–1206

Ross, R.K., Bernstein, L., Judd, H., Hanisch, R., Pike, M.C. & Henderson, B.E. (1986) Serum testosterone levels in healthy young black and white men. *J. Natl Cancer Inst.*, **76**, 45–48

Ross, R.K., Bernstein, L., Lobo, R.A., Shimizu, H., Stanczyk, F.Z., Pike, M.C. & Henderson, B.E. (1992) 5-Alpha reductase activity and risk of prostate cancer among Japanese and US White and Black males. *Lancet*, **339**, 887–889

Ross, R.K., Coetzee, G.A., Reichardt, J., Skinner, E. & Henderson, B.E. (1995) Does the racial-ethnic variation in prostate cancer risk have a hormonal basis? *Cancer*, **75**, 1778–1782

Ross, R.K., Pike, M.C., Coetzee, G.A., Reichardt, J.K.V., Yu, M.C., Feigelson, H., Stanczyk, F.Z., Kolonel, L.N. & Henderson, B.E. (1998) Androgen metabolism and prostate cancer: establishing a model of genetic susceptibility. *Cancer Res.*, **58**, 4497–4504

Ross, R.K., Reichardt, J.K., Ingles, S.A. & Coetzee, G.A. (2000) The genetic epidemiology of prostate cancer: closing in on a complex disease. In: Chung, L.W., Isaacs, W.B. & Simons, J.W., eds, *Prostate Cancer in the 21st Century,* Totowa, NJ, Humana Press (in press)

Smith, J.R., Freije, D., Carpten, J.D., Gronberg, H., Xu, J., Isaacs, S.D., Brownstein, M.J., Bova, G.S., Guo, H., Bujnovszky, P., Nusskern, D.R., Damber, J.E., Bergh, A., Emanuelsson, M., Kallioniemi, O.P., Walker-Daniels, J., Bailey-Wilson, J.E., Beaty, T.H., Meyers, D.A., Walsh, P.C., Collins, F.S., Trent, J.M. & Isaacs, W.B. (1996) Major susceptibility locus for prostate cancer on chromosome 1 suggested by a genome-wide search. *Science,* **274**, 1371–1374

Stanford, J.L., Just, J.J., Gibbs, M., Wicklund, K.G., Neal, C.L., Blumenstein, B.A. & Ostrander, E.A. (1997) Polymorphic repeats in the androgen receptor gene: molecular markers of prostate cancer risk. *Cancer Res.*, **57**, 1194–1198

Taylor, J.A., Hirvonen, A., Watson, M., Pittman, G., Mohler, J.L. & Bell, D.A. (1996) Association of prostate cancer with vitamin D receptor gene polymorphism. *Cancer Res.*, **56**, 4108–4110

Thigpen, A.E., Davis, D.L., Gautier, T., Imperato-McGinley, J. & Russell, W. (1992) The molecular basis of steroid 5-alpha-reductase deficiency in a large Dominican kindren. *New Engl. J. Med.*, **327**, 1216–1219

Xu, J., Meyers, D., Freije, D., Isaacs, S., Wiley, K., Nusskern, D., Ewing, C., Wilkens, E., Bujnovszky, P., Bova, G.S., Walsh, P., Isaacs, W., Schleutker, J., Matikainen, M., Tammela, T., Visakorpi, T., Kallioniemi, O.P., Berry, R., Schaid, D., French, A., McDonnell, S., Schroeder, J., Blute, M., Thibodeau, S. & Trent, J. (1998) Evidence for a prostate cancer susceptibility locus on the X chromosome. *Nature Genet.*, **20**, 175–179

Yoshizawa, K., Willett, W.C., Morris, S.J., Stampfer, M.J., Spiegelman, D., Rimm, E.B. & Giovannucci, E. (1998) Study of prediagnostic selenium level in toenails and the risk of advanced prostate cancer. *J. Natl Cancer Inst.*, **90**, 1219–1224

Corresponding author:

R.K. Ross
Flora Thornton Chairman,
Preventive Medicine,
Catherine & Josephine Aresty Professor of
Preventive Medicine and Urology,
1441 Eastlake Avenue Rm 8302B,
Los Angeles, CA 90089-9181,
USA

Biomarkers in Cancer Chemoprevention
Miller, A.B., Bartsch, H., Boffetta, P., Dragsted, L. and Vainio, H. eds
IARC Scientific Publications No. 154
International Agency for Research on Cancer, Lyon, 2001

Exposure biomarkers in chemoprevention studies of liver cancer

C.P. Wild and P.C. Turner

Hepatocellular carcinoma (HCC) is the most common type of liver cancer, the major risk factors being hepatitis B and C viruses and aflatoxins; other factors such as alcohol are also of importance in some populations. Aflatoxin exposure biomarkers include urinary aflatoxin metabolites and aflatoxin–albumin adducts in peripheral blood. These biomarkers are well validated and have been applied in studies of many populations worldwide. They are proving to be valuable end-points in intervention studies, including chemoprevention studies. The biomarkers permit assessment of primary prevention measures to reduce aflatoxin intake. In addition, the determination of individual urinary aflatoxin metabolite profiles means that the effectiveness of chemopreventive agents designed to modulate aflatoxin metabolism can also be evaluated. Both aflatoxin–albumin adducts and urinary aflatoxin metabolites have been associated with increased HCC risk in prospective studies, indicating the predictive value of these biomarkers at the group level. However, given the multifactorial and multistep nature of HCC, it is unlikely that these exposure biomarkers will be predictive at the individual level or be of value as surrogate end-points in longer-term intervention trials aimed at reducing disease incidence. Aflatoxin-related mutations at codon 249 of the *p53* gene in plasma may be more relevant in this regard but their application requires further understanding of the temporal appearance of this biomarker in relation to the natural history of the disease.

Introduction

There are estimated to be half a million new cases of liver cancer each year worldwide. The major risk factors for hepatocellular carcinoma (HCC) are hepatitis B and C viruses (HBV and HCV), aflatoxins and alcohol. In developing countries, HBV and HCV are estimated to be associated with 67% and 24% of cases respectively, while the corresponding figures for developed countries are 29 and 22% (Pisani *et al.*, 1997). Aflatoxins contribute significantly to HCC incidence in regions of high HBV prevalence, where the two factors appear to act synergistically (Wild & Hall, 1999). Tobacco, oral contraceptives and schistosome infection also contribute to HCC incidence in some parts of the world, although they are quantitatively of less importance (Stuver, 1998). In the other major form of liver cancer, cholangiocarcinoma, the major risk factor identified is infection with liver flukes (e.g., *Opisthorchis viverrini*) (Parkin *et al.*, 1991), although *N*-nitrosamines may also play a role (Srivatanakul *et al.*, 1991). Occupational exposure to vinyl chloride is associated with development of another type of liver cancer, angiosarcoma (IARC, 1987). This chapter deals briefly with exposure biomarkers to the risk factors for HCC, but focuses primarily on aflatoxins, for which most experience with chemoprevention has been accrued.

Hepatitis viruses

Exposure to hepatitis B and C viruses can be accurately assessed by measuring either viral proteins or circulating antibodies to those proteins in the peripheral blood. Resolved infections and persistent active infections can also be differentiated (IARC, 1994). For example, persistent hepatitis B surface antigen (HBsAg) in the blood indicates chronic carriage and the presence of the 'e' antigen and HBV DNA indicates active viral replication. Positivity for antibody to HBsAg (anti-HBs) indicates resolved infection and immunity. Presence of antibody to hepatitis B core antigen (anti-HBc) indicates exposure to the virus, with acute, chronic or resolved infection depending on the presence of other viral markers. In cross-sectional studies in areas of endemic HBV infection, such as Guinea-Conakry in west Africa, >80% of individuals are anti-HBs-positive and 15–20% HBsAg-positive (Sylla *et al.*, 1999). The availability of these biomarkers of hepatitis exposure has permitted the strong and specific association between HBV exposure and HCC to be established in both prospective and case–control studies (IARC, 1994). In addition, the markers can be used in evaluation of the effectiveness of HBV vaccination programmes in the prevention of both chronic carriage and HCC.

Alcohol

A number of biomarkers of exposure to alcohol consumption are being developed. One of the most promising is the detection of acetaldehyde bound to peripheral blood proteins (Conduah Birt *et al.*, 1998). Alcohol is metabolized to acetaldehyde by alcohol dehydrogenase and can be further metabolized to acetate by acetaldehyde dehydrogenase. Acetaldehyde is genotoxic and may be important in the carcinogenic action of alcohol (IARC, 1988). This is supported by recent observations that risk of alcohol-related cancer of the oral cavity and pharynx is increased in individuals with alcohol dehydrogenase genotypes associated with fast metabolism of alcohol to acetaldehyde; the risk is limited to heavy drinkers (Harty *et al.*, 1997). Consequently a biomarker of acetaldehyde–DNA or –protein adducts would be valuable. Conduah Birt *et al.* (1998) have described the covalent binding of acetaldehyde to globin and shown that the level of this biomarker is elevated in heavy drinkers. This biomarker could potentially be used to assess the impact of chemoprevention or lifestyle changes on alcohol-related cellular damage.

Aflatoxins

Measuring individual exposure to aflatoxins by either dietary questionnaire or food analysis is problematic. First, aflatoxins can contaminate a variety of cereals and oilseeds, so that measured intakes of one or two specific food commodities may not be useful as a surrogate for aflatoxin exposure, particularly in countries with a varied diet. Further, in countries with limited variety in the staple diet, the similarity of diet between individuals makes questionnaire-based assessments uninformative. Second, aflatoxin contamination is heterogeneous in nature, affecting perhaps one groundnut or maize kernel in a given batch. This hampers food analysis because it makes representative sampling extremely difficult to achieve. In the light of these difficulties, biomarkers of aflatoxin exposure have been developed and subsequently widely applied in studies of exposed human populations (Montesano *et al.*, 1997). In many ways, the aflatoxin biomarkers have become a paradigm for other chemical carcinogens and have been employed in chemoprevention studies to a degree not yet achieved in other cases (Groopman & Kensler, 1999).

The exposure biomarkers for aflatoxins have been developed on the basis of an understanding of their mechanisms of action, in particular their activation and detoxification by phase I and II enzymes (Guengerich *et al.*, 1998). The principal biomarkers of exposure include urinary metabolites and adducts, specifically albumin adducts in blood and DNA adducts in urine. These are considered briefly below; their application to chemoprevention studies is then discussed in the light of the properties of these biomarkers.

Urinary aflatoxin biomarkers

The hydroxylated aflatoxin metabolite, aflatoxin M_1 (AFM_1), was detected in human urine samples 30 years ago (Campbell *et al.*, 1970). Significant advances in analytical sensitivity and specificity were achieved with the application of immunoaffinity columns to purify aflatoxins from urine samples (Groopman *et al.*, 1985; Wild *et al.*, 1986b). This approach permitted a number of aflatoxins to be detected in a single urine sample (Groopman *et al.*, 1985); these included metabolites considered to be detoxification products such as AFM_1, but also a nucleic acid adduct of aflatoxin B_1 (AFB_1), AFB_1-N7-guanine. More recently, the AFB_1–mercapturic acid conjugate has been detected in human urine (Wang *et al.*, 1999).

Urinary levels of aflatoxin metabolites have been shown to correlate with measurements of dietary intake at the individual level (Zhu *et al.*, 1987; Groopman *et al.*, 1992). While this is true for a number of metabolites, it is not the case for all, aflatoxin P_1 being one exception. This is probably because this metabolite can be excreted in both the bile and urine and hence measurements in only one compartment can be misleading (Groopman *et al.*, 1993). However, excellent correlations between dietary intake, AFB_1-N7-guanine and total urinary aflatoxin levels have been found, particularly when exposure and urinary levels are integrated over a number of days by collecting consecutive 24-hour urine samples (Groopman *et al.*, 1992). This latter approach overcomes problems due to the rapid excretion of aflatoxins in the urine, which occurs over some 24–48 hours following ingestion and can lead to rapidly fluctuating levels. For example, in a study in The Gambia, 24-hour urine samples were collected on four consecutive days from 20 individuals and the day-to-day fluctuation in urinary levels was marked, exceeding two orders of magnitude in some cases (Groopman *et al.*, 1992; Hall & Wild, 1994). This could clearly lead to misclassification of exposure status if only a single day's level were taken into account. However, when the mean daily urinary aflatoxin levels over four days were compared with the mean daily aflatoxin intake for each individual over one week, excellent correlations were observed for both total urinary aflatoxins and AFB_1-N7-guanine (Groopman *et al.*, 1992).

In a prospective cohort study in Shanghai, urinary levels of aflatoxin metabolites were positively associated with HCC risk and a synergistic interaction between HBV and aflatoxins was observed (Qian *et al.*, 1994). The association was strongest with AFB_1-N7-guanine adduct, but was also positive with AFM_1. Similar data have been obtained from a prospective cohort study in Taiwan using urinary aflatoxins and aflatoxin–albumin adducts as exposure markers (Wang *et al.*, 1996). To date, the biomarker data in these studies have been used as a categorical rather than continuous variable, with simple dichotomization into positive or negative or high and low levels. Thus no quantitative dose–response data are yet available.

Aflatoxin–albumin adducts

Aflatoxin–albumin adducts are detected in peripheral blood of both animals and humans exposed to aflatoxin. As with DNA adducts, there is a linear dose–response relationship following aflatoxin exposure in rats (Wild *et al.*, 1986a, 1996). Furthermore, with multiple exposures, adducts accumulate and reach a plateau; in different strains of rats, mice, hamsters and guinea-pigs, the albumin adducts are correlated with hepatic DNA adduct levels and reflect qualitatively the susceptibility of these four species to aflatoxin toxicity and carcinogenicity (Wild *et al.*, 1986a; 1996). In rats, the aflatoxin–albumin conjugate appears to turn over with the half-life of albumin itself (Sabbioni *et al.*, 1987; Wild *et al.*, 1986a). This has not been formally demonstrated in humans, but assuming it to be the case, a single measurement should reflect an integration of exposure over the previous two to three months. Certainly, the albumin adduct measurement provides a more stable biomarker than urinary metabolites (Hall & Wild, 1994). Aflatoxin–albumin adducts were associated with increased HCC risk in HBV carriers in the Taiwan cohort mentioned above (Wang *et al.*, 1996).

Aflatoxin–albumin adducts have proved to be an excellent marker for assessing exposure in many countries worldwide (Montesano *et al.*, 1997). We detected the highest levels and prevalence of exposure in west Africa (The Gambia, Guinea, Senegal, Burkina Faso) where groundnuts, a frequent source of aflatoxin exposure, are consumed as a dietary staple. High levels are also observed in maize-consuming countries such as Kenya and parts of southern China. Lower levels of exposure occur where the diet is more varied (e.g., Thailand, Nepal). In contrast, in Europe, the United States and Canada, almost no human sera have been found to be aflatoxin-positive in our studies (Montesano *et al.*, 1997; Wild *et al.*, unpublished data).

If adducts are to be used as outcome markers in intervention studies, it is important to understand the parameters which can influence their level. In a study of environmental and genetic determinants of aflatoxin–albumin adduct levels in The Gambia, we observed that season and geography were major factors associated with the levels of adducts measured (Wild *et al.*, 2000). The seasonal variations were consistent with previous observations (Wild *et al.*, 1990) but were probably further complicated by annual variations in toxin levels dependent on climatic conditions. In contrast, genetic polymorphisms in aflatoxin-metabolizing enzymes or phenotypic expression of cytochrome P450 3A4 (measured by urinary cortisol:6-hydroxy-cortisol ratio) were not strong determinants of adduct levels (Wild *et al.*, 2000). In adults, HBV status did not appear to be associated with higher adduct levels, although there was a suggestion of this in children, particularly at the time of acute infection (Allen *et al.*, 1992; Wild *et al.*, 1993; P.C. Turner, M. Mendy, A.J. Hall, M. Fortuin, H. Whittle & C.P. Wild, unpublished data).

p53 mutations

The AFB1-N7-guanine adduct can give rise to G to T transversion mutations (Bailey *et al.*, 1996). In HCC from regions of high aflatoxin exposure, *p53* mutations are common and specifically a high prevalence of an AGG (Arg) to AGT (Ser) mutation at codon 249 has been detected (Montesano *et al.*, 1997). This mutation is also seen in non-tumour liver tissue from regions of the world where aflatoxin exposure is elevated (Aguilar *et al.*, 1994), but is extremely rare in HCC from countries where aflatoxin exposure is low, even though HBV infection is present in association with many of the tumours. Harris has concluded that the weight of evidence, according to the Bradford Hill criteria, supports the causality of the relationship between aflatoxin exposure and the codon 249 mutation (Hussain & Harris, 1998). Kirk *et al.* (2000) have reported the presence of this same mutation in the plasma of HCC cases in The Gambia. The prevalence in cases (36%) was significantly higher than in cirrhotics (15%) and controls without liver disease (6%). Furthermore, the mutation was not detected in plasma from HCC cases from France. It remains to be determined whether the codon 249 mutations in plasma DNA observed in The Gambia are related specifically to the presence of HCC, in which case the cirrhotics and controls who were positive for the biomarker may have occult cancer, or whether the mutation also reflects heavy aflatoxin exposure with frequent somatic mutation and clonal expansion of individual hepatocytes. A prospective study would provide an answer to this question.

Chemoprevention

HBV vaccination is a priority for reducing the global burden of HCC. Currently, however, only about 1% of African children receive the vaccine. There are 360 million HBV carriers worldwide and the continuing restricted access to the vaccine means that the number of carriers will remain high for at least several decades. Given the high numbers of HBV carriers and the synergistic interaction between aflatoxins and HBV, intervention to reduce aflatoxin exposure is also merited.

Interventions to reduce aflatoxin-related disease involve initiatives at the individual level or community level (Wild & Hall, 2000). The community level approach can involve either pre- or post-harvest measures, while at the individual level the intervention can comprise a change in diet to avoid intake of frequently contaminated foods or chemoprevention to reduce the toxicity of aflatoxins once ingested.

Aflatoxin exposure biomarkers have been developed based on an understanding of the metabolism of these compounds in humans (Figure1 A). A number of these biomarkers have been validated and subsequently widely applied in human populations. Modulation of the level of a biomarker can be used to assess the effectiveness of interventions (Figure 1 B). For example, primary prevention measures aimed at reducing aflatoxin intake should lead to a reduction in all biomarker levels. In addition, the ability to examine individual aflatoxin metabolite profiles means that the effectiveness of chemopreventive agents designed to modulate aflatoxin metabolism can also be evaluated (Figure 1 B).

Most progress has been made using the urinary aflatoxin metabolites and aflatoxin–albumin adducts in relation to chemoprevention with oltipraz in the People's Republic of China, but the latter biomarker is also being applied in a post-harvest intervention in Guinea-Conakry (Sylla *et al.*, 1999). The use of biomarkers in these studies is discussed briefly below.

Animal species show marked differences in sensitivity to aflatoxin–DNA and –protein adduct formation and susceptibility to aflatoxin carcinogenesis (Wild *et al.*, 1996). Induction of glutathione *S*-transferases (GST) and aflatoxin aldehyde reductase decreases aflatoxin–DNA and –protein adduct formation and blocks aflatoxin carcinogenicity in rats (Judah *et al.*, 1993; Groopman & Kensler, 1999). Therefore a similar modulation of the balance between aflatoxin activation and detoxification in humans has been sought. The drug used, oltipraz, is one originally prescribed to treat schistosomiasis.

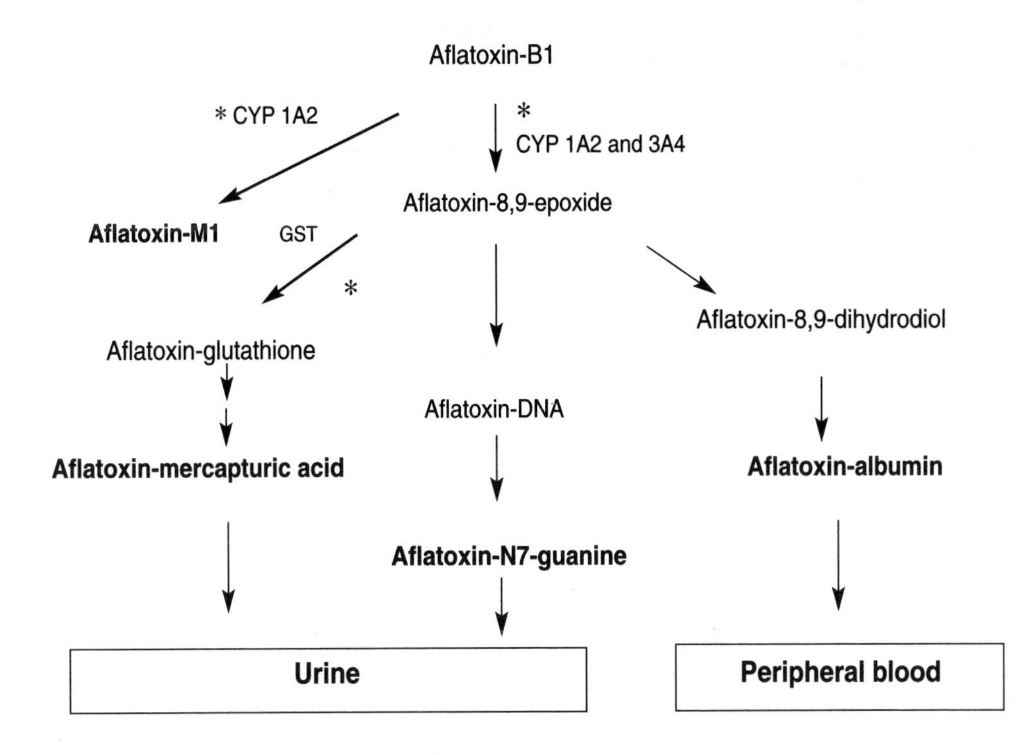

B

Aflatoxin biomarker	Primary intervention:	Secondary intervention:	
	Reduce intake	Oltipraz	Chlorophyllin
Aflatoxin–albumin	**Decrease**	**Decrease**	**Decrease**
Aflatoxin M_1	**Decrease**	**Decrease**	**Decrease**
Aflatoxin-N7-guanine	**Decrease**	**Decrease**	**Decrease**
Aflatoxin–mercapturic acid	**Decrease**	**Increase**	**Decrease**

Figure 1. Aflatoxin metabolism and biomarkers

A: Aflatoxin biomarkers in bold, oltipraz effects indicated by stars. Oltipraz reduces CYP 1A2 and increases glutathione *S*-transferase (GST) activity (Wang *et al.*, 1999) (see text).

B: Effects of various aflatoxin intervention strategies on biomarker levels. Primary intervention to reduce aflatoxin intake, secondary intervention by (*a*) oltipraz to modify aflatoxin metabolism, (*b*) chlorophyllin to reduce gastrointestinal absorption (see text for details).

In China, Kensler and colleagues have demonstrated that oltipraz can modulate aflatoxin metabolism, probably by inhibiting activity of cytochrome P450 (CYP) 1A2, an enzyme which activates AFB_1 to AFB_1-8,9-epoxide and by increasing the level of GST-mediated conjugation of the epoxide to glutathione (Jacobson *et al.*, 1997; Kensler *et al.*, 1998, 1999; Wang *et al.*, 1999; see also Kensler *et al.*, this volume). These effects were demonstrated by assay of urinary AFM_1 (a product of CYP1A2 metabolism of AFB_1), peripheral blood aflatoxin–albumin adducts and the urinary aflatoxin-mercapturic acid conjugate. In a phase IIb clinical trial in Qidong County, People's Republic of China, 250 or 500 mg oltipraz was administered weekly over one year (Kensler *et al.*, 1999). While the phase II trials establish an effect of oltipraz on aflatoxin metabolism, a further trial would be required to evaluate the chemopreventive action of oltipraz against aflatoxin-induced carcinogenesis, with disease incidence as an outcome. Unless aflatoxin exerts a hepatocarcinogenic effect late in the natural history of the disease, a long follow-up would be required to detect the effect of the intervention. Oltipraz has been reported to inhibit HBV replication in cells *in vitro* by induction of a p53-mediated effect on the HBV reverse transcriptase (Chi *et al.*, 1998). If this also occurs *in vivo*, markers of HBV replication could be used to monitor the effect in oltipraz intervention trials.

If any of the above aflatoxin biomarkers are demonstrated to be strong predictors of cancer risk, they could be used as surrogate measures of disease outcome in future studies. It is unlikely that the transient aflatoxin adducts (with DNA or albumin) will fulfil this requirement at the individual level; this is suggested indirectly in rats where a correlation between adducts and liver cancer was seen at the group but not individual level (Kensler *et al.*, 1997). However, the specific *p53* codon 249 mutation related to aflatoxin exposure may be more predictive of individual risk. In this respect, the above-mentioned recent identification of this mutation in the plasma of Gambians with liver cancer or cirrhosis is encouraging (Kirk *et al.*, 2000).

In addition to oltipraz, a number of other chemopreventive agents are being developed with respect to aflatoxin (Kelly *et al.*, 2000). One of these is chlorophyllin, which can inhibit aflatoxin adduct formation and carcinogenicity by non-covalent complex formation between the two compounds (Hayashi *et al.*, 1999). The biomarkers of aflatoxin exposure mentioned above would be equally applicable to evaluating effects of this type of chemopreventive agent.

In summary, both aflatoxin–albumin adducts and urinary aflatoxin metabolites have been associated with increased HCC risk in prospective studies, indicating the predictive value of these biomarkers at the group level. However, given the multifactorial and multistep nature of HCC, it is unlikely that these exposure biomarkers will be predictive for HCC risk at the individual level or be of value as surrogate end-points in longer-term intervention trials aimed at reducing disease incidence. The aflatoxin-related codon 249 *p53* mutations in plasma may be more relevant in this regard, but further understanding of the temporal relationship between the appearance of this biomarker and the natural history of the disease is needed.

Acknowledgement

The authors are grateful for the support from the National Institute of Environmental Health Sciences, USA, Grant No. ES06052.

References

Aguilar, F., Harris, C.C., Sun, T., Hollstein, M. & Cerutti, P. (1994) Geographic variation of p53 mutational profile in nonmalignant human liver. *Science*, **264**, 1317–1319

Allen, S.J., Wild, C.P., Wheeler, J.G., Riley, E.M., Montesano, R., Bennett, S., Whittle, H.C. & Hall, A.J. (1992) Aflatoxin exposure, malaria and hepatitis B infection in rural Gambian children. *Trans. R. Soc. Trop. Med. Hyg.*, **86**, 426–430

Bailey, E.A., Iyer, R.S., Stone, M.P., Harris, T.M. & Essigmann, J.M. (1996) Mutational properties of the primary aflatoxin B1-DNA adduct. *Proc. Natl Acad. Sci. USA*, **93**, 1535–1539

Campbell, T.C., Caedo, J.P., Jr, Bulatao-Jayme, J., Salamat, L. & Engel, R.W. (1970) Aflatoxin M1 in human urine. *Nature*, **227**, 403–404

Chi, W.J., Doong, S.L., Lin-Shiau, S.Y., Boone, C.W., Kelloff, G.J. & Lin, J.K. (1998) Oltipraz, a novel inhibitor of hepatitis B virus transcription through elevation of p53 protein. *Carcinogenesis*, **19**, 2133–2138

Conduah Birt, J.E., Shuker, D.E. & Farmer, P.B. (1998) Stable acetaldehyde–protein adducts as biomarkers of alcohol exposure. *Chem. Res. Toxicol.*, 11, 136–142

Groopman, J.D. & Kensler, T.W. (1999) The light at the end of the tunnel for chemical-specific biomarkers: daylight or headlight? *Carcinogenesis*, **20**, 1–11

Groopman, J.D., Donahue, P.R., Zhu, J.Q., Chen, J.S. & Wogan, G.N. (1985) Aflatoxin metabolism in humans: detection of metabolites and nucleic acid adducts in urine by affinity chromatography. *Proc. Natl Acad. Sci. USA*, **82**, 6492–6496

Groopman, J.D., Hall, A.J., Whittle, H., Hudson, G.J., Wogan, G.N., Montesano, R. & Wild, C.P. (1992) Molecular dosimetry of aflatoxin-N7-guanine in human urine obtained in the Gambia, West Africa. *Cancer Epidemiol. Biomarkers Prev.*, **1**, 221–227

Groopman, J.D., Wild, C.P., Hasler, J., Junshi, C., Wogan, G.N. & Kensler, T.W. (1993) Molecular epidemiology of aflatoxin exposures: validation of aflatoxin-N7-guanine levels in urine as a biomarker in experimental rat models and humans. *Environ. Health Perspect.*, **99**, 107–113

Guengerich, F.P., Johnson, W.W., Shimada, T., Ueng, Y.F., Yamazaki, H. & Langouet, S. (1998) Activation and detoxication of aflatoxin B1. *Mutat. Res.*, **402**, 121–128

Hall, A.J. & Wild, C.P. (1994) Epidemiology of aflatoxin related disease. In: Eaton, D.A. & Groopman, J.D., eds, *The Toxicology of Aflatoxins: Human Health, Veterinary and Agricultural Significance,* New York, Academic Press, pp. 233–258

Harty, L.C., Caporaso, N.E., Hayes, R.B., Winn, D.M., Bravo-Otero, E., Blot, W.J., Kleinman, D.V., Brown, L.M., Armenian, H.K., Fraumeni, J.F., Jr & Shields, P.G. (1997) Alcohol dehydrogenase 3 genotype and risk of oral cavity and pharyngeal cancers. *J. Natl Cancer Inst.*, **89**, 1698–1705

Hayashi, T., Schimerlik, M. & Bailey, G. (1999) Mechanisms of chlorophyllin anticarcinogenesis: dose-responsive inhibition of aflatoxin uptake and biodistribution following oral co-administration in rainbow trout. *Toxicol. Appl. Pharmacol.*, **158**, 132–140

Hussain, S.P. & Harris, C.C. (1998) Molecular epidemiology of human cancer: contribution of mutation spectra studies of tumor suppressor genes. *Cancer Res.*, **58**, 4023–4037

IARC (1987) *IARC Monographs on the Evaluation of Carcinogenic Risks to Humans,* Suppl. 7, *Overall Evaluations of Carcinogenicity: An Updating of IARC Monographs Volumes 1 to 42*, Lyon, IARC

IARC (1988) *IARC Monographs on the Evaluation of Carcinogenic Risks to Humans*, Volume 44, *Alcohol Drinking*, Lyon, IARC

IARC (1994) *IARC Monographs on the Evaluation of Carcinogenic Risks to Humans*, Volume 59, *Hepatitis Viruses*, Lyon, IARC

Jacobson, L.P., Zhang, B.C., Zhu, Y.R., Wang, J.B., Wu, Y., Zhang, Q.N., Yu, L.Y., Qian, G.S., Kuang, S.Y., Li, Y.F., Fang, X., Zarba, A., Chen, B., Enger, C., Davidson, N.E., Gorman, M.B., Gordon, G.B., Prochaska, H.J., Egner, P.A., Groopman, J.D., Munoz, A., Helzlsouer, K.J. & Kensler, T.W. (1997) Oltipraz chemoprevention trial in Qidong, People's Republic of China: study design and clinical outcomes. *Cancer Epidemiol. Biomarkers Prev.*, **6**, 257–265

Judah, D.J., Hayes, J.D., Yang, J.C., Lian, L.Y., Roberts, G.C., Farmer, P.B., Lamb, J.H. & Neal, G.E. (1993) A novel aldehyde reductase with activity towards a metabolite of aflatoxin B1 is expressed in rat liver during carcinogenesis and following the administration of an anti-oxidant. *Biochem. J.*, **292**, 13–18

Kelly V.P., Ellis, E.M., Manson, M.M., Chanas, S.A., Moffat, G.J., McLeod, R., Judah, D.J., Neal, G.E. & Hayes, J.D. (2000) Chemoprevention of aflatoxin B1 hepatocarcinogenesis by coumarin, a natural benzopyrone that is a potent inducer of aflatoxin B1-aldehyde reductase, the glutathione S-transferase A5 and P1 subunits, and NAD(P)H:quinone oxidoreductase in rat liver. *Cancer Res.*, **60**, 957–969

Kensler, T.W., Gange, S.J., Egner, P.A., Dolan, P.M., Munoz, A., Groopman, J.D., Rogers, A.E. & Roebuck, B.D. (1997) Predictive value of molecular dosimetry: individual versus group effects of oltipraz on aflatoxin-albumin adducts and risk of liver cancer. *Cancer Epidemiol. Biomarkers Prev.*, **6**, 603–610

Kensler, T.W., He, X., Otieno, M., Egner, P.A., Jacobson, L.P., Chen, B., Wang, J.-S., Zhu, Y.-R., Zhang, B.-C., Wang, J.-B., Wu, Y., Zhang, Q.-N., Qian, G.-S., Kuang, S.-Y., Fang, S., Li, Y.-F., Yu, L.-Y., Prochaska, H.J., Davidson, N.E., Gordon, G.B., Gorman, M.B., Zarba, A., Enger, C., Muñoz, A., Helzlsouer, K.J. & Groopman, J.D. (1998) Oltipraz chemoprevention trial in Qidong, People's Republic of China: modulation of serum aflatoxin albumin adduct biomarkers. *Cancer Epidemiol. Biomarkers Prev.*, **7**, 127–134

Kensler, T.W., Groopman, J.D., Sutter, T.R., Curphey, T.J. & Roebuck, B.D. (1999) Development of cancer chemopreventive agents: oltipraz as a paradigm. *Chem. Res. Toxicol.*, **12**, 113–126

Kirk, G.D., Camus-Randon, A.-M., Mendy, M., Goedert, J.J., Merle, P., Trepo, C., Brechot, C., Hainaut, P. & Montesano, R. (2000) 249ser p53 mutations in plasma DNA from hepatocellular carcinoma patients from The Gambia, West Africa. *J. Natl Cancer Inst.*, **92**, 148–153

Montesano, R., Hainaut, P. & Wild, C.P. (1997) Hepatocellular carcinoma: from gene to public health. *J. Natl Cancer Inst.*, **89**, 1844-1851

Parkin, D.M., Srivatanakul, P., Khlat, M., Chenvidhya, D., Chotiwan, P., Insiripong, S., L'Abbe, K.A. & Wild, C.P. (1991) Liver cancer in Thailand. I. A case-control study of cholangiocarcinoma. *Int. J. Cancer*, **48**, 323–328

Pisani, P., Parkin, D.M., Munoz, N. & Ferlay, J. (1997) Cancer and infection: estimates of the attributable fraction in 1990. *Cancer Epidemiol. Biomarkers Prev.*, **6**, 387–400

Qian, G.S., Ross, R.K., Yu, M.C., Yuan, J.M., Gao, Y.T., Henderson, B.E., Wogan, G.N. & Groopman, J.D. (1994) A follow-up study of urinary markers of aflatoxin exposure and liver cancer risk in Shanghai, People's Republic of China *Cancer Epidemiol. Biomarkers Prev.*, **3**, 3–10

Sabbioni, G., Skipper, P.L., Buchi, G. & Tannenbaum, S.R. (1987) Isolation and characterization of the major serum albumin adduct formed by aflatoxin B1 in vivo in rats. *Carcinogenesis*, **8**, 819–824

Srivatanakul, P., Ohshima, H., Khlat, M., Parkin, M., Sukaryodhin, S., Brouet, I. & Bartsch, H. (1991) Opisthorchis viverrini infestation and endogenous nitrosamines as risk factors for cholangiocarcinoma in Thailand. *Int. J. Cancer*, **48**, 821–825

Stuver, S. (1998) Towards global control of liver cancer. *Seminars Cancer Biol.*, **8**, 299–306

Sylla, A., Diallo, M.S., Castegnaro, J. & Wild, C.P. (1999) Interactions between hepatitis B virus infection and exposure to aflatoxins in the development of hepatocellular carcinoma; a molecular epidemiological approach. *Mutat. Res.*, **428**, 187–196

Wang, J.S., Shen, X., He, X., Zhu, Y.R., Zhang, B.C., Wang, J.B., Qian, G.S., Kuang, S.Y., Zarba, A., Egner, P.A., Jacobson, L.P., Munoz, A., Helzlsouer, K.J., Groopman, J.D. & Kensler, T.W. (1999) Protective alterations in phase 1 and 2 metabolism of aflatoxin B1 by oltipraz in residents of Qidong, People's Republic of China. *J. Natl Cancer Inst.*, **91**, 347–354

Wang, L.Y., Hatch, M., Chen, C.J., Levin, B., You, S.L., Lu, S.N., Wu, M.H., Wu, W.P., Wang, L.W., Wang, Q., Huang, G.T., Yang, P.M., Lee, H.S. & Santella, R.M. (1996) Aflatoxin exposure and risk of hepatocellular carcinoma in Taiwan. *Int. J. Cancer*, **67**, 620–625

Wild, C.P. & Hall, A.J. (1999) Hepatitis B virus and liver cancer: unanswered questions. *Cancer Surveys*, **33**, 35–54

Wild, C.P. & Hall, A.J. (2000) Primary prevention of hepatocellular carcinoma in developing countries. *Mutat. Res.*, **462**, 381–393

Wild, C.P., Garner, R.C., Montesano, R. & Tursi, F. (1986a) Aflatoxin B1 binding to plasma albumin and liver DNA upon chronic administration to rats. *Carcinogenesis*, **7**, 853–858

Wild, C.P., Umbenhauer, D., Chapot, B., & Montesano, R. (1986b) Monitoring of individual human exposure to aflatoxins (AF) and N-nitrosamines (NNO) by immunoassays. *J. Cell Biochem.*, 30, 171–179

Wild, C.P., Jiang, Y.Z., Allen, S.J., Jansen, L.A.M., Hall, A.J. & Montesano, R. (1990) Aflatoxin-albumin adducts in human sera from different regions of the world. *Carcinogenesis*, **11**, 2271–2274

Wild, C.P., Fortuin, M., Donato, F., Whittle, H.C., Hall, A.J., Wolf, C.R. & Montesano, R. (1993) Aflatoxin, liver enzymes, and hepatitis B virus infection in Gambian children. *Cancer Epidemiol. Biomarkers Prev.*, **2**, 555–561

Wild, C.P., Hasegawa, R., Barraud, L., Chutimataewin, S., Chapot, B., Ito, N. & Montesano, R. (1996) Aflatoxin albumin adducts – a basis for comparative carcinogenesis between animals and humans. *Cancer Epidemiol. Biomarkers Prev.*, **5**, 179–189

Wild, C.P., Fen, Y., Turner, P.C., Chemin, I., Chapot, B., Mendy, M., Whittle, H., Kirk, G.D. & Hall, A.J. (2000) Environmental and genetic determinants of aflatoxin-albumin adducts in The Gambia. *Int. J. Cancer*, **86**, 1–7

Zhu, J.Q., Zhang, L.S., Hu, X., Xiao, Y., Chen, J.S., Xu, Y.C., Fremy, J. & Chu, F.S. (1987) Correlation of dietary aflatoxin B1 levels with excretion of aflatoxin M1 in human urine. *Cancer Res.*, **47**, 1848–1852

Corresponding author:

C.P. Wild
Molecular Epidemiology Unit,
Epidemiology and Health Services Research,
School of Medicine,
Algernon Firth Building,
University of Leeds,
Leeds LS2 9JT,
UK

Biomarkers in Cancer Chemoprevention
Miller, A.B., Bartsch, H., Boffetta, P., Dragsted, L. and Vainio, H. eds
IARC Scientific Publications No. 154
International Agency for Research on Cancer, Lyon, 2001

Significance of hepatic preneoplasia for cancer chemoprevention

P. Bannasch, D. Nehrbass and A. Kopp-Schneider

Hepatic preneoplasia represents an early stage in neoplastic development, preceding both benign and malignant neoplasia. This applies particularly to foci of altered hepatocytes (FAH), that precede the manifestation of hepatocellular adenomas and carcinomas in all species investigated. Morphological, microbiochemical and molecular biological approaches *in situ* have provided evidence for striking similarities in specific changes of the cellular phenotype of preneoplastic FAH emerging in experimental and human hepatocarcinogenesis, irrespective of whether this was elicited by chemicals, hormones, radiation, viruses or, in animal models, by transgenic oncogenes or *Helicobacter hepaticus*. Different types of FAH have been distinguished and related to three main preneoplastic hepatocellular lineages: (1) the glycogenotic-basophilic cell lineage, (2) its xenomorphic-tigroid cell variant, and (3) the amphophilic-basophilic cell lineage. The predominant glycogenotic-basophilic and tigroid cell lineages develop especially after exposure to DNA-reactive chemicals, radiation, hepadnaviridae, transgenic oncogenes and local hyperinsulinism, their phenotype indicating initiation by insulin or insulinomimetic effects of the oncogenic agents. In contrast, the amphophilic cell lineage of hepatocarcinogenesis has been observed mainly after exposure of rodents to peroxisome proliferators that are not directly DNA-reactive or to hepadnaviridae, the biochemical pattern mimicking an effect of thyroid hormone, including mitochondrial proliferation and activation of mitochondrial enzymes. Hepatic preneoplastic lesions are increasingly used as end-points in carcinogenicity testing, particularly in medium-term carcinogenesis bioassays. This has been complemented more recently by the use of FAH as indicators of chemoprevention, although possible pitfalls of this approach have to be considered carefully. Our ever-increasing knowledge on the metabolic and molecular changes that characterize preneoplastic lesions and their progression to neoplasia provides a new basis for rational approaches to chemoprevention by drugs, hormones or components of the diet.

Introduction

Hepatocellular carcinoma (HCC) is one of the most frequent malignant neoplasms in humans, and has a very poor prognosis. Primary and secondary prevention appear to be the most promising approaches in the fight against this fatal disease. Chronic infection with the hepatitis B (HBV) and C (HCV) viruses, ingestion of foodstuffs contaminated with chemical hepatocarcinogens, particularly the naturally occuring mycotoxin aflatoxin B_1, and abuse of alcoholic beverages have been identified as major risk factors for the development of HCC (Bosch, 1997; Montesano *et al.*, 1997; Stuver, 1998). At least some of these factors may act synergistically, as suggested particularly for HBV and aflatoxins by epidemiological observations. Vaccination against HBV has been introduced in several high-risk areas for HCC and the first results are promising (Chang *et al.*, 1997). However little, if any, progress has been made in prevention of HCC due to other risk factors.

For most of the risk factors for human HCC, appropriate animal models, including chronic infection of woodchucks with the woodchuck hepatitis virus (WHV) which is closely related to HBV, have been established (Okuda & Tabor, 1997). These models were instrumental in the analysis of the mechanism of hepatocarcinogenesis, and especially that of hepatic preneoplasia (Bannasch, 1996). Preneoplastic foci of altered hepatocytes (FAH) precede the manifestation of both benign (adenoma) and malignant hepatocellular neoplasms by long lag periods, which may vary between months and years depending on the cause of neoplastic development and the life span of the species affected. FAH were discovered more than three decades ago in rodents treated with nitrosamines (Bannasch, 1968; Friedrich-Freksa *et al.*, 1969) and have since been observed in a large number of species, including non-human primates and humans, after exposure to hepatocarcinogenic agents of virtually all known classes, such as various chemicals, hormones, HBV, HCV, WHV, *Helicobacter hepaticus* and radiation with X-rays, neutrons or α-particles from Thorotrast. FAH have also been found in transgenic rodent strains and a mutant (LEC) rat strain suffering from hereditary hepatitis, which are prone to develop a high incidence of HCC (Grisham, 1996; Bannasch & Schröder, 2001). FAH have been studied most extensively in rodent models of chemical hepatocarcinogenesis (Hasegawa & Ito, 1994; Pitot & Dragan, 1994; Farber, 1996; Bannasch & Zerban, 1997), and have been increasingly used as endpoints in carcinogenicity testing (Bannasch, 1986; US National Institute of Environmental Health Sciences, 1989), particularly in risk identification by medium-term carcinogenesis bioassays (Ito *et al.*, 1998; Williams & Enzmann, 1998). More recently, FAH have also been used as biomarkers for studying chemopreventive effects in experimental hepatocarcinogenesis. Although this approach appears to be attractive, it is evident that only detailed knowledge of the pathobiolgy of the preneoplastic lesions and of the various animal models of hepatocarcinogenesis employed can avoid pitfalls in the evaluation of possible chemopreventive effects.

Cellular origin of HCC and definition of preneoplasia

There is continuing debate on the existence of potential stem-like liver cells which might be identical with, or closely related to, the so-called oval cells derived from the cholangioles, and might give rise to both cholangiocellular and hepatocellular carcinomas (Kitten & Ferry, 1998; Lazaro *et al.*, 1998; Steinberg *et al.*, 1999). While there is general agreement that oval cells may be precursors of cholangiocellular neoplasms, their role in the evolution of hepatocellular neoplasms remains controversial (Bannasch & Zerban, 1997). In experimental chemical hepatocarcinogenesis, it has been clearly shown that the dose determines whether the carcinogenic process is accompanied by oval cell proliferation. Only after exposure to high doses that lead to pronounced toxic damage of the liver parenchyma does oval cell proliferation occur frequently early during hepatocarcinogenesis. A similar dose-dependence has been observed for development of liver fibrosis and cirrhosis. These findings show clearly that neither oval cell proliferation nor liver cirrhosis is an obligatory prerequisite for development of HCC. In contrast, FAH appear at all dose levels that lead to HCC, irrespective of whether cirrhotic changes or oval cell proliferation occur. Evidence for the preneoplastic nature of FAH has been provided by a number of laboratories (Hasegawa & Ito, 1994; Pitot & Dragan, 1994; Farber, 1996; Bannasch & Zerban, 1997; Williams & Enzmann, 1998), hepatic preneoplasia being defined as phenotypically altered cell populations that have no obvious neoplastic nature but indicate an increased risk for the development of both benign and malignant neoplasms (Bannasch, 1986). The earliest-emerging FAH are composed of differentiated hepatocytes, which show specific morphological, metabolic and molecular aberrations, and gradually dedifferentiate, while progressing through various intermediate forms to the malignant phenotype.

Pathomorphology of preneoplastic hepatocellular lineages

Experimental chemical hepatocarcinogenesis

For a long time, hepatic preneoplasia as defined above was almost exclusively studied during experimental hepatocarcinogensis in rodents. On the basis of cytomorphological and simple cytochemical criteria, resulting mainly from staining of alcohol-fixed serial sections with haematoxylin and eosin (H&E) and the periodic acid–Schiff reaction (PAS) to reveal glycogen, at least eight different types of preneoplastic FAH have been distinguished (Bannasch & Zerban, 1992; Goodman *et al.*, 1994). Comprehensive sequential morphological, stereological and biochemical studies *in situ* revealed that the different phenotypes of FAH are integral parts of preneoplastic hepatocellular lineages leading from highly differentiated hepatocellular phenotypes to poorly differentiated neoplastic phenotypes. Three main hepatocellular lineages have been distinguished (Figure 1): (1) the glycogenotic-basophilic, (2) the xenomorphic-

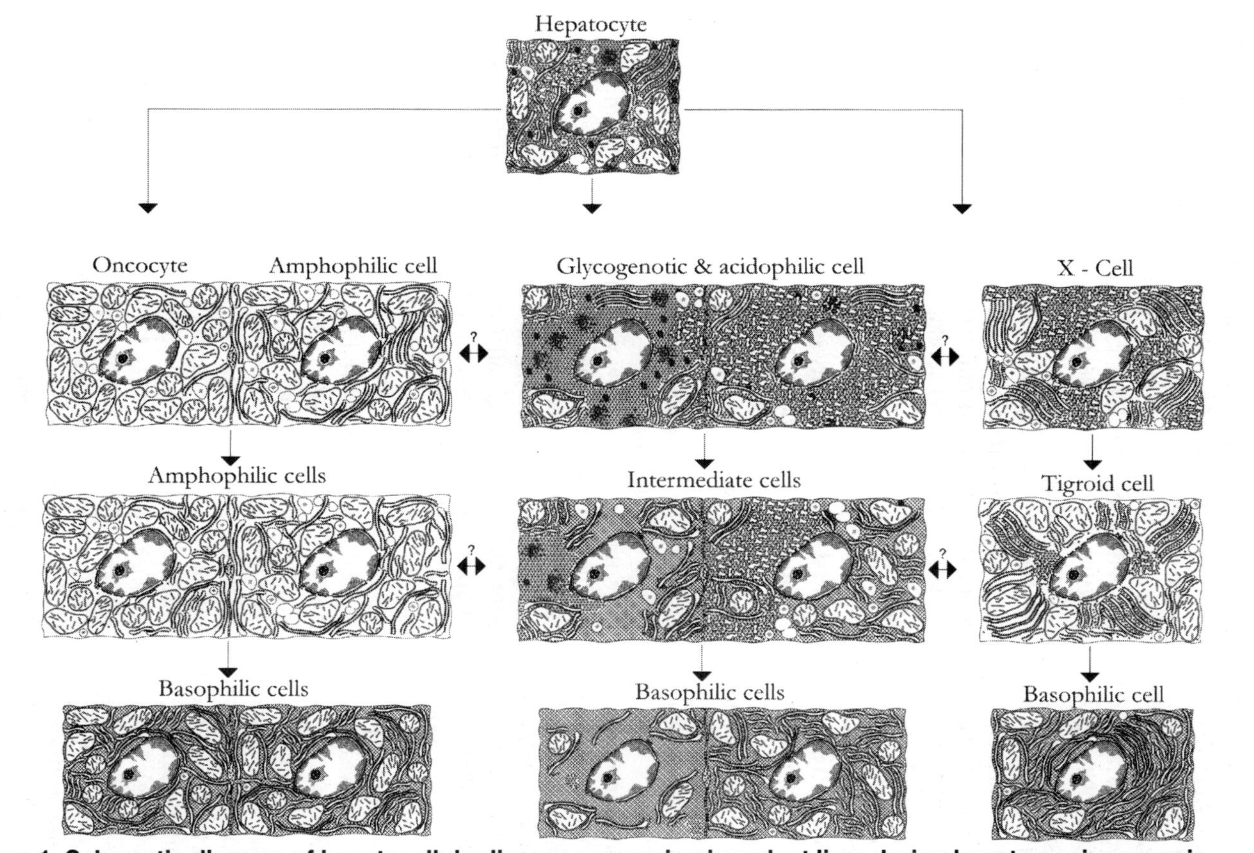

Figure 1. Schematic diagram of hepatocellular lineages emerging in rodent liver during hepatocarcinogenesis.
The predominant sequence of cellular changes (centre) starts with glycogenotic clear and acidophilic (smooth endoplasmic reticulum-rich) hepatocytes and progresses through intermediate phenotypes in mixed cell populations to glycogen-poor, homogeneously basophilic (ribosome-rich) cellular phenotypes prevailing in undifferentiated hepatocellular carcinomas. The tigroid basophilic cell lineage (to the right), originating from xenomorphic hepatocytes (X-cells), is characterized by cells with abundant highly ordered stacks of the rough endoplasmic reticulum and apparently represents a less altered variant of the glycogenotic-basophilic cell lineage. The amphophilic cell lineage (to the left), which has hitherto been described mainly in rats treated with nongenotoxic peroxisome proliferators, and may include oncocytes in woodchucks chronically infected with the woodchuck hepatitis virus, consists of cells with a glycogen-poor cytoplasm containing both abundant granular-acidophilic (mitochondria and peroxisomes) and basophilic (ribosomes) components (from Bannasch, 1998).

tigroid, and (3) the amphophilic-basophilic cell lineage (Bannasch, 1996). After exposure of rats to the majority of hepatocarcinogenic chemicals, especially DNA-reactive compounds, the glycogenotic-basophilic lineage prevails. The sequence of cellular changes in this lineage starts with the appearance of glycogenotic clear and acidophilic cell foci and passes through mixed cell populations before formation of glycogen-poor, basophilic (ribosome-rich) neoplastic lesions. This progression-linked phenotypic instability is associated with a gradual reduction of the glycogen initially stored in excess, a multiplication of ribosomes resulting in increased cytoplasmic basophilia, and an ever-increasing cell proliferation and expansion from small to large foci (Bannasch, 1968; Moore & Kitagawa, 1991; Bannasch & Zerban, 1997). Although a minor but significant increase of cell proliferation is seen in the earliest glycogenotic foci, there is an inverse relationship between the gradual reduction of glycogen and the pronounced increase in cell proliferation during neoplastic development at later time points (Zerban *et al.*, 1994). The xenomorphic-tigroid cell lineage represents a less altered variant of the glycogenotic-basophilic cell lineage (Weber & Bannasch, 1994c; Ströbel *et al.*, 1998).

In contrast, the amphophilic cell lineage is characterized by a completely different phenotype, frequently produced in rat liver by hepatocarcinogens that are not directly DNA-reactive, of the peroxisome proliferator type, including several hypolipidaemic drugs and the adrenal steroid hormone dehydroepiandrosterone (Weber *et al.*, 1988; Metzger *et al.*, 1995). The amphophilic cell foci, synonyms of which are 'atypical eosinophilic foci' (Harada *et al.*, 1989), 'weakly basophilic foci' (Marsman & Popp, 1994), and 'large-cell basophilic foci' (Christensen *et al.*, 1999), are not preceded by glycogenotic foci under most experimental conditions, but are poor in glycogen from the beginning. At the ultrastructural level, the amphophilic cells exhibit a proliferation of mitochondria wrapped by profiles of rough endoplasmic reticulum, and sometimes also an increase in peroxisomes (Metzger *et al.*, 1995). During progression to the malignant phenotype, the number of ribosomes and, consequently, the cytoplasmic basophilia usually increase.

A series of stereological studies based on the morphological classification of FAH confirmed the progression-linked phenotypic instability of the predominant preneoplastic hepatocellular lineage (Table 1), and revealed that the number, size and phenotype of FAH is dose- and time-dependent, correlating with the appearance of hepatocellular adenomas and carcinomas (Moore & Kitagawa, 1986; Enzmann & Bannasch, 1987; Weber & Bannasch, 1994c). The results of these investigations also support previous observations (Moore & Kitagawa, 1986; Farber & Sarma, 1987; Bannasch & Zerban, 1992, 1997) that a reversion-linked phenotypic instability of FAH may occur under certain experimental conditions, especially after repeated but limited administration of high sublethal doses of a single or several carcinogenic chemicals, leading to reappearance of less altered phenotypes after withdrawal of these agents (Weber & Bannasch, 1994a,b,c). The cause of reversion-linked phenotypic instability is poorly understood, but it may be due mainly to the cessation of a proliferative stimulus elicited during carcinogen exposure by severe toxic parenchymal damage.

Experimental physical, viral and hormonal hepatocarcinogenesis

In the past few years, it has been shown that the glycogenotic-basophilic and tigroid-xenomorphic lineages develop not only in experimental hepatocarcinogenesis induced by chemicals such as nitrosamines, aflatoxin B_1 and phenobarbital, but also in animal models of physical, viral and hormonal hepatocarcinogenesis (Table 2). In addition to neutrons and α-particles from Thorotrast (Ober *et al.*, 1994), X-rays have been reported to produce the glycogenotic-basophilic cell lineage (Oehlert, 1978). Particularly rewarding for the understanding of human hepatocarcinogenesis are the findings in the woodchuck model of hepadnaviral hepatocarcinogenesis (Toshkov *et al.*, 1990; Radaeva *et al.*, 2000) and in the transgenic mouse models established by Chisari (Toshkov *et al.*, 1994) and Kim *et al.* (1991), in which subgenomic fragments of HBV, coding for the large envelope polypeptide and the X protein, respectively, are expressed and trigger the development of hepatocellular carcinomas.

Table 1. Chemical hepatocarcinogensis: characteristics of preneoplastic cellular changes

- **Early emergence** of foci of altered hepatocytes (FAH) in all models of chemical hepatocarcinogensis
- **Phenotypic diversity and instability**
- Integration of certain types of FAH in **three main hepatocellular lineages**
- **Dose- and time-dependence** of number, size, and phenotype of FAH correlates with appearance of hepatocellular neoplasms
- **Progression-linked phenotype instability** is associated with ever-increasing **cell proliferation** (and apoptosis) and **FAH expansion**
- **Reversion-linked phenotypic instability** results in reappearance of early preneoplastic phenotypes but not in normal liver parenchyma
- Lack of consistent **genotypic changes** in FAH
- **Field effects** rather than repeated clonal selections characterize progression of FAH
- Mathematical modelling of phenotypic cellular changes as **epigenetic events**

Table 2. Hepatocellular lineages in animal models of hepatocarcinogensis

Glycogenotic-basophilic/xenomorphic-tigroid cell lineages

- Chemicals: nitrosamines, aflatoxin B_1, thioacetamide, phenobarbital
- Radiation: X-rays, neutrons, α-particles of Thorotrast
- Viruses: woodchuck hepatitis virus, subgenomic fragments of hepatitis B virus and simian virus 40
- Insulin: intrahepatic transplantation of pancreatic islets in diabetic rats

Amphophilic-basophilic cell lineage

- Chemicals: peroxisome proliferators including dehydroepiandrosterone
- Virus: woodchuck hepatitis virus

The amphophilic-basophilic cell lineage has also been observed in experimental chemical, hormonal and hepadnaviral hepatocarcinogenesis (Bannasch *et al.*, 1995; Metzger *et al.*, 1995; Dombrowski *et al.*, 2000; Radaeva *et al.*, 2000). However, while the glycogenotic-basophilic and amphophilic-basophilic cell lineages are produced by different types of chemicals in the rat, they frequently coexist in hepadnaviral hepatocarcinogenesis in woodchucks.

Human hepatocarcinogenesis

In resected livers from humans suffering from liver cell cancer and cirrhosis as a consequence of a variety of chronic liver diseases predisposing to HCC (Table 3), FAH comparable to those observed in

Table 3. Hepatocellular lineages in human hepatocarcinogenesis

Glycogenotic-basophilic cell lineage
- HCC-bearing livers with and without cirrhosis
- Posthepatic cirrhosis (hepatitis B virus or hepatitis C virus)
- Alcoholic cirrhosis
- Biliary cirrhosis
- Cryptogenic cirrhosis
- Inborn hepatic glycogenosis

Amphophilic cell populations
- Frequent appearance under similar conditions, except inborn hepatic glycogenosis

animal models are often found (Altmann, 1994; Bannasch, 1996). We have evidence for the preneoplastic nature of the glycogenotic-basophilic cell lineage (Bannasch *et al.*, 1997b; Su *et al.*, 1997), but this remains to be demonstrated for the amphophilic cell population in human hepatocarcinogenesis. Cases of inborn hepatic glycogenosis, which result from a genetically fixed defect of glucose-6-phosphatase and are associated with a high risk of developing hepatocellular neoplasms when the patients pass through adolescence, seem to be of particular heuristic value (Bannasch *et al.*, 1984; Bianchi, 1993).

Pathobiochemistry of preneoplastic hepatocellular lineages

The abnormal morphology of FAH is associated with a variety of biochemical and molecular aberrations, as demonstrated by cytochemical, microbiochemical and molecular biological methods (Moore & Kitagawa, 1986; Farber & Sarma, 1987; Schwarz *et al.*, 1989; Pitot, 1990; Farber, 1996; Bannasch *et al.*, 1997a; Mayer *et al.*, 1998a; Feo *et al.*, 2000a). Aberrations in energy and drug metabolism have attracted the most attention, but other metabolic pathways may also be affected. Resistance to experimentally induced haemosiderosis was introduced as a marker of various types of FAH (including glycogen storage foci) in rodents (Williams *et al.*, 1976) and has been successfully applied to the detection of FAH in human hereditary haemochromatosis (Deugnier *et al.*, 1993). Similarly, excessive storage of glycogen (glycogenosis) and reduced activity of glucose-6-phosphatase, which were the first biochemical markers of FAH discovered in rodents (Bannasch, 1968; Friedrich-Freksa *et al.*, 1969), have been valuable for the identification of corresponding focal lesions in human liver (Bannasch *et al.*, 1997b). Among the enzymes involved in drug metabolism, γ-glutamyltranspeptidase (γGT) (Kalengayi *et al.*, 1975) and the placental form of glutathione *S*-transferase (GSTP) (Sato *et al.*, 1984) have been widely used as markers for FAH (Sato, 1989). It is important to realize, however, that the three preneoplastic hepatocellular lineages differ fundamentally in not only their morphological but also their biochemical phenotype. Thus, while γGT and GST-P are reasonable markers for FAH of the glycogenotic-basophilic cell lineage, they fail to reveal the majority of tigroid cell foci (Ströbel *et al.*, 1998), and are completely absent from amphophilic cell foci (Rao *et al.*, 1982; Mayer *et al.*, 1998a). Changes in the expression of certain genes, including those coding for various growth factors, and molecular genetic alterations, most of which were not studied in specific types of FAH and showed considerable interspecies variations, are considered by Grisham (1996), Bannasch & Schröder (2001) and Feo *et al.* (2000a).

We have studied the biochemical phenotype of the two main preneoplastic hepatocellular lineages, the glycogenotic-basophilic and the amphophilic-basophilic lineages, using the nitrosamine-induced stop model of rat hepatocarcinogenesis (Bannasch, 1968) and rat liver continuously exposed to dehydroepiandrosterone (Metzger *et al.*, 1995; Mayer *et al.*, 1998a). The preneoplastic FAH occupy a maximum of 10% of the total liver volume, precluding the application of conventional biochemical or molecular biological approaches in tissue homogenates. We have, therefore, adopted enzyme histochemical, immunohistochemical, microbiochemical and molecular biological methods *in situ* (Bannasch *et al.*, 1984, 1997a). Based on these approaches, metabolic and molecular patterns in the predominant types of preneoplastic hepatic foci have been outlined.

Phenotypes mimicking a response to insulin

In the glycogenotic foci, several metabolic changes apparently act in concert favouring glycogen accumulation (Bannasch *et al.*, 1997a; Mayer *et al.*, 1998b). In addition to inactivation of the adenylate cyclase-mediated signalling pathway, resulting in disturbance of phosphorylytic glycogen breakdown, the hydrolytic lysosomal degradation of glycogen by α-glucosidase is reduced. Decreased activity of glucose-6-phosphatase and expression of the glucose transporter protein GLUT2 indicate a downregulation of gluconeogenesis (Grobholz *et al.*, 1993). In contrast, increased activities of the key enzymes pyruvate kinase and glucose-6-phosphate dehydrogenase point to upregulation of glycolysis and the pentose phosphate pathway, providing precursors and energy for nucleic acid synthesis associated with increased cell proliferation (Hacker *et al.*, 1982, 1998; Klimek *et al.*, 1984). This metabolic pattern is consistent with an insulinomimetic effect of the oncogenic agents (Klimek & Bannasch, 1993; Bannasch *et al.*, 1997a). Direct evidence for such an effect in the early stages of hepatocarcinogenesis has been provided by a new animal model of hormonal hepatocarcinogenesis. Low-number intraportal pancreatic islet transplantation in streptozotocin-diabetic rats results in rapid development of proliferative focal lesions, the morphological and biochemical phenotype of which is similar to that induced by a variety of oncogenic agents (Dombrowski *et al.*, 1994, 1997). Within 1–2 years, the early-emerging glycogenotic foci gradually undergo metamorphosis towards a glycogen-poor, basophilic phenotype and give rise to hepatocellular adenomas and carcinomas, which regularly contain pancreatic islet cells. It may be relevant that an excess risk of primary liver cancer in human patients with diabetes mellitus has been repeatedly reported (Adami *et al.*, 1996; Moore *et al.*, 1998; Stuver, 1998); implications of the hyperinsulinaemia–diabetes–cancer link for preventive efforts have been considered (Moore *et al.*, 1998).

These observations and considerations prompted us to investigate the expression of several components of the insulin signalling cascade (Figure 2) in FAH that emerge in the stop model of chemical hepatocarcinogenesis. We chose to study the insulin-receptor (IR), the receptor of the insulin-like growth factor I (IGF-RI), the insulin receptor substrates-1 and -2 (IRS-1, IRS-2) and the mitogen-activated extracellular signal-regulated kinase-1 (MEK-1) by immunohistochemistry (Nehrbass *et al.*, 1998, 1999; Nehrbass, 2000), and the proto-oncogenes c-*raf*-kinase and c-*myc* by *in situ* hybridization (Bannasch, 1996). The proto-oncogene c-*raf* holds a central position in several intracellular signalling cascades (Slupsky *et al.*, 1998). The product of the c-*myc* proto-oncogene acts as a transcription factor that, according to studies in transgenic mice, regulates hepatic glycolysis (Valera *et al.*, 1995). In early glycogenotic foci, all components of the insulin signalling cascade studied were upregulated, as demonstrated particularly convincingly for IRS-1, which is a multi-site docking protein acting as a principal intracellular substrate of the insulin receptor tyrosine kinase (Table 4). These findings suggest that activation of the insulin-stimulated raf-MAP kinase signal transduction pathway elicits preneoplastic hepatic glycogenosis (Nehrbass *et al.*, 1998, 1999). In hepadnaviral hepatocarcinogenesis, the insulinomimetic effect may also be responsible for the downregulation of expression of the viral surface antigen in glycogenotic FAH, as observed in HBV transgenic mice (Toshkov *et al.*, 1994), WHV-infected woodchucks (Bannasch *et al.*, 1995; Radaeva *et al.*, 2000) and human HBV carriers (Su *et al.*, 1998), since it has been shown in studies on a human hepatoma cell line that insulin may indeed suppress the expression of the surface antigen (Chou *et al.*, 1989). *In situ* investigations on the glycogenotic-basophilic cell lineage of human hepatocarcinogenesis have revealed that in HBV carriers, preneoplastic FAH of any type preferentially albeit rarely express the X protein of HBV, while p53 accumulation is invariably negative in FAH but correlates with neoplastic progression in HCC, irrespective of the risk factors involved in regions with low exposure to aflatoxins (Su *et al.*, 1998, 2000).

Progression-linked downregulation of insulin signalling during cellular dedifferentiation

As previously shown in experimental chemical hepatocarcinogenesis for the c-*raf*-kinase (Bannasch, 1996), the overexpression of most of the proteins of the insulin signalling cascade studied in the early glycogenotic cell populations presenting a high grade of differentiation is only

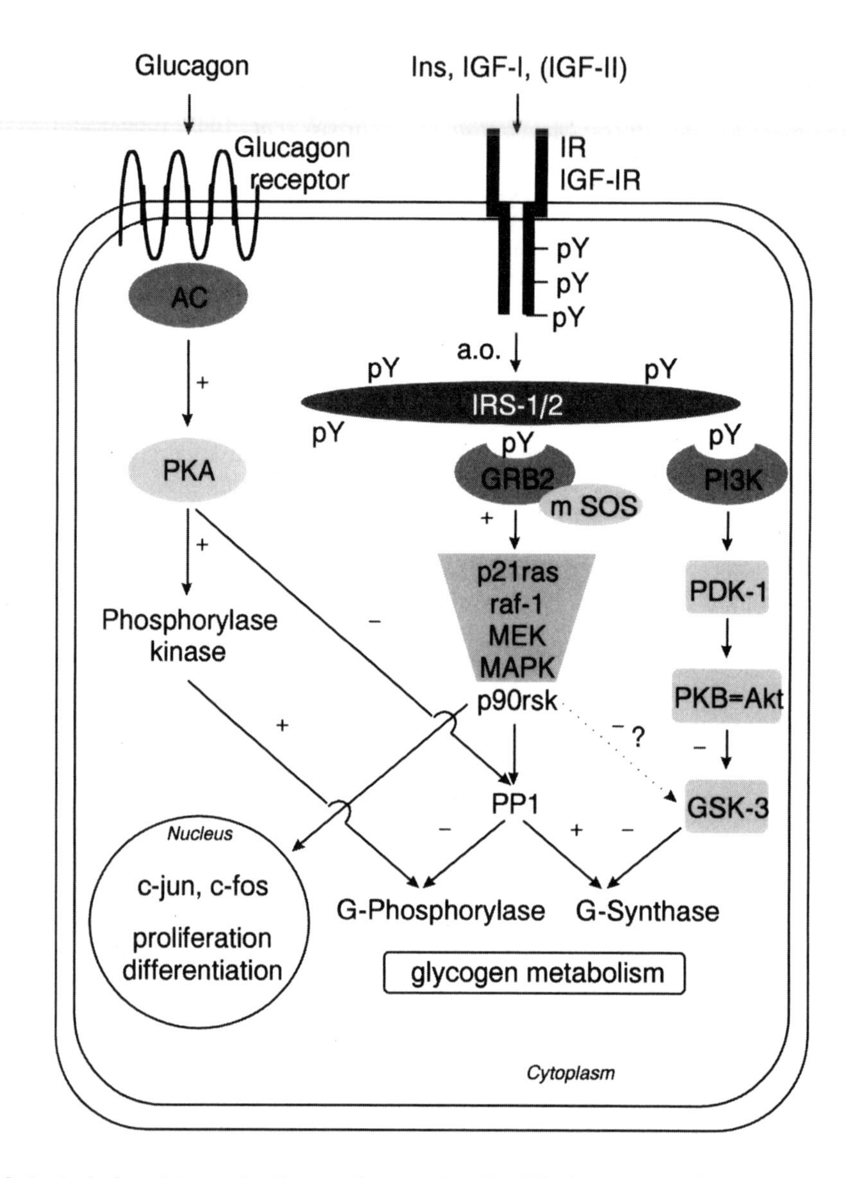

Figure 2. Selected signal transduction pathways involved in hepatocarcinogenesis, particularly the insulin-stimulated *ras*-, *raf*-, mitogen-activated signalling cascade (centre) and the glucagon-stimulated, adenylate cyclase-mediated pathway (to the left).

AC, adenylate cyclase; GRB2, growth factor receptor binding protein-2; GSK-3, glycogen synthase kinase-3; ins, insulin; IGF-I, insulin-like growth factor-I; IGF-II, insulin-like growth factor-II; IGF-IR, insulin-like growth factor-I receptor; IR, insulin receptor; IRS-1/2, insulin receptor substrate-1 and -2; MAPK, mitogen-activated protein kinase; MEK, mitogen-activated extracellular signal-regulated kinase; PDK1, phosphoinositide-dependent protein kinase-1; PI3K, phosphatidyl-inositol-3-kinase; PKA, protein kinase A; PKB (=Akt), protein kinase B; PP1, protein phosphatase 1; mSOS, mammalian son of sevenless; py, phosphotyrosine.

Table 4. Hormone-like effects of hepatocarcinogenic agents – I

Insulinomimetic effect elicits preneoplastic glycogenotic phenotype

- Activation of insulin-signalling pathway: e.g. overexpression of insulin receptor, insulin-growth factor-I receptor, insulin receptor substrate-1, insulin receptor substrate-2, mitogen-activated extracellular signal-regulated kinase-1, and c-*raf*; increased synthesis of glycogen and/or fat; reduced activities of glucose-6-phosphatase and α-glucosidase; increased activity of glucose-6-phosphate dehydrogenase and cell proliferation
- Inactivation of glucagon-signalling pathway: e.g. reduced activities of adenylate cyclase and glycogen phosphorylase; increased liver type pyruvate kinase activity

Progression from preneoplastic glycogenotic to neoplastic basophilic phenotype

- Downregulation of insulin-signalling pathway
- Isoenzyme shift (e.g. glucokinase/hexokinase; liver type pyruvate kinase/fetal pyruvate kinase) stimulating glycolysis
- Further increase in pentose phosphate pathway (glucose-6-phosphate dehydrogenase) and cell proliferation
- Gradual reduction of gluconeogenesis and glycogenesis

transient, and is gradually downregulated during progression-linked dedifferentiation in later stages of hepatocarcinogenesis (Nehrbass *et al.*, 1998, 1999; Nehrbass, 2000). This event is closely related to the reduction in the glycogen initially stored in excess (Table 4), a further increase in the expression and activity of the key enzyme of the pentose phosphate pathway, elevation of cytoplasmic basophilia due to an increase in the number of ribosomes, and an ever-increasing cell proliferation and expression of c-*myc* (Bannasch, 1996; Bannasch *et al*, 1997a). At about the same time, a shift from adult to fetal glycolytic isoenzymes, for example from glucokinase to hexokinase, and the liver-specific L-pyruvate kinase to the fetal M_2-pyruvate kinase takes place (Klimek & Bannasch, 1990, 1993; Hacker *et al.*, 1998; Steinberg *et al.*, 1999). In addition, the fetal glucose transporter protein GLUT1 emerges, while the earlier downregulation of the liver-specific adult glucose transporter protein GLUT2 is maintained (Grobholz *et al.*, 1993). This pronounced shift from anabolic to catabolic glucose metabolism is probably a prerequisite for a more effective energy supply favouring the increase in cell proliferation. It remains to be clarified, however, by which growth factors cell proliferation is further stimulated when the insulin signalling pathway is downregulated. We have speculated that the relatively weak early proliferative stimulus mediated by insulin may be replaced by alternative growth factors (Nehrbass *et al.*, 1998), particularly IGF-II, which has been shown to be frequently overexpressed in late stages of hepatocarcinogenesis (Rogler *et al.*, 1995). Growth stimulation by IGF-II is preferentially or exclusively mediated by pathways that have not been completely clarified, but may stimulate cell proliferation without exerting insulinomimetic effects on glycogen metabolism.

Phenotypes mimicking responses to thyroid and ovarian hormones

In contrast to the glycogenotic foci, the preneoplastic amphophilic cell foci induced in rats by the peroxisome proliferator dehydroepiandrosterone are characterized by downregulation of IRS-1 and show a completely different histochemical pattern (Weber *et al.*, 1988; Mayer *et al.*, 1998a; Nehrbass *et al.*, 1999). Whereas a reduction in the activity of enzymes of glycogen metabolism is associated with an early loss of glycogen, the gluconeogenic enzyme glucose 6 phosphatase, several mitochondrial enzymes including cytochrome c oxidase and glycerol-3-phosphate dehydrogenase, and some peroxisomal enzymes are usually increased in their amount or activity (Table 5), suggesting a thyromimetic effect of dehydroepiandrosterone and other peroxisome proliferators (Bannasch *et al.*, 1997a; Mayer *et al.*, 1998a). A thyromimetic action of several peroxisome proliferators such as clofibrate and acetylsalicylic acid on rat liver, including changes in messenger RNA levels of certain genes involved in mitochondrial biogenesis has been reported (Cai *et al.*, 1996). In addition, a thyromimetic effect of peroxisome proliferators on the activities of several enzymes such as glycerol-3-phosphate dehydrogenase, malic enzyme and glucose-6-phosphatase has been found in rat liver homogenates and cultured hepatocytes (Hertz *et al.*, 1993, 1996).

In rodents, a number of the biological actions of peroxisome proliferators including dehydroepiandrosterone have been shown to be mediated by the peroxisome proliferator-activated receptor α (PPARα), a member of the superfamily of nuclear steroid receptors (Green & Wahli, 1994; Schoongans *et al.*, 1997), which also mediate effects (possibly including peroxisome proliferation) of the thyroid hormone 3,3,5-triiodo-L-thyronine (T_3) (Francavilla *et al.*, 1994; Ledda-Columbano *et al.*, 1999). PPARα is apparently responsible for peroxisome proliferation, activation of target genes encoding fatty acid-metabolizing enzymes, mitogenesis and ultimately hepatocarcinogenesis (Gonzalez *et al.*, 1998). In contrast to wild-type mice, PPARα-null mice that were treated with a potent peroxisome proliferator developed neither hepatocellular neoplasms nor preneoplastic hepatocellular foci (Peters *et al.*, 1997). PPARα is also required for gene induction by the less potent peroxisome proliferator dehydroepiandrosterone in mice (Peters *et al.*, 1996). However, according to Hertz and Bar-Tana (1998), the biological effects exerted by peroxisome proliferators in the human liver may be mediated by transduction pathways independent of PPARα.

In the rat, the hypothesis of a thyromimetic effect of peroxisome proliferators eliciting the amphophilic preneoplastic phenotype has been substantiated by the recent observation of focal hyperproliferative hepatic lesions with a similar morphological and biochemical phenotype after intrahepatic thyroid tissue transplantation in thyroidectomized animals (Dombrowski *et al.*, 2000). Labelling of these rapidly emerging lesions with bromodeoxyuridine showed considerable proliferation within the transplants and in the surrounding amphophilic cell populations. Eighteen months after thyroid tissue transplantation, large amphophilic lesions were found, but frank hepatocellular neoplasms were not observed in this animal model. Hyperproliferative focal lesions resembling in some respects amphophilic cell foci were also induced in rat liver by intraportal transplantation of ovarian tissue in ovariectomized rats (Klotz *et al.*, 2000). In a preliminary long-term experiment, four of six animals developed hepato-

Table 5. Hormone-like effects of hepatocarcinogenic agents - II

Thyromimetic effect elicits preneoplastic amphophilic phenotype

- Downregulation of component (insulin receptor substrate-1) of insulin-signalling pathway
- Increased activity of glucose-6-phosphatase and glycogen loss
- Mitochondrial proliferation
- Increase in mitochondrial enzymes (glycerol-3-phosphate dehydrogenase, succinate dehydrogenase, cytochrome c oxidase) and malic enzyme

cellular neoplasms including three HCC, the phenotype of which was similar to that of the amphophilic-like FAH.

Although the glycogenotic and the amphophilic cell lineages are very different at first glance, there is circumstantial evidence from some experiments that they may transform into each other (Bannasch *et al.*, 1997a, Mayer *et al*, 1998a; Radaeva *et al.*, 2000). This interconversion is difficult to understand, but crosstalk between two or several disturbed signal transduction pathways related to insulinomimetic or thyromimetic actions of the oncogenic agents might be involved.

Hepatic preneoplastic lesions as biomarkers in chemoprevention studies

Tissue specificity and phenotypic instability

The consistent development of preneoplastic FAH in all animal models of hepatocarcinogenesis and their apparent similarity in human hepatocarcinogenesis favour the use of these lesions as biomarkers in chemoprevention studies. A number of compounds such as several antioxidants (Thamavit *et al.*, 1985; Ito *et al.*, 1992) and oltipraz when administered to animals exposed to aflatoxin B_1 (Kensler *et al.*, 1987; Roebuck *et al.*, 1991) reduce the formation of both GST-P-positive FAH and HCC in rodents. However, some of the antioxidants (e.g., butylated hydroxyanisole, butylated hydroxytoluene) which inhibit carcinogenesis in the liver may enhance carcinogenesis at other sites (Imaida *et al.*, 1983; Ito *et al.*, 1988), indicating a possible limitation of preventive effects to certain target tissues. In addition, it is evident from the discussion in the preceding sections that preneoplastic FAH are neither uniform nor stable. Their phenotypic instability is an outstanding feature of their biological behaviour, be it related to progression or to reversion of neoplastic development. As discussed previously, the reversion-linked phenotypic instability of FAH may seriously hamper the interpretation of studies on carcinogenesis (Bannasch, 1986; Bannasch & Zerban, 1992). This may be even more critical in chemoprevention studies, since a reduction in the number and size of preneoplastic FAH inherent in the animal model used may be mistaken as a positive chemopreventive effect. Animal models largely avoiding this complication are available (Bannasch & Zerban, 1992, 1997).

The methods applied for identification of preneoplastic FAH in tissue sections are also critical in studies of both carcinogenesis and chemoprevention. Thus, the progression-linked phenotypic instability of FAH characterizing particularly the predominant glycogenotic-basophilic preneoplastic cell lineage implies that certain markers such as the activation of components of the insulin signalling cascade (e.g., IRS-1) are only useful for the detection of early-appearing FAH, while other markers such as cellular hyperproliferation and overexpression of c-*myc* or M_2-pyruvate kinase may only help to detect more advanced types of FAH and hepatocellular adenomas. Several of the standard markers used in many laboratories (e.g., glycogen, glucose-6-phosphate dehydrogenase, GSTP, γGT) are suitable for the demonstration of a large proportion of FAH integrated into the glycogenotic-basophilic cell lineage, but largely or completely fail to show FAH with an amphophilic phenotype (Bannasch & Zerban, 1992, 1997; Mayer *et al.*, 1998a). There is not a single biochemical or molecular marker for all types of FAH. A number of comparative studies have clearly shown, however, that the vast majority of preneoplastic FAH are readily identifiable in H&E-stained tissue sections (complemented by serial sections treated with the PAS reagent in some specific cases) without any additional biochemical marker. The application of H&E-staining may be of particular advantage in chemoprevention studies, since both the morphological and the biochemical phenotype of FAH induced by hepatocarcinogens in a variety of rodent models is often modulated by additional exposure to other chemicals (Bannasch & Zerban, 1997).

Phenotypic modulation

In the context of chemoprevention, the phenotypic modulation produced by peroxisome proliferators is of special interest. Several peroxisome proliferators including nafenopin, clofibrate, ciprofibrate and dehydroepiandrosterone inhibit the expression of γGT and GST-P, and cause a rapid loss of glycogen from glycogenotic FAH when given after DNA-reactive hepatocarcinogens (e.g., Numoto *et al.*, 1984; Hosokawa *et al.*, 1989; Gerbracht *et al.*, 1990; Tsuda *et al.*, 1992; Mayer *et al.*, 1998a). Gerbracht *et al.* (1990) emphasized that there was no increase in apoptosis within FAH under these conditions, which might have been an alternative explanation for the reduction in the number and size of enzyme-altered foci. After additional administration of clofibrate to rats pretreated with *N*-nitrosodiethylamine, Hosokawa *et al.* (1989) found that FAH positive and negative for GST-P could be identified morphologically in H&E-stained sections. The total number of FAH, both positive and negative for GST-P, was higher in rats treated with *N*-nitrosodiethylamine followed by clofibrate than in those exposed to *N*-nitrosodiethylamine alone, indicating an enhancing rather than a reducing effect of clofibrate on the development of FAH. A higher incidence of hepatocellular carcinomas was also seen after additional treatment with clofibrate. An enhancing effect of nafenopin on rat hepatocarcinogenesis was found, although preneoplastic FAH were negative for γGT and showed only low levels or absence of several GST isoenzymes (Grasl-Kraupp *et al.*, 1993a,b).

Dehydroepiandrosterone, which had been proposed as a possible chemopreventive agent because it inhibited the focal expression of GST-P in rat liver (Moore *et al.*, 1986; Garcea *et al.*, 1987), later turned out to be a complete hepatocarcinogen of the peroxisome proliferator type (Rao *et al.*, 1992; Hayashi *et al.*, 1994; Metzger *et al.*, 1995) and to enhance carcinogenesis in the liver (Metzger *et al.*, 1998) as well as in several other tissues (Feo *et al.*, 2000b). In Fischer-344 rats initiated with a single dose of *N*-nitrosodiethylamine, feeding a diet supplemented by the thyroid hormone T_3 led to a 70% reduction in the number of GST-P-positive FAH per cm^2, although this hormone exerts strong mitogenic effects on the liver parenchyma (Ledda-Columbano *et al.*, 1999). This seems to indicate a chemopreventive potential of the thyroid hormone under these experimental conditions despite its mitogenic activity, since the authors found a 50% reduction in the incidence of HCC when similarly pretreated rats were exposed to seven cycles of T_3-supplemented diet, as compared to rats undergoing the pretreatment procedure alone (Ledda-Columbano *et al.*, 2000). Green and black tea, the consumption of which has been shown to be associated with both negative and positive effects on human cancer incidence at various sites (Steele *et al.*, 2000), have been reported to induce hepatic peroxisome proliferation (Bu-Abbas *et al.*, 1999). Thus, it is conceivable that peroxisome proliferators, thyroid hormone and possibly other chemicals may exert both carcinogenic and chemopreventive effects on the liver parenchyma and other target tissues depending on the prevailing biological conditions and the dosing and time schedules employed.

Quantitative assessment of hepatic preneoplasia

The heterogeneity and instability of the phenotypic cellular changes that characterize FAH have serious implications for quantitative assessment (Pitot *et al.*, 1989; Schwarz *et al.*, 1989; Bannasch & Zerban, 1992). Using glycogen retention after starvation as a marker for FAH induced in rat liver by an initiation–promotion protocol, Kaufmann *et al.* (1985, 1987) estimated that only one carcinoma developed for every 1000 to 10 000 focal lesions that were observed concurrent with the appearance of neoplasms. Although this discrepancy may in part be a consequence of the experimental approach used, the large number of preneoplastic FAH implies a great advantage for the detection of early stages of neoplastic development in diagnostic pathology and interventional approaches for secondary prevention. The observations discussed in this review are not readily compatible with the mathematical standard multistage model of carcinogenesis based on the assumption of repeated clonal selections. Therefore, Kopp-Schneider *et al.* (1998) have elaborated a new mathematical model of hepatocarcinogenesis called the colour-shift model. In this model, phenotypic changes in focal hepatocellular lesions are treated as epigenetic events, possibly due to alterations in a number of cells from the very beginning, and progressing to hepatocellular carcinomas by changes running parallel in larger cell populations rather than repeated clonal selections as suggested by previous stereological studies on the dose- and time-dependence of the development of FAH in rats exposed to *N*-nitrosomorpholine (Enzmann & Bannasch, 1987; Weber & Bannasch, 1994a,b,c). Comparative investigations, in which the same morphometric data are being applied to both mathematical models, are under way and should help to further elucidate the sequence of cellular and molecular changes during hepatocarcinogenesis.

Depending on the dose and duration of the carcinogenic treatment, the lag period between the first appearance of FAH and neoplasms may vary widely. In *N*-nitrosomorpholine-treated rats, lag periods for the occurrence of adenomas were between 15 and more than 50 weeks (Weber & Bannasch, 1994 a, b, c). There is at present no established marker which would permit us to predict precisely from the appearance of FAH at what time hepatocellular adenomas and carcinomas will develop. However, for many experimental situations, the assumption of a lag period of 6–12 months (about 15–30% of the average life span of the rat) seems to be reasonable. It is interesting to note that in relation to the average life span, these figures correspond closely to the 15–30 years which pass in the majority of children suffering from inborn hepatic glycogenosis type I until multiple hepatocellular neoplasms occur.

The main shortcoming of preneoplastic FAH for their use in early detection and secondary prevention of human HCC is the small size of the lesions and their location in an organ which is not easily accessible. Thus, most of the early preneoplastic FAH are smaller than a liver lobule, which has an average diameter of 1–2 mm in both rodents and humans. This small size precludes a non-invasive identification by any of the imaging procedures available at present. However, there is some hope that at least certain types of FAH can be diagnosed in patients by fine-needle biopsies. The further elucidation of the molecular and metabolic aberrations associated with neoplastic cell conversion in the liver should eventually provide a rational basis for early diagnosis, chemoprevention and, perhaps, also chemotherapy of HCC.

Acknowledgements

This work was supported by grant 0311834 from the Bundesministerium für Bildung und Forschung.

References

Adami, H.O., Chow, W.H., Nyren, O., Berne, C., Linet, M.S., Ekbom, A., Wolk, A., McLaughlin, J.K. & Fraumeni, J.F., Jr (1996) Excess risk of primary liver cancer in patients with diabetes mellitus. *J. Natl Cancer. Inst.*, **88**, 1472–1477

Altmann, H.W. (1994) Hepatic neoformations. *Pathol. Res. Pract.*, **190**, 513–577

Bannasch, P. (1968) The cytoplasm of hepatocytes during carcinogenesis. *Recent Results. Cancer Res.*, **19**, 1–100

Bannasch, P. (1986) Preneoplastic lesions as end points in carcinogenicity testing. I. Hepatic preneoplasia. *Carcinogenesis*, **7**, 689–695

Bannasch, P. (1996) Pathogenesis of hepatocellular carcinoma: sequential cellular, molecular, and metabolic changes. *Prog. Liver Dis.*, **14**, 161–197

Bannasch, P. (1998) Evolution of liver cell cancer – interaction of viruses and chemicals. In: Deutsches Krebsforschungszentrum, ed., *Current Topics in Cancer Research*, New York, Springer, pp. 61–67

Bannasch, P. & Schröder, C. (2001) Pathogenesis of primary liver tumours. In: MacSween, R.N.M., Anthony, P.P., Scheuer, P.J., Burt, T.A.D. & Portman, B.C., eds, *Pathology of the Liver*, New York, Churchill Livingstone (in press)

Bannasch, P. & Zerban, H. (1992) Predictive value of hepatic preneoplastic lesions as indicators of carcinogenic response. In: Vainio, H., Magee, P.N., McGregor, D.B. & McMichael, A.J., eds, *Mechanisms of Carcinogenesis in Risk Identification*, Lyon, IARC, pp. 389–427

Bannasch, P. & Zerban, H. (1997) Experimental chemical hepatocarcinogenesis. In: Okuda K. & Tabor E., eds, *Liver Cancer*, New York, Churchill Livingstone, pp. 213–253

Bannasch, P., Hacker, H.J., Klimek, F. & Mayer, D. (1984) Hepatocellular glycogenosis and related pattern of enzymatic changes during hepatocarcinogenesis. *Adv. Enzyme Regul.*, **22**, 97–121

Bannasch, P., Imani Koshkou, N., Hacker, H.J., Radaeva, S., Mrozek, M., Zillmann, U., Kopp-Schneider, A., Haberkorn, U., Elgas, M., Tolle, T., Roggendorf, M. & Toshkov, I. (1995) Synergistic hepatocarcinogenic effect of hepadnaviral infection and dietary aflatoxin B_1 in woodchucks. *Cancer Res.*, **55**, 3318–3330

Bannasch, P., Klimek, F. & Mayer, D. (1997a) Early bioenergetic changes in hepatocarcinogenesis: preneoplastic phenotypes mimic responses to insulin and thyroid hormone. *J. Bioenerg. Biomemb.*, **29**, 303–313

Bannasch, P., Jahn, U.-R., Hacker, H.J., Su, Q., Hofmann, W., Pichlmayr, R. & Otto, G. (1997b) Focal hepatic glycogenosis: a putative preneoplastic lesion associated with neoplasia and cirrhosis in explanted human livers. *Int. J. Oncol.*, **10**, 261–268

Bianchi, L. (1993) Glycogen storage disease I and hepatocellular tumours. *Eur. J. Pediatr.*, **152** (Suppl. 1), S63–S70

Bosch, F.X. (1997) Global epidemiology of hepatocellular carcinoma. In: Okuda K. & Tabor E., eds, *Liver Cancer*, New York, Churchill Livingstone, pp. 13–28

Bu-Abbas, A., Dobrota, M., Copeland, E., Clifford, M.N., Walker, R. & Ioannides, C. (1999) Proliferation of hepatic peroxisomes in rats following the intake of green or black tea. *Toxicol. Lett.*, **109**, 69–76

Cai, Y., Nelson, B.D., Li, R., Luciakova, K. & DePierre, J.W. (1996) Thyromimetic action of the peroxisome proliferators clofibrate, perfluorooctanoic acid, and acetylsalicylic acid includes changes in mRNA levels for certain genes involved in mitochondrial biogenesis. *Arch. Biochem. Biophys.*, **325**, 107–112

Chang, M.-H., Chen, C.-J., Lai, M.-S., Hsu, H.-M., Wu, T.-C., Kong, M.-S., Liang, D.-C., Shau, W.-Y. & Chen, D.-S. (1997) Universal hepatitis B vaccination in Taiwan and the incidence of hepatocellular carcinoma in children. *New Engl. J. Med.*, **336**, 1855–1859

Chou, C.-K., Su, T.-S., Chang, C.-P., Hu, C., Huang, M.-Y., Suen, C.-S., Chou, N.-W. & Ting, L.-P. (1989) Insulin suppresses hepatitis B surface antigen expression in human hepatoma cells. *J. Biol. Chem.*, **264**, 15304–15308

Christensen, J.G., Romach, E.H., Healy, L.N., Gonzales, A., Anderson, S.P., Malarkey, D.E., Corton, J.C., Fox, T.R., Cattley, R.C. & Goldsworthy, T.L. (1999) Altered bcl-2 family expression during non-genotoxic hepatocarcinogenesis in mice. *Carcinogenesis*, **20**, 1583–1590

Deugnier, Y.M., Charalambous, P., Le Quilleuc, D., Turlin, B., Searle, J., Brissot, P., Powell, L.W. & Halliday, J.W. (1993) Preneoplastic significance of hepatic iron-free foci in genetic hemochromatosis: a study of 185 patients. *Hepatology*, **18**, 1363–1369

Dombrowski, F., Lehringer-Polzin, M. & Pfeifer, U. (1994) Hyperproliferative liver acini after intraportal islet transplantation in streptozotocin-induced diabetic rats. *Lab. Invest.*, **71**, 688–699

Dombrowski, F., Bannasch, P. & Pfeifer, U. (1997) Hepatocellular neoplasms induced by low-number pancreatic islet transplants in streptozotocin diabetic rats. *Am. J. Pathol.*, **150**, 1071–1087

Dombrowski, F., Klotz, L., Hacker, H.J., Li, Y., Klingmüller, D., Brix, K., Herzog, V. & Bannasch, P. (2000) Hyperproliferative hepatocellular alterations after intra-portal transplantation of thyroid follicles. *Am. J. Pathol.*, **156**, 99–113

Enzmann, H. & Bannasch, P. (1987) Potential significance of phenotypic heterogeneity of focal lesions at different stages in hepatocarcinogenesis. *Carcinogenesis*, **8**, 1607–1612

Farber, E. (1996) The step-by-step development of epithelial cancer: from phenotype to genotype. *Adv. Cancer Res.*, **70**, 21–48

Farber, E. & Sarma, D.S. (1987) Hepatocarcinogenesis: a dynamic cellular perspective. *Lab. Invest.*, **56**, 4–22

Feo, F., Pascale, R.M., Simile, M.M., De Miglio, M.R., Muroni, M.R. & Calvisi, D. (2000a) Genetic alterations in liver carcinogensis: implications for new preventive and therapeutic strategies. *Crit. Rev. Oncogenesis*, **11**, 19–62

Feo, F., Pascale, R.M., Simile, R.M. & De Miglio, M.R. (2000b) Role of dehydroepiandrosterone in experimental and human carcinogenesis. In: Kalimi, M. & Regelson, W., eds, *Dehydroepiandrosterone (DHEA)*, Berlin, de Gruyter, pp. 215–236

Francavilla, A., Carr, B.I., Azzarone, A., Polimeno, L., Wang, Z., Van Thiel, D.H., Subbotin, V., Prelich, J.G. & Starzl, T.E. (1994) Hepatocyte proliferation and gene expression induced by triiodothyronine *in vivo* and *in vitro*. *Hepatology*, **20**, 1237–1241

Friedrich-Freksa, H., Papadopulu, G. & Gössner, W. (1969) Histochemische Untersuchungen der Cancerogenese in der Rattenleber nach zeitlich begrenzter Verabfolgung von Diethylnitrosamin. *Z. Krebsforsch.*, **72**, 240–253

Garcea, R., Daino, L., Pascale, R.M., Frasetto, S., Cozzolino, P., Ruggiu, M.E. & Feo, F. (1987) Inhibition by dehydroepiandrosterone of liver preneoplastic foci formation in rats after initiation-selection in experimental carcinogenesis. *Toxicol. Pathol.*, **15**, 164–168

Gerbracht, U., Bursch, W., Kraus, P., Putz, B., Reinacher, M., Zimmermann-Trosiener, I. & Schulte-Hermann, R. (1990) Effects of hypolipidemic drugs nafenopin and clofibrate on phenotypic expression and cell death (apoptosis) in altered foci of rat liver. *Carcinogenesis*, **11**, 617–624

Gonzalez, F.J., Peters, J.M. & Cattley, R. (1998) Mechanism of action of nongenotoxic peroxisome proliferators: role of the peroxisome proliferator-activated receptor α. *J. Natl Cancer Inst.*, **90**, 1702–1709

Goodman, D.G., Maronpot, R.R., Newberne, P.M., Popp, J.A. & Squire, R.A. (1994) Proliferative and selected lesions in the liver of rats. In: Street, C.S., Burek, J.D., Hardisty, J.F., Garner, F.M., Leininger, J.R., Pletscher, J.M. & Moch, R.W., eds, *Guides for Toxicologic Pathology*, Washington, Society of Toxicologic Pathologists/-American Registry of Pathology/Armed Forces Institute of Pathology, pp. G1-5, 1–24

Grasl-Kraupp, B., Huber, W., Just, W., Gibson, G. & Schulte-Hermann, R. (1993a) Enhancement of peroxisomal enzymes, cytochrome P-452 and DNA synthesis in putative preneoplastic foci of rat liver treated with the peroxisome proliferator nafenopin. *Carcinogenesis*, 14, 1007–1012

Grasl-Kraupp, B., Waldhör, T., Huber, W. & Schulte-Hermann, R. (1993b) Glutathione S-transferase isoenzyme patterns in different subtypes of enzyme-altered rat liver foci treated with the peroxisome proliferator nafenopin or with phenobarbital. *Carcinogenesis*, **14**, 2407–2412

Green, S. & Wahli, W. (1994) Peroxisome proliferator-activated receptors: finding the orphan a home. *Mol. Cell. Endocrinol.*, **100**, 149–153

Grisham, J.W. (1996) Interspecies comparison of liver carcinogenesis: implications for cancer risk assessment. *Carcinogenesis*, **18**, 59–81

Grobholz, R., Hacker, H.J., Thorens, B. & Bannasch, P. (1993) Reduction in the expression of glucose transporter protein GLUT2 in preneoplastic and neoplastic hepatic lesions and reexpression of GLUT1 in late stages of hepatocarcinogenesis. *Cancer Res.*, **53**, 4204–4211

Hacker, H.J., Moore, M.A., Mayer, D. & Bannasch, P. (1982) Correlative histochemistry of some enzymes of carbohydrate metabolism in preneoplastic and neoplastic lesions in the rat liver. *Carcinogenesis*, **3**, 1265–1272

Hacker, H.J., Steinberg, P. & Bannasch, P. (1998) Pyruvate kinase isoenzyme shift from L-type to M_2-type is a late event in hepatocarcinogenesis induced in rats by a choline-deficient/D,L-ethionine-supplemented diet. *Carcinogenesis*, **19**, 99–107

Harada, T., Maronpot, R.R., Morris, R.W. & Boorman, G.A. (1989) Observations on altered hepatocellular foci in National Toxicology Program two-year carcinogenicity studies in rats. *Toxicol. Pathol.*, **17**, 690–708

Hasegawa, R. & Ito, N. (1994) Hepatocarcinogenesis in the rat. In: Waalkes, M.P. & Ward, J.M., eds, *Carcinogenesis*, New York, Raven, pp. 39–65

Hayashi, F., Tamura, H., Yamada, J. Kasai, H. & Suga, T. (1994) Characterstics of the hepatocarcinogenesis caused by dehydroepiandrosterone, a peroxisome proliferator, in male F-344 rats. *Carcinogenesis*, **15**, 2215–2219

Hertz, R. & Bar-Tana, J. (1998) Peroxisome proliferator-activated receptor (PPAR) alpha activation and its consequences in humans. *Toxicol. Lett.*, **102–103**, 85–90

Hertz, R., Kalderon, B. & Bar-Tana, J. (1993) Thyromimetic effect of peroxisome proliferators. *Biochimie*, **75**, 257–261

Hertz, R., Nikodem, V., Ben-Ishai, A., Berman, I. & Bar-Tana, J. (1996) Thyromimetic mode of action of peroxisome proliferators: activation of "malic" enzyme gene transcription. *Biochem. J.*, **319**, 241–248

Hosokawa, S., Tatematsu, M., Aoki, T., Nakanowatari, J., Igarashi, T. & Ito, N. (1989) Modulation of diethylnitrosamine-initiated placental glutathione positive preneoplastic and neoplastic lesions by clofibrate, a hepatic peroxisome proliferator. *Carcinogenesis*, **10**, 2237–2241

Imaida, K., Fukushima, S., Shirai, T., Ohtani, M., Nakanishi, K. & Ito, N. (1983) Promoting activities of butylated hydroxyanisole and butylated hydroxytoluene on 2-stage urinary bladder carcinogenesis and inhibition of γ-glutamyltranspeptidase-positive foci development in the liver of rats. *Carcinogenesis*, **4**, 895–899

Ito, N., Tsuda, H., Hasegawa, R., Tatematsu, M., Imaida, K. & Asamoto, M. (1988) Medium-term bioassays for environmental carcinogens – two-step liver and multi-organ carcinogenesis protocols. In: Travis, C.C., ed., *Biologically Based Methods for Cancer Risk Assessment*, New York, London, Plenum, pp. 209–230

Ito, N., Shirai, T. & Hasegawa, R. (1992) Medium-term bioassays for carcinogens. In: Vainio, H., Magee, P.N., McGregor, D.B. & McMichael, A.J., eds, *Mechanisms of Carcinogenesis in Risk Identification*, Lyon, IARC, pp. 353–388

Ito, N., Imaida, K., Tamano, S., Hagiwara, A. & Shirai, T. (1998) Medium-term bioassays as alternative carcinogenicity tests. *J. Toxicol. Sci.*, **23**, 103–106

Kalengayi, M.M.R., Ronchi, G. & Desmet, V.J. (1975) Histochemistry of gamma-glutamyl transpeptidase in rat liver during aflatoxin B_1-induced carcinogenesis. *J. Natl Cancer Inst.*, **55**, 579–588

Kaufmann, W.K., MacKenzie, S.A. & Kaufman, D.G. (1985) Quantitative relationship between hepatocytic neoplasms and islands of cellular alteration during hepatocarcinogenesis in the male F344 rat. *Am. J. Pathol.*, **119**, 171–174

Kaufmann, W.K., Rahija, R.J. MacKenzie, S.A. & Kaufman, D.G. (1987) Cell cycle-dependent initiation of hepatocarcinogenesis in rats by (±) 7r, 8t-dihydroxy-9t,10t-epoxy-7,8,9,10-tetrahydrobenzo(a)pyrene. *Cancer Res.*, **47**, 3771–3775

Kensler, T.W., Egner, P.A., Dolan, P.M., Groopman, J.D. & Roebuck, B.D. (1987) Mechanism of protection against aflatoxin tumorigenicity in rats fed 5-(2-pyrazinyl)-4-methyl-1,2-dithiol-3-thione (oltipraz) and related 1,2-dithiol-3-thiones and 1,2-dithiol-3-ones. *Cancer Res.*, **47**, 4271–4277

Kim, C.M., Koike, K., Saito, J., Miyamura, T. & Jay, G. (1991) HBx gene of hepatitis B virus induces liver cancer in transgenic mice. *Nature*, **351**, 317–320

Kitten, O. & Ferry, N. (1998) Mature hepatocytes actively divide and express gamma-glutamyl transpeptidase after D-galactosamine liver injury. *Liver*, **18**, 398–404

Klimek, F. & Bannasch, P. (1990) Biochemical microanalysis of pyruvate kinase activity in preneoplastic and neoplastic liver lesions induced in rat by N-nitrosomorpholine. *Carcinogenesis*, **11**, 1377–1380

Klimek, F. & Bannasch, P. (1993) Isoenzyme shift from glucokinase to hexokinase is not an early but a late event in hepatocarcinogenesis. *Carcinogenesis*, **14**, 1857–1861

Klimek, F., Mayer, D. & Bannasch, P. (1984) Biochemical microanalysis of glycogen content and glucose-6-phosphate dehydrogenase activity in focal lesions of the rat liver induced by N-nitrosomorpholine. *Carcinogenesis*, **5**, 265–268

Klotz, L., Hacker, H.J., Klingmüller, D., Bannasch, P., Pfeifer, U. & Dombrowski, F. (2000) Hepatocellular alterations after intraportal transplantation of ovarian tissue in ovariectomized rats. *Am. J. Pathol.*, **156**, 1613–1626

Kopp-Schneider, A., Portier, C. & Bannasch, P. (1998) A model for hepatocarcinogenesis treating phenotypical changes in focal hepatocellular lesions as epigenetic events. *Math. Biosci.*, **148**, 181–204

Lazaro, C.A., Rhim, J.A., Yamada, Y. & Fausto, N. (1998) Generation of hepatocytes from oval cell precursors in culture. *Cancer Res.*, **58**, 5514–5522

Ledda-Columbano, G.M., Perra, A., Piga, R., Pibiri, M., Loi, R., Shinozuka, H. & Columbano, A. (1999) Cell proliferation induced by 3,3,5-triiodo-L-thyronine is associated with a reduction in the number of preneoplastic hepatic lesions. *Carcinogenesis*, 20, 2299–2304

Ledda-Columbano, G.M., Perra, A., Loi, R., Shinozuka, H. & Columbano, A. (2000) Cell proliferation induced by triiodo-thyronine in rat liver is associated with nodule regression and reduction of hepatocellular carcinomas. *Cancer Res.*, **60**, 603–609

Marsman, S.D. & Popp, J.A. (1994) Biological potential of basophilic hepatocellular foci and hepatic adenoma induced by the peroxisome proliferator, Wy-14,643. *Carcinogenesis*, **15**, 111–117

Mayer, D., Metzger, C., Leonetti, P., Beier, K., Benner, A. & Bannasch, P. (1998a) Differential expression of key enzymes of energy metabolism in preneoplastic and neoplastic rat liver lesions induced by N-nitrosomorpholine and dehydroepiandrosterone. *Int. J. Cancer (Pred. Oncol.)*, **79**, 232–240

Mayer, D., Klimek, F. & Bannasch, P. (1998b) Cytochemical and biochemical studies on adenylate cyclase activity in preneoplastic and neoplastic liver tissue and cultured liver cells. *Microsc. Res. Techn.*, **40**, 463–472

Metzger, C., Mayer, D., Hoffmann, H., Bocker, T., Hobe, G., Benner, A. & Bannasch, P. (1995) Sequential appearance and ultrastructure of amphophilic cell foci, adenomas and carcinomas in the liver of male and female rats treated with dehydroepiandrosterone (DHEA). *Toxicol. Pathol.*, **23**, 591–605

Metzger, C., Bannasch, P. & Mayer, D. (1998) Enhancement and phenotypic modulation of *N*-nitrosomorpholine-induced hepatocarcinogenesis by dehydroepiandrosterone. *Cancer Lett.*, **121**, 125–131

Montesano, R., Hainaut, P. & Wild, C.P. (1997) Hepatocellular carcinoma: from gene to public health. *J. Natl Cancer Inst.*, **89**, 1844–1851

Moore, M.A. & Kitagawa, T. (1986) Hepatocarcinogenesis in the rat: the effect of the promoters and carcinogens in vivo and in vitro. *Int. Rev. Cytol.*, **101**, 125–173

Moore, M.A., Thamavit, N., Tsuda, H., Sato, K., Ichira, A. & Ito, N. (1986) Modifying influence of dehydroepiandrosterone on the development of dihydroxy-di-n-propylnitrosamine-initiated lesions in the thyroid, lung and liver of F344 rats. *Carcinogenesis*, **7**, 311–316

Moore, M.A., Park, C.B. & Tsuda, H. (1998) Implications of hyperinsulinaemia-diabetes-cancer link for preventive efforts. *Eur. J. Cancer Prev.*, **7**, 89–107

Nehrbass, D. (2000) *Veränderungen in der Insulin-Signaltransduktionskaskade während der Hepatocarcinogenese der Ratte*. University of Giessen, Thesis

Nehrbass, D., Klimek, F. & Bannasch, P. (1998) Overexpression of insulin receptor substrate-1 emerges early in hepatocarcinogenesis and elicits preneoplastic hepatic glycogenosis. *Am. J. Pathol.*, **152**, 341–345

Nehrbass, D., Klimek, F., Bannasch, P. & Mayer, D. (1999) Insulin receptor substrate-1 is over-expressed in glycogenotic but not in amphophilic preneoplastic hepatic foci induced in rats by *N*-nitrosomorpholine and dehydroepiandrosterone. *Cancer Lett.*, **140**, 75–79

Numoto, S., Furukawa, K., Furuya, K. & Williams, G.M. (1984) Effects of the hepatocarcinogenic peroxisome-proliferating hypolipidemic agents clofibrate and nafenopin on the rat liver cell membrane enzymes gamma-glutamyl transpeptidase and alkaline phosphatase and on the early stages of liver carcinogenesis. *Carcinogenesis*, **5**, 1603–1611

Ober, S., Zerban, H., Spiethoff, A., Wegener, K., Schwarz, M. & Bannasch, P. (1994) Preneoplastic foci of altered hepatocytes induced in rats by irradiation with α-particles of Thorotrast and neutrons. *Cancer Lett.*, **83**, 81–88

Oehlert, W. (1978) Radiation-induced liver cell carcinoma in the rat. In: Remmer, H., Bolt, H.M., Bannasch, P. & Popper, H., eds, *Primary Liver Tumors,* Lancaster, MTP, pp. 217–225

Okuda, K. & Tabor, E., eds (1997) *Liver Cancer*, New York, Churchill Livingstone

Peters, J.M., Zhou, Y.C., Ram, P.A., Lee, S.S.T., Gonzalez, F.J. & Waxman, D.J. (1996) Peroxisome proliferator-activated receptor alpha required for gene induction by dehydroepiandrosterone-3 beta-sulfate. *Mol. Pharmacol.*, **50**, 67–74

Peters, J.M., Cattley, R.C. & Gonzalez, F.J. (1997) Role of PPAR alpha in the mechanism of action of the nongenotoxic carcinogen and peroxisome proliferator Wy-14,643. *Carcinogenesis*, **18**, 2029–2033

Pitot, H.C. (1990) Altered hepatic foci: their role in murine hepatocarcinogenesis. *Ann. Rev. Pharmacol. Toxicol.*, **30**, 465–500

Pitot, H.C. & Dragan, Y.P. (1994) Chemical induction of hepatic neoplasia. In: Fausto, N., Jakoby, W.B. & Schachter, D.A., eds, *The Liver, Biology and Pathobiology*, 3rd Ed., New York, Raven, pp. 1467–1495

Pitot, H.C., Campbell, H.A., Maronpot, R., Bawa, N., Rizvi, T.A., Xu, Y.-H., Sargent, L., Dragan, Y. & Pyron, M. (1989) Critical parameters in the quantitation of the stages of initiation, promotion, and progression in one model of hepatocarcinogenesis in the rat. *Toxicol. Pathol.*, **17**, 594–612

Radaeva, S., Li, Y., Hacker, H.J., Burger, V., Kopp-Schneider, A. & Bannasch, P. (2000) Hepadnaviral hepatocarcinogenesis: *in situ* visualization of viral antigens, cytoplasmic compartmentation, enzymic patterns, and cellular proliferation in preneoplastic hepatocellular lineages in woodchucks. *J. Hepatol.*, **33**, 580–600

Rao, M.S., Lalwani, N.D., Scarpelli, D.G. & Reddy, J.K. (1982) The absence of γ-glutamyltranspeptidase activity in putative preneoplastic lesions and in hepatocellular carcinomas induced in rats by the hypolipidemic peroxisome proliferator Wy-14,643. *Carcinogenesis*, **3**, 1231–1233

Rao, M.S., Subbaro, V., Yeldani, A.V. & Reddy, J.K. (1992) Hepatocarcinogenicity of dehydroepiandrosterone in the rat. *Cancer Res.*, **52**, 2977–2979

Roebuck, B.D., Liu, Y.-L., Rogers, A.E., Groopman, J.D. & Kensler, W. (1991) Protection against aflatoxin B_1-induced hepatocarcinogenesis in F344 rats by 5-(2-pyrazinyl)-4-methyl-1,2-dithiole-3-thione (oltipraz): predictive role for short-term molecular dosimetry. *Cancer Res.*, **51**, 5501–5506

Rogler, C.E., Rogler, L.E., Yang, D., Breiteneder-Geleef, S., Gong, S. & Wang H. (1995) Contributions of hepadnavirus research to our understanding of hepatocarcinogenesis In: Jirtle, R.L., ed., *Liver Regeneration and Carcinogenesis*, San Diego, Academic Press, pp. 113–140

Sato, K. (1989) Glutathione transferases as markers of preneoplasia and neoplasia. *Adv. Cancer Res.*, **52**, 205–255

Sato, K., Kitahara, A., Satoh, K., Ishikawa, T., Tatematsu, M. & Ito, N. (1984) The placental form of glutathione S-transferase as a new marker protein for preneoplasia in rat chemical carcinogenesis. *Gann*, **75**, 199–202

Schoongans, K., Martin, G., Staels, B. & Auwerx, J. (1997) Peroxisome proliferator-activated receptors, orphans with ligands and function. *Curr. Opin. Lipidol.*, **8**, 159–166

Schwarz, M., Buchmann, A., Schulte, M., Pearson, D. & Kunz, W. (1989) Heterogeneity of enzyme-altered foci in rat liver. *Toxicol. Lett.*, **49**, 297–317

Slupsky, J.R., Weber, C.K., Ludwig, S. & Rapp, U.R. (1998) Raf-dependent signaling pathways in cell growth and differentiation. In: Bannasch, P., Kanduc, D., Papa, S. & Tager, J.M., eds, *Cell Growth and Oncogenesis*, Basel, Birkhäuser, pp. 75–95

Steele, V.E., Kelloff, G.J., Balentine, D., Boone, C.W., Mehta, R., Bagheri, D., Sigman, C.C., Zhu, S. & Sharma, S. (2000) Comparative chemopreventive mechanisms of green tea, black tea and selected polyphenol extracts measured by *in vitro* bioassays. *Carcinogenesis*, **21**, 63–67

Steinberg, P., Klingelhöffer, A., Schäfer, A., Wüst, G., Weiße, G., Oesch, F. & Eigenbrodt, E. (1999) Expression of pyruvate kinase M_2 in preneoplastic hepatic foci of N-nitroso-morpholine-treated rats. *Virchows Arch.*, **434**, 213–220

Ströbel, P., Klimek, F., Kopp-Schneider, A. & Bannasch, P. (1998) Xenomorphic hepatocellular precursors and neoplastic progression of tigroid cell foci induced in rats with low doses of N-nitrosomorpholine. *Carcinogenesis*, **19**, 2069–2080

Stuver, S.O. (1998) Towards global control of liver cancer. *Seminars Cancer Biol.*, **18**, 299–306

Su, Q., Benner, A., Hofmann, W.J., Otto, G., Pichlmayr, R. & Bannasch, P. (1997) Human hepatic preneoplasia: phenotypes and proliferation kinetics of foci and nodules of altered hepatocytes and their relationship to liver cell dysplasia. *Virchows Arch.*, **431**, 391–406

Su, Q., Schröder. C.H., Hofmann, W.J., Otto, G., Pichlmayr, R. & Bannasch, P. (1998) Expression of hepatitis B virus X protein in HBV-infected human livers and hepatocellular carcinomas. *Hepatology*, **26**, 1109–1120

Su, Q., Schröder, C.H., Otto, G. & Bannasch, P. (2000) Overexpression of p53 is not directly related to hepatitis B x protein expression and is associated with neoplastic progression rather than hepatic preneoplasia. *Mutation Res.*, **462**, 365–380

Thamavit, W., Tatematsu, M., Ogiso, T., Mera, Y., Tsuda, H. & Ito, N. (1985) Dose-dependent effects of butylated hydroxyanisole, butylated hydroxytoluene, and ethoxyquin in induction of foci of rat liver cells containing the placental form of glutathione S-transferase. *Cancer Lett.*, **27**, 295–305

Toshkov, I., Hacker, H.J., Roggendorf, M. & Bannasch, P. (1990) Phenotypic patterns of preneoplastic and neoplastic hepatic lesions in woodchucks infected with woodchuck hepatitis virus. *J. Cancer Res. Clin. Oncol.*, **116**, 581–90.

Toshkov, I., Chisari, F.V. & Bannasch, P. (1994) Hepatic preneoplasia in hepatitis B virus transgenic mice. *Hepatology*, **20**, 1162–1172

Tsuda, H., Ozaki, K., Uwagawa, S., Yamaguchi, S., Hakoi, K., Aoki, T., Kato, T., Sato, K. & Ito, N. (1992) Effects of modifying agents on conformity of enzyme phenotype and proliferative potential in focal preneoplastic and neoplastic liver cell lesions in rats. *Jpn. J. Cancer Res.*, **83**, 1154–1165

US National Institute of Environmental Health Sciences (1989) Significance of foci of cellular alteration. A symposium. *Toxicol. Pathol.*, **17**, 557–735

Valera, A., Pujol, A., Gregori, X., Riu, E., Visa, J. & Bosch, F. (1995) Evidence from transgenic mice that myc regulates hepatic glycolysis. *FASEB J.*, **9**, 1067–1078

Weber, E. & Bannasch P. (1994a) Dose and time dependence of the cellular phenotype in rat hepatic preneoplasia and neoplasia induced by single oral exposures to N-nitrosomorpholine. *Carcinogenesis*, **15**, 1219–1226

Weber, E. & Bannasch, P. (1994b) Dose and time dependence of the cellular phenotype in rat hepatic preneoplasia and neoplasia induced in stop experiments by oral exposure to N-nitroso-morpholine. *Carcinogenesis*, **15**, 1227–1234

Weber, E. & Bannasch, P. (1994c) Dose and time dependence of the cellular phenotype in rat hepatic preneoplasia and neoplasia induced by continuous oral exposure to N-nitroso-morpholine. *Carcinogenesis*, **15**, 1235–1242

Weber, E., Moore, M.A. & Bannasch, P. (1988) Enzyme histochemical and morphological phenotype of amphophilic foci and amphophilic/tigroid cell neoplastic nodules in rat liver after combined treatment with dehydroepiandrosterone and N-nitrosomorpholine. *Carcinogenesis*, **9**, 1049–1054

Williams, G.M. & Enzmann, H. (1998) The rat liver hepatocellular-altered focus-limited bioassay for chemicals with carcinogenic activity. In: Kitten, K.T., ed., *Carcinogenicity*, New York, Marcel Dekker, pp. 361–394

Williams, G.M., Klaiber, M., Parker, S.E. & Farber, E. (1976) Nature of early appearing carcinogen-induced liver lesions resistant to iron accumulation. *J. Natl Cancer Inst.*, **57**, 157–165

Zerban, H., Radig, S., Kopp-Schneider, A. & Bannasch, P. (1994) Cell proliferation and cell death (apoptosis) in hepatic preneoplasia and neoplasia are closely related to phenotypic cellular diversity and instability. *Carcinogenesis*, **15**, 2467–2473

Corresponding author:

P. Bannasch
Division of Cell Pathology - C0100
Deutsches Krebsforschungszentrum
Im Neuenheimer Feld 280
69120 Heidelberg,
Germany

Biomarkers in Cancer Chemoprevention
Miller, A.B., Bartsch, H., Boffetta, P., Dragsted, L. and Vainio, H. eds
IARC Scientific Publications No. 154
International Agency for Research on Cancer, Lyon, 2001

Hepatocellular carcinoma: susceptibility markers

H.E. Blum

Genetic polymorphisms of the carcinogen-metabolizing enzymes cytochrome P450 (CYP), glutathione *S*-transferase (GST) M1 and *N*-acetyltransferase (NAT2) as well as *p53* polymorphisms have been studied experimentally as susceptibility markers for hepatocellular carcinoma development in hepatocellular carcinoma cell lines and in mouse hepatocellular carcinomas. In addition, these susceptibility markers have been studied in hepatocellular carcinoma patients, in the context of coexisting alcohol consumption, smoking and/or HBV infection. To date, there is no clear evidence that susceptibility markers have an overall impact on hepatocarcinogenesis, but in subgroups of individuals, such as smokers, susceptibility markers are emerging indicators for hepatocellular carcinoma risk definition.

Introduction

Hepatocellular carcinoma (HCC) is one of the most common tumours in the world and usually represents a late complication of chronic progressive liver disease (Schafer & Sorrell, 1999). Although less frequent in the United States and Europe, these tumours have an annual incidence of up to 500 cases per 100 000 population in certain regions of Asia and sub-Saharan Africa. The reasons for this high incidence are chronic infections with hepatitis B virus (HBV), hepatitis C virus (HCV) and the delta virus (HDV), as well as HBV–HCV coinfections (Di Bisceglie, 1998). The clinical course of HBV and HCV infection depends in part on molecular characteristics of the viruses, in part on the patients' HLA haplotype (Kuzushita *et al.*, 1998) and in part on other coexisting risk factors. Well recognized non-viral exogenous agents associated with HCC development are alcohol and aflatoxin B_1 (AFB_1). In the western countries, alcohol-induced liver injury is a leading cause of liver cirrhosis and the most important HCC risk factor (Donato *et al.*, 1997). In southern China and Africa, dietary ingestion of high levels of AFB_1 may present a special environmental hazard, particularly in chronically HBV-infected individuals. Other exogenous factors have also been incriminated and include dietary iron overload (Mandishona *et al.*, 1998), long-term use of oral contraceptives (Mant & Vessey, 1995; Waetjen & Grimes, 1996) and high-dose anabolic steroids. The development of hepatic cirrhosis, particularly in association with inherited genetic diseases, such as α-1-antitrypsin deficiency or haemochromatosis, places the individual at a greatly increased risk for HCC development. The major clinical HCC risk factor is liver cirrhosis, since 70–90% of HCCs develop in a macronodular cirrhosis. In addition, HCCs are more frequent in males than in females and the incidence generally increases during the last decades of life and with age (Poynard *et al.*, 1997; Taylor-Robinson *et al.*, 1997; De Vos Irvine *et al.*, 1998; Deuffic *et al.*, 1998; El-Serag & Mason, 1999). The etiological risk factors may exist either alone or in combination, e.g., HCV infection and alcohol use (Corrao & Arico, 1998) or HBV infection and exposure to AFB_1 (Sun *et al.*, 1999), further increasing HCC risk.

As susceptibility markers of HCC development, carcinogen-metabolizing enzymes have been analysed in HCC cell lines and in mouse HCCs as well as in HCC patients. In HCC patients, susceptibility markers have been studied in the context of coexisting alcohol consumption, smoking and/or HBV infection. In the following, genetic polymorphisms of cytochrome P450 (*CYP*), glutathione *S*-transferase (*GST*) M1, *N*-acetyltransferase (*NAT2*) and *p53* are discussed as susceptibility markers of HCC development.

Cytochrome P450

The expression and polymorphisms of cytochrome P450 (CYP) isozymes CYP1A1, CYP2A6, CYP2B6, CYP2D6, CYP2E1, CYP3A4, CYP3D4 and CYP1A or CYP3D7 as carcinogen-metabolizing enzymes have been studied *in vitro* as well as in animal models and HCC patients.

The status of CYP enzymes was studied in a transgenic mouse model (ATX mice) carrying the HBV X gene under the control of the α-1-antitrypsin regulatory elements (Chomarat *et al.*, 1998). The HBx protein is suspected to play a central role in hepatocarcinogenesis through its *trans*-activation of cellular genes, including oncogenes, as well as through *cis*-acting elements. Indeed, the HBV X gene has been shown to *trans*-activate the expression of specific CYP isozymes. In ATX mice, however, no *trans*-activation of CYP1A or CYP2A5 was observed. These data indicate that HBx expression alone is insufficient to induce *trans*-activation of *CYP* genes (Chomarat *et al.*, 1998). As detailed below, there is evidence, however, that *p53* polymorphism at codon 72 interacts with CYP1A1 and carotenoid levels in smoking-related hepatocarcinogenesis (Yu *et al.*, 1999b).

Epidemiological evidence indicates that dietary AFB_1 and mutations in the *p53* tumour-suppressor gene are involved in hepatocarcinogenesis. However, the correlation between AFB_1–DNA adduct formation or *p53* mutations and their activation pathways has not been elucidated to date. In this context, Mace *et al.* (1997) established SV40-immortalized human liver epithelial cell lines stably expressing CYP1A2, CYP2A6, CYP2B6 and CYP3A4 proteins. These cell lines were highly sensitive to the cytotoxic effects of AFB_1, with formation of DNA adducts and CYP-dependent mutations of the *p53* gene. These findings indicate that the differential expression of specific *CYP* genes in human hepatocytes can modulate AFB_1-induced cytotoxicity, DNA adduct levels and the frequency of *p53* mutations (Mace *et al.*, 1997).

Since cigarette smoking has been associated with increased HCC risk in some epidemiological studies, and *CYP1A1* is involved in biotransformation of tobacco-derived polycyclic aromatic hydrocarbons (PAHs), *CYP1A1* polymorphisms have been studied in HBV-infected patients (Yu *et al.*, 1999a). *CYP1A1* genotypes were associated with increased HCC risk in smokers but not in non-smokers, indicating that tobacco-derived PAHs can increase HCC risk among HBV-infected patients and that *CYP1A1* polymorphism is an important modulator of the hepatocarcinogenic effect of PAHs (Yu *et al.*, 1999a).

In mouse liver tumours, the hepatic CYP2A6 enzyme is invariably overexpressed, suggesting that the oxidative metabolism of procarcinogens in the liver may be involved in hepatocarcinogenesis. In contrast, in human HCCs, CYP2A6 overexpression has been found not to be an invariable phenotype (Raunio *et al.*, 1998). In another study of HCC patients and healthy individuals, however, subjects with two active *CYP2D6* genes (rapid metabolizers) were found to be at increased risk for HCC development, especially in patients without evidence of HBV or HCV infection (Agundez *et al.*, 1996).

Susceptibility to HCC development has also been studied in relation to *CYP2E1* genotype. In cigarette smokers but not in non-smokers, homozygosity for the c1/c1 genotype was associated with significantly increased risk for HCC, indicating that tobacco-derived PAHs may play a role in HCC development (Yu *et al.*, 1995). For *CYP2E1* genotypes, no such association was found in Korean HCC patients, irrespective of HBV and HCV infection or chronic alcohol consumption (Lee *et al.*, 1997).

Glutathione *S*-transferase M1

Genetic polymorphisms of glutathione *S*-transferases (GST) are involved in carcinogen metabolism. GST status has been studied in a transgenic mouse model (ATX mice) carrying the HBV X gene under the control of the α-1-antitrypsin regulatory elements (Chomarat *et al.*, 1998). In this model, no *trans*-activation of GST was observed, indicating that HBx expression alone is insufficient to induce *trans*-activation of *GST* genes (Chomarat *et al.*, 1998).

Susceptibility to HCC development was further studied by *GSTM1* genotyping of HCC patients as well as healthy individuals. *GSTM1*-null genotype was not associated with increased HCC risk (Yu *et al.*, 1995). As detailed below, however, there is a significantly increased HCC risk in patients with the *p53* Pro allele at codon 72 in smokers with the *GSTM1*-null phenotype (Yu *et al.*, 1999a).

N-Acetyltransferase 2

N-Acetyltransferase 2 (NAT2) is a polymorphic enzyme which is expressed in the liver in a genotype-dependent manner and is involved in activation and inactivation of carcinogens through *N*-acetylation. In a study of HCC patients and healthy individuals, subjects with inactivating mutations of NAT2 (slow acetylators) were found to be at increased risk for HCC development, especially in patients without evidence of HBV or HCV infection (Agundez *et al.*, 1996).

p53

An analysis of the Pro variant allele of the *p53* polymorphism at codon 72 in HBV-infected patients with or without HCC revealed no overall increased HCC risk in one study (Yu *et al.*, 1999a). However, there were synergistic effects on HCC development for the *p53* Pro allele at codon 72 and chronic liver disease as well as for HCC family history in first-degree relatives. A significantly increased HCC risk in patients with the Pro allele was further observed in smokers with the *GSTM1*-null phenotype. *p53* polymorphism at codon 72 also interacts with CYP1A1 and carotenoid levels in smoking-related hepatocarcinogenesis (Yu *et al.*, 1999a).

Conclusion

There is no clear evidence to date that susceptibility markers have an overall impact on hepatocarcinogenesis. In subgroups of individuals, however, such as smokers, susceptibility markers are emerging indicators for hepatocellular carcinoma risk definition. The establishment of new hepatocellular carcinoma animal models and larger-scale studies in hepatocellular carcinoma patients stratified according to individual characteristics, such as smoking, alcohol ingestion, HBV or HCV infection and others, should allow clearer definition of the role of susceptibility in the management of individuals at risk of hepatocellular carcinoma development.

References

Agundez, J.A., Olivera, M., Ladero, J.M., Rodriguez-Lescure, A., Ledesma, M.C., Diaz-Rubio, M., Meyer, U.A. & Benitez, J. (1996) Increased risk for hepatocellular carcinoma in NAT2-slow acetylators and CYP2D6-rapid metabolizers. *Pharmacogenetics*, **6**, 501–512

Chomarat, P., Rice, J.M., Slagle, B.L. & Wild, C.P. (1998) Hepatitis B virus-induced liver injury and altered expression of carcinogen metabolising enzymes: the role of the HBx protein. *Toxicol. Lett.*, **102–103**, 595–601

Corrao, G. & Arico, S. (1998) Independent and combined action of hepatitis C virus infection and alcohol consumption on the risk of symptomatic liver cirrhosis. *Hepatology*, **27**, 914–919

De Vos Irvine, H., Goldberg, D., Hole, D.J. & McMenamin, J. (1998) Trends in primary liver cancer. *Lancet*, **351**, 215–216

Deuffic, S., Poynard, T., Buffat, L. & Valleron, A.-J. (1998) Trends in primary liver cancer. *Lancet*, **351**, 214–215

Di Bisceglie, A.M. (1998) Hepatitis C. *Lancet*, **351**, 351–355

Donato, F., Tagger, A., Chiesa, R., Ribero, M.L., Tomasoni, V., Fasola, M., Gelatti, U., Portera, G., Boffetta, P. & Nardi, G. (1997) Hepatitis B and C virus infection, alcohol drinking and hepatocellular carcinoma: a case-control study in Italy. *Hepatology*, **26**, 579–584

El-Serag, H.B. & Mason, A.C. (1999) Rising incidence of hepatocellular carcinoma in the United States. *New Engl. J. Med.*, **340**, 745–750

Kuzushita, N., Hayashi, N., Moribe, T., Katayama, K., Kanto, T., Nakatani, S., Kaneshige, T., Tatsumi, T., Ito, A., Mochizuki, K., Sasaki, Y., Kasahara, A. & Hori, M. (1998) Influence of HLA haplotypes on the clinical courses of individuals infected with hepatitis C virus. *Hepatology*, **27**, 240–244

Lee, H.S., Yoon, J.H., Kamimura, S., Iwata, K., Watanabe, H. & Kim, C.Y. (1997) Lack of association of cytochrome P450 2E1 genetic polymorphisms with the risk of human hepatocellular carcinoma. *Int. J. Cancer*, **71**, 737–740

Mace, K., Aguilar, F., Wang, J.S., Vautravers, P., Gomez-Lechon, M., Gonzalez, F.J., Groopman, J., Harris, C.C. & Pfeifer, A.M. (1997) Aflatoxin B1-induced DNA adduct formation and p53 mutations in CYP450-expressing human liver cell lines. *Carcinogenesis*, **18**, 1291–1297

Mandishona, E., MacPhail, A.P., Gordeux, V.R., Kedda, M.-A., Paterson, A.C., Rouault, T.A. & Kew, M.C. (1998) Dietary iron overload as a risk factor for hepatocellular carcinoma in black Africans. *Hepatology*, **27**, 1563–1566

Mant, J.W.F. & Vessey, M.P. (1995) Trends in mortality from primary liver cancer in England and Wales. *Br. J. Cancer*, **72**, 800–803

Poynard, T., Bedossa, P., Opolon, P., for the METAVIR, CLINIVIR and DOSVIRC Groups (1997) Natural history of liver fibrosis progression in patients with chronic hepatitis C. *Lancet*, **349**, 825–832

Raunio, H., Juvonen, R., Pasanen, M., Pelkonen, O., Paakko, P. & Soini, Y. (1998) Cytochrome P4502A6 (CYP2A6) expression in human hepatocellular carcinoma. *Hepatology*, **27**, 427–432

Schafer, D.F. & Sorrell, M.F. (1999) Hepatocellular carcinoma. *Lancet*, **353**, 1253–1257

Sun, Z., Lu, P., Gail, M.H., Pee, D., Zhang, Q., Ming, L., Wang, J., Wu, Y., Liu, G., Wu, Y. & Zhu, Y. (1999) Increased risk of hepatocellular carcinoma in male hepatitis B surface antigen carriers with chronic hepatitis who have detectable urinary aflatoxin metabolite M1. *Hepatology*, **30**, 379–383

Taylor-Robinson, S.D., Foster, G.R., Arora, S., Hargreaves, S. & Thomas, H.C. (1997) Increase in primary liver cancer in the UK, 1979–1994. *Lancet*, **350**, 1142–1143

Waetjen, L.E. & Grimes, D.A. (1996) Oral contraceptives and primary liver cancer: temporal trends in three countries. *Obstet. Gynecol.*, **88**, 945–949

Yu, M.W., Gladek-Yarborough, A., Chiamprasert, S., Santella, R.M., Liaw, Y.F. & Chen, C.J. (1995) Cytochrome P450 2E1 and glutathione S-transferase M1 polymorphisms and susceptibility to hepatocellular carcinoma. *Gastroenterology*, **109**, 1266–1273

Yu, M., Chiu, Y., Yang, S., Santella, R., Chern, H., Liaw, Y. & Chen, C. (1999a) Cytochrome P450 1A1 genetic polymorphisms and risk of hepatocellular carcinoma among chronic hepatitis B carriers. *Br. J. Cancer*, **80**, 598–603

Yu, M.W., Yang, S.Y., Chiu, Y.H., Chiang, Y.C., Liaw, Y.F. & Chen, C.J. (1999b) A p53 genetic polymorphism as a modulator of hepatocellular carcinoma risk in relation to chronic liver disease, familial tendency, and cigarette smoking in hepatitis B carriers. *Hepatology*, **29**, 697–702

Corresponding author:

H. E. Blum,
Department of Medicine II,
University of Freiburg,
Hugstetter Strasse 55,
D-79106 Freiburg,
Germany

Biomarkers in Cancer Chemoprevention
Miller, A.B., Bartsch, H., Boffetta, P., Dragsted, L. and Vainio, H. eds
IARC Scientific Publications No. 154
International Agency for Research on Cancer, Lyon, 2001

Carcinogen biomarkers for lung or oral cancer chemoprevention trials

S.S. Hecht

The potential applicability of specific carcinogen-derived biomarkers in chemoprevention trials against lung and oral cancer is discussed. At present, there are no examples of the use of these biomarkers in chemoprevention trials, but the principle has been established in chemoprevention trials directed at aflatoxin B_1-induced liver cancer. Polycyclic aromatic hydrocarbons (PAHs) and tobacco-specific nitrosamines are among the most important carcinogens invoked as causes of lung and oral cancer. Biomarkers that are potentially practical for current application in chemoprevention trials are 7,8-dihydroxy-9,10-epoxy-7,8,9,10-tetrahydrobenzo[*a*]pyrene–DNA adducts, as determined by HPLC with fluorescence detection, nitrosamino acids in urine, 4-(methylnitrosamino)-1-(3-pyridyl)-1-butanol and its glucuronides in urine, nicotine metabolites in urine, and metabolites of cytochrome P450 substrates in urine. Biomarkers that need further development or exploration before application in trials include 7-methylguanine in DNA, tobacco-specific nitrosamine–DNA adducts, acrolein/crotonaldehyde–DNA adducts, PAH–protein adducts, acetaldehyde–protein adducts, pyrene metabolites in urine and benzo[*a*]pyrene metabolites in urine. Such carcinogen derived-biomarkers could be applied in chemoprevention trials to test the hypothesis that chemopreventive agents alter carcinogen metabolic activation and detoxification and, ultimately, risk for cancer.

Introduction

Carcinogen-derived biomarkers are quantifiable compounds that are formed from specific carcinogens. Examples are adducts with DNA, haemoglobin or albumin, and metabolites in blood or urine. No examples have yet been reported of the use of these biomarkers in lung or oral cancer chemoprevention trials in humans. However, they are being effectively employed in studies of liver cancer prevention by oltipraz in individuals with high exposure to aflatoxin B_1 (Kensler *et al.*, 1997, 1998; Groopman & Kensler, 1999; see also Wild & Turner in this volume). There is every reason to believe that similar strategies can be used in studies of lung and oral cancer. This chapter discusses specific carcinogen-derived biomarkers potentially suitable for application in chemoprevention trials against cancers of the lung and oral cavity in humans.

Carcinogen involvement in lung cancer

Cigarette smoking causes 87% of lung cancer (American Cancer Society, 2000). There are 55 carcinogens in cigarette smoke that have been evaluated by IARC and for which there is sufficient evidence for carcinogenicity in either laboratory animals or humans (Hecht, 1999). Of these, 20 compounds have been found convincingly to induce lung tumours in at least one animal species and have been positively identified in cigarette smoke (Table 1) (Hecht, 1999). The potential contributions of each of these compounds, as well as free radicals and oxidative damage, to lung cancer induction in humans have been evaluated (Hecht, 1999). The evidence is strongest for specific polycyclic aromatic hydrocarbons (PAHs), typified by benzo[*a*]pyrene (B[a]P) and the tobacco-specific nitrosamine 4-(methylnitrosamino)-1-(3-pyridyl)-1-butanone (NNK). Figure 1 presents a scheme linking nicotine addiction and lung cancer via B[a]P, NNK and other carcinogens of cigarette smoke (Hecht, 1999). Nicotine addiction is the reason that people continue to smoke. While nicotine itself is not considered to be carcinogenic, each cigarette contains small doses of B[a]P, NNK and other carcinogens. Although each individual dose is small, the overall carcinogen dose in years of smoking is substantial. Most carcinogens in cigarette smoke require metabolic activation to exert their carcinogenic effects via formation of DNA adducts. Several competing detoxification processes protect

Table 1. Pulmonary carcinogens in cigarette smoke[a]

Carcinogen class	Compound	Amount in mainstream cigarette smoke (ng/cig)	Sidestream/mainstream ratio	Representative lung tumorigenicity in species
PAHs	Benzo[*a*]pyrene (B[a]P)	20–40	2.5–3.5	Mouse, rat, hamster
	Benzo[*b*]fluoranthane	4–22		Rat
	Benzo[*j*]fluoranthane	6–21		Rat
	Benzo[*k*]fluoranthane	6–12		Rat
	Dibenzo[*a,i*]pyrene	1.7–3.2		Hamster
	Indeno[1,2,3-*cd*]pyrene	4–20		Rat
	Dibenz[*a,h*]anthracene	4		Mouse
	5-Methylchrysene	0.6		Mouse
Aza-arenes	Dibenz[*a,h*]acridine	0.1		Rat
	7H-Dibenzo[*c,g*]carbazole	0.7		Hamster
N-Nitrosamines	N-Nitrosodiethylamine	ND–2.8	< 40	Hamster
	4-(Methylnitrosamino)-1-(3-pyridyl)-1-butanone (NNK)	80–770	1–4	Mouse, rat, hamster
Miscellaneous organic compounds	1,3-Butadiene	$20–70 \times 10^3$		Mouse
	Ethyl carbamate	20–38		Mouse
Inorganic compounds	Nickel	0–510	13–30	Rat
	Chromium	0.2–500		Rat
	Cadmium	0–6670	7.2	Rat
	Polonium-210	0.03–1.0 pCi	1.0–4.0	Hamster
	Arsenic	0–1400		None
	Hydrazine	24–43		Mouse

[a] From Hecht (1999)

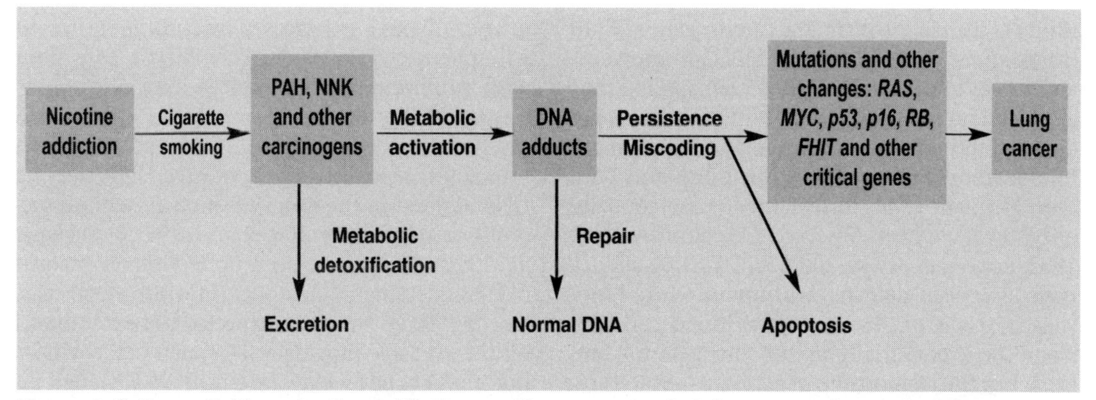

Figure 1. Scheme linking nicotine addiction and lung cancer via tobacco smoke carcinogens

against carcinogenesis. The balance between metabolic activation and detoxification is a target for chemopreventive agents. If metabolic activation can be decreased or detoxification increased, then DNA adduct formation should decrease. This should result in a decreased number of permanent mutations in critical regions of oncogenes and tumour-suppressor genes, and a consequent decrease in loss of normal cellular growth control mechanisms, with the ultimate result being decreased cancer incidence.

The biomarkers considered here are those that would specifically relate to changes in metabolic activation and detoxification of lung or oral carcinogens.

Carcinogen involvement in oral cancer

Cigarette smoking causes 90% of oral cancer in males and 60% in females (Shopland, 1995). Unlike lung cancer, the risk of oral cancer is multiplicatively enhanced by alcohol consumption (Blot *et al.*, 1996). The mechanistic framework illustrated in Figure 1 can also be applied to oral cancer, although the genetic changes have not been characterized as extensively. There is good evidence from animal studies that PAHs and tobacco-specific nitrosamines such as NNK and *N'*-nitrosonornicotine (NNN) play a significant role as causes of cancer of the oral cavity (Hecht & Hoffmann, 1989; Hoffmann & Hecht, 1990). Other nitrosamines may be involved, particularly in combination with alcohol intake (Chhabra *et al.*, 1996). Acetaldehyde, the major metabolite of ethanol, and other aldehydes in tobacco smoke may also contribute (Vaca *et al.*, 1998).

Snuff-dipping causes oral cancer (IARC, 1985). Tobacco-specific nitrosamines are the most prevalent strong carcinogens present in snuff and are likely to play a significant role as causative agents in people who use these products (Hecht, 1998). Betel-quid chewing, with tobacco, is the major cause of cancer of the oral cavity in India and other parts of southern Asia. Tobacco-specific nitrosamines, areca-specific nitrosamines and oxidative damage are believed to be causative factors (IARC, 1985; Hecht, 1998; Hoffmann *et al.*, 1994; Bartsch *et al.*, 1999).

DNA adducts as biomarkers

PAHs

A large number of studies have quantified 'PAH–DNA adducts' by immunochemical methods. These will not be considered further here because they do not quantify specific PAH adducts, but rather measure them as a group (Santella, 1999). Individual PAHs differ widely in carcinogenic activity, limiting the utility of this approach for chemoprevention trials. Similarly, many studies have employed ^{32}P-postlabelling to investigate 'hydrophobic-DNA adducts', some of which are probably PAH–DNA adducts, in humans (Kriek *et al.*, 1998). None of these studies has quantified specific PAH–DNA adducts. Consequently, the same limitations apply as to the immunochemical studies.

More specific methods including synchronous fluorescence spectroscopy, phosphorescence, and HPLC with fluorescence detection (HPLC-fluorescence) of released tetraols have been developed for quantitation of B[a]P–DNA adducts formed via 7,8-dihydroxy-9,10-epoxy-7,8,9,10-tetrahydroB[a]P

(BPDE), the major ultimate carcinogen of B[a]P (Kriek *et al.*, 1998). Of these, HPLC-fluorescence analysis of released tetraols has been applied most extensively. BPDE–DNA adducts have been detected in human lung, and relationships to aryl hydrocarbon hydroxylase inducibility and polymorphisms in genes for carcinogen-metabolizing enzymes have been observed (Alexandrov *et al.*, 1992; Rojas *et al.*, 1998, 2000). BPDE–DNA adducts have also been detected in human white blood cells (Rojas *et al.*, 1995). White blood cell DNA would be a potentially useful surrogate for lung DNA, but the relationship of adduct levels to those in the lung has not been extensively investigated.

In mice, BPDE–DNA adduct levels measured by HPLC-fluorescence of released tetraols are decreased in lung and liver by pretreatment with benzyl isothiocyanate (BITC), an inhibitor of B[a]P-induced lung tumorigenesis (Sticha *et al.*, 2000). However, there may be limitations to the application of this assay in chemoprevention trials in humans. In one study of adduct levels in coke-oven workers, the levels in 50% of the samples were below the detection limit (Rojas *et al.*, 1995). Among smokers in the same study, 65% of the samples had undetectable adduct levels. Similar results were obtained in a recent larger investigation; overall 52% of exposed individuals including coke-oven workers and smokers had detectable levels of the adduct (Rojas *et al.*, 2000). Remarkably, adduct levels were detected in 93% of subjects with *GSTM1*-null genotype, but not in individuals who were *GSTM1*-positive (Rojas *et al.*, 2000). Therefore, this biomarker may have utility in chemoprevention trials with smokers, provided they are preselected as *GSTM1*- null. Otherwise, the relatively frequent occurrence of non-detectable adduct levels would impose practical limitations on the use of this biomarker in chemoprevention studies.

Nitrosamines

Acid or enzymatic hydrolysis of DNA isolated from tissues of rodents treated with NNK or NNN releases 4-hydroxy-1-(3-pyridyl)-1-butanone (HPB) (Hecht, 1998). Studies *in vitro* and *in vivo* have demonstrated conclusively that this arises via pyridyloxobutylation of DNA (Hecht, 1998). HPB-releasing adducts in DNA can be quantified by gas chromatography/mass spectrometry (GC/MS) (Foiles *et al.*, 1991). In rats, DNA adduct levels in lung cell types are decreased by chronic treatment with phenethyl isothiocyanate (PEITC), consistent with inhibition of NNK-induced lung carcinogenesis by PEITC (Staretz *et al.*, 1997). The GC/MS method has been applied in limited studies in humans. One investigation detected HPB-releasing DNA adducts in the lungs of smokers, while a second was negative (Foiles *et al.*, 1991; Blomeke *et al.*, 1996). There have been no reports on analysis of HPB-releasing DNA adducts in white blood cells, but their levels would be expected to be low, based on the available lung data. This methodology does not appear to have adequate sensitivity at present for application in chemoprevention trials in humans.

Methylation of DNA by nitrosamines results in the formation of O^6- and 7-methylguanine (O^6- and 7-mG). These adducts may be produced upon exposure to NNK or other methylating nitrosamines such as *N*-nitrosodimethylamine. Their levels in human lung have been examined in several studies (Hecht & Tricker, 1999). The average level of O^6-mG is reported to be 27 adducts per 10^8 nucleotides in peripheral lung. Levels of 7-mdG in lung DNA are reported to range from 2.1 to 42.7 adducts per 10^7 dG; one study has reported higher levels of these adducts in smokers than in non-smokers. Based on limited data, levels of 7-mdG in lymphocyte and bronchial DNA are correlated (Mustonen *et al.*, 1993). Highly sensitive methods have been developed for detection of O^6-mG in human DNA (Kyrtopoulos, 1998). However, this adduct was non-detectable in 95% of 407 peripheral blood leukocyte DNA samples analysed (EUROGAST Study Group, 1994).

Other DNA adduct biomarkers

Adducts of acrolein and crotonaldehyde have been detected in human leukocyte DNA (Nath *et al.*, 1996). In gingival tissue DNA, levels were higher in smokers than in non-smokers (Nath *et al.*, 1998). This assay has not been tested with exfoliated oral cells. Acetaldehyde–DNA adducts have been detected in peripheral white blood cells of alcohol abusers, but rarely in control subjects (Fang & Vaca, 1997). 8-Oxo-dG is commonly detected in exfoliated oral cells (Yarborough *et al.*, 1996), leukocyte DNA (van Zeeland *et al.*, 1999; Asami *et al.*, 1996) and various tissues including lung (Asami *et al.*, 1997) and urine (Prieme *et al.*, 1998). Some conflicting results have been obtained, perhaps due to methodological difficulties (Collins, 1999). With the use of adequate techniques, 8-oxo-dG measurements could be useful particularly in studies of oral cancer associated with betel quid use, in cases where tobacco is not involved. Nevertheless, the significance of this adduct with respect to carcinogenesis is still unclear.

Protein adducts as biomarkers

PAHs

Non-specific immunochemical methods have been used to quantify PAH–albumin adducts, with the same limitations as discussed above (Tang *et al.*, 1999). Specific methods have been developed for quantitation of BPDE–haemoglobin and albumin adducts, with detection by GC/MS or fluorescence. In a study of newspaper vendors in high-density traffic areas, 60% of the subjects had detectable BPDE–haemoglobin adduct levels (Pastorelli *et al.*, 1996). BPDE–haemoglobin adducts were detectable in 14% of lung cancer patients, while BPDE–albumin adducts were detectable in 55% (Pastorelli *et al.*, 1998). In 65 employees with no occupational exposure, BPDE–haemoglobin adducts were detectable in 11% of the subjects in summer and 16% in winter (Pastorelli *et al.*, 1999). In another study, levels of these adducts were higher in smokers than in non-smokers and were detectable in all subjects. Adducts of chrysene-1,2-diol-,3,4-epoxide were also detected by GC/MS (Melikian *et al.*, 1997). Laser-induced fluorescence detection of BPDE–albumin adducts promises to be highly sensitive, but limited human data are available (Ozbal *et al.*, 1999). The overall utility of BPDE–protein adducts in chemoprevention studies is unclear at present, as there are discordant data with respect to detectability.

Nitrosamines

The tobacco-specific nitrosamines NNK and NNN form adducts with haemoglobin by pyridyloxobutylation of globin (Hecht, 1998). Mild base treatment of this haemoglobin releases HPB, which can be reliably quantified by GC/MS. In rats, levels of HPB-releasing haemoglobin adducts, which are esters, correlate with the corresponding pulmonary levels of DNA adducts formed by NNK and are decreased significantly by the chemopreventive agent PEITC (Murphy *et al.*, 1990; Hecht *et al.*, 1996). In initial studies, this adduct was detected in 73% of smokers, but later investigations found detectable levels in only 10–15% of subjects (Carmella *et al.*, 1990; Hecht, Carmella & Murphy, unpublished data). Adduct levels in smokers are generally close to background levels, limiting the utility of this assay as a biomarker in chemoprevention studies (Hecht, 1998). Higher adduct levels have been found in snuff users than in smokers (Hecht, 1998).

Methylation of globin by alkylating nitrosamines does not appear to be useful because of the high background levels resulting from endogenous methylation processes (Tornqvist *et al.*, 1988). Globin ethylation has been investigated only superficially to date (Kautiainen *et al.*, 1989).

Other protein biomarkers

Of the compounds considered here, acetaldehyde has attracted the most attention with respect to protein adducts. Acetaldehyde forms stable imidazolidinone adducts by reaction with *N*-terminal valine (Conduah Birt *et al.*, 1998; de Jersey *et al.*, 1992). Formation of such adducts in human haemoglobin can be monitored by mass spectrometry and may be applicable as a biomarker of acetaldehyde uptake (Conduah Birt *et al.*, 1998). Malondialdehyde-acetaldehyde protein adducts have also been characterized (Kearley *et al.*, 1999). Immunoassay techniques have been employed to quantify acetaldehyde–haemoglobin adducts in alcoholic patients (Lin *et al.*, 1993). Other protein adducts of acetaldehyde have been described, but there are few quantitative studies in humans (de Jersey *et al.*, 1992; IARC, 1999).

Urinary metabolites as biomarkers

PAHs

Urinary metabolite analysis could be used to test the hypothesis that intervention with a chemopreventive agent alters carcinogen metabolism either by enhancement of detoxification or inhibition of metabolic activation. 1-Hydroxypyrene is the most widely used biomarker of PAH uptake (Jongeneelen, 1997). Levels of 1-hydroxypyrene glucuronide exceed those of free 1-hydroxypyrene (Strickland *et al.*, 1994). Most studies have measured total 1-hydroxypyrene, after hydrolysis of conjugates (Jongeneelen, 1997). Levels of this biomarker increase in response to smoking, consumption of charcoal-broiled meat and occupational or medicinal exposure to PAHs (Jongeneelen, 1997;

Sithisarankul *et al.*, 1997; Wu *et al.*, 1998). The effects of chemopreventive agents have not been examined, but presumably in situations of constant exposure, inhibition or induction of cytochrome P450 1A1 could modulate levels of total 1-hydroxypyrene. Evidence for this was found in the study of Wu *et al.* (1998), who observed effects of a CYP1A1 polymorphism on 1-hydroxypyrene levels after adjusting for PAH exposure in coke-oven workers.

We are now developing methods to analyse B[a]P metabolites in human urine, with the ultimate goal of developing B[a]P activation/detoxification metabolite profiles which could be applicable in chemoprevention studies. The first step in this work was development of a method for analysis of *r*-7,*t*-8,9,*c*-10-tetrahydroxy-7,8,9,10-tetrahydro-benzo[*a*]pyrene (*trans-anti*-B[a]P-tetraol, the major hydrolysis product of *anti*-BPDE) in human urine. This metabolite has been detected and quantified by GC/MS in the urine of coke-oven workers, psoriasis patients treated with a coal tar-containing ointment, and cigarette smokers (Simpson *et al.*, 2000). There are also limited published data on urinary metabolite profiles in workers exposed to phenanthrene, chrysene and B[a]P (Grimmer *et al.*, 1997).

Nitrosamines

4-(Methylnitrosamino)-1-(3-pyridyl)-1-butanol (NNAL) and its glucuronides (NNAL-Gluc) are major urinary metabolites of NNK (Hecht, 1998). These metabolites have been quantified in the urine of smokers, snuff-dippers, ex-smokers, individuals exposed to environmental tobacco smoke and newborns of smoking mothers (Hecht, 1998; Lackmann *et al.*, 1999; Hecht *et al.*, 1999a). They are also sensitive to the effects of chemopreventive agents. In rats treated with PEITC, levels of urinary NNAL plus NNAL-Gluc increased 4–6-fold in tandem with decreased haemoglobin adduct formation (Hecht *et al.*, 1996). These results are consistent with inhibition by PEITC of hepatic NNK metabolism. In smokers who consumed watercress as a source of PEITC, levels of NNAL-Gluc as well as of NNAL plus NNAL-Gluc increased significantly, suggesting similar effects of PEITC in rats and humans (Hecht *et al.*, 1995). In mice treated with indole-3-carbinol, an inhibitor of NNK-induced lung tumorigenesis, levels of urinary NNAL decreased due to increased hepatic clearance of NNK (Morse *et al.*, 1990). Similar effects were observed in smokers treated with indole-3-carbinol (Taioli *et al.*, 1997).

Quantification of NNAL and NNAL-Gluc is a practical method for obtaining information on the effects of chemopreventive agents on NNK metabolism in smokers. NNAL and NNAL-Gluc are readily quantified in the urine of all smokers, providing an excellent biomarker of lung carcinogen uptake (Hecht, 1998, 1999). Provided that there is a specific hypothesis to be tested with respect to the chemopreventive agent employed, this biomarker should be widely applicable in chemoprevention studies.

Endogenous formation of *N*-nitroso compounds can be monitored by the quantification of urinary nitrosamino acids (Bartsch & Spiegelhalder, 1996). This methodology is highly practical and has been widely applied in studies of endogenous nitrosation and its inhibition, related to cancers of the oral cavity, oesophagus and other sites. Thus, chemopreventive agents which inhibit nitrosation can be evaluated in this way. However, effects on metabolism could not be evaluated since these compounds are in general excreted unchanged.

3-Methyladenine is a compound excreted in the urine that could result in part from metabolic activation of nitrosamines. However, there are multiple sources of urinary 3-methyladenine, including the diet (Fay *et al.*, 1997). In controlled studies, levels of 3-methyladenine in the urine of smokers were elevated. Similar results have been obtained with 3-ethyladenine, but the source of the ethylating agent is unknown (Kopplin *et al.*, 1995; Prevost & Shuker, 1996; Fay *et al.*, 1997).

Other urinary biomarkers

Various drugs have been used as non-invasive biomarkers of specific cytochrome P450 activities (Guengerich *et al.*, 1997). Examples include coumarin for CYP2A6, debrisoquine for CYP2D6, caffeine for CYP1A2 and chlorzoxazone for CYP2E1. Assays for metabolites of these drugs could be useful in certain chemoprevention studies if modulation of a specific cytochrome P450 is proposed. As examples, several studies have examined the effects of watercress, a rich source of PEITC, on drug metabolism in humans. Watercress caused a decrease in levels of oxidative metabolites of acetaminophen, which was attributed to inhibition of oxidative metabolism by CYP2E1 (Chen *et al.*, 1996). The area under the chlorzoxazone plasma concentration–time curve was significantly increased after watercress ingestion, indicating inhibition of hydroxylation of chlorzoxazone (Leclercq *et al.*, 1998). Watercress, however, had no effect on CYP2D6 activity as monitored with debrisoquine or CYP2A6 activity as assessed by coumarin metabolism (Caporaso *et al.*, 1994; Hecht & Murphy, unpublished data).

Effects on nicotine metabolism may also provide support for modulation of drug-metabolizing enzymes by chemopreventive agents. Nicotine is a naturally abundant substrate in smokers who would be on these trials. In a recently completed study, we found that watercress consumption induced glucuronidation of nicotine, cotinine and 3′-hydroxycotinine, suggesting that PEITC induces UDP-glucuronosyl transferase activity in smokers (Hecht *et al.*, 1999b).

Conclusions

Table 2 summarizes information on the potential utility of carcinogen-derived biomarkers in lung and oral cancer chemoprevention trials. The only specific adduct measurement which is practical and useful for current application is HPLC-fluorescence detection of BPDE–DNA adducts. There are some limitations to this assay. The analyte is detectable in leukocytes of a large proportion of PAH-exposed individuals only when they are *GSTM1*-null. Moreover, the relationship between leukocyte and lung or oral tissue levels of BPDE–DNA adducts has not been established using this assay. Urinary nitrosamino acids have already been widely applied in human trials, but these specifically relate to endogenous formation of nitrosamines, not to activation and detoxification. Measurement of urinary NNAL plus NNAL-Gluc is highly practical in smokers and can be applied in chemoprevention trials provided there is a specific *a priori* hypothesis. Nicotine metabolites and P450 substrates could also find potential use in specific chemoprevention trials.

Several types of biomarker are potentially applicable, but further research is necessary to improve detection methods and to test them in larger groups of individuals. Nitrosamine–DNA adducts, PAH–protein adducts, and urinary metabolites of PAH could be particularly important. Several biomarkers have already been examined in fairly extensive studies and do not appear to be suitable for chemoprevention trials because they are generally undetectable or detectable at relatively low levels; examples include O^6-mG in DNA, acetaldehyde–DNA adducts and tobacco-specific nitrosamine adducts with haemoglobin. Other biomarkers may have numerous endogenous and/or exogenous sources and their relevance to cancer induction is unclear; examples are methylated haemoglobin, alkyladenines in urine and 8-oxo-dG in DNA.

Other specific carcinogen-derived biomarkers as well as less specific tests, e.g. immunoassay and ^{32}P-postlabelling, may ultimately be applicable in chemoprevention studies. Biomarkers of the uptake and metabolism of chemopreventive agents (e.g., total isothiocyanates in urine) are becoming available and the application of these is also important for ensuring compliance and assessing individual differences in the metabolism of these agents (Chung *et al.*, 1998).

Table 2. Application of carcinogen-derived biomarkers in lung and oral cancer chemoprevention trials

Practical for current application	Need development and further exploration	Unlikely to be useful for chemoprevention trials
BPDE–DNA adducts by HPLC-fluorescence	7-mG in DNA	O^6-mG in DNA
Nitrosamino acids in urine	Tobacco-specific nitrosamine adducts in DNA	Acetaldehyde-DNA
NNAL plus NNAL-Gluc in urine	Acrolein/crotonaldehyde–DNA adducts	8-Oxo-dG in DNA
Nicotine metabolites in urine	Acrolein–protein adducts	Tobacco-specific nitrosamine Hb adducts
P450 substrate metabolites in urine	PAH–protein adducts	Methylated Hb
	Pyrene metabolites in urine	Alkyladenines in urine
	B[a]P metabolites in urine	

Acknowledgements

Our work on chemoprevention is supported by U.S. National Cancer Institute Grant CA-46535. Dr Hecht is an American Cancer Society Research Professor.

References

Alexandrov, K., Rojas, M., Geneste, O., Castegnaro, M., Camus, A.-M., Petruzzelli, S., Giuntini, C. & Bartsch, H. (1992) An improved fluorometric assay for dosimetry of benzo[*a*]pyrene diol-epoxide-DNA adducts in smokers' lung: comparisons with total bulky adducts and aryl hydrocarbon hydroxylase activity. *Cancer Res.*, **52**, 6248–6253

American Cancer Society (2000) *Cancer Facts and Figures 2000*, Atlanta, GA, American Cancer Society, pp. 28–31

Asami, S., Hirano, T., Yamaguchi, Y., Itoh, H. & Kasai, H. (1996) Increase of a type of oxidative DNA damage, 8-hydroxyguanine, and its repair activity in human leukocytes by cigarette smoking. *Cancer Res.*, **56**, 2546–2549

Asami, S., Manabe, H., Miyake, J., Tsurudome, Y., Hirano, T., Yamaguchi, R., Itoh, H. & Kasai, H. (1997) Cigarette smoking induces an increase in oxidative DNA damage, 8-hydroxydeoxyguanosine, in a central site of the human lung. *Carcinogenesis*, **18**, 1763–1766

Bartsch, H. & Spiegelhalder, B. (1996) Environmental exposure to N-nitroso compounds (NNOC) and precursors: an overview. *Eur. J. Cancer Prev.*, **5**, 11–18

Bartsch, H., Rojas, M., Nair, V., Nair, J. & Alexandrov, K. (1999) Genetic cancer susceptibility and DNA adducts: studies in smokers, tobacco chewers, and coke oven workers. *Cancer Detect. Prevent.*, **23**, 445–453

Blomeke, B., Greenblatt, M.J., Doan, V.D., Bowman, E.D., Murphy, S.E., Chen, C.C., Kato, S. & Shields, P.G. (1996) Distribution of 7-alkyl-2'-deoxyguanosine adduct levels in human lung. *Carcinogenesis*, **17**, 741–748

Blot, W.J., McLaughlin, J.K., Devesa, S.S. & Fraumeni, J., Jr (1996) Cancers of the oral cavity. In: Schottenfeld, D. & Fraumeni, J., eds, *Cancer Epidemiology and Prevention*, New York, Oxford University Press, pp. 666–680

Caporaso, N., Whitehouse, J., Monkman, S., Boustead, C., Issaq, H., Fox, S., Morse, M.A., Idle, J.R. & Chung, F.-L. (1994) *In vitro* but not *in vivo* inhibition of CYP2D6 by phenethyl isothiocyanate (PEITC), a constituent of watercress. *Pharmacogenetics*, **4**, 275–280

Carmella, S.G., Kagan, S.S., Kagan, M., Foiles, P.G., Palladino, G., Quart, A.M., Quart, E. & Hecht, S.S. (1990) Mass spectrometric analysis of tobacco-specific nitrosamine hemoglobin adducts in snuff dippers, smokers, and non-smokers. *Cancer Res.*, **50**, 5438–5445

Chen, L., Mohr, S.N. & Yang, C.S. (1996) Decrease of plasma and urinary oxidative metabolites of acetaminophen after consumption of watercress by human volunteers. *Clin. Pharmacol. Ther.*, **60**, 651–660

Chhabra, S.K., Souliotis, V.L., Kyropoulos, S.A. & Anderson, L.M. (1996) Nitrosamines, alcohol, and gastrointestinal tract cancer: recent epidemiology and experimentation. *In Vivo*, **10**, 265–284

Chung, F.-L., Jiao, D., Getahun, S.M. & Yu, M.C. (1998) A urinary biomarker for uptake of dietary isothiocyanates in humans. *Cancer Epidemiol. Biomarkers Prev.*, **7**, 103–108

Collins, A.R. (1999) Oxidative DNA damage, antioxidants, and cancer. *Bioessays*, **21**, 238–246

Conduah Birt, J.E., Shuker, D.E.G.& Farmer, P.B. (1998) Stable acetaldehyde–protein adducts as biomarkers of alcohol exposure. *Chem. Res. Toxicol.*, **11**, 136–142

de Jersey, J., Worrall, S. & Wilce, P. (1992) Modification of proteins and other biological molecules by acetaldehyde: adduct structure and functional significance. *Int. J. Biochem.*, **24**, 1899–1906

EUROGAST Study Group (1994) O^6-Methylguanine in blood leucocyte DNA: an association with the geographic prevalence of gastric cancer and with low levels of serum pepsinogen A, a marker of severe chronic atrophic gastritis. *Carcinogenesis*, **15**, 1815–1820

Fang, J.-L. & Vaca, C.E. (1997) Detection of DNA adducts of acetaldehyde in peripheral white blood cells of alcohol abusers. *Carcinogenesis*, **18**, 627–632

Fay, L.B., Leaf, C.D., Germaud, E., Aeschlimann, J.-M., Steen, C., Shuker, D.E.G. & Turesky, R.J. (1997) Urinary excretion of 3-methyladenine after consumption of fish containing high levels of dimethylamine. *Carcinogenesis*, **18**, 1039–1044

Foiles, P.G., Akerkar, S.A., Carmella, S.G., Kagan, M., Stoner, G.D., Resau, J.H. & Hecht, S.S. (1991) Mass spectrometric analysis of tobacco-specific nitrosamine-DNA adducts in smokers and nonsmokers. *Chem. Res. Toxicol.*, **4**, 364–368

Grimmer, G., Jacob, J., Dettbarn, G. & Naujack, K.-W. (1997) Determination of urinary metabolites of polycyclic aromatic hydrocarbons (PAH) for the risk assessment of PAH-exposed workers. *Int. Arch. Occup. Environ. Health*, **69**, 231–239

Groopman, J.D. & Kensler, T.W. (1999) The light at the end of the tunnel for chemical-specific biomarkers: daylight or headlight? *Carcinogenesis*, **20**, 1–11

Guengerich, F.P. (1997) Cytochrome P450 enzymes. In: Guengerich, F.P., ed., *Comprehensive Toxicology: Biotransformation*, Vol. 3, New York, Elsevier Science, pp. 37–68

Hecht, S.S. (1998) Biochemistry, biology, and carcinogenicity of tobacco-specific *N*-nitrosamines. *Chem. Res. Toxicol.*, **11**, 559–603

Hecht, S.S. (1999) Tobacco smoke carcinogens and lung cancer. *J. Natl Cancer Inst.*, **91**, 1194–1210

Hecht, S.S. & Hoffmann, D. (1989) The relevance of tobacco-specific nitrosamines to human cancer. *Cancer Surv.*, **8**, 273–294

Hecht, S.S. & Tricker, A.R. (1999) Nitrosamines derived from nicotine and other tobacco alkaloids. In: Gorrod, J.W. & Jacob, P. III, eds, *Analytical Determination of Nicotine and Related Compounds and their Metabolites*, Amsterdam, Elsevier Science, pp. 421–488

Hecht, S.S., Chung, F.-L., Richie, J.P., Jr, Akerkar, S.A., Borukhova, A., Skowronski, L. & Carmella, S.G. (1995) Effects of watercress consumption on metabolism of a tobacco-specific lung carcinogen in smokers. *Cancer Epidemiol, Biomarkers Prev.*, **4**, 877–884

Hecht, S.S., Trushin, N., Rigotty, J., Carmella, S.G., Borukhova, A., Akerkar, S.A. & Rivenson, A. (1996) Complete inhibition of 4-(methylnitrosamino)-1-(3-pyridyl)-1-butanone induced rat lung tumorigenesis and favorable modification of biomarkers by phenethyl isothiocyanate. *Cancer Epidemiol. Biomarkers Prev.*, **5**, 645–652

Hecht, S.S., Carmella, S.G., Chen, M., Koch, J.F.D., Miller, A.T., Murphy, S.E., Jensen, J.A., Zimmerman, C.L. & Hatsukami, D.K. (1999a) Quantitation of urinary metabolites of a tobacco-specific lung carcinogen after smoking cessation. *Cancer Res.*, **59**, 590–596

Hecht, S.S., Carmella, S.G. & Murphy, S.E. (1999b) Effects of watercress consumption on urinary metabolites of nicotine in smokers. *Cancer Epidemiol. Biomarkers Prev.*, **8**, 907–913

Hoffmann, D., Brunnemann, K.D., Prokopczyk, B. & Djordjevic, M.V. (1994) Tobacco-specific *N*-nitrosamines and *areca*-derived *N*-nitrosamines: chemistry, biochemistry, carcinogenicity, and relevance to humans. *J. Toxicol. Environ. Health*, **41**, 1–52

Hoffmann, D. & Hecht, S.S. (1990) Advances in tobacco carcinogenesis. In: Cooper, C.S. & Grover, P.L., eds, *Handbook of Experimental Pharmacology*, Heidelberg, Springer-Verlag, 94/I, pp. 63–102

IARC (1985) *IARC Monographs on the Evaluation of the Carcinogenic Risk of Chemicals to Humans*, Vol. 37, *Tobacco Habits other than Smoking; Betel-Quid and Areca-Nut Chewing; and some Related Nitrosamines*, Lyon, IARC

IARC (1999) *IARC Monographs on the Evaluation of Carcinogenic Risks to Humans*, Vol. 71, *Re-evaluation of some Organic Chemicals, Hydrazine and Hydrogen Peroxide* (Part Two), Lyon, IARC, pp. 318–335

Jongeneelen, F.J. (1997) Methods for routine biological monitoring of carcinogenic PAH-mixtures. *Sci. Total Environ.*, **199**, 141–149

Kautiainen, A., Tornqvist, M., Svensson, K. & Osterman-Golkar, S. (1989) Adducts of malonaldehyde and a few other aldehydes to hemoglobin. *Carcinogenesis*, **10**, 2123–2130

Kearley, M.L., Patel, A., Chien, J. & Tuma, D.J. (1999) Observation of a new non-fluorescent malondialdehyde-acetaldehyde-protein adduct by ^{13}C NMR spectroscopy. *Chem. Res. Toxicol.*, **12**, 100–105

Kensler, T., Gange, S., Egner, P., Dolan, P., Munoz, A., Groopman, J., Rogers, A. & Roebuck, B. (1997) Predictive value of molecular dosimetry: individual versus group effects of oltipraz on aflatoxin-albumin adducts and risk of liver cancer. *Cancer Epidemiol. Biomarkers Prev.*, **6**, 603–610

Kensler, T.W., He, X., Otieno, M., Egner, P.A., Jacobson L.P., Chen, B., Wang, J.-S., Zhu, Y.-R., Zhang, B.-C., Wang, J.-B., Wu, Y., Zhang, Q.-N., Qian, G.-S., Kuang, S.-Y., Fang, X., Li, Y.-F., Yu, L.-Y., Prochaska, H.J., Davidson, N.E., Gordon, G.B., Gorman, M.B., Zarba, A., Enger, C., Munoz, A., Helzsouer, K.J. & Groopman, J.D. (1998) Oltipraz chemoprevention trial in Qidong, P.R. China: modulation of serum aflatoxin albumin adduct biomarkers. *Cancer Epidemiol. Biomarkers Prev.*, **7**, 127–134

Kopplin, A., Eberle-Adamkiewicz, G., Glüsenkamp, K.-H., Nehls, P. & Kirstein, U. (1995) Urinary excretion of 3-methyladenine and 3-ethyladenine after controlled exposure to tobacco smoke. *Carcinogenesis*, **16**, 2637–2641

Kriek, E., Rojas, M., Alexandrov, K. & Bartsch, H. (1998) Polycyclic aromatic hydrocarbon-DNA adducts in humans: relevance as biomarkers for exposure and cancer risk. *Mutat. Res.*, **400**, 215–231

Kyrtopoulos, S.A. (1998) DNA adducts in humans after exposure to methylating agents. *Mutat. Res.*, **405**, 135–143

Lackmann, G.M., Salzberger, U., Tollner, U., Chen, M., Carmella, S.G. & Hecht, S.S. (1999) Metabolites of a tobacco-specific carcinogen in the urine of newborns. *J. Natl Cancer Inst.*, **91**, 459–465

Leclercq, I., Desager, J.-P. & Horsmans, Y. (1998) Inhibition of chlorzoxazone metabolism, a clinical probe for CYP2E1, by a single ingestion of watercress. *Clin. Pharmacol. Ther.*, **64**, 144–149

Lin, R.C., Shahidi, S., Kelly, T.J., Lumeng, C. & Lumeng, L. (1993) Measurement of hemoglobin-acetaldehyde adduct in alcoholic patients. *Alcohol. Clin. Exp. Res.*, **17**, 669–674

Melikian, A.A., Sun, P., Pierpont, C., Coleman, S. & Hecht, S.S. (1997) Gas chromatography-mass spectrometric determination of benzo[*a*]pyrene and chrysene diol epoxide globin adducts in humans. *Cancer Epidemiol. Biomarkers Prev.*, **6**, 833–839

Morse, M.A., LaGreca, S.D., Amin, S.G. & Chung, F.-L. (1990) Effects of indole-3-carbinol on lung tumorigenesis and DNA methylation induced by 4-(methylnitrosamino)-1-(3-pyridyl)-1-butanone (NNK) and on the metabolism and disposition of NNK in A/J mice. *Cancer Res.*, **50**, 2613–2617

Murphy, S.E., Palomino, A., Hecht, S.S. & Hoffmann, D. (1990) Dose-response study of DNA and hemoglobin adduct formation by 4-(methylnitrosamino)-1-(3-pyridyl)-1-butanone in F344 rats. *Cancer Res.*, **50**, 5446–5452

Mustonen, R., Schoket, B. & Hemminki, K. (1993) Smoking-related DNA adducts: ^{32}P-postlabeling analysis of 7-methylguanine in human bronchial and lymphocyte DNA. *Carcinogenesis*, **14**, 151–154

Nath, R.G., Ocando, J.E. & Chung, F.-L. (1996) Detection of 1,N^2-propanodeoxy-guanosine adducts as potential endogenous DNA lesions in rodent and human tissues. *Cancer Res.*, **56**, 452–456

Nath, R.G., Ocando, J.E., Guttenplan, J.B. & Chung, F.-L. (1998) 1,N^2-Propanodeoxy-guanosine adducts: Potential new biomarkers of smoking-induced DNA damage in human oral tissue. *Cancer Res.*, **58**, 581–584

Ozbal, C.C., Skipper, P.L., Dasai, R.R. & Tannenbaum, S.R. (1999) Detection of human serum albumin adducts of benzo[*a*]pyrene-diol-epoxides by HPLC with laser-induced fluorescence detection. *Proc. Am. Assoc. Cancer Res.*, **40**, 293

Pastorelli, R., Restano, J., Guanci, M., Maramonte, M., Magagnotti, C., Allevi, R., Lauri, D., Fanelli, R. & Airoldi, L. (1996) Hemoglobin adducts of benzo[*a*]pyrene diol-epoxide in newspaper vendors: association with traffic exhaust. *Carcinogenesis*, **17**, 2389–2394

Pastorelli, R., Guanci, M., Cerri, A., Negri, E., La Vecchia, C., Fumagalli, F., Mezzetti, M., Cappelli, R., Panigalli, T., Fanelli, R. & Airoldi, L. (1998) Impact of inherited polymorphisms in glutathione *S*-transferase M1, microsomal epoxide hydrolase, cytochrome P450 enzymes on DNA, and blood protein adducts of benzo[*a*]pyrene-diolepoxide. *Cancer Epidemiol. Biomarkers Prev.*, **7**, 703–709

Pastorelli, R., Guanci, M., Restano, J., Berri, A., Micoli, G., Minoia, C., Alcini, D., Carrer, P., Negri, E., La Vecchia, C., Fanelli, R. & Airoldi, L. (1999) Seasonal effect on airborne pyrene, urinary 1-hydroxypyrene, and benzo[*a*]pyrene diol epoxide-hemoglobin adducts in the general population. *Cancer Epidemiol. Biomarkers Prev.*, **8**, 561–565

Prevost, V. & Shuker, D.E.G. (1996) Cigarette smoking and urinary 3-alkyladenine excretion in man. *Chem. Res. Toxicol.*, **9**, 439–444.

Prieme, H., Loft, S., Kharlund, M., Gronbaek, K., Tonnesen, P. & Paulsen, H.E. (1998) Effect of smoking cessation on oxidative DNA modification estimated by 8-oxo-7,8-dihydro-2′-deoxyguanosine excretion. *Carcinogenesis*, **19**, 347–351

Rojas, M., Alexandrov, K., Auburtin, G., Wastiaux-Denamur, A., Mayer, L., Mahieu, B., Sebastien, P. & Bartsch, H. (1995) Anti-benzo[*a*]pyrene diolepoxide-DNA adduct levels in peripheral mononuclear cells from coke oven workers and the enhancing effect of smoking. *Carcinogenesis*, **16**, 1373–1376

Rojas, M., Alexandrov, K., Cascorbi, I., Brockmoller, J., Likhachev, A., Pozharisski, K., Bouvier, G., Auburtin, G., Mayer, L., Koop-Schneider, A., Roots, I. & Bartsch, H. (1998) High benzo[*a*]pyrene diol-epoxide DNA adduct levels in lung and blood cells from individuals with combined CYP1A1 MspI/MspI-GSTM1*0/*0 genotypes. *Pharmacogenetics*, **8**, 109–118

Rojas, M., Cascorbi, I., Alexandrov, K., Kriek, E., Auburtin, G., Mayer, L., Kopp-Schneider, A., Roots, I. & Bartsch, H. (2000) Modulation of benzo[*a*]pyrene diol-epoxide-DNA adduct levels in human white blood cells by *CYP1A1, GSTM1,* and *GSTT1* polymorphism. *Carcinogenesis*, **21**, 35–41

Santella, R.M. (1999) Immunological methods of detection of carcinogen-DNA damage in humans. *Cancer Epidemiol. Biomarkers Prev.*, **8**, 733–739

Shopland, D.R. (1995) Tobacco use and its contribution to early cancer mortality with a special emphasis on cigarette smoking. *Environ. Health Perspect.*, **103** (suppl. 8), 131–142

Simpson, C.D., Wu, M.-T., Christiani, D.C., Santella, R.M., Carmella, S.G. & Hecht, S.S. (2000) Determination of *r*-7,*t*-8,9,*c*-10-tetrahydroxy-7,8,9,10-tetrahydrobenzo-[*a*]pyrene in human urine by gas chromatography-negative ion chemical ionization-mass spectrometry. *Chem. Res. Toxicol.*, **13**, 271–280

Sithisarankul, P., Vineis, P., Kang, D., Rothman, N., Caporaso, N. & Strickland, P. (1997) The association of 1-hydroxypyrene-glucuronide in human urine with cigarette smoking and broiled or roasted meat consumption. *Biomarkers*, **2**, 217–221

Staretz, M.E., Foiles, P.G., Miglietta, L.M. & Hecht, S.S. (1997) Evidence for an important role of DNA pyridyloxobutylation in rat lung carcinogenesis by 4-(methylnitrosamino)-1-(3-pyridyl)-1-butanone: effects of dose and phenethyl isothiocyanate. *Cancer Res.*, **57**, 259–266

Sticha, R.K., Staretz, M.E., Liang, H., Kenney, P.M.J. & Hecht, S.S. (2000) Effects of benzyl isothiocyanate and phenethyl isothiocyanate on benzo[*a*]pyrene metabolism and DNA adduct formation in the A/J mouse. *Carcinogenesis* (in press)

Strickland, P.T., Kang, D., Bowman, E.D., Fitzwilliam, A., Downing, T.E., Rothman, N., Groopman, J.D. & Weston, A. (1994) Identification of 1-hydroxypyrene glucuronide as a major pyrene metabolite in human urine by synchronous fluorescence spectroscopy and gas chromatography-mass spectrometry. *Carcinogenesis*, **15**, 483–487

Taioli, E., Garbers, S., Bradlow, H.L., Carmella, S.G., Akekar, S. & Hecht, S.S. (1997) Effects of indole-3-carbinol on the metabolism of 4-(methylnitrosamino)-1-(3-pyridyl)-1-butanone (NNK) in smokers. *Cancer Epidemiol. Biomarkers Prev.*, **6**, 517–522

Tang, D., Warburton, D., Tannenbaum, S.R., Skipper, P., Santella, R.M., Cereijido, G.S., Crawford, F.G. & Perera, F.P. (1999) Molecular and genetic damage from environmental tobacco smoke in young children. *Cancer Epidemiol. Biomarkers Prev.*, **8**, 427–431

Tornqvist, M., Osterman-Golkar, S., Kautiainen, A., Naslund, M., Calleman, C.J. & Ehrenberg, L. (1988) Methylations in human hemoglobin. *Mutat. Res.*, **204**, 521–529

Corresponding author:

S. Hecht
University of Minnesota Cancer Center,
Mayo Mail Code 806,
420 Delaware St., SE,
Minneapolis,
MN 55455,
USA

Vaca, C.E., Nilsson, J.A., Fang, J.L. & Grafstrom, R.C. (1998) Formation of DNA adducts in human buccal epithelial cells exposed to acetaldehyde and methylglyoxal *in vitro*. *Chem-Biol. Interact.*, **108**, 197–208

van Zeeland, A.A., de Groot, A.J.L., Hall, J. & Donato, F. (1999) 8-Hydroxydeoxy-guanosine in DNA from leukocytes of healthy adults: relationship with cigarette smoking, environmental tobacco smoke, alcohol and coffee consumption. *Mutat. Res.–Gen. Toxicol. Environ. Mutag.*, **439**, 249–257

Wu, M.-J., Huang, S.-L., Ho, C.-K., Yeh, Y.-F. & Christiani, D.C. (1998) Cytochrome *P450 1A1 Msp*I polymorphism and urinary 1-hydroxypyrene concentrations in coke-oven workers. *Cancer Epidemiol, Biomarkers Prev.*, **7**, 823–829

Yarborough, A., Zhang, Y.-J., Hsu, T.-M. & Santella, R.M. (1996) Immunoperoxidase detection of 8-hydroxydeoxyguanosine in aflatoxin B1-treated rat liver and human oral mucosal cells. *Cancer Res.*, **56**, 683–688

Biomarkers in Cancer Chemoprevention
Miller, A.B., Bartsch, H., Boffetta, P., Dragsted, L. and Vainio, H. eds
IARC Scientific Publications No. 154
International Agency for Research on Cancer, Lyon, 2001

Lung cancer: chemoprevention and intermediate effect markers

M.S. Tockman

Even after smoking cessation, genetic damage in the airways epithelium may lead to focal progression of lung carcinogenesis. Some centres now report as many new lung cancer cases among former smokers as among current smokers. Chemoprevention is a potential approach to diminish the progression of pre-clinical genetic damage. The most intensively studied lung cancer chemoprevention agents are the retinoids, including vitamin A and its synthetic analogues and precursors. While effective in suppressing lung carcinogenesis in animal models, retinoids have failed to inhibit carcinogenesis in human chemoprevention trials with premalignant end-points (sputum atypia, bronchial metaplasia). In trials with lung cancer end-points, administration of retinoids either was ineffective or, in the case of β-carotene, led to greater lung cancer incidence and mortality. In view of these findings, markers of specific retinoid effect (i.e., levels of RAR-β) become less relevant. Other markers of genetic instability and proliferation may be useful for both early detection and potentially as intermediate-effect markers for new chemoprevention trials. Cytological atypia, bronchial metaplasia, protein (hnRNP A2/B1) overexpression, *ras* oncogene activation and tumour-suppressor gene deletion, genomic instability (loss of heterozygosity, microsatellite alterations), abnormal methylation, helical CT detection of atypical adenomatous hyperplasia and fluorescent bronchoscopic detection of angiogenic squamous dysplasia offer great promise for molecular diagnosis of lung cancer far in advance of clinical presentation. These end-points can now be evaluated as monitors of response to chemoprevention as potential intermediate-effect markers.

Chemoprevention

Lung cancer continues to be the leading cause of cancer death in most developed countries. In the United States in 1998, 171 500 new cases and 160 100 deaths maintained lung cancer as the leading cancer killer of both men (exceeding prostate cancer) and women (exceeding breast cancer) (Landis *et al.*, 1998). The persistence of grim lung cancer incidence and mortality figures over three decades demands new approaches to control this disease. Tobacco smoking is generally accepted to be responsible for 85–90% of lung cancer. Yet even after cessation of smoking, unrepaired genetic damage in the airways epithelium may lead to focal progression of carcinogenesis. New cases of lung cancer are now as common among former smokers as in current smokers in the United States (Kurie *et al.*, 1995). Chemoprevention is a potential approach to mitigate the progression of preclinical genetic damage. As originally termed by Sporn *et al.* (1976), chemoprevention is the interruption of the carcinogenic process through application of natural or pharmacological agents.

The potential for lung cancer chemoprevention is built upon two observations: field cancerization and multistep carcinogenesis (Lippman *et al.*, 1994). Slaughter *et al.* (1953) first noted the widespread damage (field cancerization) that persists in the aerodigestive tract of current and former smokers. From completely sectioned autopsy specimens, Auerbach *et al.* (1961, 1979) recognized three progressive grades of histological abnormality in the epithelium of 'uninvolved bronchi' surrounding lung tumours: hyperplasia (an increase in the number of cell rows), metaplasia (loss of cilia) and dysplasia (presence of atypical cells). They also reported an increased frequency of carcinoma *in situ* associated with a higher frequency of smoking, from 0% in nonsmokers to 11% in men who smoked two or more packs per day. Field cancerization implies that detection of premalignant or malignant lesions in one area of the pulmonary epithelium identifies an epithelium at increased risk of developing neoplastic lesions in other areas (Khuri & Lippman, 2000).

Observation of progressive histopathological abnormality preceding cancer reflects the underlying multistep carcinogenesis. Tumours develop through a series of specific genetic changes in proto-oncogenes and tumour-suppressor genes (Bishop, 1987; Fearon & Vogelstein, 1990; Weinberg, 1989). Changes that cannot be repaired and do not trigger apoptosis may lead to a cellular growth advantage. Many of these genetic changes are acquired before and during the earliest stages of clonal expansion and are retained by daughter cells through the course of carcinogenesis and malignancy. Wistuba *et al.* (1999) observed that increasing severity of histopathological changes in lung squamous carcinoma is associated with a progression of genetic changes (increasing frequency of loss of heterozygosity; LOH). Regions of chromosomal loss are suspected to have contained tumour-suppressor genes, the loss of which would be advantageous for growth. This genetic damage is widespread throughout the airway even in areas of normal-appearing epithelium, and persists long after removal of the insult. If detected during the premalignant period, these genetic changes could serve as markers of carcinogenesis and indicate the need for chemoprevention.

Retinoids, including vitamin A and its synthetic analogues, are the most intensively studied lung cancer chemoprevention agents. Retinoids promote the differentiation of airway epithelium *in vitro* (Jetten *et al.*, 1992) and suppress lung carcinogenesis *in vivo* (Moon *et al.*, 1994). Most effects of retinoids on gene expression are mediated by nuclear retinoic acid receptors RARs (α, β and γ) and retinoid X receptors (RXR α, β and γ), which function as retinoid-activated transcription factors. Like the other members of the steroid receptor superfamily, each of these receptors is thought to bind to specific response elements (retinoic acid response elements, RAREs) which govern the expression of genes and modify post-translational mechanisms that regulate cell growth and differentiation (Chambon, 1996; Mangelsdorf *et al.*, 1994). In the case of all-*trans*-retinoic acid, these post-translational mechanisms also may include degradation of CDK-4 through the ubiquitin–proteasome pathway (Sueoka *et al.*, 1999) and binding of E_2F-4 to (the RB family member) p107 leading to transcriptional suppression (Lee *et al.*, 1998). Reduced RAR-β expression is observed in non-small-cell lung cancer (NSCLC) compared to paired normal tissues. Lotan (1997) found that the messenger RNA (mRNA) expression of RAR-β was suppressed in more than 50% of oral and lung premalignant lesions in individuals without cancer (e.g., oral leukoplakia and squamous metaplasia), in dysplastic lesions adjacent to cancer, and in malignant oral and lung carcinomas. Picard *et al.* (1999) measured reduced RAR-β protein levels in tumour compared to normal tissues resected from 76 NSCLC patients using immunohistochemistry. Decreased expression of RAR-β in lung cancer tissue and cell lines suggests that loss of RAR-β expression may be important in the development of lung cancer (Xu *et al.*, 1997).

Metaplasia and dysplasia end-points

Metaplasia and dysplasia of the bronchial epithelium have been monitored as a general index of progression towards lung cancer. Alternatively, RAR-β expression might be considered a specific index of retinoid activity. Examination of these markers might indicate a chemopreventive effect in advance of clinical lung cancer development (intermediate end-points). Synchronous evaluations of these intermediate end-points, however, do not show similar results. Ayoub *et al.* (1999) examined the effects of 13-*cis*-retinoic acid on both bronchial epithelium and RAR-β expression. Decreased baseline expression of RAR-β mRNA was found in bronchial brushings in 44 (23%) of 188 smokers. These 44 were randomly assigned to receive a placebo or 13-*cis*-retinoic acid (30 mg/day) for six months. While only 18 of 44 (41%, eight 13-*cis*-retinoic acid treated and ten placebo) completed the follow-up bronchoscopy, the 13-*cis*-retinoic acid group showed a significant upregulation of RAR-β expression at the end of 13-*cis*-retinoic acid treatment ($p = 0.001$). Corresponding cytological changes in bronchial brushings were uncommon. Thus, RAR-β expression may be upregulated by 13-*cis*-retinoic acid treatment even in the absence of cytological improvement. Xu *et al.* (1999) evaluated the effects of 13-*cis*-retinoic acid on bronchial mucosal biopsies of heavy smokers and found a significant increase in RAR-β mRNA in the biopsies of the treated group after six months of treatment without a reversal of squamous metaplasia.

Khuri and Lippman (2000) and Siegfried (1998) have summarized the published results of randomized clinical trials of lung cancer chemoprevention. Lung premalignancy (sputum atypia or bronchial metaplasia) was the end-point of five trials (Table 1). All three randomized trials that studied the effect of retinoids on lung premalignancy in smokers gave negative results. In their small (n = 150), placebo-controlled trial of etretinate to reverse sputum cytology metaplasia, Arnold *et al.* (1992) found that metaplasia was reversed in 32% of etretinate-treated subjects and 30% of placebo subjects. In a placebo-controlled trial of isotretinoin (n = 87), Lee *et al.* (1994) found similar reductions in metaplasia index regardless of treatment (54% isotretinoin vs. 60% placebo). Smoking cessation during this trial showed a better correlation with reduction in metaplasia index than did chemopreventive treatment. In a similarly designed trial of fenretinide in 68 subjects, there was also a negative result (Kurie *et al.*, 1999).

Sputum atypia was the premalignancy end-point of two randomized studies of chemoprevention. Treatment with β-carotene (50 mg per day) plus retinol (25 000 IU on alternate days), led to no significant reduction in the prevalence of sputum atypia or in cytological progression in 755 asbestos workers (McLarty *et al.*, 1995). Although Heimburger *et al.* (1988) reported a significant improvement in sputum atypia in a folic acid–vitamin B_{12} treatment group after a four-month, randomized, placebo-controlled trial (n = 73), a reanalysis of these data using standard analytical methods found no significant difference in sputum atypia between the placebo and treatment groups (Lippman *et al.*, 1994).

Primary lung cancer end-point

Two large randomized trials sponsored by the US National Cancer Institute have evaluated chemoprevention of primary lung cancer. In the Alpha-Tocopherol, Beta-Carotene (ATBC) Prevention Study, α-tocopherol and β-carotene were examined in a randomized, placebo-controlled factorial

Table 1. Results of randomized aerodigestive clinical/translational trials

Intervention	End-point	Number	Result	Reference/Year
Etretinate	Metaplasia	150	Negative	Arnold *et al.* (1992)
Isotretinoin	Metaplasia	87	Negative	Lee *et al.* (1994)
Fenretinide	Metaplasia	68	Negative	Kurie *et al.* (1999)
β-Carotene + retinol	Sputum atypia	755	Negative	McLarty *et al.* (1995)
Vitamin B_{12} + folic acid	Sputum atypia	73	Negative	Heimburger *et al.* (1988)
β-Carotene	Lung cancer	29 133	Harmful	Alpha-Tocopherol, Beta Carotene Cancer Prevention Study Group (1994)
β-Carotene + retinol	Lung cancer	18 314	Harmful	Omenn *et al.* (1996)
β-Carotene	Lung cancer	22 071	Negative	Hennekens *et al.* (1996)
β-Carotene	Lung cancer	39 876	Negative	Lee *et al.* (1999)
Vitamin E	Lung cancer	29 133	Negative	Alpha-Tocopherol, Beta Carotene Cancer Prevention Study Group (1994)
Retinyl palmitate	Second primary LC	307	Negative	Pastorino *et al.* (1993)
Retinyl palmitate	Second primary LC	2592	Negative	de Vries *et al.* (1999)
N-Acetylcysteine	Second primary LC	2592	Negative	de Vries *et al.* (1999)

Adapted from Khuri & Lippman (2000)

Table 2. Intermediate effect markers

Marker	Reported in pre-malignant/ malignant specimens	Reference
RAR-β mRNA	Pre-malignant	Lotan (1997), Ayoub *et al.* (1999)
hnRNP A2/B1	Pre-malignant	Tockman *et al.* (1997), Fielding *et al.* (1999)
ras oncogene activation	Pre-malignant	Rodenhius & Slebos (1992), Mao *et al.* (1994a)
p53 deletion	Pre-malignant	Hollstein *et al.* (1991), Mao *et al.* (1994a)
Loss of heterozygosity at 3p, 9p	Pre-malignant	Mao *et al.* (1994b, 1997)
Promoter CpG island methylation	Pre-malignant	Herman *et al.* (1996), Belinsky *et al.* (1998)
Helical CT detection of atypical adenomatous hyperplasia	Malignant Malignant	Henschke *et al.* (1999), Kitimara *et al.* (1999)
Fluorescent bronchoscopy detection of angiogenic squamous dysplasia	Pre-malignant	Lam *et al.* (1993)

design among 29 133 Finnish male cigarette smokers (Alpha-Tocopherol, Beta-Carotene Cancer Prevention Study Group, 1994). After 5–8 years receiving either (1) β-carotene (20 mg/day), (2) α-tocopherol (50 mg/day), (3) the combination or (4) placebo, the group receiving β-carotene had a statistically significant 18% increase in lung cancer incidence and an 8% increase in total mortality compared with the placebo group. This result was in sharp contrast with the epidemiological data, which show an inverse association of dietary and serum β-carotene with lung cancer (IARC, 1998). Smokers who consume more fruits and vegetables are reported to have a lower risk for lung cancer, but the nutrients responsible for this protection remain unknown. Carotenoids in fruits and vegetables were postulated to have a protective action by quenching singlet oxygen and free radicals that could lead to lipid peroxidation.

The Beta-Carotene and Retinol Efficacy Trial (CARET) confirmed the harmful effect of supplemental β-carotene upon cigarette smokers (Omenn *et al.*, 1996). CARET was a multicentre, randomized, double-blind, placebo-controlled primary prevention trial in the United States involving a total of 18 314 smokers, former smokers and workers exposed to asbestos. This trial was stopped 21 months early when interim analysis showed that a combination of 30 mg of β-carotene and 25 000 IU of retinol per day had no benefit and may have had an adverse effect on the incidence of lung cancer and on the risk of death. The active treatment group had statistically significant excesses of lung cancer incidence (28%), lung cancer mortality (46%), cardiovascular mortality (26%) and all-cause mortality (17%) compared with the placebo group.

The reasons for the harmful effects of β-carotene supplementation are neither clear nor universal. No significant harmful (or beneficial) effect was observed on rates of lung cancer, total cancer or cardiovascular disease after 12 years of β-carotene supplementation (50 mg, alternate days) among 22 071 males in the Physicians' Health Study (Hennekens *et al.*, 1996). While many physicians were nonsmokers, there were no differences between current and former smokers in any of these end-points. Similarly, the Women's Health Study found neither positive nor negative effects of β-carotene (50 mg, alternate days) on the incidence of cancer or of cardiovascular disease (Lee *et al.*, 1999). This randomized, double-blind, placebo-controlled trial among 39 876 women aged 45 years or older tested the effects of aspirin, vitamin E and β-carotene in prevention of cancer and cardiovascular disease. The β-carotene component was terminated early after a median treatment duration of 2.1 years in response to the publication of the Finnish and CARET trials. While no mechanism satisfactorily explains the effects of β-carotene on lung cancer, the absence of beneficial effects and the potential for harm suggests that this agent should not be used for chemoprevention (IARC, 1998; Hong, 1999).

Second primary lung cancer end-point

Perhaps as a result of widespread epithelial injury (field cancerization), patients with a successfully resected first primary lung cancer have a high annual incidence (2–5%) of second primary lung cancer (Grover & Piantadosi, 1989). The adjuvant effect of vitamin A (as retinyl palmitate, 300 000 IU per day for 12 months) was evaluated in 307 patients with stage I NSCLC, randomly assigned to treatment or placebo after curative surgery. After a median follow-up of 46 months, the treatment group developed fewer second primary tumours (37% vs. 48%) and showed a significantly greater time to new primary (Pastorino *et al.*, 1993). These encouraging results have been followed by the EUROSCAN trial, the largest chemoprevention study in curatively treated early-stage oral cancer, laryngeal cancer and lung cancer (n = 2595) (de Vries *et al.*, 1999). Initiated by the European Organization for Reseach and Treatment of Cancer (EORTC) in 1988, EUROSCAN was an open-label, multicentre, two-year trial of retinyl palmitate and *N*-acetylcysteine (NAC) administered in a 2×2 factorial design. Preliminary results show no difference between treatment and control arms for second primary tumours of the head and neck and the lung, local/regional recurrence and distant metastases, or long-term survival rates.

In trials designed to look at primary aerodigestive tract end-points therefore, chemoprevention of premalignancy (sputum atypia or bronchial metaplasia) has been unsuccessful and chemoprevention of primary or second primary lung cancer has either been unsuccessful or harmful (Table 1). The intermediate end-point of specific retinoid (13-*cis*-retinoic acid) effect, RAR-β mRNA expression has differed from the more general lung cancer premalignancy markers (sputum atypia and bronchial metaplasia). The absence of benefit from chemoprevention with 13-*cis*-retinoic acid in reducing lung cancer incidence indicates that a marker of retinoid effect does not necessarily provide a good intermediate index of lung cancer chemoprevention.

Lung cancer as secondary end-point

In terms of lung cancer as a secondary end-point, vitamin E and selenium show some promise for chemoprevention. In a placebo-controlled trial to evaluate the effect of selenium (200 μg in brewer's yeast per day) on the incidence of skin basal-cell or squamous-cell carcinomas, Clark *et al.* (1996) found a reduction of second primary lung cancers. Seven dermatology clinics in the eastern United States treated 1312 former non-melanoma skin cancer patients with selenium or placebo for an average of 4.5 years. After follow-up of 8271 person-years, selenium treatment did not significantly affect the incidence of basal-cell or squamous-cell skin cancer. However, compared with controls, patients treated with selenium had significant reductions in total cancer mortality, total cancer incidence and incidence of lung, colorectal and prostate cancers. The trial was stopped early because of the apparent reductions in total cancer mortality and total cancer incidence in the selenium group. No case of selenium toxicity occurred. The potential lung cancer risk of low selenium levels is supported by follow-up data from a Finnish serum bank (Knekt *et al.*, 1998). The selenium levels of 95 cases of lung cancer arising during a 20-year follow-up of 9101 cancer-free individuals were compared with 190 matched controls. After adjustment, Knekt *et al.* (1998) found a significant excess lung cancer relative risk among those in the lowest compared with the highest tertiles of serum selenium levels. The low selenium–lung cancer association was even stronger at lower levels (<5.9 mg/litre) of α-tocopherol.

A slight chemopreventive effect of vitamin E was observed in the α–tocopherol arm of the ATBC study (Alpha-Tocopherol, Beta-Carotene Cancer Prevention Study Group, 1994). In contrast to the significant increase in lung cancer incidence with β-carotene, vitamin E gave no adverse effect on lung cancer incidence, but a trend towards reduction in lung cancer incidence in participants with longer α-tocopherol exposure. The epidemiological dietary and baseline serum α-tocopherol levels from the ATBC study support the protective effect of vitamin E against lung cancer (Woodson *et al.*, 1999), as do data from China. An occupational cohort of Chinese tin miners was followed for development of lung cancer. Although there was no significant overall association between prospectively collected serum α-tocopherol, γ-tocopherol or selenium levels and incidence of lung cancer, results from this study suggest that higher α-tocopherol levels may be protective in men less than 60 years old and in those who do not drink alcohol (Ratnasinghe *et al.*, 2000).

In summary, while natural and synthetic vitamin A metabolites and analogues (retinoids) have been found to suppress head and neck and lung carcinogenesis in animal models, chemopreventive interventions have failed to show inhibition of carcinogenesis in individuals with premalignant lesions and were either neutral or harmful in terms of lung cancer end-points. With no demonstrated effectiveness of retinoid chemoprevention, discussion of markers of specific retinoid effect (i.e., RAR-β) becomes moot.

Intermediate-effect markers

A number of genetic instability and proliferation markers have been proposed as general markers of carcinogenesis. These markers may be useful both for early detection and potentially as intermediate-effect markers for chemoprevention trials.

Abnormal protein (hnRNP A2/B1) overexpression

Sputum cells can now be probed for altered gene expression to assess the preneoplastic state of the airway. During the Johns Hopkins Lung Project (JHLP), we developed an archive of dysplastic sputum specimens and associated clinical data linking specimens to lung cancer outcome. These archived specimens were examined with two monoclonal antibodies (MoAbs 703D4 and 624H12; raised at the US National Cancer Institute) to NSCLC and SCLC cell lines (respectively). Antibody differential display was used to identify biomarkers of lung cancer in archived sputum specimens two years before clinical detection of lung cancer, with a sensitivity of 91% and a specificity of 88% (Tockman *et al.*, 1988).

Monoclonal antibody 703D4 recognizes an epitope of hnRNP A2 and its splice variant, hnRNP B1 (Zhou *et al.*, 1996). The target antigen for 703D4 was purified, leading to detection of three sequences, including one across a site of alternate exon splicing, which all identified a single protein, heterogeneous nuclear ribonucleoprotein A2 (hnRNP A2). The splice variant hnRNP-B1 was a minor co-purifying immunoreactive protein. hnRNPs are members of a family of ribonuclear proteins which are generally thought to regulate the shuttling of nascent RNA transcripts between the nucleus and cytoplasm. The hnRNP A2/B1 family of antigens is frequently observed in transformed bronchial epithelium (Zhou *et al.*, 1996) and its increased expression is associated with a critical phase of fetal lung development for three mammalian systems, suggesting an oncofetal role for this protein (Montuenga *et al.*, 1998).

We are conducting a clinical trial to evaluate the performance of the hnRNP A2/B1 protein as a biomarker for early detection of second primary lung cancer (SPLC) (Tockman *et al.*, 1994). Individuals at risk for SPLC have the highest annual incidence of lung cancer (2–5%) among asymptomatic populations (Grover & Piantadosi, 1989). The Lung Cancer Early Detection Working Group (LCEDWG), thoracic surgeons and medical oncologists from leading medical centres throughout the United States and Canada, and the US National Cancer Institute are collaborating in this trial.

Monoclonal antibody 703D4 binds hnRNP A2/B1 within selected epithelial cells that exfoliate in the sputum (Tockman *et al.*, 1998). In all cells correctly diagnosed by immunocytochemistry, we recognize at least a proplastic morphology. These morphological criteria reflect proliferative changes in nuclear morphology and a level of cytoplasmic immaturity. When such cells bind monoclonal antibody, we consider them sentinel cells for preclinical lung cancer (Tockman *et al.*, 1998). Sentinel cells expressing upregulated levels of hnRNP A2/B1 are found infrequently (1/5000) among cells which normally express low levels of this protein. By developing a cell-based diagnostic approach, rather than a traditional mass extraction assay, we preserve in these isolated cells the natural upregulated signal compared to background noise.

After the first year of the LCEDWG trial, 13 SPLCs were identified (Tockman *et al.*, 1997). The sensitivity and specificity of the hnRNP A2/B1 biomarker for later SPLC were 77–82% and 65–81%, respectively. Among the cases called positive by immunocytochemistry and image cytometry, 67% developed SPLC within one year. This diagnostic accuracy exceeds that commonly found in cancer screening tests with prostate-specific antigen (PSA) (Tockman, 1996). Sueoka *et al.* (1999) and Fielding *et al.* (1999) have independently confirmed that this epitope can be used to detect preclinical lung cancer. More than 6000 individuals have been screened with this biomarker in ongoing clinical trials in North America, the United Kingdom, China and Japan. Detection of hnRNP A2/B1 overexpression in sputum epithelial cells with proplastic morphology appears to be the basis of a cytotest that could be the basis for a strategy of preclinical lung cancer diagnosis.

Specific oncogene activation (ras) and tumour-suppressor gene deletion (p53, 3p, 9p)

Three closely related genes, H-*ras*, N-*ras* and K-*ras* make up the *ras* family of oncogenes. The highly conserved 21-kDa protein products of these genes are important signal transduction elements which participate in cell cycle regulation, controlling proliferation. Mutation of the *KRAS2* oncogene is one of the most commonly occurring genetic lesions in colorectal cancer (Bos *et al.*, 1987), and is frequently seen in lung cancer (Rodenhuis & Slebos, 1992). Using the JHLP archive of preclinical sputum linked to tumour outcome data, we have demonstrated that specific mutations can be detected in non-malignant sputum specimens in advance of clinical lung cancer (Mao *et al.*, 1994a).

In this pilot study, we selected 15 participants in the JHLP with no malignancy in sputum cytology who went on to develop adenocarcinoma or large-cell carcinoma of the lung. These histological cell types were selected because they have a higher incidence of K-*ras* mutations (30%) than other lung tumours (Rodenhuis *et al.*, 1988). We also looked for *p53* gene mutations because these are among the most common genetic alterations in lung cancers (and other cancers) (Takahashi *et al.*, 1989; Hollstein *et al.*, 1991). The first exon of K-*ras* or exons 5–8 of the *p53* gene were amplified by polymerase chain reaction (PCR) from DNA extracted from the paraffin-embedded primary lung tumour. After cloning, the K-*ras* gene was sequenced to detect mutations. Tumours not containing K-*ras* mutations were sequenced for *p53* mutations to detect tumour-specific markers. Once mutations specific for each tumour were identified, oligonucleotide probes were prepared, specific for the wild-type sequence or individual mutant K-*ras* and *p53*. These probes were hybridized to sputum DNA that had been amplified by PCR, cloned into a phage vector and transferred to nylon membranes. Ten of the 15 patients had primary tumours which contained either a K-*ras* or a *p53* gene mutation. Identical mutations were detected in nonmalignant sputum cells from 8 of 10 patients who had tumours containing oncogene mutations. Patients whose tumours did not contain mutations and control patients without cancer were negative for sputum mutations by this assay.

This study demonstrated that 8 of 15 patients (53%) with adenocarcinoma or large-cell carcinoma of the lung had detectable mutations in sputum cells from 1 to 13 months before clinical diagnosis. The identification of specific gene abnormalities is less sensitive than the protein marker described above and its applicability is further limited by the need to know the specific mutation sequence with which to probe the sputum specimens. In this pilot study, the mutation sequence was determined from the resected tumour. This approach is obviously not practical for screening undiagnosed individuals. However, with future advances in gene chip technology, it may become feasible to probe for all possible mutations of common oncogenes and tumour-suppressor genes in sputum specimens of asymptomatic individuals.

Genomic instability (loss of heterozygosity, microsatellite alterations)

Microsatellite markers are small repeated DNA sequences found in the introns (non-coding regions) of a gene. PCR amplification of these repeat sequences provides a rapid method for assessment of loss of heterozygosity (LOH) and facilitates mapping of tumour-suppressor genes

(Nawroz *et al.*, 1994; Ruppert *et al.*, 1993). Yet microsatellites can provide additional information. Expansions or deletions of these repeating elements are called microsatellite alterations. These alterations, acquired during division of a single transformed cell, are passed onto daughter cells during clonal expansion. Since they are not transcribed, microsatellite alterations provide no growth advantage to the cell. However, their detection in histological specimens is equivalent to the detection of neoplastic (clonal) cell populations. Although detection of microsatellite alterations does not indicate the specific genetic change in the tumour, detection of clonal cell populations might serve as a marker for use in cancer screening (Mao *et al.*, 1994b).

Widespread microsatellite instability was first reported in colorectal tumors (Peinado *et al.*, 1992). In hereditary nonpolyposis colorectal carcinoma (HNPCC), mutations of mismatch repair genes are probably responsible for microsatellite alterations at multiple locations in the genome (Leach *et al.*, 1993). However, in non-HNPCC-associated tumours, including lung cancer, there is not a similar widespread loss of mismatch repair, indicating that another, as yet unknown, mechanism is responsible for somatic alterations of repeat sequences (Merlo *et al.*, 1994).

The pattern of microsatellite alterations and LOH may be specific for different types of cancer. The high incidence of these changes on chromosomes 3, 5, 8, 9, 10, 11, 17 and 20 has been described in lung cancer specimens (Merlo *et al.*, 1994; Xu *et al.*, 1997b), although the role of these changes in carcinogenesis is not yet known. Perhaps it is the cumulative effect of these genetic injuries that is important. We have already shown that microsatellite alterations are clonal markers for detection of human lung cancer, and microsatellite alterations at selected loci can be recognized in sputum cells before clinical lung cancer (Mao *et al.*, 1994b). When microsatellite alterations occurred at more than one locus, there was a significant association with hnRNP A2/B1 overexpression (Zhou *et al.*, 1999). We tested 41 paired tumour and normal DNA specimens from NSCLC patients surgically resected at the Moffitt Cancer Center. Eleven di- and tetranucleotide repeat markers were selected for their high frequency of loss (LOH) or alteration in lung cancer. In 41 paired tumour/normal samples, 19 patients (46%) had more than two loci of microsatellite instability (loss or alteration), while 13 out of 41 patients (32%) had only one locus of microsatellite instability. The total frequency for microsatellite alteration was 78%, suggesting that this panel of markers may have a sensitivity comparable to that of hnRNP protein markers for detection of lung cancer. We are now evaluating whether this performance is maintained in sputum specimens of high-risk individuals.

Abnormal methylation

The *p16* gene is located on the short arm of chromosome 9 (at 9p21) and is frequently mutated or inactivated in tumours and cell lines derived from lung cancer (Hamada *et al.*, 1998; Shapiro *et al.*, 1995). This gene codes for a protein that binds to the cyclin-dependent kinases 4 or 6 (CDK4 or CDK6) and prevents the kinase from phosphorylating (activating) cyclin D1. When phosphorylated, the activated cyclin D1 phosphorylates the retinoblastoma protein to allow release of E_2F transactivators and progression through the cell cycle (Lukas *et al.*, 1995). *p16*, therefore, acts as a tumour-suppressor gene, inhibiting mitosis and cell proliferation. Downregulation or loss of *p16* expression could contribute to the loss of cell-cycle control and provide a cellular growth advantage. Mao *et al.* (1997) have shown that nearly two thirds of former smokers show some genetic alteration in their bronchial cells. The frequency of LOH at the tumour-suppressor sites 9p21(*p16*) and 17p13(*p53*) was nearly the same for smokers and for former smokers. In contrast, lifetime nonsmokers showed no LOH at 9p21, suggesting that loss of 9p21 sequences might be an early event in the development of lung cancer.

Loss of *p16* expression through gene deletion or expression of an altered protein through gene mutation are established causes of loss of tumour-suppressor function (Shapiro *et al.*, 1995). Merlo *et al.* (1995) have described inhibition of *p16* gene transcription by promoter region hypermethylation and Herman *et al.* (1996) have described a novel PCR assay for detection of this condition. Briefly, the addition of methyl groups to a sequence motif (CpG islands) in the gene promoter region results in failure of gene transcription (Gonzalez-Zulueta *et al.*, 1995). For *p16*, these sequence motifs start at the promoter and extend into exon (transcribed region) 1α (Merlo *et al.*, 1995). Myohanen *et al.* (1998) have shown that transcriptional repression induced by CpG island methylation could be at least partly reversed by cell culture treatment with the demethylating agent 5-aza-2′-deoxycytidine.

Belinsky *et al.* (1998) have measured hypermethylation of the CpG islands of the *p16* gene in sputum from lung cancer patients and found a high correlation with the early stages of NSCLC. They suggested that detection of *p16* CpG island hypermethylation might be useful in the prediction of individuals who might develop lung cancer. As yet, however, no prospective studies have assessed the performance of the hypermethylation assay on samples from individuals at risk for developing lung cancer.

Helical CT detection of atypical adenomatous hyperplasia (AAH)

A new radiographic approach has been pioneered in the United States and Japan (Kaneko *et al.*, 1996; Henschke *et al.*, 1999) using low-dose, single-breath, helical computed tomography (CT). Henschke *et al.* (1999), in their Early Lung Cancer Action Project (ELCAP), screened 1000 smokers and former smokers aged 60 years and older with CT and detected 27 lung cancers, over 80% of which were stage 1a tumours. However, the histological distribution of these lung lesions was skewed; 78% were adenocarcinoma and a further 11% were mixed adenosquamous cancers. The very small 'ground glass opacities' detected by CT are frequently characterized histologically as atypical adenomatous hyperplasia (AAH), appearing as type II pneumocytes proliferating along the alveolar septa. Kitamura *et al.* (1999) have characterized AAH as alveolar intraepithelial neoplasia. Certain populations of AAH cells exhibit active proliferation, aneuploidy, 3p and 9p deletions, K-*ras* codon 12 mutation and disruption of cell-cycle control, but *p53* gene aberrations are rare and telomerase activation is absent (Kitamura *et al.*, 1999). AAH and AAH-like carcinomas may constitute a population of heterogeneous lesions representing progressive steps towards overt bronchiolo-alveolar lung carcinoma (Kitamura *et al.*, 1995). Although the treatment and prognostic implications of the lesion have not yet been determined, helical CT-detected AAH could be characterized and evaluated as a surrogate end-point biomarker in future chemoprevention trials.

Fluorescent bronchoscopy detection of angiogenic squamous dysplasia

Experience in other epithelial organs (i.e., cervix, oesophagus, colon) has shown that if the neoplastic lesion can be detected and treated at the earliest, intraepithelial stage, the rate of cure can be significantly improved (Anderson *et al.*, 1988; Jiang *et al.*, 1989; Winawer *et al.*, 1993). Lam *et al.* (1998) showed that for the early detection of bronchogenic carcinoma, a combination of conventional white-light bronchoscopy and light-induced fluorescence endoscopy significantly increased the ability of the bronchoscopist to recognize and diagnose small invasive cancers, carcinoma *in situ* and intraepithelial lesions. The treatment and survival implications of fluorescence-detected lesions have not yet been studied.

In one study of 54 patients with known or suspected lung cancer, 15% of patients diagnosed with invasive carcinoma also harboured carcinoma *in situ* (Lam *et al.*, 1993). The relative sensitivity of fluorescence bronchoscopy, with regard to identification of moderate/severe dysplasia or carcinoma *in situ*, was 50% greater than that of conventional white-light bronchoscopy. To our knowledge, no study has evaluated whether lesions detected by fluorescence bronchoscopy have enhanced significance due to altered gene expression or the presence of angiogenic squamous dysplasia.

In the biopsies from fluorescence bronchoscopies of smokers at high risk for lung cancer, Keith *et al.* (2000) observed a lesion consisting of capillary blood vessels closely juxtaposed to and projecting into metaplastic or dysplastic squamous bronchial epithelium, which they termed angiogenic squamous dysplasia. This lesion represents a qualitatively distinct form of angiogenesis in which there is architectural rearrangement of the capillary microvasculature. Genetic analysis of surface epithelium in a random subset of lesions revealed loss of heterozygosity at chromosome 3p in 53% of angiogenic squamous dysplasia lesions. No confirmed *p53* mutations were identified. Compared with normal epithelium, proliferative activity was markedly elevated in angiogenic squamous dysplasia lesions. The lesion was not present in biopsies from 16 normal nonsmoker control subjects. While neither the treatment nor prognostic implications of this lesion have been determined, it is another potential intermediate-effect marker for trials of chemoprevention.

Conclusion

These potential intermediate-effect markers offer great promise to establish a molecular diagnosis of lung cancer far in advance of clinical presentation and to follow the response to chemoprevention. Any or all of these tests could be incorporated into the routine management of individuals at risk of developing primary or second primary lung cancer. However, several issues must be considered before these tests are ready for clinical application. First, test performance characteristics must be confirmed in prospective trials. For several of these markers, such trials are already in progress. Second, a management and intervention strategy must be developed that will be appropriate to the stage at which lung cancer is diagnosed. The ability to detect lung cancer at the stage of clonal expansion, well in advance of malignant invasion of the basement membrane, suggests that a non-invasive chemoprevention approach might be the primary therapeutic intervention in such cases. Preliminary studies of chemopreventive agents are now under way at the US National Cancer Institute. Several of these agents could be delivered by inhaler to place a maximum dose directly on the transformed epithelium. The next stage will be clinical trials to evaluate combined diagnostic and therapeutic approaches to assess their impact on the incidence of clinical lung cancer.

References

Alpha-Tocopherol, Beta Carotene Cancer Prevention Study Group. The effect of vitamin E and beta carotene on the incidence of lung cancer and other cancers in male smokers. *New Engl. J. Med.*, **330**, 1029–1035

Anderson, G.H., Boyes, D.A., Benedet, J.L., Le Riche, J.C., Matisic, J.P., Suen, K.C., Worth, A.J., Millner, A. & Bennett, O.M. (1988) Organization and results of the cervical cytology screening program in British Columbia, 1955-85. *Br. Med. J.*, **296**, 975–978

Arnold, A.M., Browman, G.P., Levine, M.N., D'Souza, T., Johnstone, B., Skingley, P., Turner-Smith, L., Cayco, R., Booker, L., Newhouse, M. & Hryniuk, W.M. (1992) The effect of the synthetic retinoid etretinate on sputum cytology: results from a randomised trial. *Br. J. Cancer*, **65**, 737–743

Auerbach, O., Stout, A.P., Hammond, E.C. & Garfinkel, L. (1961) Changes in bronchial epithelium in relation to cigarette smoking and in relation to lung cancer. *New Engl. J. Med.*, **265**, 253–267

Auerbach, O., Hammond, E.C. & Garfinkel, L. (1979) Changes in bronchial epithelium in relation to cigarette smoking, 1955-1960 vs. 1970-1977. *New Engl. J. Med.*, **300**, 381–385

Ayoub, J., Jean-Francois, R., Cormier, Y., Meyer, D., Ying, Y., Major, P., Desjardins, C. & Bradley, W.E. (1999) Placebo-controlled trial of 13-cis-retinoic acid activity on retinoic acid receptor-beta expression in a population at high risk: implications for chemoprevention of lung cancer. *J. Clin. Oncol.*, **17**, 3546–3552

Belinsky, S.A., Nikula, K.J., Palmisano,W.A., Michels, R., Saccamano, G., Gabrielson, E., Baylin, S.B. & Herman, J.G. (1998) Aberrant methylation of *p16*(INK4a) is an early event in lung cancer and a potential biomarker for early diagnosis. *Proc. Natl Acad. Sci. USA*, **95**, 11891–11896

Bishop, J.M. (1987) The molecular genetics of cancer. *Science*, **235**, 305–311

Bos, J.L., Fearon, E.R., Hamilton, S.R., Verlaan-de Vries, M., Van Boom, J.H., Van der Eb, A.J. & Vogelstein, B. (1987) Prevalence of ras gene mutations in human colorectal cancers. *Nature*, **327**, 293–297

Chambon, P. (1996) A decade of molecular biology of retinoic acid receptors. *FASEB J.*, **10**, 940–954

Clark, L.C., Combs, G.F. Jr, Turnbull, B.W., Slate, E.H., Chalker, D.K., Chow, J., Davis, L.S., Glover, R.A., Graham, G.F., Gross, E.G., Krongrad, A., Lesher, J.L., Jr, Park, H.K., Sanders, B.B, Jr, Smith, C.L. & Taylor, J.R. (1996) Effects of selenium supplementation for cancer prevention in patients with carcinoma of the skin. A randomized controlled trial. Nutritional Prevention of Cancer Study Group. *J. Am. Med. Ass.*, **276**, 1957–1963

de Vries, N., van Zandwijk, N. & Pastorino, U. (1999) Chemoprevention of head and neck and lung (pre)cancer. *Recent Results Cancer Res.*, **151**, 13–25

Fearon, E.R. & Vogelstein, B. (1990) A genetic model for colorectal tumorigenesis. *Cell*, **61**, 759–767

Fielding, P., Turnbull, L., Prime, W., Walshaw, M. & Field, J.K. (1999) Heterogeneous nuclear ribonucleoprotein A2/B1 up-regulation in bronchial lavage specimens: a clinical marker of early lung cancer detection. *Clin. Cancer Res.*, **5**, 4048–4052

Gonzalez-Zulueta, M., Bender, C.M., Young, A.S., Nguyen, T., Beart, R.W., van Tornout, J.M. & Jones, P.A. (1995) Methylation of the 5´CpG island of the *p16/CDKN2* tumor suppressor gene in normal and transformed human tissues correlates with gene silencing. *Cancer Res.*, **55**, 4531–4535

Grover, F.L. & Piantadosi, S. (1989) Recurrence and survival following resection of bronchioloalveolar carcinoma of the lung—the Lung Cancer Study Group experience. *Ann. Surg.*, **209**, 779–790

Hamada, K., Kohno, T., Kawanishi, M., Ohwada, S. & Yokota, J. (1998) Association of CDKN2A (p16)/CDKN2B(p15) alterations and homozygous chromosome arm 9p deletions in human lung carcinoma. *Genes Chromosomes Cancer*, **22**, 232–240

Heimburger, D.C., Alexander, C.B., Birch, R., Butterworth, C.E., Jr, Bailey, W.C. & Krumdieck, C.L. (1988) Improvement in bronchial squamous metaplasia in smokers treated with folate and vitamin B12. Report of a preliminary randomized, double-blind intervention trial. *J. Am. Med. Ass.*, **259**, 1525–1530

Hennekens, C.H., Buring, J.E., Manson, J.E., Stampfer, M., Rosner, B., Cook, N.R., Belanger, C., LaMotte, F., Gaziano, J.M., Ridker, P.M., Willett, W. & Peto, R. (1996) Lack of effect of long-term supplementation with beta carotene on the incidence of malignant neoplasms and cardiovascular disease. *New Engl. J. Med.*, **334**, 1145–1149

Henschke, C.I., McCauley, D.I., Yankelevitz, D.F., Naidich, D.P., McGuinness, G., Miettinen, O.S., Libby, D.M., Pasmantier, M.W., Koizumi, J., Altorki, N.K. & Smith, J.P. (1999) Early Lung Cancer Action Project: overall design and findings from baseline screening. *Lancet*, **354**, 99–105

Herman, J.G., Graff, J.R., Myohanen, S., Nelkin, B.D. & Baylin, S.B. (1996) Methylation-specific PCR: a novel PCR assay for methylation status of CpG islands. *Proc. Natl Acad. Sci. USA*, **93**, 9821–9826

Hollstein, M., Sidransky, D., Vogelstein, B. & Harris, C.C. (1991) *P53* mutations in human cancers. *Science*, **253**, 49–53

Hong, W.K. (1999) Chemoprevention of lung cancer. *Oncology*, **13** (Suppl. 5), 135–141

IARC (1998) *IARC Handbooks of Cancer Prevention*, Volume 2, *Carotenoids*, Lyon, IARC

Jetten, A.M., Nervi, C. & Vollberg, T.M. (1992) Control of squamous differentiation in tracheobronchial and epidermal epithelial cells: role of retinoids. *J. Natl Cancer Inst. Monogr.*, **13**, 93–100

Jiang, T., Yan, S. & Zhao, L. (1989) Preventing effect of linwei dihuang decoction on esophageal carcinoma. *Jpn. J. Cancer Chemother.*, **16**, 1511–1518

Kaneko, M., Eguchi, K., Ohmatsu, H., Kakinuma, R., Naruke, T., Suemasu, K. & Moriyama, N. (1996) Peripheral lung cancer: screening and detection with low-dose spiral CT versus radiography. *Radiology*, **201**, 789–802

Keith, R.L., Miller, Y.E., Gemmill, R.M., Drabkin, H.A., Dempsey, E.C., Kennedy, T.C., Prindiville, S. & Franklin, W.A. (2000) Angiogenic squamous dysplasia in bronchi of individuals at high risk for lung cancer. *Clin. Cancer Res.*, **6**, 1616–1625

Khuri, F.R. & Lippman, S.M. (2000) Lung cancer chemoprevention. *Semin. Surg. Oncol.*, **18**, 100–105

Kitamura, H., Kameda, Y., Nakamura, N., Nakatani, Y., Inayama, Y., Iida, M., Noda, K., Ogawa, N., Shibagaki, T. & Kanisawa, M. (1995) Proliferative potential and p53 overexpression in precursor and early stage lesions of bronchioloalveolar lung carcinoma. *Am. J. Pathol.*, **146**, 876–887

Kitamura, H., Kameda, Y., Ito, T. & Hayashi, H. (1999) Atypical adenomatous hyperplasia of the lung. Implications for the pathogenesis of peripheral lung adenocarcinoma. *Am. J. Clin. Pathol.*, **111**, 610–622

Knekt, P., Marniemi, J., Teppo, L., Heliovaara, M. & Aromaa, A. (1998) Is low selenium status a risk factor for lung cancer? *Am. J. Epidemiol.*, **148**, 975–982

Kurie, J.M., Spitz, M.R. & Hong, W.K. (1995) Lung cancer chemoprevention: targeting former rather than current smokers. *Cancer Prevent. Int.*, **2**, 55–64

Kurie, J.M., Lee, J.S., Khuri, F.R., Morice, R.C., Walsh, G.L., Broxson, A., Ro, J.Y., Kemp, B.L., Lee, J.K.J., Lotan, R., Xu, X., Lu, D. & Hong, W.K. (1999) 4-Hydroxyphenylretinamide (4-HPR) in the reversal of bronchial metaplasia and dysplasia in smokers [abstract]. *Proc. Am. Soc. Clin. Oncol.*, **18**, 473a

Lam, S., MacAulay, C., Hung, J., LeRiche, J., Profio, A.E. & Palcic, B. (1993) Detection of dysplasia and carcinoma in situ with a lung imaging fluorescence endoscope device. *J. Thor. Cardiovasc. Surg.*, **105**, 1035–1040

Lam, S., Kennedy, T., Unger, M., Miller, Y.E., Gelmont, D., Rusch, V., Gipe, B., Howard, D., le Riche, J.C., Coldman, A.C. & Gazdar, A.F. (1998) Localization of bronchial intraepithelial neoplastic lesions by fluorescence bronchoscopy. *Chest*, **113**, 696–702

Landis, S.H., Murray, T., Bolden, S. & Wingo, P.A. (1998) Cancer statistics [published errata appear in *CA Cancer J. Clin.*, 1998, **48**, 192 and 1998, **48**, 329] *CA Cancer J. Clin.*, **48**, 6–9

Leach, F., Nicolaides, N.C., Papadopoulos, N., Liu, B., Jen, J., Parsons, R., Peltomaki, P., Sistonen, P., Aaltonen, L.A., Nystrom-Lahti, M., Guan, X.Y., Zhang, J., Meltzer, P.S., Yu, J.W., Kao, F.T., Chen, D.J., Cerosaletti, K.M., Fournier, R.E.K., Todd, S., Lewis, T., Leach, R.J., Naylor, S.L.,

Weissenbach, J., Meeklin, J.P., Jarvinen, H., Petersen, G.M., Hamilton, S.R., Green, J., Jass, J., Watson, P., Lynch, H.T., Trent, J.M., de la Chapelle, A., Kinzler, K.W. & Vogelstein, B. (1993) Mutation of a *mutS* homolog in hereditary nonpolyposis colorectal cancer. *Cell*, 75, 1215–1225

Lee, J.S., Lippman, S.M., Benner, S.E., Lee, J.J., Ro, J.Y., Lukeman, J.M., Morice, R.C., Peters, E.J., Pang, A.C., Fritsche, H.A. Jr & Hong, W.K. (1994) Randomized placebo-controlled trial of isotretinoin in chemoprevention of bronchial squamous metaplasia. *J. Clin. Oncol.*, **12**, 937–945

Lee, H.Y., Dohi, D.F., Kim, Y.H., Walsh, G.L., Consoli, U., Andreeff, M., Dawson, M.I., Hong, W.K. & Kurie, J.M. (1998) All-trans retinoic acid converts E2F into a transcriptional suppressor and inhibits the growth of normal human bronchial epithelial cells through a retinoic acid receptor-dependent signaling pathway. *J. Clin. Invest.*, **101**, 1012–1019

Lee, I.M., Cook, N.R., Manson, J.E., Buring, J.E. & Hennekens, C.H. (1999) Beta-carotene supplementation and incidence of cancer and cardiovascular disease: the Women's Health Study *J. Natl Cancer Inst.*, **91**, 2102–2106

Lippman, S.M., Benner, S.E. & Hong, W.K. (1994) Cancer chemoprevention. *J Clin. Oncol.*, **12**, 851–873

Lotan, R. (1997) Retinoids and chemoprevention of aerodigestive tract cancers. *Cancer Metastasis Rev.*, **16**, 349–356

Lukas, J., Parry, D., Aagaard, L., Mann, D.J., Bartkova, J., Strauss, M., Peters, G. & Bartek, J. (1995) Retinoblastoma protein-dependent cell-cycle inhibition by the tumour suppressor p16. *Nature*, **375**, 503–506

Mangelsdorf, D.J., Umesono, K. & Evans, R.M. (1994) The retinoid receptors. In: Sporn, M.B., Roberts, A.B. & Goodman, D.S., eds, *The Retinoids*, New York, Raven Press, pp. 319–350

Mao, L., Hruban, R.H., Boyle, J.O., Tockman, M. & Sidransky, D. (1994a) Detection of oncogene mutations in sputum precedes diagnosis of lung cancer. *Cancer Res.*, **54**, 1634–1637

Mao, L., Lee, D.J., Tockman, M.S., Erozan, Y.S., Askin, F. & Sidransky, D. (1994b) Microsatellite alterations as clonal markers for the detection of human cancer. *Proc. Natl Acad. Sci. USA*, **91**, 9871–9875

Mao, L., Lee, J.G., Kurie, J.M., Fan, Y.H., Lippman, S.M., Lee, J., Ro, J.Y., Broxson, A., Yu, R., Morice, R.C., Kemp, B.L., Khuri, F.R., Walsh, G.L., Hittleman, W.N. & Hong, W.K. (1997) Clonal genetic alterations in the lungs of current and former smokers. *J. Natl Cancer Inst.*, **89**, 857–862

McLarty, J.W., Holiday, D.B., Girard, W.M., Yanagihara, R.H., Kummet, T.D. & Greenberg, S.D. (1995) Beta-carotene, vitamin A, and lung cancer chemoprevention: results of an intermediate endpoint study. *Am. J. Clin. Nutr.*, **62**, 1431S–1438S

Merlo, A., Herman, J.G., Mao, L., Lee, D.J., Gabrielson, E., Burger, P.C., Baylin, S.B. & Sidransky, D. (1995) 5′CpG island methylation is associated with transcriptional silencing of the tumour suppressor p16/CDKN2/MTS1 in human cancers. *Nat. Med.*, **1**, 686–692

Merlo, A., Mabry, M., Gabrielson, E., Vollmer, R., Baylin, S.B. & Sidransky, D. (1994) Frequent microsatellite instability in primary small cell lung cancer. *Cancer Res.*, **54**, 2098–2101

Montuenga, L., Zhou, J., Avis, I., Martinez, A., Vos, M., Sunday, M. & Mulshine, J.L. (1998) The early lung cancer marker heterogeneous nuclear ribonucleoprotein (hnRNP) A2/B1 is an oncofetal antigen. *Am. J. Resp. Cell Mol. Biol.*, **19**, 554–562

Moon, R.C., Mehta, R.G. & Rao, K.J. (1994) Retinoids and cancer in experimental animals. In: Sporn, M.B., Roberts, A.B. & Goodman, D.S., eds, *The Retinoids*, New York, Raven Press, pp. 573–595

Myohanen, S.K., Baylin, S.B. & Herman, J.G. (1998) Hypermethylation can selectively silence individual $p16^{INK4A}$ alleles in neoplasia. *Cancer Res.*, **58**, 591–593

Nawroz, H., van der Riet, P., Hruban, R.H., Koch, W., Ruppert, J.M. & Sidransky, D. (1994) Allotype of head and neck squamous cell carcinoma. *Cancer Res.*, **54**, 1152–1155

Omenn, G.S., Goodman, G.E., Thornquist, M.D., Balmes, J., Cullen, M.R., Glass, A., Keogh, J.P., Meyskens, F.L., Valanis, B., Williams, J.H., Barnhart, S. & Hammar, S. (1996) Effects of a combination of beta carotene and vitamin A on lung cancer and cardiovascular disease. *New Engl. J. Med.*, **334**, 1150–1155

Pastorino, U., Infante, M., Maioli, M., Chiesa, G., Buyse, M., Firket, P., Rosmentz, N., Clerici, M., Soresi, E., Valente, M., Belloni, P.A. & Ravasi, G. (1993) Adjuvant treatment of stage I lung cancer with high-dose vitamin A. *J. Clin. Oncol.*, **11**, 1216–1222

Peinado, M.A., Malkhosyan, S., Velazquez, A. & Perucho, M. (1992) Isolation and characterization of allelic losses and genes in colorectal tumors by arbitrarily primed polymerase chain reaction. *Proc. Natl Acad. Sci. USA*, **89**, 10065–10069

Picard, E., Seguin, C., Monhoven, N., Rochette-Egly, C., Siat, J., Borrelly, J., Martinet, Y., Martinet, N. & Vignaud, J.M. (1999) Expression of retinoid receptor genes and proteins in non-small-cell lung cancer. *J. Natl Cancer Inst.*, **91**, 1059–1066

Ratnasinghe, D., Tangrea, J.A., Forman, M.R., Hartman, T., Gunter, E.W., Qiao, Y.L., Yao, S,X., Barett, M.J., Giffen, C.A., Erozan, Y., Tockman, M.S. & Taylor, P.R. (2000) Serum tocopherols, selenium and lung cancer risk among tin miners in China. *Cancer Causes Control*, **11**, 129–135

Rodenhuis, S. & Slebos, R.J.C. (1992) Clinical significance of ras oncogene activation in human lung cancer. *Cancer Res.*, **52**, 2665s–2669s

Rodenhuis, S., Slebos, R.J., Boot, A.J., Evers, S.G., Mooi, W.J., Wagenaar, S.S., Van Bodegom, P.C. & Bos, J.L. (1988) Incidence and possible clinical significance of K-*ras* oncogene activation in adenocarcinoma of the human lung. *Cancer Res.*, **48**, 5738–5741

Ruppert, J.M., Tokino, K. & Sidransky, D. (1993) Evidence for two bladder cancer suppressor loci on human chromosome 9. *Cancer Res.*, **53**, 5093–5094

Shapiro, G.I., Park, J.E., Edwards, C.D., Mao, L., Merlo, A., Sidransky, D., Ewen, M.E. & Rollins, B.J. (1995) Multiple mechanisms of $p16^{INK4A}$ inactivation in non-small cell lung cancer cell lines. *Cancer Res.*, **55**, 6200–6209

Siegfried, J.M. (1998) Biology and chemoprevention of lung cancer. *Chest*, **113** (1 Suppl.), 40S–45S

Slaughter, D.P., Southwick, H.W. & Smejkal, W. (1953) "Field cancerization" in oral stratified squamous epithelium. *Cancer*, **6**, 963–968

Sporn, M.B., Dunlop, N.M., Newton, D.L. & Smith, J.M. (1976) Prevention of chemical carcinogenesis by vitamin A and its synthetic analogs (retinoids). *Fed. Proc.*, **35**, 1332–1338

Sueoka, E., Goto, Y., Sueoka, N., Kai, Y., Kozu, T. & Fujiki, H. (1999) Heterogeneous nuclear ribonucleoprotein B1 as a new marker of early detection for human lung cancers. *Cancer Res.*, **59**, 1404–1407

Sueoka, N., Lee, H.Y., Walsh, G.L., Hong, W.K. & Kurie, J.M. (1999) Posttranslational mechanisms contribute to the suppression of specific cyclin:CDK complexes by all-*trans* retinoic acid in human bronchial epithelial cells. *Cancer Res.*, **59**, 3838–3844

Takahashi, T., Nau, M.M., Chiba, I., Levitt, M., Pass, M., Gazdar, A.F. & Minna, J.D. (1989) *P53*: a frequent target for genetic abnormalities in lung cancer. *Science*, **246**, 491–494

Tockman, M.S., Gupta, P.K., Myers, J.D., Frost, J.K., Baylin, S.B., Gold, E.B., Chase, A.M., Wilkinson, P.H. & Mulshine, J.L. (1988) Sensitive and specific monoclonal antibody recognition of human lung cancer antigen on preserved sputum cells: a new approach to early lung cancer detection. *J. Clin. Oncol.*, **6**, 1685–1693

Tockman, M.S., Erozan, Y.S., Gupta, P., Piantadosi, S., Mulshine, J.L. & Ruckdeschel, J.C. (1994) The early detection of second primary lung cancers by sputum immunostaining. *Chest*, **106**, 385S–390S

Tockman, M.S. (1996) Monoclonal antibody detection of premalignant lesions of the lung. In: Fortner, J.G. & Sharp, P.A., eds, *Accomplishments in Cancer Research*, Philadelphia, Lippincott-Raven, pp. 169–177

Tockman, M.S., Mulshine, J.L., Piantadosi, S., Erozan, Y.S., Gupta, P.K., Ruckdeschel, J.C., Taylor, P.R., Zhukov, T., Zhou, W.H., Qiao, Y.L. & Yen, S.X. (1997) Prospective detection of preclinical lung cancer; results from two studies of heterogeneous nuclear ribonucleoprotein A2/B1 overexpression. *Clin. Cancer Res.*, **3**, 2237–2246

Tockman, M.S., Zhukov, T.A., Erozan, Y.S., Westra, W.H., Zhou, J. & Mulshine, J.L. (1998) Antigen retrieval improves hnRNP A2/B1 immunohistochemical localization in premalignant lesions of the lung. In: Martinet, Y., Hirsch, F.R., Martinet, N., Vignaud, J.M. & Mulshine, J.L., eds, *Clinical and Biological Basis of Lung Cancer Prevention*, Basel, Birkhauser Verlag, pp. 239–246

Weinberg, R.A. (1989) Oncogenes, antioncogenes, and the molecular basis of multistep carcinogenesis. *Cancer Res.*, **49**, 3713–3721

Winawer, S.J., Zamber, A.G., Ho, M.N., O'Brien, M.J., Gottleib, L.S., Sternberg, S.S., Waye, J.D., Schapiro, M., Bond, J.H., Panish, J.F., Ackroyd, F., Shike, M., Kurtz, R.C., Hornsby-Lewis, L., Gerdes, H., Stewart, E.T., and the National Polyp Study Workgroup (1993) Prevention of colorectal cancer by colonoscopic polypectomy. *New Engl. J. Med.*, **329**, 1977–1981

Wistuba, I.I., Behrens, C., Milchgrub, S., Bryant, D., Hung, J., Minna, J.D. & Gazdar, A.F. (1999) Sequential molecular abnormalities are involved in the multistage development of squamous cell lung carcinoma. *Oncogene*, **18**, 643–650

Woodson, K., Tangrea, J.A., Barrett, M.J., Virtamo, J., Taylor, P.R. & Albanes, D. (1999) Serum alpha-tocopherol and subsequent risk of lung cancer among male smokers. *J. Natl Cancer Inst.*, **91**, 1738–1743

Xu, X.C., Sozzi, G., Lee, J.S., Lee, J.J., Pastorino, U., Pilotti, S., Kurie, J.M., Hong, W.K. & Lotan, R. (1997a) Suppression of retinoic acid receptor beta in non-small-cell lung cancer in vivo: implications for lung cancer development. *J. Natl Cancer Inst.*, **89**, 624–629

Xu, L.H., Bonacum, J., Wu, L., Ahrent, S., Westra, W.H., Yang, S., Tockman, M., Sidransky, D. & Jen, J. (1997b) Identification of frequently altered microsatellite markers for clinical detection of non-small cell lung cancer. *Proc. Am. Ass. Cancer Res.*, **38**, 329

Xu, X.C., Lee, J.S., Lee, J.J., Morice, R.C., Liu, X., Lippman, S.M., Hong, W.K. & Lotan, R. (1999) Nuclear retinoid acid receptor beta in bronchial epithelium of smokers before and during chemoprevention. *J. Natl Cancer Inst.*, **91**, 1317–1321

Zhou, J., Mulshine, J.L., Unsworth, E.J., Scott, F.M., Avis, I.M., Vos, M.D. & Treston, A.M. (1996) Purification and characterization of a protein that permits early detection of lung cancer. *J. Biol. Chem.*, **271**, 10760–10766

Zhou, J., Nong, L., Wloch, M., Zhukov, T. & Tockman, M.S. (1999) Expression of early lung cancer detection marker: hnRNP A2/B1 and its relation to microsatellite instability in non small cell lung cancer. *Proc. Am. Ass. Cancer Res.*, **40**, 140–141

Corresponding author:

M. S. Tockman
Professor of Medicine and Program Leader,
Molecular Screening,
H. Lee Moffitt Cancer Center and Research Institute,
Tampa, FL,
USA

Biomarkers in Cancer Chemoprevention
Miller, A.B., Bartsch, H., Boffetta, P., Dragsted, L. and Vainio, H. eds
IARC Scientific Publications No 154
International Agency for Research on Cancer, Lyon, 2001

Metabolic polymorphisms as susceptibility markers for lung and oral cavity cancer

U. Nair and H. Bartsch

Lung and oral cavity cancers are causally associated with tobacco use. Alcohol is an independent risk factor for oral cavity cancer. Major classes of carcinogens present in tobacco and tobacco smoke are converted into DNA-reactive metabolites by cytochrome P450 (CYP)-related enzymes, several of which display genetic polymorphism. Individual susceptibility to cancer is likely to be modified by the genotype for enzymes involved in the activation or detoxification of carcinogens in tobacco and repair of DNA damage. Molecular epidemiological studies to assess the risk associated with metabolic polymorphisms for cancers of the lung and head and neck have shown that the overall effect of common polymorphisms is moderate in terms of penetrance and relative risk. However, some gene combinations like mutated *CYP1A1/GSTM1*-null genotype seem to predispose the lung and oral cavity of smokers to an even higher risk for cancer or DNA damage, although these results require confirmation in larger well defined studies that take into account the existence of ethnic variations even within the commonly defined groups. Retinoids, isothiocyanates and tea polyphenols have been identified as possible chemopreventive agents for cancers of the lung and oral cavity. While a number of trials have been conducted with retinoids or β-carotene, the results were ambiguous and the causes are still being debated. The possible interaction of chemopreventive agents with metabolic polymorphisms as biomarkers in chemoprevention trials is discussed.

Introduction

Analytical and molecular epidemiology studies have identified two types of population at elevated risk of cancer: (*a*) groups exposed to particular environmental or lifestyle risk factors, and (*b*) carriers of mutated cancer-determining genes that confer a very high cancer risk (Caporaso & Goldstein, 1995). Compared to the highly penetrant polymorphic genes, low-penetrance susceptibility genes (such as those involved in carcinogen metabolism and DNA repair) modestly increase the risk for cancer in exposed individuals, perhaps at low doses of carcinogens (Vineis *et al.*, 1994; Vineis, 1997). An important mechanism by which carcinogens cause DNA mutations is by covalent binding to DNA to form adducts, which if not repaired can result in a mutation when the DNA is copied during cell division. Thus DNA adducts are sensitive markers of exposure and, subject to certain conditions, also of disease risk. However, most carcinogens need to be metabolized by enzymes that may differ between individuals due to either inherited polymorphic DNA sequence variations or up-regulation or down-regulation of genes for metabolic enzymes by other external agents such as diet, alcohol, medication or lifestyle exposures. This entails complex gene–environment and gene–gene interactions. In principle, understanding the molecular basis of disease can facilitate prevention by allowing the identification of individuals who are at increased risk and of susceptible populations so that preventive strategies can be designed for the greatest impact. It is therefore important to identify candidate genes for which biologically plausible mechanisms for effects on cancer risk can be proposed.

Tobacco exposure is the largest single cause of lung and oral cavity cancers (Doll, 1998). Alcohol use is the second major risk factor for the development of oral cancer. The major classes of carcinogens from tobacco, polycyclic aromatic hydrocarbons (PAHs) and tobacco-specific nitrosamines, are causally associated with lung and oral cancer (Hoffmann & Hecht, 1990; Hecht, 1999a; McClellan, 1996; Bergen & Caporaso, 1999). Most tobacco carcinogens are converted to DNA-reactive metabolites by oxidative, mainly cytochrome P450-related, enzymes (CYPs) and are further metabolized by phase II enzymes. PAHs are converted into phenolic metabolites; thus benzo[*a*]pyrene (B[a]P) is oxidized to B[a]P-7,8-diol by a CYP-mediated process. Secondary metabolism, mainly involving epoxide hydrolase and other CYP isoforms, leads to formation of the highly reactive (+)-*anti*-B[a]P diolepoxide (BPDE) which is a relatively good substrate for glutathione *S*-transferase (GST) enzymes GSTM1, M2, M3 and P1 (Coles & Ketterer, 1990). Among the tobacco-specific nitrosamines, 4-(methylnitrosamino)-1-(3-pyridyl)-1-butanone (NNK) and *N*′-nitrosonornicotine (NNN) are the most important. NNK is a powerful lung carcinogen in all species tested and human exposure levels are comparable to the doses that cause tumours in laboratory animals. The metabolism of NNK includes α-methyl-hydroxylation, α-methylene-hydroxylation and pyridine-*N*-oxidation by CYP-mediated reactions and reduction to 4-(methylnitrosamino)-1-(3-pyridyl)-1-butanol (NNAL) and its conjugation as a glucuronide (Hecht, 1999a,b). Tobacco smoke also contains reactive oxygen species and reactive nitrogen species that impose oxidative stress on smokers' tissue. As a consequence, oxidative DNA-base damage has been detected in respiratory tract tissue of smokers, with formation of lipid peroxidation products such as malondialdehyde, crotonaldehyde and *trans*-4-hydroxy-2-nonenal, the last of which can be further epoxidized by CYP-mediated reactions to form promutagenic exocyclic DNA adducts (Chung *et al.*, 1996; Nair *et al.*, 1999a) that could contribute to carcinogenesis in the upper aerodigestive tract (Nath *et al.*, 1998). Chewing of tobacco alone or with betel quid also results in generation of large amounts of reactive oxygen species in the mouth; these and the tobacco-specific nitrosamines are the major genotoxic agents implicated in oral cancer related to chewing (Nair *et al.*, 1996). Since tobacco carcinogens, reactive oxygen species and lipid peroxidation products are likely to be substrates for metabolizing enzymes, the extent of DNA damage and ultimately the cancer risk may be affected by polymorphic CYP and GST enzymes (Brockmöller *et al.*, 1998; Nair *et al.*, 1999b). In many instances, the association between a particular genetic risk factor and cancer appears to be race- or ethnicity-specific. Susceptibility markers that are on the pathway to the development of cancer have often been suggested as appropriate intermediate biomarkers for use in chemoprevention studies. This review focuses on metabolic polymorphisms implicated in susceptibility to lung and oral cancer, with a view to integrating them as biomarkers in chemoprevention strategies.

Polymorphic metabolic genes and cancer risk for oral and lung cancer

Genetic polymorphisms occur at a population frequency of more than 1% and can result in marked inter-individual differences. Numerous alleles that cause defective, qualitatively altered, diminished or enhanced rates of drug metabolism have been identified for many of the phase I and phase II enzymes and underlying molecular mechanisms have been elucidated.

Cytochrome P450

The cytochrome P450 enzymes are a large multigene family and are important in phase I detoxification/activation reactions. In a review of the results of all genotype-based case–control studies published in the 1990s on the effect of genetic variants of CYPs alone or in combination as risk modifiers of tobacco-related cancers (Bartsch *et al.*, 2000), the overall effects of common CYP polymorphisms were found to be modest, with odds ratios ranging from 2 to 10. Extensive studies have been conducted with the aim of identifying risk genotypes for lung cancer, mostly focussing on *CYP1A1*, *CYP2D6* and *CYP2E1*, while studies on oral cavity cancer have appeared more recently.

CYP1A1: The human enzyme CYP1A1 is involved in the activation of major classes of tobacco procarcinogens, such as PAHs and aromatic amines, and is present in many epithelial tissues. About 10% of the Caucasian population has a highly inducible form of this enzyme (termed B[a]P-hydroxylase or previously arylhydrocarbon hydroxylase), which is associated with an increased risk for bronchial, laryngeal and oral cavity tumours in smokers (Nebert *et al.*, 1996). The induction of *CYP1A1* is initiated by the specific binding of aromatic inducer compounds to the aromatic hydrocarbon (Ah) receptor. An Ah receptor nuclear translocator (*Arnt*) gene is further involved in the *CYP1A1* induction pathway. So far, no relationship has been found between Ah-receptor polymorphism and lung cancer risk (Kawajiri *et al.*, 1995; Micka *et al.*, 1997). Beginning with the pioneering work by Kellerman *et al.* (1973) on B[a]P-hydroxylase inducibility and bronchogenic carcinoma, studies on the association of the genetic polymorphism of *CYP1A1* and cancer started after co-segregation of the CYP1A1 high-inducibility phenotype and polymorphism of the MspI restriction site (Petersen *et al.*, 1991). At the present time, the *CYP1A1* gene is known to contain four important sequence polymorphisms. The *CYP1A1* Ile-Val (m2) mutation in the haem-binding region results in a twofold increase in microsomal enzyme activity and is in complete linkage disequilibrium in Caucasians with the *CYP1A1 Msp*I (m1) mutation, which has also been associated experimentally with increased catalytic activity (Landi *et al.*, 1994). Although the Ile-Val mutation in the *CYP1A1* allele does not increase activity *in vitro* (Zhang *et al.*, 1996; Persson *et al.*, 1997), it may be linked to other functional polymorphisms, for example in the regulatory region important for *CYP1A1* inducibility. Significant ethnic differences in the frequency of homozygous *CYP1A1 Msp*I alleles have been observed, and both the *Msp*I and *Val* alleles are rarer in Caucasian than in Japanese populations (Cascorbi *et al.*, 1996a).

Lung cancer: The various *CYP1A1* gene polymorphisms are differentially associated with risk of lung cancer. As the prevalence of both *m1* and *m2* alleles is higher in Japanese, studies have mainly investigated difference in cancer risk between homozygous mutant versus other genotypes, while in other populations, the effect is determined using pooled homo- and heterozygous genotypes. Data on lung cancer among Japanese suggest an increased risk associated with both the *m1* allele (Okada *et al.*, 1994) and *m2* allele (Hayashi *et al.*, 1992) polymorphism; the risk is particularly elevated among light smokers and for development of squamous-cell carcinoma (SCC). In studies on Caucasians, these findings were not confirmed (Hirvonen *et al.*, 1993; Alexandrie *et al.*, 1994; Bouchardy *et al.*, 1997), possibly due to the much lower prevalence of the *m1* allele in Caucasians. However, Xu *et al.* (1996) did show an association for the *m1* allele in Caucasians. An African-American-specific *m3* allele does not seem to confer increased risk for lung cancer overall, but an association with increased risk for adenocarcinoma has been seen (Taioli *et al.*, 1998). A recently described *m4* mutation in close proximity to *m1* has been investigated in one study on Caucasians, and no correlation with lung cancer risk was found (Cascorbi *et al.*, 1996b).

Oral cancer: The CYP1A1 enzyme is present in oral tissue (Romkes *et al.*, 1996), and the link between *CYP1A1* variants and oral cancer risk has been investigated in a number of studies. An overrepresentation of the *CYP1A1* Ile-Val variant occurred among Caucasian patients with oral cancer and this was significantly higher in nonsmokers than smokers (Park *et al.*, 1997). Similarly, an increased prevalence of *CYP1A1* Val-Val variant was found among Japanese patients with head-and-neck cancers and especially those with pharyngeal cancer (Morita *et al.*, 1999). Individuals with the homozygous *CYP1A1 Msp*I (m1/m1) variant were at significantly increased risk for oral SCC, in particular after exposure to low concentrations of PAHs (Tanimoto *et al.*, 1999) and in combination with *GSTM1*-null, risk was also significantly elevated for various subsites, especially buccal mucosa.

Combined effect of CYP1A1 and GSTM1 genotypes on cancer susceptibility

Lung cancer: Smokers with the exon-7 Ile-Val mutation were found to have more PAH–DNA adducts in their white blood cells than smokers without the variant (Mooney *et al.*, 1997). In lung parenchymal tissue of smokers, levels of BPDE and bulky (PAH)–DNA adducts were positively correlated with CYP1A1 enzyme activity (Alexandrov *et al.*, 1992). Risk of lung cancer, especially SCC, is increased in individuals with a combination of a homozygous rare allele of the *CYP1A1* gene (either *m2* or *m1*) and the null *GSTM1* gene compared with those having other combinations of genotypes in Japanese populations (Hayashi *et al.*,

1992). Subsequently, a remarkably high risk for SCC at low cigarette dose (OR 41, 95% confidence interval (CI) 8.7–193.6) was demonstrated in Japanese (Nakachi *et al.*, 1993). Significant increases were seen in B[a]P adduct levels in lung tissue from smokers with an 'at-risk' genotype, with a multiplicative effect at a lower level of exposure (Bartsch *et al.*, 1995, 1999). BPDE adduct levels in lung and leukocytes of Caucasian smokers were correlated with *CYP1A1* genotype, most strongly in *GSTM1*-deficient smokers (Bartsch, 1996; Rojas *et al.*, 2000). These findings provide a mechanistic basis for the correlation of high-risk genotypes with increased risks for tobacco-related lung cancer, even at low levels of cigarette smoking (Kihara & Noda, 1995). Significantly more *p53* mutations were seen in lung tumours of Japanese smokers having the susceptible *CYP1A1* genotype. Individuals with the combination of *CYP1A1* m2/m2 and *GSTM1*-null genotypes had an eightfold greater frequency of *p53* mutations than persons with neither genotype (Kawajiri *et al.*, 1996). Operated lung cancer patients with high pulmonary CYP1A1 enzyme inducibility (Bartsch *et al.*, 1990) and/or a high-risk genotype combination (Goto *et al.*, 1996) had shorter survival.

Oral cancer: In Japanese patients, the high-risk combination *CYP1A1* MspI (m1/m1) and *GSTM1*-null genotype conferred a high risk for oral SCC (Sato *et al.*, 1999) and a similar combination increased significantly the risk for cancer of the buccal mucosa and upper gingiva (Tanimoto *et al.*, 1999). In Caucasians, while no effect was reported by several authors (Matthias *et al.*, 1998; Oude Ophius *et al.*, 1998), Park *et al.* (1997) reported a significant effect, especially in non-smoker cases. But in one report, the highest prevalence of *p53* mutation was observed in oral tumours from patients with the *CYP1A1*(Val)/*GSTM1* active genotype (Lazarus *et al.*, 1998).

Overall, there is increasing evidence that individuals with the homozygous *CYP1A1* MspI and Ile-Val genotypes are at higher risk for contracting smoking-associated lung (SCC) and oral cancers, as particularly seen in Asian study populations where these high-risk allele frequencies are 8–18 times higher than in Caucasians (Bartsch *et al.*, 2000). The cancer risk is further increased in carriers of the combined high-risk *CYP1A1*/*GSTM1*-null genotypes.

CYP2D6: *CYP2D6* has received particular attention as a genetic susceptibility factor since the first studies on the link between lung cancer risk and extensive metabolizer (EM) status (Ayesh *et al.*, 1984). The poor metabolizer (PM) phenotype, inherited as an autosomal recessive trait, is due to several defective allelic *CYP2D6* variants, three of which account for more that 90% of all poor metabolizers. An allele associated with 2–12-fold amplification of the *CYP2D6* gene is found in carriers known as 'ultra-rapid metabolizers' (UM) (Daly *et al.*, 1996). Conflicting data exist as to whether *CYP2D6* is expressed in human lung (Raunio *et al.*, 1999). This isozyme can activate the tobacco-specific nitrosamine NNK and also nicotine, but other P450s are more active in this respect. Associations have been found between nicotine dependence and PM phenotype (Boustead *et al.*, 1997) and between UM and smoking addiction (Saarikoski *et al.*, 2000). Higher DNA adduct levels have been reported among homozygous and heterozygous extensive metabolizers compared with people classified as poor metabolizers.

Lung cancer: Among the nine genotyping studies that have been reported, a significant association was found between lung cancer and the EM genotype in three studies, and in one study an association with UM genotype and lung cancer risk was found in African Americans (Bartsch *et al.*, 2000). However, in two meta-analyses, there was no association or one of borderline significance between the EM genotype and increased risk (Christensen *et al.*, 1997; Rostami Hodjegan *et al.*, 1998).

Oral cancer: In Caucasians, a significantly higher frequency of the PM genotype among cases was reported in one study, while the time for lymph node metastasis was shorter in PM compared with EM subjects (Worrall *et al.*, 1998). No effect was found in other studies (Gonzales *et al.*, 1998; Matthias *et al.*, 1998).

Overall, evidence for a role of *CYP2D6* polymorphisms as a risk factor for lung or oral cancer is weak, conflicting and inconclusive.

CYP2E1: The ethanol-inducible *CYP2E1* metabolizes many known procarcinogens, including NNN, NNK and other volatile nitrosamines found in tobacco smoke. Wide inter-individual variation in expression of the *CYP2E1* gene in humans has been reported, which is possibly due to gene–environment interactions. *CYP2E1* is induced in mice exposed to cigarette smoke by inhalation (Villard *et al.*, 1998). Its regulation involves complex transcriptional and post-transcriptional mechanisms. Several polymorphic alleles occurring at low frequency have been identified and the most studied are the *Rsa*I $G_{-1259}C$ or *Pst*I $C_{-1091}T$ restriction fragment length polymorphisms (RFLP); these appear to be in complete linkage disequilibrium with each other (c1: common and c2: rare allele). Although the primary sequence of the enzyme does not appear to be altered, increased gene transcription has been suggested (Watanabe *et al.*, 1994). A second allele of the *CYP2E1* gene is revealed by a *Dra*I RFLP (C: minor and D: common allele). While in Caucasians no relationship was found between genotype and the activity of this enzyme *in vivo*, in Japanese the presence of the variant c2 alleles resulted in a significant reduction in oral clearance of chlorzoxazone, after adjustment for age and sex. The mean activity in individuals with the c2/c2 genotype was significantly lower than that in individuals with either the homozygous wild-type or the heterozygous genotype. Body weight and dietary factors were the major modulators of inter-individual variation (Le Marchand *et al.*, 1999). As the frequencies of variant alleles are very low in Caucasians and African Americans, the statistical power of these studies was low.

Lung cancer: In a number of studies of Caucasians, no significant association was found, but the wild-type *Dra*I genotype was associated with an increased risk for lung cancer in studies in Japanese (Uematsu *et al.*, 1994), Mexican Americans (Wu *et al.*, 1998) and a mixed Hawaiian population (Le Marchand *et al.*, 1998). More conflicting results have been published concerning the *Rsa*I/*Pst*I mutation. The rare *Pst*I/*Rsa*I c2 allele has been associated with decreased risk for cancer in some studies (Wu *et al.*, 1997; Le Marchand *et al.*, 1998), and in one study the c2 allele frequency was significantly lower among cases than controls (Persson *et al.*, 1993). The homozygous c2 genotype correlated positively with *p53* mutations in Japanese (Oyama *et al.*, 1997a); in a small study of Caucasians, the c2 genotype conferred a significant risk especially for lung adenocarcinoma (El Zein *et al.*, 1997).

Oral cancer : No association between head-and-neck cancer and *CYP2E1* variants was reported in two studies (Matthias *et al.*, 1998; Morita *et al.*, 1999), but a higher prevalence of the c2 allele was reported in non-betel quid-chewing Chinese patients (Hung *et al.*, 1997). As the frequencies of variant alleles are very low in Caucasians and African Americans, the statistical power of the studies was low.

Glutathione S-transferases

The glutathione S-transferase genes (*GST*s) form a superfamily of consisting of four distinct families, named Alpha, Mu, Pi and Theta. The *GSTM1*, *GSTT1* and *GSTP1* genes are polymorphic in humans. *GSTM1* is expressed at high levels in liver but not in lung. *GSTM1* has three alleles: *GSTM1 0/0* (null) is gene deletion in homozygotes, while *GSTM1**A* and *GSTM1**B* differ by a single base in exon 7 and encode enzyme monomers that form active homo- and heterodimeric forms. The prevalence of the *GSTM1*-null genotype shows ethnic differences, with reported ranges of 22–35% in Africans, 38–67% in Caucasians and 33–63% in East Asian populations (Rebbeck, 1997).

Lung cancer: As an illustration of the potential population impact of these genes, it has been estimated that 17% of lung cancers may be attributed to *GSTM1* genotypes (McWilliams *et al.*, 1995). Recent meta-analyses of 19 studies indicate that the *GSTM1*-null genotype confers a modest increased risk of lung cancer, the relative risk among Caucasians being 1.21 (95% CI, 1.06–1.39), and among Asians 1.45 (95% CI, 1.23–1.70). For lung cancer, risk seems to be related particularly to squamous and small cell histologies. Estimates for Asians show higher consistency and risk is greater in Kreyberg I histologies (d'Errico *et al.*, 1999). Ten studies have examined the relationship between *GSTM1* polymorphism and lung cancer risk according to tobacco consumption. In five of the ten studies, the OR was higher among heavy smokers than among light smokers and in the other five studies was higher among light smokers (Stucker *et al.*, 1999). There is evidence that female smokers with the *GSTM1*-null genotype are at higher risk for lung cancer than males (Tang *et al.*, 1998). Although *GSTM1* is not expressed in human lung, GSTM3 activity is found, which seems to be co-regulated with the *GSTM1* form. Thus, individuals with the nulled *GSTM1* genotype suffer from impaired detoxification of tobacco carcinogens, both qualitatively because of the absence of *GSTM1* in the body and low expression of *GSTM3* in the lung, and quantitatively because of the overall lower GST activity (Nakajima *et al.*, 1995).

Oral cancer: A significant association with the null genotype was seen in some Japanese studies, one at low dose of cigarettes (Sato, 1999) and another in alcohol-drinking cases (Nomura, 2000), while a lack of association has also been reported (Tanimoto *et al.*, 1999). An association was reported among French smokers (Jourenkovo-Mironova *et al.*, 1999b), while other studies among Caucasians have reported a lack of association (Deakin *et al.*, 1996; Park *et al.*, 1997). *GSTM1*-null conferred a significant increased risk for oral leukoplakia among Indian betel-quid chewers, which was further increased in combination with *GSTT1*-null genotypes (Nair *et al.*, 1999b).

GSTT1: Ethnic differences in prevalence exist. The frequency of the *GSTT1*-null genotype varies from 10–18% in Caucasians (Rebbeck, 1997) to 58% in Chinese (Lee *at al.*, 1995). The overall biological effect of this polymorphism is difficult to predict, as the enzyme is involved in both detoxification (monohalomethanes, ethylene oxides) and metabolic activation reactions (methylene chloride).

Lung cancer: A number of studies have shown no association of *GSTT1* with lung cancer risk (Harries *et al.*, 1997; Ryberg et *al.*, 1997). However, a significant association was observed with a concurrent lack of *GSTM1* and *GSTT1* genes and susceptibility to squamous cell lung cancers, although individually neither genotype showed any association (Saarikoski *et al.*, 1998).

Oral cancer: The *GSTT1*-null genotype was shown to increase the risk for oral and pharyngeal cancers among French smokers (Jourenkova-Mironova *et al.*, 1999b), but most other studies did not find an associated risk (Deakin *et al.*, 1996; Mathias *et al.*, 1998; Oude Ophius *et al.*, 1998; Worrall *et al.*, 1998).

GSTP1: Although genetic polymorphisms have been reported, only the exon 5 polymorphism (G to A, valine to isoleucine) in the *GSTP1* gene is linked to changed enzyme activity. The 105Val variant is more active than the 105Ile variant in conjugation of diol epoxides of some PAHs, suggesting that the allele 105Ile variant may be more susceptible to the carcinogenic effects of diol epoxides of PAHs.

Lung cancer: The frequency of the homozygous *GG* genotype was significantly higher in male lung cancer patients (Ryberg *et al.*, 1997), whereas no association was reported in other studies (Butkiewicz & Chorazy, 1999; Harries *et al.*, 1997). The combination of the three risk genotypes of *GSTM3 AA*, *GSTP1* (AG or GG) and *GSTM1*-null conferred an increased risk of lung cancer in heavy smokers (Jourenkova-Mironova *et al.*, 1998).

Oral cancer: The GSTP1 variant genotype conferred an increased risk for oral and/or pharyngeal cancer in a number of studies (Mathias *et al.*, 1998; Jourenkova-Mironovo *et al.*, 1999b; Katoh *et al.*, 1999; Park *et al.*, 1999). In an oral cancer case–control study among ethnic Indians, we observed a statistically significant risk associated with a homozygous variant genotype of the *GSTP1* gene (Nair *et al.*, unpublished).

N-Acetyltransferase

N-Acetyltransferase-1 (*NAT1*) and *N*-acetyltransferase-2 (*NAT2*) genes are polymorphically expressed in a variety of tissues. NAT2 may either detoxify or activate aromatic amines found in tobacco smoke, such as 4-aminobiphenyl (Hengstler *et al.*, 1998). Both phenotypic assays and genotypic assays for NAT2 can be used to classify individuals as rapid or slow acetylators. Genetic variants of the *NAT* genes have been cloned and at least 19 rare alleles for *NAT2* have been detected (Vatsis *et al.*, 1995) but NAT*5, *6 and *7 alleles account for most of the slow-acetylator allele in Caucasian populations, providing high concordance between genotype and phenotype. For NAT1, eight alleles have been identified. The distribution of NAT1 and NAT2 alleles differs widely between racial and ethnic groups.

Lung cancer: Information about the role of NAT enzymes as risk factors for lung cancer is ambiguous. No association with NAT2 slow or rapid genotypes was seen in a Spanish study (Martinez *et al.*, 1995). In contrast, the homozygous rapid NAT2 genotype was associated with increased risk in a German study (Cascorbi *et al.*, 1996b) and the slow acetylator with an increased risk in a Japanese study (Oyama *et al.*, 1997b). In a Swedish study, an increased risk for NAT2 slow acetylators among never-smokers but an increased risk for rapid acetylators among smokers was seen (Nyberg *et al.*, 1998). A significantly increased risk of developing pulmonary disorders among asbestos-exposed subjects with a combined *GSTM1*-null and *NAT2* slow genotype has also been reported (Hirvonen *et al.*, 1995, 1996). In a meta-analysis of estimates for NAT2 slow acetylators, an OR of 0.96 (95% CI, 0.82–1.10) was computed. A significant association was reported between *NAT1* genotype and lung cancer risk in smokers (Bouchardy *et al.*, 1998; Abdel-Rahman *et al.*, 1998), whereas no association between lung cancer risk and NAT2 was observed. Recently, individuals with *GSTM1*-null and/or NAT2 slow genotypes were reported to be at increased risk of contracting non-operable lung cancer at young age (Hou *et al.*, 2000).

Oral cancer: Only a few studies have been conducted. The NAT1*10 slow metabolizer genotype was associated with oral cancer risk in Japanese, especially in non-smokers, but not the NAT2 rapid-acetylator genotype (Katoh *et al.*, 1998). In another Japanese study, the NAT2 slow-metabolizer genotype predisposed towards higher risk (Morita *et al.*, 1999), reflecting the tendencies observed among Caucasians (Gonzales *et al.*, 1998; Jourenkova-Mironova *et al.*, 1999a).

Other metabolic polymorphisms

There are a number of other metabolic polymorphisms which, given their substrates and functions, could be expected to have some effect on the risk for lung and oral cancers. However, only a few studies testing these hypotheses for some isoforms have appeared.

CYP1A2: CYP1A2 activates many dietary and tobacco procarcinogens, notably aromatic and heterocyclic amines and nitrosamines, and also metabolizes nicotine. In contrast to extrahepatic CYP1A1, CYP1A2 appears to be expressed mainly in the liver and only weakly in the peripheral lung (Mace *et al.*, 1998). Like CYP1A1, CYP1A2 is regulated in part by the Ah-receptor system and induced in humans by a variety of chemicals. The activity of this enzyme can be determined in a non-invasive assay involving measurement of caffeine 3-demethylation. Recently, two genetic polymorphisms of the human *CYP1A2* gene have been identified, one in the 5′ flanking region affecting enzyme inducibility (Nakajima *et al.*, 1999) and another in intron 1 which is associated with high catalytic activity of the enzyme, when subjects are exposed to tobacco smoke (MacLeod *et al.*, 1998; Sachse *et al.*, 1999). About 45% of healthy Caucasians are homozygous for the high-inducibility genotype. A subgroup of smokers had a 1.6-fold increase in caffeine demethylation ratio (ratio of paraxanthine : caffeine in serum) over that in nonsmokers. Interactions between *GSTM1* status and CYP1A2 and CYP1A1 enzyme induction have been observed in smokers: *GSTM1* deficiency was associated not only with increased hepatic CYP1A2 activity in current smokers but also with significantly increased levels of bulky PAH–DNA adducts in the lung parenchyma of smokers and ex-smokers, over that in individuals with wild-type *GSTM1* (Bartsch *et al.*, 1995; Bartsch & Hietanen, 1996). CYP1A2 activity was higher in *GSTM1*-null subjects after exposure to cigarette smoke and heterocyclic amines from cooked meat. Exposed individuals with *CYP1A1* Ile–Val alleles had greater CYP1A2 activity than those with wild-type *CYP1A1* (MacLeod *et al.*, 1997). *GSTM1*-null was associated with higher levels of 4-aminobiphenyl–haemoglobin adducts in smokers (Yu *et al.*, 1995).

CYP1B1: A key enzyme in the production of potentially carcinogenic estrogen metabolites, CYP1B1 also activates many PAH-dihydrodiols, aromatic amines and other groups of procarcinogens. CYP1B1 is also induced by Ah-receptor ligands. Several genetic polymorphisms have been identified (Bailey *et al.*, 1998), but the role of *CYP1B1* in lung/oral cancer has not been investigated in epidemiological studies.

CYP2A6: In humans, CYP2A6 is involved in the metabolism of several carcinogens, mediates 7-hydroxylation of coumarin, a component of cigarette smoke, and activates several nitrosamines in tobacco smoke, including NNK (Hecht, 1999b; Tiano *et al.*, 1994). The catalytic selectivity of CYP2A6 appears to overlap with that of CYP2E1. The location of CYP2A6 and 2E1 in extrahepatic tissues such as lung, nasal and pharyngeal areas is of interest. Two *CYP2A6* variant alleles have been identified (*2 and *3). The prevalence of the Leu^{160}His variant allele in Caucasians is about 2% and it is associated with lower coumarin 7-hydroxylation activity. A new allele has been described in which exons 5–9 are deleted (Nunoya *et al.*, 1998). Individuals lacking functional CYP2A6 have impaired nicotine metabolism and may thus be protected against tobacco dependence. No association of *CYP2A6**2 with lung cancer was detected in a single report (London *et al.*, 1999).

CYP2C9: The levels of all smoking-related DNA adducts in the larynx were correlated with the presence of P4502C protein, suggesting a role of *CYP2C9* in DNA adduction of PAH-type tobacco carcinogens (Degawa *et al.*, 1994). The level of bulky DNA adducts in normal bronchial tissue of smokers was higher in individuals with the homozygous *CYP2C9**3/*3 genotype (Ozawa *et al.*, 1999). In African Americans and Caucasians, the *2 allele was associated with borderline-increased risk for lung cancer (London *et al.*, 1996, 1997).

CYP2C19: Several defective *CYP2C19* alleles are the basis for the (*S*)-mephenytoin 4′-hydroxylase polymorphism. The most common variant allele, *2, has an aberrant splice site in exon 5 (DeMorais *et al.*, 1994a). The premature stop codon mutant *3 allele has so far been found only in Asians (DeMorais *et al.*, 1994b). There is evidence that CYP2C19 expressed by yeast has a major role in the accumulation of the proximate mutagen B[a]P-7,8-dihydrodiol. A very small study among Japanese patients revealed a significant association of the poor-metabolizer genotype with lung SCC (Tsuneoka *et al.*, 1996).

CYP3A4: CYP3A4 can activate numerous procarcinogenic PAH dihydrodiols, such as B[a]P-dihydrodiol, and also metabolizes NNN (Patten *et al.*, 1997). Whether genetic or solely environmental factors are responsible for the wide variation in human CYP3A4 activity is unknown. Although the three *CYP3A* genes, *3A4*, *3A5*, and *3A7*, are expressed at widely different levels, polymorphism has been found only for *CYP3A4* and *CYP3A5* to date. Several allelic variants of the *CYP3A4* gene have been reported (Peyronneau *et al.*, 1993), but none was apparently related to catalytic activity in the liver samples from which the DNA was derived. No extensive studies on *CYP3A4* polymorphism have been reported.

NQO1: NAD(P)H:quinone oxidoreductase (NQO1) is a flavoprotein that catalyses the reduction of quinones, quinone amines and azo dyes, thereby protecting cells from reactive oxygen species generated from these compounds by the activity of reducing enzymes such as cytochrome P450 reductase. Two alleles have been identified. *NQO1* Pro-Ser, the less common, encodes an inactive protein and is termed null (Rosvold *et al.*, 1995). Significant ethnic variations in the frequency of the variant allele have been reported. The *NQO1* Pro allele is associated with increased risk for lung cancer in non-Hispanic whites and African Americans (Wienke *et al.*, 1997). Allelic variants at *NQO1* have been associated with susceptibility to lung tumours.

Other enzymes: Microsomal epoxide hydrolases (mEH), UDP-glucuronyl transferases (UDPGTs) catalysing glucuronidation of *N*-hydroxyarenes, heterocyclic amines and aromatic dihydrodiols, phenols, quinols and NNK, and myeloperoxidase (MPO) which can activate B[a]P and aromatic amine in cigarette smoke, all display genetic polymorphisms, and would be candidate susceptibility genes to modify the risk for cancers of the lung and oral cavity.

Genetic polymorphism in metabolic enzymes as biomarkers in chemoprevention

As a number of xenobiotic oxidations by cytochrome P450 enzymes are affected by endobiotic chemicals, endobiotic–xenobiotic interactions as well as drug–drug interactions may be of great importance in relation to the chemopreventive, pharmacological and toxicological actions of chemopreventive agents. Three classes of chemopreventive agents, retinoids, tea polyphenols and isothiocyanates, are reviewed in the context of their possible modulation of and interaction with metabolic enzymes, to evaluate the importance of integrating individual metabolic genotype/phenotype in designing effective chemopreventive strategies.

Retinoids

Vitamin A, synthetic and naturally occuring retinoids and β-carotene have attracted wide interest as possible chemopreventive agents against lung and oral cavity cancer. β-Carotene is directly absorbed by the intestine and a proportion of it is then converted to retinol. Additionally, β-apo-8´-carotenal, an excentric cleavage product of β-carotene, is a strong inducer of cytochrome P4501A1 and 1A2 enzymes in mice and rats (Gradelet *et al.*, 1996). Such induction of P450 enzymes might occur with high doses of β-carotene supplementation, as reported in ferrets, where the formation of β-apo-8´-carotenal was 2.5 times higher in lung extracts (Wang *et al.*, 1999).

The major established pathway of retinol activation involves mobilization of retinyl esters, reversible conversion of released retinol into retinal and irreversible conversion of retinal into the key functional all-*trans*-retinoic acid. In adult mammalian hepatic tissue, biosynthesis of all-*trans*-retinoic acid is catalysed primarily by ADH and ALDH, but other enzymes including P450 have also been reported to catalyse biosynthesis of all-*trans*-retinoic acid from precursor retinoids (Roos *et al.*, 1998). Rat liver microsomes in the presence of NADPH converted retinol to polar metabolites, including 4-hydroxyretinol (Huang *et al.*, 1999). This activity was also shown in a reconstituted monooxygenase system containing purified forms of rat P450 enzymes including CYP1B1. More recently CYP1A1 has been shown to oxidize retinal to retinoic acid. In human skin, CYP1A1 and CYP1A2 convert all-*trans*- and 9-*cis*-retinoic acid into corresponding isomers. The basal expression of *CYP1A1* and *CYP1A2* can be inhibited by all-*trans*-retinoic acid (Li *et al.*, 1995).

3-Methylcholanthrene(3-MC) and B[a]P can increase all-*trans*-retinoic acid catabolism in human skin or induce local tissue depletion. This is caused primarily by xenobiotic-mediated induction of CYP1A1, which also is involved in inactivation of all-*trans*-retinoic acid to 4-hydroxy-retinoic acid. Competitive inhibitory effects of vitamin A, all-*trans*-retinol, all-*trans*-retinal, all-*trans*-retinoic acid and retinyl palmitate on rat CYP1A1-dependent monooxygenase activity were observed in a reconstituted system containing the microsomal fraction prepared from recombinant *Saccharomyces cerevisiae* cells producing rat CYP1A1 and yeast NADPH-P450 reductase (Inouye et *al.*, 1999). Retinol and retinal decreased the mutagenicity of heterocyclic amines in the Salmonella/reversion assay, behaved as competitive inhibitors of isoquinoline-induced mutagenesis and strongly inhibited CYP1A1- and CYP1A2-dependent monooxygenases activities (Edenharder *et al.*, 1999). Retinol and retinoic acid were strong competitive inhibitors for xenobiotic oxidations catalysed by recombinant human CYP1A1 and CYP2C19 (Yamazaki & Shimada, 1999).

Ethanol and its major oxidative metabolite, acetaldehyde, both inhibit the generation of all-*trans*-retinoic acid (Deltour *et al.*, 1996). Concurrently, the CYP2E1-catalysed oxidation of ethanol can initiate lipid peroxidation via generation of a variety of free radicals. The lipid peroxides thus formed could then be converted via CYP2E1-catalysed reactions to alcohols and aldehydes, including *trans*-4-hydroxy-2-nonenal, that act as potent inhibitors of all-*trans*-retinoic acid synthesis. (Khalighi *et al.*, 1999). Prolonged use of alcohol, drugs or both accelerates the breakdown of retinol through cross-induction of degradative enzymes. There is also competition between ethanol and retinoic acid precursors. Depletion ensues, with associated hepatic and extrahepatic pathology, including carcinogenesis. Ethanol also interferes with the conversion of β-carotene to retinol (Leo & Lieber, 1999). Thus ethanol, while promoting a deficiency of vitamin A, also enhances its toxicity as well as that of β-carotene.

In conclusion, the role of microsomal cytochrome P450 in vitamin A metabolism and maintenance of vitamin A homeostasis should be considered in formulating chemopreventive strategies with retinoids, especially in the presence of exogenous exposure.

Human studies: The prevalence of an array of polymorphic genes was determined in a cohort of male smokers who participated in the α-tocopherol, β-carotene (ATBC) study among a fairly genetically homogeneous Caucasian population in Finland. Unlike *CYP1A1* and *CYP2E1* mutant frequencies, which in keeping with studies among other

Caucasian populations were low, most of the genes studied (including the *GSTM1*-null allele, *NQO1* Ser linked to loss of enzyme activity and *ADH3-2* lower enzyme activity) had a sufficiently high frequency in this population to allow investigation of gene–environment interactions (Woodson *et al.*, 1999a). In a nested case–control study within the ATBC cohort, *GSTM1*-null genotype was not associated with lung cancer risk in male smokers, but may have conferred a higher susceptibility with cumulative tobacco exposure. The association was attenuated by α-tocopherol but not by β-carotene supplementation (Woodson *et al.*, 1999b). One possible mechanism for this interaction may be the reported activation of carcinogens, as β-carotene has been shown to induce several carcinogen-metabolizing enzymes including CYP1A1/2 in ferrets (Wang *et al.*, 1999).

Tea polyphenols

Tea (black, green or oolong) is produced from the tea plant (*Camellia sinensis*) by various processing conditions. Black tea is produced by fermenting the leaves, green tea leaves are not fermented, while oolong teas are semi-fermented followed by a heating process to halt fermentation.

The primary catechins in green tea are epicatechin, epicatechin-3-gallate, epigallocatechin and epigallocatechin-3-gallate. Other polyphenols include flavanols and their glycosides and depsides, such as chlorogenic acid, coumaroylquinic acid and one that is unique to tea, theogallin (3-galloylquinic acid). Also present are quinic acids, carotenoids, trigalloylglucose, lignin, protein, chlorophyll, minerals, caffeine and very small amounts of other methylxanthines such as theophylline, theobromine and theanine. Depending on the amount of oxidation and condensation of catechins, green tea may also contain constituents commonly found in black tea, such as theaflavin, theaflavic acids, volatile compounds and thearubigens (Graham, 1992).

Tea polyphenols, specifically the catechins epigallocatechin-3-gallate, epigallocatechin and epicatechin-3-gallate, which account for 30–40% of the extractable solids of green tea leaves, are believed to mediate many of the cancer-chemopreventive effects. The chemopreventive potential of tea has been extensively reviewed (Yang & Wang, 1993; Weisburger, 1999), but the mode of action is still not clear. There appear to be a whole spectrum of activities with which these tea components interfere. Several plausible mechanisms have been put forward, including inhibition of ultraviolet radiation- and tumour promoter-induced ornithine decarboxylase, cyclo-oxygenase and lipoxygenase activities; antioxidant- and free radical-scavenging activity; enhancement of antioxidant (glutathione peroxidase, catalase and quinone reductase) and phase II (GST) enzyme activities; inhibition of lipid peroxidation; and anti-inflammatory activity. These properties of tea polyphenols make them effective chemopreventive agents against the initiation, promotion and progression stages of multistage carcinogenesis (Katiyar & Mukhtar, 1997). The pharmacokinetic properties of tea polyphenols are largely unknown. Tea catechins are rapidly absorbed and the addition of milk does not impair their bioavailability.

Inhibition of tumorigenesis by tea and tea polyphenols has been demonstrated in several rodent models, for sites including skin, lung, oesophagus, forestomach, duodenum, small intestine, colon and liver (Yang & Wang, 1993; Fujiki, 1999). After oral administration of tea preparations to animals, the activities of glutathione peroxidase, catalase, GST, NADPH-quinone reductase, UDPGT and methoxyresorufin *O*-dealkylase were moderately enhanced. Theaflavins have been shown to have inhibitory action against NNK-induced pulmonary hyperproliferation and tumorigenesis (Yang *et al.*, 1997, 1998). The inhibitory effect of tea on NNK-induced tumorigenesis has been explained in part by increased metabolism of NNK in the rat liver and decreased bioavailability in the lung (Chung *et al.*, 1998). The induction of hepatic cytochrome P450 enzymes such as CYPA2, 1A1, and 2B1 has been described in rats given either green or black tea (Sohn *et al.*, 1994). Caffeine has been identified as the active component in tea responsible for enzyme induction (Chen *et al.*, 1996). Glucuronidation and sulfation of tea polyphenols are the major elimination pathways, and competition among tea polyphenols for glucuronosyltransferase and sulfotransferase could inhibit epigallocatechin-3-gallate elimination. Oral administration of lyophilized green tea to female CD-1 mice stimulated liver microsomal glucuronidation of estrone, estradiol and 4-nitrophenol (Zhu *et al.*, 1998). The effects of green tea flavonoids on 7-ethoxyresorufin-*O*-deethylase (CYP1A2), *p*-nitrophenol hydroxylase (CYP2E1), erythromycin-*N*-demethylase (CYP3A) and nifedipine oxidase (CYP3A4) were examined in human liver microsomes. Epicatechin-3-gallate was the most potent inhibitor of 7-ethoxyresorufin-*O*-deethylase in human liver microsomes. The effect of the green tea flavonoids on 7-ethoxyresorufin-*O*-deethylase was complex; in addition to inhibition at high concentrations of flavonoid, moderate activation was seen at lower concentrations (Obermeier *et al.*, 1995). Using standardized cell cultures, the green and black tea extracts and tea polyphenols were shown to inhibit B[a]P adduct formation with human DNA and induce GST and quinone reductase (Steele *et al.*, 2000). Activation of the mitogen-activated protein kinase pathway by green tea polyphenols (Lin *et al.*, 1999) might be responsible for the regulation of the antioxidant-responsive element which is believed to mediate the induction of phase II enzymes by many drugs, and may be stimulated by green tea polyphenols in the transcription of phase II detoxifying enzymes (Yu *et al.*, 1997).

Human studies: In contrast to the consistently observed inhibition of tumorigenesis by tea in many animal models, studies concerning the effects of tea on the incidence of human cancers have been inconclusive. Some epidemiological studies on the effect of tea ingestion on cancer risk have suggested an inhibitory effect, others an enhancing effect, and still others a lack of effect (Katiyar & Mukhtar, 1997; Yang, 1999; Goldbohm *et al.*, 1996).

A catechin esterase which converts epigallocatechin-3-gallate to epigallocatechin has been found in saliva. Holding a tea solution in the mouth for a few minutes without swallowing produced high salivary catechin levels, which were two orders of magnitude higher than in plasma. This suggests that slow drinking of green tea could be an effective way to deliver high concentrations of catechins to the oral cavity and oesophagus for prevention studies (Yang *et al.*, 1999). Human oral precancerous mucosal lesions have been reported to respond to tea (Li *et al.*, 1999).

Isothiocyanates

Glucosinolates are stable precursors of isothiocyanates, which are typically present in plants at high concentrations. They are hydrolysed to isothiocyanates by the coexisting but physically segregated enzyme myrosinase, which is released upon wounding (cutting, chewing) plant cells. Isothiocyanates and related substances, indoles, are responsible for the sharp taste in cruciferous (brassica) vegetables (broccoli, cabbage, Brussels sprouts, cauliflower, collards, kale, kohlrabi, mustard greens, rutabaga, turnips, bok choy). In the cruciferous family, the inducer activity is principally due to the highly reactive isothiocyanates (R–N=C=S; mustard oils). More than 20 isothiocyanates have been shown to inhibit the formation of carcinogen-induced tumours of several animal target organs. Of the indole glucosinolates which are predominant in brassica vegetables, glucobrassicin forms an unstable isothiocyanate which degrades into indole-3-carbinol. The isothiocyanate sulforaphane is the most potent naturally occurring inducer of phase II enzymes. A CYP2E1-mediated effect has been associated with the anti-genotoxicity of the broccoli constituent sulforaphane (Barcelo *et al.*, 1996). Young sprouts of broccoli and cauliflower contain much higher levels of glucoraphanin, the glucosinolate precursor of sulforaphane, than do the mature counterparts. Isothiocyanates prevent carcinogenesis in laboratory animals by blocking carcinogen activation or enhancing detoxification. Many isothiocyanates have one or the other type of activity or both in various test systems. Some isothiocyanates are inhibitors of specific cytochrome P450 isozymes in rodent tissues. Numerous isothiocyanates are potent inducers of GST and NADPH quinone reductase. Several effectively inhibit carcinogenesis and their effect can be remarkably specific. While benzyl isothiocyanate (BITC) inhibits lung tumour induction by B[a]P in A/J mice, phenethyl isothiocyanate (PEITC) does not. In contrast, induction of lung tumours by NNK is inhibited by PEITC but not by BITC. PEITC inhibits the metabolic activation of NNK in rat lung and is a selective inhibitor of certain P450 enzymes in rat liver, although it has limited effect on phase II enzymes (Verhoeven *et al.*, 1997; Hecht, 1999c).

Human studies: Several metabolic experiments in humans have consistently shown increased CYP1A2 activity after consumption of cruciferous vegetables (Verhoeven *et al.*, 1997). A number of studies to test the effect of cruciferous vegetables in humans in the context of metabolic polymorphisms have been reported. Among Chinese in Singapore, urinary isothiocyanate levels were not dependent on either *GSTM1* or *GSTP1* genotypes, but urinary excretion was significantly higher in *GSTT1*-positive than in *GSTT1*-null subjects in the highest tertile of cruciferous vegetable intake. These results suggest the possible presence of inducers of GSTT1 in cruciferous vegetables (Seow *et al.*, 1998). *GSTM1* apparently indirectly plays a role in the induction of CYP1A2 activity. Among weekly consumers of cruciferous vegetables in a study among non-Hispanic whites, *GSTM1*-null individuals showed significantly higher CYP1A2 activity relative to *GSTM1*-non-null individuals. Cruciferous vegetables also induce CYP1A2 activity, and CYP1A2 activity was induced in *GSTM1*-null subjects, suggesting that cruciferous vegetables contain CYP1A2 inducers, which are probably deactivated in the presence of GSTM1 (Probst-Hensch *et al.*, 1998). GSTM1 rapidly conjugates isothiocyanates to GSH, leading to excretion (Zhang *et al.*, 1995). Lin *et al.* (1998) reported that although subjects in the highest quartile of broccoli intake had a low risk of colorectal adenomas, when stratified by the *GSTM1* genotype a protective effect was observed only among subjects with the *GSTM1*-null genotype. Among smoking volunteers, Hecht *et al.* (1995) showed that PEITC (watercress) inhibited the oxidative metabolism of NNK in humans similarly to rodents; 7 of 11 subjects had increased levels of urinary metabolites of NNK (NNAL and NNAL glucuronide). As PEITC can induce UDP glucuronosyltransferase activity and as that is the path of NNK excretion, it would be interesting to study whether polymorphisms in these enzymes would affect NNK detoxification.

Careful consideration should be given in future research to the fact that the plant polyphenols act synergistically in complex ways with other constituents of the plant. These unknown and poorly understood interactions could play a significant role in the anticarcinogenic efficacy of the polyphenol constituents. A lesson can be drawn from pharmaceutical agents available today: isolated extracts of one primary active constituent are generally not without side-effects.

Perspectives: metabolic polymorphisms and chemoprevention in humans

The evaluation of the net effect of chemopreventive agents in real life is difficult, as exposure to many different carcinogens, drugs, dietary factors and endobiotics occurs simultaneously. Several of these use the same metabolic pathways. Induction of a pathway that is protective against one group of compounds may potentiate the toxic or carcinogenic effects of another class or *vice versa*. Although bifunctional inducers such as indole-3-carbinol can be directly beneficial, as shown by the increased metabolism of aflatoxin B_1 to less toxic and carcinogenic metabolites due to induction of P450 enzymes, such a protective effect may not extend to PAHs. Further, prophylaxis based on inhibition of cytochrome P450, known to play a role in primary metabolism of a wide range of dietary and physiological compounds, carries a high risk of inducing adverse responses. A balance between activation and detoxification is the determining factor for the net effect. In contrast to bifunctional inducers (Prochaska & Talalay, 1988) which induce both phase II as well as selected phase I activities, monofunctional inducers elevate phase II metabolism without significantly affecting phase I activity. They appear to be the preferred agents for producing resistance against a wide range of chemical insults without adverse side-effects. The possibility of changed organotropy should also be considered; for example, decreased carcinogen metabolism in the liver could affect peripheral organs such as the lung, or increased metabolism in the liver could affect the kidney or bladder. A detailed understanding of mechanism of action and host genotype/phenotype profile is required before setting up chemoprevention trials.

Molecular epidemiological studies attempting to identify gene–carcinogen interactions and other mechanistic aspects in humans require comprehensive integration of genotype and phenotype biomarkers. IARC (Toniolo *et al.*, 1997; Vineis *et al.*, 1999) proposed that the design and analysis of molecular epidemiological studies should include:

- a clear definition of representative study populations and controls;

- a sample size adequate to provide the necessary statistical power;
- proper documentation (or measurement) of exposure to carcinogens or protective agents;
- avoidance of confounding due to mixed ethnic background of study subjects;
- study only of gene polymorphisms that have been shown to lead to altered phenotypic expression.

Future investigations should include both single and joint risk effects of multiple polymorphism combinations and should assess interaction between susceptibility genes and other endogenous (e.g., hormones) and exogenous (e.g., tobacco smoke, vitamin supplementation) risk factors. For appropriate and rigorous hypothesis testing, these studies will require a reasonable prevalence of relevant alleles. The data generated from the Human Genome Project and the Cancer Gene Anatomy Programme together with gene-chip technology have provided high-throughput means to look at general genetic damage as well as specific genetic changes at the molecular level. Such chips may be designed to evaluate subjects at risk, for example those carrying genetic polymorphisms, for determining appropriate target population for intervention strategies. Knowledge of the prevalence and distribution of common genetic susceptibility factors and the ability to identify susceptible individuals or subgroups will have substantial preventive implications, in particular if more data are collected showing that people with certain 'at-risk' genotypes are more susceptible to low levels of carcinogens. It is conceivable that such subjects could be (i) more easily persuaded to avoid hazardous exposures such as tobacco, (ii) targeted for intensive smoking cessation programmes, (iii) be enrolled in chemoprevention trials and (iv) be involved in cancer screening programmes that would not be appropriate for the general population.

References

Abdel Rahman, S.Z., El Zein, R.A., Zwischenberger, J.B. & Au, W.W. (1998) Association of the NAT1*10 genotype with increased chromosome aberrations and higher lung cancer risk in cigarette smokers. *Mutat. Res.*, **398**, 43–54

Alexandrie, A.K., Sundberg, M.I., Seidegard, J., Tornling, G. & Rannug, A. (1994) Genetic susceptibility to lung cancer with special emphasis on CYP1A1 and GSTM1: a study on host factors in relation to age at onset, gender and histological cancer types. *Carcinogenesis*, **15**, 1785–1790

Alexandrov, K., Rojas, M., Geneste, O., Castegnaro, M., Camus, A.M., Petruzzelli, S., Giuntini, C. & Bartsch, H. (1992) An improved fluorometric assay for dosimetry of benzo(a)pyrene diol-epoxide-DNA adducts in smokers' lung: comparisons with total bulky adducts and aryl hydrocarbon hydroxylase activity. *Cancer Res.*, **52**, 6248–6253

Ayesh, R., Idle, J.R., Ritchie, J.C., Crothers, M.J. & Hetzel, M.R. (1984) Metabolic oxidation phenotypes as markers for susceptibility to lung cancer. *Nature*, **312**, 169–170

Bailey, L.R., Roodi, N., Dupont,W.D. & Parl, F.F. (1998) Association of cytochrome P450 1B1 (CYP1B1) polymorphism with steroid receptor status in breast cancer. *Cancer Res.*, **58**, 5038–5041

Barcelo, S., Gardiner, J.M., Gescher, A. & Chipman, J.K. (1996) CYP2E1-mediated mechanism of anti-genotoxicity of the broccoli constituent sulforaphane. *Carcinogenesis*, **17**, 277–282

Bartsch, H., Nair, U., Risch, A., Rojas, M., Wikman, H. & Alexandrov, K. (2000) Genetic polymorphism of CYP genes, alone or in combination, as a risk modifier of tobacco-related cancers. *Cancer Epidemiol. Biomarkers Prev.*, **9**, 3–28

Bergen, A.W. & Caporaso, N.(1999) Cigarette smoking. *J. Natl Cancer Inst.*, **91**, 1365–1375

Bouchardy, C., Mitrunen, K., Wikman, H., Husgafvel-Pursiainen, K., Dayer, P., Benhamou, S. & Hirvonen, A. (1998) N-Acetyltransferase NAT1 and NAT2 genotypes and lung cancer risk. *Pharmacogenetics*, **8**, 291–298

Bouchardy, C., Wikman, H., Benhamou, S., Hirvonen, A., Dayer, P. & Husgafvel-Pursiainen, K. (1997) CYP1A1 genetic polymorphisms, tobacco smoking and lung cancer risk in a French Caucasian population. *Biomarkers*, **2**, 131–134

Boustead, C., Taber, H., Idle, J.R. & Cholerton, S. (1997) CYP2D6 genotype and smoking behaviour in cigarette smokers. *Pharmacogenetics*, **7**, 411–414

Brockmöller, J., Cascorbi, I., Kerb, R., Sachse, C. & Roots, I. (1998) Polymorphisms in xenobiotic conjugation and disease predisposition. *Toxicol. Lett.*, **102–103**, 173–183

Butkiewicz, D. & Chorazy, M. (1999) Individual predisposition to lung neoplasm – the role of genes involved in metabolism of carcinogens [in Polish]. *Postepy. Hig. Med Dosw.*, **53**, 655–673

Caporaso, N. & Goldstein, A. (1995) Cancer genes: single and susceptibility: exposing the difference. *Pharmacogenetics*, **5**, 59–63

Cascorbi, I., Brockmoller, J. & Roots, I. (1996a) A C4887A polymorphism in exon 7 of human CYP1A1: population frequency, mutation linkages, and impact on lung cancer susceptibility. *Cancer Res.*, **56**, 4965–4969

Cascorbi, I., Brockmoller, J., Mrozikiewicz, P.M., Bauer, S., Loddenkemper, R. & Roots, I. (1996b) Homozygous rapid arylamine N-acetyltransferase (NAT2) genotype as a susceptibility factor for lung cancer. *Cancer Res.*, **56**, 3961–3966

Chen, L., Bondoc, F.Y., Lee, M.J., Hussin, A.H., Thomas, P.E. & Yang, C.S. (1996) Caffeine induces cytochrome P4501A2: induction of CYP1A2 by tea in rats. *Drug Metab. Disp.*, **24**, 529–533

Christensen, P.M., Gotzsche, P.C. & Brosen, K. (1997) The sparteine/debrisoquine (CYP2D6) oxidation polymorphism and the risk of lung cancer: a meta-analysis. *Eur. J. Clin. Pharmacol.*, **51**, 389–393

Chung, F.-L., Chen, H.-J.C. & Nath, R.G.(1996) Lipid peroxidation as a potential endogenous source for the formation of exocyclic DNA adducts. *Carcinogenesis*, **17**, 2105–2111

Chung, K.T., Wong, T.Y., Wei, C.I., Huang, Y.W. & Lin, Y. (1998) Tannins and human health: a review. *Crit. Rev. Food Sci. Nutr.*, **38**, 421–464

Coles, B. & Ketterer, B. (1990) The role of glutathione and glutathione transferases in chemical carcinogenesis. *Crit. Rev. Biochem. Mol. Biol.*, **25**, 47–70

Daly, A.K., Brockmoller, J., Broly, F., Eichelbaum, M., Evans, W.E., Gonzalez, F.J., Huang, J.D., Idle, J.R., Ingelman Sundberg, M., Ishizaki, T., Jacqz Aigrain, E., Meyer, U.A., Nebert, D.W., Steen,V.M., Wolf, C.R. & Zanger, U.M. (1996) Nomenclature for human CYP2D6 alleles. *Pharmacogenetics*, **6**, 193–201

Deakin, M., Elder, J., Hendrickse, C., Peckham, D., Baldwin, D., Pantin, C., Wild, N., Leopard, P., Bell, D.A., Jones, P., Duncan, H., Brannigan, K., Alldersea, J., Fryer, A.A. & Strange, R.C. (1996) Glutathione S-transferase GSTT1 genotypes and susceptibility to cancer: studies of interactions with GSTM1 in lung, oral, gastric and colorectal cancers. *Carcinogenesis*, **17**, 881–884

Degawa, M., Stern, S.J., Martin, M.V., Guengerich, F.P., Fu, P.P., Ilett, K.F., Kaderlik, R.K. & Kadlubar, F.F. (1994) Metabolic activation and carcinogen-DNA adductdetection in human larynx. *Cancer Res.*, **54**, 4915–4919

Deltour, L., Ang, H.L. & Duester, G. (1996) Ethanol inhibition of retinoic acid synthesis as a potential mechanism for fetal alcohol syndrome. *FASEB J.*, **10**, 1050–1057

DeMorais, S.M.F., Wilkinson, G.R., Blaisdell, J., Meyer, U.A., Nakamura, K. & Goldstein, J.A. (1994a) Identification of a new genetic defect responsible for the polymorphism of (S)-mephenytoin metabolism in Japanese. *Mol. Pharmacol.*, **46**, 594–598

DeMorais, S.M.F., Wilkinson, G.R., Blaisdell, J., Nakamura, K., Meyer, U.A. & Goldstein, J.A. (1994b) The major genetic defect responsible for the polymorphism of S-mephenytoin metabolism in humans. *J. Biol. Chem.*, **269**, 15419–15422

d'Errico, A., Malats, N., Vineis, P. & Boffetta, P. (1999) Review of studies of selected metabolic polymorphisms and cancer. In: Vineis, P., Malats, M., Lang, M., d'Errico, A., Caporaso, N., Cuzick, J., Boffetta, P., eds, *Metabolic Polymorphisms and Susceptibility to Cancer* (IARC Scientific Publications No. 148), Lyon, IARC, pp. 323–393

Doll, R. (1998) Uncovering the risks of smoking: historical perspective. *Stat. Meth. Med. Res.*, **7**, 87–117

Edenharder, R., Worf Wandelburg, A., Decker, M. & Platt, K.L. (1999) Antimutagenic effects and possible mechanisms of action of vitamins and related compounds against genotoxic heterocyclic amines from cooked food. *Mutat. Res.*, **444**, 235–248

El Zein, R.A., Zwischenberger, J.B., Abdel Rahman, S.Z., Sankar, A.B. & Au, W.W. (1997) Polymorphism of metabolizing genes and lung cancer histology: prevalence of CYP2E1 in adenocarcinoma. *Cancer Lett.*, **112**, 71–78

Fujiki, H. (1999) Two stages of cancer prevention with green tea. *J. Cancer Res. Clin. Oncol.*, **125**, 589–597

Goldbohm, R.A., Hertog, M.G., Brants, H.A., van Poppel, G. & van den Brandt, P.A. (1996) Consumption of black tea and cancer risk: a prospective cohort study. *J. Natl Cancer Inst.*, **88**, 93–100

Gonzalez, M.V., Alvarez, V., Pello, M.F., Menendez, M.J., Suarez, C. & Coto, E.(1998) Genetic polymorphism of N-acetyltransferase-2, glutathione S-transferase-M1, and cytochromes P450IIE1 and P450IID6 in the susceptibility to head and neck cancer. *J. Clin. Pathol.*, **51**, 294–298

Goto, I., Yoneda, S., Yamamoto, M. & Kawajiri, K. (1996) Prognostic significance of germ line polymorphisms of the CYP1A1 and glutathione S-transferase genes in patients with non-small cell lung cancer. *Cancer Res.*, **56**, 3725–3730

Gradelet, S., Leclerc, J., Siess, M.H. & Astorg, P.O. (1996) beta-Apo-8′-carotenal, but not beta-carotene, is a strong inducer of liver cytochromes P4501A1 and 1A2 in rat. *Xenobiotica*, **26**, 909–919

Graham, H.N. (1992) Green tea composition, consumption, and polyphenol chemistry. *Prev. Med.*, **21**, 334–350

Harries, L.W., Stubbins, M.J., Forman, D., Howard, G.C. & Wolf, C.R. (1997) Identification of genetic poly-morphisms at the glutathione S-transferase Pi locus and association with susceptibility to bladder, testicular and prostate cancer. *Carcinogenesis*, **18**, 641–644

Hayashi, S., Watanabe, J. & Kawajiri, K. (1992) High susceptibility to lung cancer analyzed in terms of combined genotypes of P450IA1 and Mu-class glutathione S-transferase genes. *Jpn. J. Cancer Res.*, **83**, 866–870

Hecht, S.S. (1999a) Tobacco smoke carcinogens and lung cancer. *J. Natl Cancer Inst.*, **91**, 1194–1210

Hecht, S.S. (1999b) DNA adduct formation from tobacco-specific N-nitrosamines. *Mutat. Res.*, **424**, 127–142

Hecht, S.S. (1999c) Chemoprevention of cancer by isothiocyanates, modifiers of carcinogen metabolism. *J. Nutr.*, **129**, 768S–774S

Hecht, S.S., Chung, F.L., Richie, J.P., Jr, Akerkar, S.A., Borukhova, A., Skowronski, L. & Carmella, S.G. (1995) Effects of watercress consumption on metabolism of a tobacco-specific lung carcinogen in smokers. *Cancer Epidemiol. Biomarkers Prev.*, **4**, 877–884

Hengstler, J.G., Arand, M., Herrero, M.E. & Oesch, F. (1998) Polymorphisms of N-acetyltransferases, glutathione S-transferases, microsomal epoxide hydrolase and sulfotransferases: influence on cancer susceptibility. *Recent Results Cancer Res.*, **154**, 47–85

Hirvonen, A., Husgafvel Pursiainen, K., Anttila, S., Karjalainen, A. & Vainio, H. (1993) Polymorphism in CYP1A1 and CYP2D6 genes: possible association with susceptibility to lung cancer. *Environ. Health Perspect.*, **101** Suppl. 3, 109–112

Hirvonen, A., Pelin, K., Tammilehto, L., Karjalainen, A., Mattson, K. & Linnainmaa, K. (1995) Inherited GSTM1 and NAT2 defects as concurrent risk modifiers in asbestos-related human malignant mesothelioma. *Cancer Res.*, **55**, 2981–2983

Hirvonen, A., Saarikoski, S.T., Linnainmaa, K., Koskinen, K., Husgafvel Pursiainen, K., Mattson, K. & Vainio, H. (1996) Glutathione S-transferase and N-acetyltransferase genotypes and asbestos-associated pulmonary disorders. *J. Natl Cancer Inst.*, **88**, 1853–1856

Hoffmann, D. & Hecht, S.S. (1990) Advances in tobacco carcinogenesis. In: Cooper C.S. & Grover P.L., eds, *Chemical Carcinogenesis and Mutagenesis* I, Berlin, Heidelberg, Springer, pp. 63–103

Hou, S.M., Ryberg, D., Falt, S., Deverill, A., Tefre, T., Borresen, A.L., Haugen, A. & Lambert, B. (2000) GSTM1 and NAT2 polymorphisms in operable and non-operable lung cancer patients. *Carcinogenesis*, **21**, 49–54

Huang, D.Y., Ohnishi, T., Jiang, H., Furukawa, A. & Ichikawa, Y. (1999) Inhibition by retinoids of benzo(a)pyrene metabolism catalyzed by 3-methylcholanthrene-induced rat cytochrome P-450 1A1. *Metabolism*, **48**, 689–692

Hung, H.C., Chuang, J., Chien, Y.C., Chern, H.D., Chiang, C.P., Kuo, Y.S., Hildesheim, A. & Chen, C.J. (1997) Genetic polymorphisms of CYP2E1, GSTM1, and GSTT1; environmental factors and risk of oral cancer. *Cancer Epidemiol. Biomarkers Prev.*, **6** , 901–905

Inouye, K., Mae, T., Kondo, S. & Ohkawa, H. (1999) Inhibitory effects of vitamin A and vitamin K on rat cytochrome P4501A1-dependent monooxygenase activity. *Biochem. Biophys. Res. Commun.*, **262**, 565–569

Jourenkova-Mironova, N., Wikman, H., Bouchardy, C., Voho, A., Dayer, P., Benhamou, S. & Hirvonen,A. (1998) Role of glutathione S-transferase GSTM1, GSTM3, GSTP1 and GSTT1 genotypes in modulating susceptibility to smoking-related lung cancer. *Pharmacogenetics*, **8**, 495–502

Jourenkova-Mironova, N., Wikman, H., Bouchardy, C., Mitrunen, K., Dayer, P., Benhamou, S. & Hirvonen, A. (1999a) Role of arylamine N-acetyltransferase 1 and 2 (NAT1 and NAT2) genotypes in susceptibility to oral/pharyngeal and laryngeal cancers. *Pharmacogenetics*, **9**, 533–537

Jourenkova-Mironova, N., Voho, A., Bouchardy, C., Wikman, H., Dayer, P., Benhamou, S. & Hirvonen, A. (1999b) Glutathione S-transferase GSTM1, GSTM3, GSTP1 and GSTT1 genotypes and the risk of smoking-related oral and pharyngeal cancers. *Int. J. Cancer*, **81**, 44–48

Katiyar, S.K. & Mukhtar, H. (1997) Tea antioxidants in cancer chemoprevention. *J. Cell Biochem. Suppl.*, **27**, 59–67

Katoh, T., Kaneko, S., Boissy, R., Watson, M., Ikemura, K. & Bell, D.A. (1998) A pilot studytesting the association between N-acetyltransferases 1 and 2 and risk of oral squamous cell carcinoma in Japanese people. *Carcinogenesis*, **19**, 1803–1807

Katoh, T., Kaneko, S., Takasawa, S., Nagata, N., Inatomi, H., Ikemura, K., Itoh, H., Matsumoto, T., Kawamoto, T. & Bell, D.A. (1999) Human glutathione S-transferase P1 polymorphism and susceptibility to smoking related epithelial cancer; oral, lung, gastric, colorectal and urothelial cancer. *Pharmacogenetics*, **9**, 165–169

Kawajiri, K., Watanabe, J., Eguchi, H., Nakachi, K., Kiyohara, C. & Hayashi, S. (1995) Polymorphisms of human Ah receptor gene are not involved in lung cancer. *Pharmacogenetics*, **5**, 151–158

Kawajiri, K., Eguchi, H., Nakachi, K., Sekiya, T. & Yamamoto, M. (1996) Association of CYP1A1 germ line polymorphisms with mutations of the p53 gene in lung cancer. *Cancer Res.*, **56**, 72–76

Kellerman, G., Shaw, C.R. & Luyten-Kellerman, M. (1973) Aryl hydrocarbon hydroxylase inducibility and bronchogenic carcinoma. *New Engl. J. Med.*, **289**, 934–937

Khalighi, M., Brzezinski, M.R., Chen, H. & Juchau, M.R. (1999) Inhibition of human prenatal biosynthesis of all-*trans*-retinoic acid by ethanol, ethanol metabolites, and products of lipid peroxidation reactions: a possible role for CYP2E1. *Biochem. Pharmacol.*, **57**, 811–821

Kihara, M. & Noda, K. (1995) Risk of smoking for squamous and small cell carcinomas of the lung modulated by combinations of CYP1A1 and GSTM1 gene polymorphisms in a Japanese population. *Carcinogenesis*, **16**, 2331–2336

Landi, M.T., Bertazzi, P.A., Shields, P.G., Clark, G., Lucier, G.W., Garte, S.J., Cosma, G. & Caporaso, N.E. (1994) Association between CYP1A1 genotype, mRNA expression and enzymatic activity in humans. *Pharmacogenetics*, **4**, 242–246

Lazarus, P., Sheikh, S.N., Ren, Q., Schantz, S.P., Stern, J.C., Richie, J.P., Jr & Park, J.Y. (1998) p53, but not p16 mutations in oral squamous cell carcinomas are associated with specific CYP1A1 and GSTM1 polymorphic genotypes and patient tobacco use. *Carcinogenesis*, **19**, 509–514

Le Marchand, L., Sivaraman, L., Pierce, L., Seifried, A., Lum, A., Wilkens, L.R. & Lau, A.F. (1998) Associations of CYP1A1, GSTM1, and CYP2E1 polymorphisms with lung cancer suggest cell type specificities to tobacco carcinogens. *Cancer Res.*, **58**, 4858–4863

Le Marchand, L., Wilkinson, G.R. & Wilkens, L.R. (1999) Genetic and dietary predictors of CYP2E1 activity: a phenotyping study in Hawaii Japanese using chlorzoxazone. *Cancer Epidemiol. Biomarkers Prev.*, **8**, 495–500

Lee, E.J., Wong, J.Y., Yeoh, P.N. & Gong, N.H. (1995) Glutathione S transferase-theta (GSTT1) genetic polymorphism among Chinese, Malays and Indians in Singapore. *Pharmacogenetics*, **5**, 332–334

Leo, M.A. & Lieber, C.S. (1999) Alcohol, vitamin A, and beta-carotene: adverse interactions, including hepatotoxicity and carcinogenicity. *Am. J. Clin. Nutr.*, **69**, 1071–1085

Li, X.Y., Astrom, A., Duell, E.A., Qin, L., Griffiths, C.E. & Voorhees, J.J. (1995) Retinoic acid antagonizes basal as well as coal tar and glucocorticoid-induced cytochrome P4501A1 expression in human skin. *Carcinogenesis*, **16**, 519–524

Li, N., Sun, Z., Han, C. & Chen, J. (1999) The chemopreventive effects of tea on human oral precancerous mucosa lesions. *Proc. Soc. Exp. Biol. Med.*, **220**, 218–224

Lin, H.J., Probst Hensch, N.M., Louie, A.D., Kau, I.H., Witte, J.S., Ingles, S.A., Frankl, H.D., Lee, E.R. & Haile, R.W. (1998) Glutathione transferase null genotype, broccoli, and lower prevalence of colorectal adenomas. *Cancer Epidemiol. Biomarkers Prev.*, **7**, 647–652

Lin, J.K., Liang, Y.C. & Lin Shiau, S.Y. (1999) Cancer chemoprevention by tea polyphenols through mitotic signal transduction blockade. *Biochem. Pharmacol.*, **58**, 911–915

London, S.J., Daly, A.K., Leathart, J.B., Navidi, W.C. & Idle, J.R. (1996) Lung cancer risk in relation to the CYP2C9*1/CYP2C9*2 genetic polymorphism among African-Americans and Caucasians in Los Angeles County, California. *Pharmacogenetics*, **6**, 527–533

London, S.J., Sullivan Klose, T., Daly, A.K. & Idle, J.R. (1997) Lung cancer risk in relation to the CYP2C9 genetic polymorphism among Caucasians in Los Angeles County. *Pharmacogenetics*, **7**, 401–404

London, S.J., Idle, J.R., Daly, A.K. & Coetzee, G.A. (1999) Genetic variation of CYP2A6, smoking, and risk of cancer [letter]. *Lancet*, **353**, 898–899

Mace, K., Bowman, E.D., Vautravers, P., Shields, P.G., Harris, C.C. & Pfeifer, A.M. (1998) Characterisation of xenobiotic-metabolising enzyme expression in human bronchial mucosa and peripheral lung tissues. *Eur. J. Cancer*, **34**, 914–920

MacLeod, S., Sinha, R., Kadlubar, F.F. & Lang, N.P. (1997) Polymorphisms of CYP1A1 and GSTM1 influence the in vivo function of CYP1A2. *Mutat. Res.*, **376**, 135–142

MacLeod, S., Tang, Y.-M., Yokoi, T., Kamataki, T., Doublin, S., Lawson, B., Massengill, J., Kadlubar, F., & Lang, N. (1998) The role of recently discovered genetic polymorphism in the regulation of the human CYP1A2 gene. *Proc. Am. Assoc. Cancer Res.*, **39**, 396

Martinez, C., Agundez, J.A., Olivera, M., Martin, R., Ladero, J.M. & Benitez, J. (1995) Lung cancer and mutations at the polymorphic NAT2 gene locus. *Pharmacogenetics*, **5**, 207–214

Matthias, C., Bockmuhl, U., Jahnke, V., Harries, L.W., Wolf, C.R., Jones, P.W., Alldersea, J., Worrall, S.F., Hand, P., Fryer, A.A. & Strange, R.C. (1998) The glutathione S-transferase GSTP1 polymorphism: effects on susceptibility to oral/pharyngeal and laryngeal carcinomas. *Pharmacogenetics*, **8**, 1–6

McClellan, R.O. (1996) Tobacco-specific N-nitrosamines: recent advances. *Crit. Rev. Toxicol.*, **26**, 119–253

McWilliams, J.E., Sanderson, B.J., Harris, E.L., Richert Boe, K.E. & Henner, W.D. (1995) Glutathione S-transferase M1 (GSTM1) deficiency and lung cancer risk. *Cancer Epidemiol. Biomarkers Prev.*, **4**, 589–594

Micka, J., Milatovich, A., Menon, A., Grabowski, G.A., Puga, A. & Nebert, D.W. (1997) Human Ah receptor (AHR) gene: localization to 7p15 and suggestive correlation of polymorphism with CYP1A1 inducibility. *Pharmacogenetics*, 7, 95–101

Mooney, L.A., Bell, D.A., Santella, R.M., Van Bennekum, A.M., Ottman, R., Paik, M., Blaner, W.S., Lucier, G.W., Covey, L., Young, T.L., Cooper, T.B., Glassman,A.H. & Perera, F.P. (1997) Contribution of genetic and nutritional factors to DNA damage in heavy smokers. *Carcinogenesis*, **18**, 503–509

Nair, J., Ohshima, H., Nair, U.J. & Bartsch, H. (1996) Endogenous formation of nitrosamines and oxidative DNA-damaging agents in tobacco users. *Crit. Rev. Toxicol.*, **26**, 149–161

Nair, J., Barbin, A., Velic, I. & Bartsch, H. (1999a) Etheno DNA-base adducts from endogenous reactive species. *Mutat. Res.*, **424**, 59–69

Nair, U.J., Nair, J., Mathew, B. & Bartsch, H. (1999b) Glutathione S-transferase M1 and T1 null genotypes as risk factors for oral leukoplakia in ethnic Indian betel quid/tobacco chewers. *Carcinogenesis*, **20**, 743–748

Nakachi, K., Imai, K., Hayashi, S. & Kawajiri, K. (1993) Polymorphisms of the CYP1A1 and glutathione S-transferase genes associated with susceptibility to lung cancer in relation to cigarette dose in a Japanese population. *Cancer Res.*, **53**, 2994–2999

Nakajima, T., Elovaara, E., Anttila, S., Hirvonen, A., Camus, A.M., Hayes, J.D., Ketterer, B. & Vainio, H. (1995) Expression and polymorphism of glutathione S-transferase in human lungs: risk factors in smoking-related lung cancer. *Carcinogenesis*, **16**, 707–711

Nakajima, M., Yokoi, T., Mizutani, M., Kinoshima, M., Funayama, M. & Kamataki, T. (1999) Genetic polymorphism in the 5′-flanking region of human CYP1A2 gene effect on the CYP1A2 inducibility in humans. *J. Biochem.*, **125**, 803–808

Nath, R.G., Ocando, J.E., Guttenplan, J.B. & Chung, F.-L. (1998) 1,N2-Propanodeoxyguanosine adducts: potential new biomarkers of smoking-induced DNA damage in human oral tissue. *Cancer Res.*, **58**, 581–584

Nebert, D.W., McKinnon, R.A. & Puga, A. (1996) Human drug-metabolizing enzyme polymorphisms: effects on risk of toxicity and cancer. *DNA Cell Biol.*, **15**, 273–280

Nomura, T., Noma, H., Shibara, T., Yokoyama, A., Muramatsu, T. & Ohmori, T. (2000) Aldehyde dehydrogenase 2 and glutathione S-transferase M1 polymorphism in relation to the risk for oral cancer in Japanese drinkers. *Oral Oncology*, **36**, 42–46

Nunoya, K., Yokoi,T., Kimura, K., Inoue, K., Kodama, T., Funayama, M., Nagashima, K., Funae, Y., Green, C., Kinoshita, M. & Kamataki, T. (1998) A new deleted allele in the human cytochrome P450 2A6 (CYP2A6) gene found in individuals showing poor metabolic capacity to coumarin and (+)-cis-3,5-dimethyl-2-(3-pyridyl)thiazolidin-4-one hydrochloride (SM-12502). *Pharmacogenetics*, **8**, 239–249

Nyberg, F., Hou, S.M., Hemminki, K., Lambert, B. & Pershagen, G. (1998) Glutathione S-transferase mu1 and N-acetyltransferase 2 genetic polymorphisms and exposure to tobacco smoke in nonsmoking and smoking lung cancer patients and population controls. *Cancer Epidemiol. Biomarkers Prev.*, **7**, 875–883

Obermeier, M.T., White, R.E. & Yang, C.S. (1995) Effects of bioflavonoids on hepatic P450 activities. *Xenobiotica*, **25**, 575–584

Okada, T., Kawashima, K., Fukushi, S., Minakuchi, T. & Nishimura, S. (1994) Association between a cytochrome P450 CYPIA1 genotype and incidence of lung cancer. *Pharmacogenetics*, **4**, 333–340

Oude Ophuis, M.B., van Lieshout, E.M., Roelofs, H.M., Peters, W.H. & Manni, J.J. (1998) Glutathione S-transferase M1 and T1 and cytochrome P4501A1 polymorphisms in relation to the risk for benign and malignant head and neck lesions. *Cancer*, **82**, 936–943

Oyama, T., Kawamoto, T., Mizoue, T., Sugio, K., Kodama, Y., Mitsudomi, T. & Yasumoto, K. (1997a) Cytochrome P450 2E1 polymorphism as a risk factor for lung cancer: in relation to p53 gene mutation. *Anticancer Res.*, **17**, 583–587

Oyama, T., Kawamoto, T., Mizoue, T., Yasumoto, K., Kodama,Y. & Mitsudomi, T. (1997b) N-Acetylation polymorphism in patients with lung cancer and its association with p53 gene mutation. *Anticancer Res.*, **17**, 577–581

Ozawa, S., Schoket, B., McDaniel, L.P., Tang, Y.M., Ambrosone, C.B., Kostic, S., Vincze, I. & Kadlubar, F.F. (1999) Analyses of bronchial bulky DNA adduct levels and CYP2C9, GSTP1 and NQO1 genotypes in a Hungarian study population with pulmonary diseases. *Carcinogenesis*, **20**, 991–995

Park, J.Y., Muscat, J.E., Ren, Q., Schantz, S.P., Harwick, R.D., Stern, J.C., Pike,V., Richie, J.P., Jr & Lazarus, P. (1997) CYP1A1 and GSTM1 polymorphisms and oral cancer risk. *Cancer Epidemiol. Biomarkers Prev.*, 6, 791–797 [published erratum appears in *Cancer Epidemiol. Biomarkers Prev.*, 1997, **6**, 1108]

Park, J.Y., Schantz, S.P., Stern, J.C., Kaur, T. & Lazarus, P. (1999) Association between glutathione S-transferase pi genetic polymorphisms and oral cancer risk. *Pharmacogenetics*, **9**, 497–504

Patten, C.J., Smith, T.J., Friesen, M.J., Tynes, R.E., Yang, C.S. & Murphy, S.E. (1997) Evidence for cytochrome P450 2A6 and 3A4 as major catalysts for N′ nitrosonornicotine α-hydroxylation by human liver microsomes. *Carcinogenesis*, **18**, 1623–1630

Persson, I., Johansson, I., Bergling, H., Dahl, M.L., Seidegard, J., Rylander, R., Rannug, A., Hogberg, J. & Sundberg, M.I. (1993) Genetic polymorphism of cytochrome P4502E1 in a Swedish population. Relationship to incidence of lung cancer. *FEBS Lett.*, **319**, 207–211

Persson, I., Johansson, I. & Ingelman Sundberg, M. (1997) In vitro kinetics of two human CYP1A1 variant enzymes suggested to be associated with interindividual differences in cancer susceptibility. *Biochem. Biophys. Res. Commun.*, **231**, 227–230

Petersen, D.D., McKinney, C.E., Ikeya, K., Smith, H.H., Bale, A.E., McBride, O.W. & Nebert, D.W. (1991) Human CYP1A1 gene: cosegregation of the enzyme inducibility phenotype and an RFLP. *Am. J. Hum. Genet.*, **48**, 720–725

Peyronneau, M.A., Renaud, J.P., Jaouen, M., Urban, P., Cullin, C., Pompon, D. & Mansuy, D. (1993) Expression in yeast of three allelic cDNAs coding for human liver P-450 3A4. Different stabilities, binding properties and catalytic activities of the yeast-produced enzymes. *Eur. J. Biochem.*, **218**, 355–361

Probst Hensch, N.M., Tannenbaum, S.R., Chan, K.K., Coetzee, G.A., Ross, R.K. & Yu, M.C. (1998) Absence of the glutathione S-transferase M1 gene increases cytochrome P4501A2 activity among frequent consumers of cruciferous vegetables in a Caucasian population. *Cancer Epidemiol. Biomarkers Prev.*, **7**, 635–638

Prochaska, H.J. & Talalay, P. (1988) Regulatory mechanisms of monofunctional and bifunctional anticarcinogenic enzyme inducers in murine liver. *Cancer Res.*, **48**, 4776–4782

Raunio, H., Hakkola, J., Hukkanen, J., Lassila, A., Päivärinta, K., Pelkonen, O., Anttila, S., Piipari, P., Boobis, A. & Edwards, R.J. (1999) Expression of xenobiotic-metabolizing CYPs in human pulmonary tissue. *Exp. Toxicol. Pathol.*, **51**, 412–417

Rebbeck, T.R. (1997) Molecular epidemiology of the human glutathione S-transferase genotypes GSTM1 and GSTT1 in cancer susceptibility. *Cancer Epidemiol. Biomarkers Prev.*, **6**, 733–743

Rojas, M., Cascorbi, I., Alexandrov, K., Kriek, E., Auburtin, G., Mayer, L., Kopp-Schneider, A., Roots, I. & Bartsch, H. (2000) Modulation of benzo[a]pyrene diolepoxide-DNA adduct levels in human white blood cells by CYPA1, GSTM1 and GSTT1 polymorphism. *Carcinogenesis*, 21, 35–41

Romkes, M., White, C., Johnson, J., Eibling, D., Landreneau, R. & Branch, R. (1996) Expression of cytochrome P450 mRNA in human lung, head and neck tumours, and normal adjacent tissues. *Proc. Am. Assoc. Cancer. Res.*, **37**, 105

Roos, T.C., Jugert, F.K., Merk, H.F. & Bickers, D.R. (1998) Retinoid metabolism in the skin. *Pharmacol. Rev.*, **50**, 315–333

Rostami Hodjegan, A., Lennard, M.S., Woods, H.F. & Tucker, G.T. (1998) Meta-analysis of studies of the CYP2D6 polymorphism in relation to lung cancer and Parkinson's disease. *Pharmacogenetics*, **8**, 227–238

Rosvold, E.A., McGlynn, K.A., Lustbader, E.D. & Buetow, K.H. (1995) Identification of an NAD(P)H:quinone oxidoreductase polymorphism and its association with lung cancer and smoking. *Pharmacogenetics*, **5**, 199–206

Ryberg, D., Skaug, V., Hewer, A., Phillips, D.H., Harries, L.W., Wolf, C.R., Ogreid, D., Ulvik, A., Vu, P. & Haugen, A. (1997) Genotypes of glutathione transferase M1 and P1 and their significance for lung DNA adduct levels and cancer risk. *Carcinogenesis*, **18**, 1285–1289

Saarikoski, S.T., Voho, A., Reinikainen, M., Anttila, S., Karjalainen, A., Malaveille, C., Vainio, H., Husgafvel Pursiainen, K. & Hirvonen, A. (1998) Combined effect of polymorphic GST genes on individual susceptibility to lung cancer. *Int. J. Cancer*, **77**, 516–521

Saarikoski, S.T., Sata, F., Husgafvel-Pursiainen, K., Rautalahti, M., Haukka, J., Impivaara, O., Järvisalo, J., Vainio, H. & Hirvonen, A. (2000) CYP2D6 ultrarapid metabolizer genotype as potential modifier of smoking behaviour. *Pharmacogenetics*, **10**, 5–10

Sachse, C., Brockmöller, J., Bauer, S. & Roots, I. (1999) Functional significance of a C A polymorphism in intron I of the cytochrome P450 CYPIA2 gene tested with caffeine. *Br. J. Clin. Pharmacol.*, **47**, 445–449

Sato, M., Sato, T., Izumo, T. & Amagasa, T. (1999) Genetic polymorphism of drug-metabolizing enzymes and susceptibility to oral cancer. *Carcinogenesis*, **20**, 1927–1931

Seow, A., Shi, C.Y., Chung, F.L., Jiao, D., Hankin, J.H., Lee, H.P., Coetzee, G.A. & Yu, M.C. (1998) Urinary total isothiocyanate (ITC) in a population-based sample of middle-aged and older Chinese in Singapore: relationship with dietary total ITC and glutathione S-transferase M1/T1/P1 genotypes. *Cancer Epidemiol. Biomarkers Prev.*, **7**, 775–781

Sohn, O.S., Surace, A., Fiala, E.S., Richie, J.P., Jr, Colosimo, S., Zang, E. & Weisburger, J.H. (1994) Effects of green and black tea on hepatic xenobiotic metabolizing systems in the male F344 rat. *Xenobiotica*, **24**, 119–127

Steele, V.E., Kelloff, G.J., Balentine, D., Boone, C.W., Mehta, R., Bagheri, D., Sigman, C.C., Zhu, S. & Sharma, S. (2000) Comparative chemopreventive mechanisms of green tea, black tea and selected polyphenol extracts measured by in vitro bioassays. *Carcinogenesis*, **21**, 63–67

Stucker, I., de Waziers, I., Cenee, S., Bignon, J., Depierre, A., Milleron, B., Beaune, P. & Hemon, D. (1999) GSTM1, smoking and lung cancer: a case-control study. *Int. J. Epidemiol.*, **28**, 829–835

Taioli, E., Ford, J., Trachman, J., Li, Y., Demopoulos, R. & Garte, S. (1998) Lung cancer risk and *CYP1A1* genotype in African Americans. *Carcinogenesis*, **19**, 813–817

Tang, D.L., Rundle, A., Warburton, D., Santella, R.M., Tsai, W.Y., Chiamprasert, S., Hsu, Y.Z. & Perera, F.P. (1998) Associations between both genetic and environmental biomarkers and lung cancer: evidence of a greater risk of lung cancer in women smokers. *Carcinogenesis*, **19**, 1949–1953

Tanimoto, K., Hayashi, S., Yoshiga, K. & Ichikawa, T. (1999) Polymorphisms of the CYP1A1 and GSTM1 gene involved in oral squamous cell carcinoma in association with a cigarette dose. *Oral Oncology*, **35**, 191–196

Tiano, H.F., Wang, R.-L., Hosokawa, M., Crespi, C., Tindall, K.R. & Langenbach, R. (1994) Human CYP2A6 activation of 4-(methylnitrosamino)-1-(3-pyridyl)-1-butanone (NNK): mutational specificity in the gpt gene of AS52 cells. *Carcinogenesis*, **15**, 2859–2866

Toniolo, P., Boffetta, P., Shuker, D., Rothman, N., Hulka, B. & Pearce, N., eds (1997) *Application of Biomarkers in Cancer Epidemiology* (IARC Scientific Publications No. 142), Lyon, IARC

Tsuneoka, Y., Fukushima, K., Matsuo, Y., Ichikawa, Y. & Watanabe, Y. (1996) Genotype analysis of the CYP2C19 gene in the Japanese population. *Life Sci.*, **59**, 1711–1715

Uematsu, F., Ikawa, S., Kikuchi, H., Sagami, I., Kanamaru, R., Abe, T., Satoh, K., Motomiya, M. & Watanabe, M. (1994) Restriction fragment length polymorphism of the human CYP2E1 (cytochrome P450IIE1) gene and susceptibility to lung cancer: possible relevance to low smoking exposure. *Pharmacogenetics*, **4**, 58–63

Vatsis, K.P., Weber, W.W., Bell, D.A., Dupret, J.M., Evans, D.A., Grant, D.M., Hein, D.W., Lin, H.J., Meyer, U.A., Relling, M.V., Sim, E., Suzuki,T. & Yamazoe, Y. (1995) Nomenclature for N-acetyltransferases. *Pharmaco-genetics*, **5**, 1–17

Verhoeven, D.T., Verhagen, H., Goldbohm, R.A., van den Brandt, P.A. & van Poppel, G. (1997) A review of mechanisms underlying anticarcinogenicity by brassica vegetables. *Chem.-Biol. Interact.*, **103**, 79–129

Villard, P.H., Seree, E.M., Re, J.L., De Meo, M., Barra, Y., Attolini, L., Dumenil, G., Catalin, J., Durand, A. & Lacarelle, B. (1998) Effects of tobacco smoke on the gene expression of the Cyp1a, Cyp2b, Cyp2e, and Cyp3a subfamilies in mouse liver and lung: relation to single strand breaks of DNA. *Toxicol. Appl. Pharmacol.*, **148**, 195–204

Vineis, P. (1997) Molecular epidemiology: low-dose carcinogens and genetic susceptibility. *Int. J. Cancer*, **71**, 1–3

Vineis, P., Bartsch, H., Caporaso, N., Harrington, A.M., Kadlubar, F.F., Landi, M.T., Malaveille, C., Shields, P.G., Skipper, P., Talaska, G. & Tannnenbaum, S.R. (1994) Genetically based N-acetyltransferase metabolic polymorphism and low-level environmental exposure to carcinogens. *Nature*, **369**, 154–156

Vineis, P., Malats, M., Lang, M., d'Errico, A., Caporaso, N., Cuzick, J. & Boffetta, P., eds (1999) *Metabolic Polymorphisms and Susceptibility to Cancer* (IARC Scientific Publications No. 148), Lyon, IARC

Wang, X.D., Liu, C., Bronson, R.T., Smith, D.E., Krinsky, N.I. & Russell, M. (1999) Retinoid signaling and activator protein-1 expression in ferrets given beta-carotene supplements and exposed to tobacco smoke. *J. Natl Cancer Inst.*, **91**, 60–66

Watanabe, J., Hayashi, S. & Kawajiri, K. (1994) Different regulation and expression of the human CYP2E1 gene due to the RsaI polymorphism in the 5′-flanking region. *J. Biochem. Tokyo*, **116**, 321–326

Weisburger, J.H. (1999) Tea and health: the underlying mechanisms. *Proc. Soc. Exp. Biol. Med.*, **220**, 271–275

Wiencke, J.K., Spitz, M.R., McMillan, A. & Kelsey, K.T. (1997) Lung cancer in Mexican-Americans and African-Americans is associated with the wild-type genotype of the NAD(P)H: quinone oxidoreductase polymorphism. *Cancer Epidemiol. Biomarkers Prev.*, **6**, 87–92

Woodson, K., Ratnasinghe, D., Bhat, N.K., Stewart, C., Tangrea, J.A., Hartman, T.J., Stolzenberg Solomon, R., Virtamo, J., Taylor, P.R. & Albanes, D. (1999a) Prevalence of disease-related DNA polymorphisms among participants in a large cancer prevention trial. *Eur. J. Cancer Prev.*, **8**, 441–447

Woodson, K., Stewart, C., Barrett, M., Bhat, N.K., Virtamo, J., Taylor, P.R. & Albanes, D. (1999b) Effect of vitamin intervention on the relationship between GSTM1, smoking, and lung cancer risk among male smokers. *Cancer Epidemiol. Biomarkers Prev.*, **8**, 965–970

Worrall, S.F., Corrigan, M., High, A., Starr, D., Matthias, C., Wolf, C.R., Jones, P.W., Hand, P., Gilford, J., Farrell, W.E., Hoban, P., Fryer, A.A. & Strange, R.C. (1998) Susceptibility and outcome in oral cancer: preliminary data showing an association with polymorphism in cytochrome P450 CYP2D6. *Pharmacogenetics*, **8**, 433–439

Wu, X., Shi, H., Jiang, H., Kemp, B., Hong, W.K., Delclos, G.L. & Spitz, M.R. (1997) Associations between cytochrome P4502E1 genotype, mutagen sensitivity, cigarette smoking and susceptibility to lung cancer. *Carcinogenesis*, **18**, 967–973

Wu, X., Amos, C.I., Kemp, B.L., Shi, H., Jiang, H., Wan, Y. & Spitz, M.R. (1998) Cytochrome P450 2E1 DraI polymorphisms in lung cancer in minority populations. *Cancer Epidemiol. Biomarkers Prev.*, **7**, 13–18

Xu, X., Kelsey, K.T., Wiencke, J.K., Wain, J.C. & Christiani, D.C. (1996) Cytochrome P450 CYP1A1 MspI polymorphism and lung cancer susceptibility. *Cancer Epidemiol. Biomarkers Prev.*, **5**, 687–692

Yamazaki, H. & Shimada, T. (1999) Effects of arachidonic acid, prostaglandins, retinol, retinoic acid and cholecalciferol on xenobiotic oxidations catalysed by human cytochrome P450 enzymes. *Xenobiotica*, **29**, 231–241

Yang, C.S. (1999) Tea and health. *Nutrition*, **15**, 946–949

Yang, C.S. & Wang, Z.Y. (1993) Tea and cancer. *J. Natl Cancer Inst.*, **85**, 1038–1049

Yang, G.Y., Liu, Z., Seril, D.N., Liao, J., Ding, W., Kim, S., Bondoc, F. & Yang, C.S. (1997) Black tea constituents, theaflavins, inhibit 4-(methylnitrosamino)-1-(3-pyridyl)- 1-butanone (NNK)-induced lung tumorigenesis in A/J mice. *Carcinogenesis*, **18**, 2361–2365

Corresponding author:

U. Nair
Division of Toxicology and Cancer Risk Factors - C0200,
Deutsches Krebsforschungszentrum,
Im Neuenheimer Feld 280,
D-69120 Heidelberg,
Germany

Yang, C.S., Yang, G.Y., Landau, J.M., Kim, S. & Liao, J. (1998) Tea and tea polyphenols inhibit cell hyperproliferation, lung tumorigenesis, and tumor progression. *Exp. Lung Res.*, **24**, 629–639

Yang, C.S., Lee, M.J. & Chen, L. (1999) Human salivary tea catechin levels and catechin esterase activities: implication in human cancer prevention studies. *Cancer Epidemiol. Biomarkers Prev.*, **8**, 83–89

Yu, M.C., Ross, R.K., Chan, K.K., Henderson, B.E., Skipper, P.L., Tannenbaum, S.R. & Coetzee, G.A. (1995) Glutathione S-transferase M1 genotype affects aminobiphenyl-hemoglobin adduct levels in white, black and Asian smokers and nonsmokers. *Cancer Epidemiol. Biomarkers Prev.*, **4**, 861–864

Yu, R., Jiao, J.J., Duh, J.L., Gudehithlu, K., Tan, T.H. & Kong, A.N. (1997) Activation of mitogen-activated protein kinases by green tea polyphenols: potential signaling pathways in the regulation of antioxidant-responsive element-mediated phase II enzyme gene expression. *Carcinogenesis*, **18**, 451–456

Zhang, Y., Kolm, R.H., Mannervik, B. & Talalay, P. (1995) Reversible conjugation of isothiocyanates with glutathione catalyzed by human glutathione transferases. *Biochem. Biophys. Res Commun.*, **206**, 748–755

Zhang, Z.Y., Fasco, M.J., Huang, L., Guengerich, F.P. & Kaminsky, L.S. (1996) Characterization of purified human recombinant cytochrome P4501A1-Ile462 and -Val462: assessment of a role for the rare allele in carcinogenesis. *Cancer Res.*, **56**, 3926–3933

Zhu, B.T., Taneja, N., Loder, D.P., Balentine, D.A. & Conney, A.H. (1998) Effects of tea polyphenols and flavonoids on liver microsomal glucuronidation of estradiol and estrone. *J. Steroid Biochem. Mol. Biol.*, **64**, 207–215

Subject Index

Aberrant crypt foci, 7, 103, 116
Acetaldehyde, 216, 247, 248, 249, 279
N-Acetylcysteine, 261
N-Acetyltransferase, 5, 36, 142, 181, 182, 243, 276
Actinic keratoses, 7, 85, 88
Adducts, *see* DNA adducts, Protein adducts
Adenoma
  colorectal, 7, 14, 15, 18, 21, 103, 105, 113–116, 133
  hepatic, 223, 226
Adenomatous atypical hyperplasia, 192, 200, 265
Aerodigestive tract, 5, 18, 34, 37, 257, 261, 272
Aflatoxins, 5
  albumin adducts, 41, 216, 217
  DNA adducts, 32, 216, 218, 242
  and liver cancer, 216, 218–220, 223, 241
  metabolites, 31, 216
  protein adducts, 31, 217
Age, 119, 131, 132, 166
Ageing, breast tissue, 163, 166
AIDS, 21, 38, 39
Albinism, 93
Albumin adducts, 31, 41, 216–218, 249
Alcohol
  and colorectal cancer, 62, 65
  and liver cancer, 5, 216, 223, 241
  metabolism, 279
  and oral cancer, 247, 271
2-Amino-1-methyl-6-phenylimidazo[4,5-*b*]-pyridine, *see* PhIP
Δ4-Androstenedione, 150, 151
Androgens, 5, 150, 157, 193, 199, 201–203, 208
Angiogenesis, 194
Antiestrogens, 16
Antioxidants, 4, 58, 97, 105, 191, 208, 232
α-1-Antitrypsin deficiency, 241
APC gene, 15, 20, 22, 33, 37, 115, 117, 134–140
Apoptosis, 15, 154, 192, 194, 258
Aromatase, 151, 157, 180
Ashkenazi Jews, 140, 144, 179
Aspirin, 37
Ascorbic acid, *see* Vitamin C
Ataxia telangiectasia, 37, 180

Barrett's oesophagus, 8, 14, 20
Benzo[*a*]pyrene, 245, 247, 272
Betel quid, 247, 249, 272
Bias, 63–65
Biomarkers, 27
  definition, 1
  exposure, 3–6, 29–32
  intermediate effect, 6–8, 13–22
  predictive value, 50
  prognostic value, 38
  surrogacy, 38–41
  susceptibility, 8–10
  validity, 2, 5, 29, 57, 63, 70
Bile acids, 102, 105
Bladder, urinary
  cancer, 18, 20, 34, 36, 58
  precancer, 14
Bloom's syndrome, 37
Bowel, large, *see* Colorectum
Bowman–Birk inhibitor, 53
Bran, 119
Breast
  ATM gene, 180
  BRCA genes, 15, 37, 62, 178–184
  cancer, 34, 37
    and diet, 149
    and hormonal factors, 5, 149–159, 163, 166, 180
  density, 7, 163–167, 173
  ductal carcinoma *in situ*, 7, 14, 35, 171–174
  intermediate-effect biomarkers, 163–167, 171–174
  mammography, 7, 163, 173, 174
  screening, 173, 174, 182
  susceptibility biomarkers, 177–184
  tamoxifen trial, 49, 184
Broccoli, 38, 281

Calcium, 101, 105, 115, 116
Calprotectin, 104
Carbohydrates, 149
Carcinoembryonic antigen, 32, 117
Carcinoid tumours, 32
β-Carotene, 5, 58, 105, 116, 119, 141, 259–262, 279, 280
Carotenoids, 9
Catechins, 280
Catechol-*O*-methyltransferase (COMT), 152–154, 180, 182
β-Catenin, 115, 137
CD44, 118
Celecoxib, 21, 22, 37
Cervix uteri, cancer of, 3, 32, 38
Children, 87, 139
Chlorophyllin, 220
Chromosome alterations, 18, 172–173
Cigarette smoking, see Tobacco
Cirrhosis, 241
Clofibrate, 232
Colorectum
  aberrant crypt foci, 7, 103, 116
  adenomas, 7, 14, 15, 18, 21
  cancer, 20, 21, 36, 37, 62, 101
    and alcohol, 65
    dietary factors, 101, 105–108, 115
  DNA adducts, 104
  exposure biomarkers, 5, 101–104
  intermediate-effect biomarkers, 7, 103, 113–120
  intervention trials, 52, 105–108
  p53 mutations, 117
  susceptibility biomarkers, 131–144
  see also Hereditary non-polyposis colorectal cancer *and* Familial adenomatous polyposis
Confounding, 65
Contraceptives, oral, 156, 241
Cotinine, 5
Cowden syndrome, 178
COX-2, 5, 10, 16, 62
Cyclins, 18, 172
Cyclobutane pyrimidine dimers, 69
Cytochrome P450 (CYP), 5, 97, 152, 180, 242, 272–275, 277–278, 281
  CYP1A1, 97, 242, 250, 272
  CYP1A2, 65, 219, 250, 277, 281
  CYP2A6, 242, 277
  CYP1B1, 277
  CYP2B6, 242
  CYP2C9, 277
  CYP2C19, 277
  CYP2D6, 97, 250, 274
  CYP2E1, 65, 250, 274
  CYP3A2, 208
  CYP3A4, 210, 217, 277
  CYP17, 208
  CYP19, 182

Dehydroepiandrosterone, 151, 157, 228, 231–234
Detoxification, 58, 96, 142, 154, 216, 245, 249, 272, 276, 281
Diabetes, 229
Dietary factors, 58
  breast cancer, 149
  colorectal cancer, 101, 105–108, 115

liver cancer, 216, 241
prostate cancer, 191
2-Difluoromethylornithine, 18, 22, 37, 51–53
DNA
adducts, 5, 6, 8, 29, 31, 57–58, 104, 271
aflatoxin, 32, 216, 218, 242
bulky, 58, 273, 277, 278
etheno, 5, 104, 108
exocyclic, 272
postlabelling analysis, 31, 58, 104, 247
in skin, 4
tobacco carcinogens, 245, 247, 273
methylation, 34, 248, 264
repair, 8, 36, 58, 61, 70, 74, 94, 103, 133, 138, 181
synthesis, unscheduled, 8
Ductal carcinoma *in situ*, 7, 14, 35, 171–174
Dysplasia, 8, 113, 258, 265

Endocrine factors, *see* Hormones
Enzyme
antioxidant, 97
carcinogen-metabolizing, 9, 35, 58, 241, 271–282
DNA repair, 35, 37, 58, 74
induction, 272, 280, 281, 282
steroid-metabolizing, 151–154
see also Cytochrome P450, Glutathione *S*-transferase
Epidermal growth factor (EGF), 16, 18
Estradiol, 150, 151, 158
Estrogens, 150–156, 166
Estrogen receptor, 16, 18
Ethanol, *see* Alcohol
Etheno adducts, 5, 104, 108
Ethical issues, 10, 143
Exercise, 115, 149, 157
Exposure biomarkers, 3, 29–32
breast, 149–159
colorectum, 5, 101–104
liver, 3, 215–220
lung, 245–251
prostate, 191
skin, 4, 69–78
validation, 5

Faecal occult blood test, 101, 104
Familial adenomatous polyposis, 9, 18, 20, 37, 103, 104, 116, 119, 131, 136–138, 142
Fanconi's anaemia, 37
Fat, 101, 115, 116, 119, 149, 191
Fenretinide, 156, 184, 259
α-Fetoprotein, 32
Fibre, 101, 105, 115, 116
Field cancerization, 257
Finasteride, 202, 211
Fish oil, 119
Flavonoids, 158
Fluorescent bronchoscopy, 265
Foci of altered hepatocytes, 223–229, 233–235
Folate, 115, 141
Food, *see* Dietary factors
Fruit, 58, 61, 107, 115, 116

Gene
amplification, 16
APC, 15, 20, 22, 33, 37, 115, 134–140
AR, 209
ATM, 180
BRCA, 15, 37, 62, 178–184
CDKN2A, 95
COMT, 154, 180, 182
DNA repair, 35, 37, 94, 133, 138
HPRT, 33
MOM1, 131, 140
MSH2, 140, 178
MTHFR, 141
NAT2, 5, 36, 142, 181, 182, 243, 276
penetrance, 8, 10, 35, 62
PTCH, 95
PTEN, 139, 178
ras, 15, 33, 95, 115, 117, 135, 263, 265
SHH, 95
SMO, 95
SRDA5A2, 209, 211
see also Cytochrome P450, Glutathione *S*-transferase *and* p53
Genetic polymorphism, *see* Polymorphism
Genomic instability, 32, 34, 263
Glucose-6-phosphate dehydrogenase, 233
Glucuronides, 30, 153
γ-Glutamyltranspeptidase, 228, 233
Glutathione *S*-transferase (GST), 38, 218, 272, 281
GSTM1, 5, 36, 63, 96, 97, 181–183, 242, 273, 275, 280, 282
GSTM3, 275, 276
GSTP1, 96, 181–183, 228, 275, 276, 282
GSTT1, 96, 181, 275, 276, 282
and breast cancer, 181–183
and liver cancer, 233
and lung cancer, 275–276
and skin cancer, 96, 97
Glycogen, 226–230
Glycophorin A assay, 33
Gorlin's syndrome, 139
Growth hormone-releasing hormone, 155

Haemochromatosis, 241
Haemoglobin adducts, 31, 249
Head and neck cancer, 16, 18, 261 (*see also* Oral cancer)
Hepatitis viruses, 3
and liver cancer, 5, 33, 215, 220, 223, 241
vaccination, 218
Hepatocellular carcinoma (HCC), *see* Liver, cancer of
Hereditary non-polyposis colorectal cancer, 9, 16, 37, 119, 131–133, 138–139, 178
Heterocyclic amines, 5, 30, 32, 36, 65, 104, 141, 277
Heterogeneous nuclear ribonucleoprotein (hnRNP), 7, 262
Homocysteine, 62
Hormones
and breast cancer, 5, 149–159, 163, 166, 180
exogenous, 149
insulin, 151, 154–158, 229
and liver cancer, 229, 232
and prostate cancer, 5, 201–203
replacement therapy, 156, 159, 166
steroids, 150–154, 157, 180
thyroid, 232, 234
Human immunodeficiency virus, 38, 39
Human papillomavirus, 3, 38
5-Hydroxyindoleacetic acid, 32
8-Hydroxydeoxyguanosine, 5, 61

Iceland, 179
Immunosuppression, 85
Inflammatory bowel disease, 134
Insulin, 151, 155, 156, 158, 229
Insulin-like growth factors, 16, 150, 154–159, 208, 231
Intermediate effect biomarkers, 6–8, 13–22, 32–35, 49
breast, 163–167, 171–174
colorectum, 7, 103, 113–120
liver, 223–235

lung, 257–266
prostate, 191–197, 200–203
skin, 81–89
Intervention studies, 16, 27
breast, 34, 155–158, 184
colorectum, 52, 105–108, 116
liver, 7, 218, 220
lung, 259–262, 266, 279
oltipraz, 31, 38
population selection, 3, 8, 9, 18, 27, 35, 283
phase I, 19
oral leukoplakia, 53
phase II, 2, 18
breast, 7, 19
colon, 52
liver, 220
oral leukoplakia, 53
prostate, 19
phase III, 2, 18
breast, 157
colon, 53
prostate, 19, 210–211
Intraepithelial neoplasia, 8, 13, 15, 35
cervical, 35
ductal carcinoma *in situ*, 7, 14, 35
prostate, 7, 14, 35, 191–197, 200–203
8-Isoprostane F-2α, 61
Isothiocyanates, 5, 31, 38, 248, 250, 281
Isotretinoin, 259

Juvenile polyposis, 139

Kidney tumours, 34
Knock-out mice, 10

Labelling index, 118, 119
Large bowel, *see* Colorectum
Leukaemia, 37
Leukoplakia, 53
Li–Fraumeni syndrome, 10, 37, 95, 178
Linoleic acid, 108
Lipid peroxidation, 5, 104, 108, 272, 279
Liver
angiosarcoma, 215
cancer of, 32, 215
and aflatoxins, 5, 7, 33, 218, 223
and alcohol, 5, 223
and hepatitis viruses, 5, 33, 215, 220, 223, 241
p53 mutations, 7, 38, 218
cholangiocarcinoma, 215
exposure biomarkers, 3, 215–220
hepatocellular carcinoma, 7, 41, 215, 223, 241
intermediate effect biomarkers, 223–235
susceptibility biomarkers, 241–243
Lobular carcinoma *in situ*, 35, 171
Loss of heterozygosity, 16, 258, 263
Lung, cancer of, 34, 37
atypical cells in sputum, 7, 259, 261
and GST polymorphism, 63
exposure biomarkers, 245–251
intermediate effect biomarkers, 257–266
non-small-cell, 34
and oxidative stress, 272
p53 mutations, 263
retinoic acid receptor β, 7, 258
second primary, 261, 262
and smoking, 245, 257
susceptibility biomarkers, 271–283
Luteinizing hormone, 157
Lycopene, 191, 208
Lymphoma, 37
Lynch syndrome, *see* Hereditary non-polyposis colorectal cancer

Mammography, 7, 163, 173, 174, 184
Measurement error, 57, 61, 113, 117
Meat, 58, 101, 104, 115, 119, 141
Melanin, 94
Melanoma, 7, 70, 81, 86–87, 93 (*see also* Skin, cancer)
Menopause, 166
Mercapturic acids, 30
Metabolism, 9
of chemopreventive agents, 38
carcinogen, 35, 219, 271–282
oxidative, 35
steroid hormone, 151, 153, 155, 180
Metabolites
carcinogenic, 61
urinary, 30–31, 216, 249
Metaplasia, bronchial, 258, 261
Methionine, 62
Methylation, DNA, 34, 248, 264
Methylenetetrahydrofolate reductase, 62, 141
4-(Methylnitrosamino)-1-(3-pyridyl)-1-butanone (NNK), 30, 245, 272, 274, 277, 282
Microsatellites, 16, 34, 134, 138, 210, 263
Mismatch repair, 37, 133, 134, 138
Monooxygenase, *see* Cytochrome P450
Muir–Torre syndrome, 178
Mutations, 8, 32, 271
ATM gene, 180
APC gene, 9, 22, 37, 115, 117, 135–140
breast cancer genes, 37, 178
β-catenin, 115
mismatch repair genes, 37, 138
MSH2, 178
p53 gene, 10, 37
in colorectal cancer, 115, 116, 117
in liver cancer, 7, 33, 218, 220, 242
in lung tumours, 263, 274
in skin tumours, 7, 82
PTEN gene, 139, 178
ras gene, 115, 117, 135, 263
screening, 179, 182
STK11, 178
tyrosinase gene, 94

NAD(P)H:quinone oxidoreductase (NQO1), 97, 278
Naevus, 4, 7, 84, 87, 89, 95, 139
NAT2 gene, 36, 142, 181, 182, 243, 276
Nicotine, 31, 245, 251, 274, 277
Nitrosamines, 32, 245, 247–250, 277
N'-Nitrosonornicotine, 272, 274, 278
Nonsteroidal anti-inflammatory drugs, 9, 18, 21, 22, 119
Nutrition, *see* Dietary factors

Obesity, 115, 149, 151, 155, 157, 191
Oesophagus, Barrett's, 8, 14, 20
Oltipraz, 31, 38, 41, 218, 232
Oncogene, 14, 18 (*see also specific genes*)
Oral cancer
leukoplakia, 53, 276
susceptibility markers, 271–283
and tobacco use, 247, 271
Ornithine decarboxylase, 18, 51
Ovary, cancer of, 178
Overtreatment, 174
Oxidative stress, 4, 5, 6, 8, 97, 272
8-Oxodeoxyguanosine, 32, 104, 248

p53, 18
and breast cancer, 18
gene mutations, 7, 10
in colorectal cancer, 117
Li–Fraumeni syndrome, 10, 37, 178
in liver cancer, 7, 33, 218, 220, 242
in lung tumours, 263, 274
in skin tumours, 7, 82
knock-out mice, 10

polymorphism, 242, 243
Papanicolaou test, 32
Parity, 165
Penetrance, gene, 8, 10, 35, 62
Peroxisome proliferators, 231–234
Peutz–Jeghers syndrome, 139, 178
Phenobarbital, 32
Phenyl isothiocyanate, 31, 248, 250, 281
PhIP, 104
Photoproducts, 32, 69
Physical activity, *see* Exercise
Phytoestrogens, 158
Polycyclic aromatic hydrocarbons, 5, 32, 245, 247, 249, 272, 273
Polycystic ovary syndrome, 150, 157
Polymorphism, genetic, 8, 36, 271–283
N-acetyltransferase, 5, 36, 142, 276
APC, 139
CYP genes, 97, 242, 272–274, 277–278
COMT, 154
glutathione *S*-transferase (GST), 36, 38, 96, 275–276
methylenetetrahydrofolate reductase, 62, 141
p53, 242, 243
SRDA5A2, 209, 211
VDR, 210
xenobiotic-metabolism, 36, 217, 271–282
Polyps, *see* Adenoma
Polyphenols, 5, 280, 282
Postlabelling, 31, 58, 69–71, 104, 247
Predictive value of biomarker, 50
Progesterone, 150, 166
Progestogens, 150
Prognostic value of biomarker, 38
Prolactin, 167
Proliferating cell nuclear antigen, 118
Proliferation, 15, 16, 103, 105, 116, 135, 154, 192, 224, 232
mucosal, 118–120, 194
Proportionate risk, 50
Prostate
cancer, 191, 199
exposure biomarkers, 5
hormonal effects, 5, 201–203
intraepithelial neoplasia, 7, 14, 191–197, 200–203
-specific antigen, 18, 32, 194, 200, 203
susceptibility biomarkers, 207–211
Protein adducts, 29, 31, 249

Quality of life, 20

Radiation, 224, 226
Raloxifene, 156, 184
ras gene, 15, 33, 95, 115, 117, 135, 200, 205
Reactive oxygen species, 94, 96, 97, 272, 278
Recommendations, 11
Renal cancer, 34
Replacement therapy, 156, 159, 166
Retinoblastoma, 37, 95
13-*cis*-Retinoic acid, 258, 261
all-*trans*-Retinoic acid, 279
Retinoids, 184, 258, 259, 279
Retinyl palmitate, 261, 279

Screening
for breast cancer, 173, 174, 182
familial adenomatous polyposis, 142
mutation, 179
for prostate cancer, 199, 201, 203
Second primary lung cancer, 261, 262
Selenium, 5, 105, 194, 208, 261
Sex hormone binding globulin, 150, 151, 155, 157, 158, 208
Skin
cancer, 36, 37, 69, 81–82, 93–98 (*see also* Melanoma)
colour, 76
exposure biomarkers, 4, 69–78
intermediate-effect biomarkers, 7
p53 mutations, 7, 82
susceptibility biomarkers, 93–98
Smoking, *see* Tobacco
Solar radiation, 69, 81
Somatostatin, 155
Soybean, 53
Sputum atypia, 7, 259, 261
Steroids, 150–154, 157, 180
Sulfate esters, 30, 153
Sulindac, 22, 37, 53, 116, 119
Sunburn, 7, 69, 82, 88, 93
Sunscreen, 75, 82, 85, 93, 97
Surrogacy, 38–41
Surrogate end-point biomarkers, 13–22, 29 (*see also* Intermediate-effect biomarkers)
Susceptibility biomarkers, 8–10, 35–38
breast cancer, 177–184
colorectal cancer, 131–144
liver cancer, 241–243
lung, 271–283
prostate cancer, 207–211
skin cancer, 93–98

Tamoxifen, 49, 62, 156, 159, 174, 184
Tanning, 70, 75
Tea, 234, 280
Testosterone, 150, 157, 201
Thymidine glycol, 32
Thyroid, 32, 232, 234
Tobacco smoking
carcinogen metabolism polymorphism, 271–283
carcinogens, 246, 272
chewing, 272
and lung cancer, 245, 257, 271, 274
and oral cancer, 271
and oxidative stress, 272
-specific nitrosamines, 32, 245, 247–250, 272, 277
Tocopherols, 5, 262 (*see also* Vitamin E)
Transforming growth factor (TGF), 16, 18, 115, 134, 138
Transgenic mice, 6, 8, 9, 226, 242
Trials, *see* Intervention studies
Tumour-suppressor genes, 15, 18, 37, 179, 263 (*see also specific genes*)
Turcot's syndrome, 139
Tyrosinase, 94

Ulcerative colitis, 134
Ultraviolet radiation, 36, 69, 81, 93
cis-Urocanic acid, 86

Vaccination, 218, 223
Validity, 2, 5, 29, 63
Vegetables, 9, 58, 61, 65, 107, 115, 116, 281, 282
Vinyl chloride, 215
Vitamins
A, 119, 258, 261, 279
C (ascorbic acid), 32, 105, 116, 119
D, 180, 208, 210
E (α-tocopherol), 58, 105, 116, 119, 194, 208, 259, 261, 280

Watercress, 30, 250, 282
Wnt signalling, 137

Xenobiotic metabolism, 9, 35, 58
Xeroderma pigmentosum, 36, 37, 70, 94